43 uwe 110
lbf 40.202

Ausgeschieden
im Jahr 2025

EMC ANALYSIS METHODS AND COMPUTATIONAL MODELS

by

Frederick M. Tesche
EMC Consultant
Dallas, Texas

Michel V. Ianoz
Ecole Polytechnique Fédérale de Lausanne
Lausanne, Switzerland

Torbjörn Karlsson
EMICON
Linköping, Sweden

A WILEY-INTERSCIENCE PUBLICATION

JOHN WILEY & SONS, INC.

New York • Chichester • Brisbane • Toronto • Singapore • Weinheim

A NOTE TO THE READER
This book has been electronically reproduced from digital information stored at John Wiley & Sons, Inc. We are pleased that the use of this new technology will enable us to keep works of enduring scholarly value in print as long as there is a reasonable demand for them. The content of this book is identical to previous printings.

This text is printed on acid-free paper.

Copyright © 1997 by John Wiley & Sons, Inc.

All rights reserved. Published simultaneously in Canada.

Reproduction or translation of any part of this work beyond that permitted by Section 107 or 108 of the 1976 United States Copyright Act without the permission of the copyright owner is unlawful. Requests for permission or further information should be addressed to the Permissions Department, John Wiley & Sons, Inc., 605 Third Avenue, New York, NY 10158-0012

Library of Congress Cataloging in Publication Data:
Tesche, Frederick M.
  EMC analysis methods and computational models / by Frederick M. Tesche, Michel V. Ianoz, Torbjörn Karlsson.
    p.   cm.
  Includes index.
  ISBN 0-471-15573-X (cloth : alk. paper)
  1. Electromagnetic compatibility--Mathematical models.   I. Ianoz, M. (Michel)   II. Karlsson, Torbjörn.   III. Title.
TK7867.2.T47 1996
621.382'24'011--dc20   96-12448

Printed in the United States of America

10 9 8 7 6 5 4 3

To Sharon, Eugenie, and Maud, who were so kind to permit us this caprice—and so wise to understand that this book was, for us, more than simply a caprice.

CONTENTS

PREFACE xix

Acknowledgments xxv

PART I PRELIMINARIES 1

1. INTRODUCTION TO MODELING AND EMC 3

  1.1 The Concept of Modeling   3

  1.2 Validation of Models   6
      1.2.1 Example of Experimental Model Validation   7
      1.2.2 Model Validation Using Nonexperimental Methods   7

  1.3 Building Models in Electromagnetics   8

  1.4 EMC Modeling: A Historical Overview   9

  1.5 Considerations for EMC Modeling   12
      1.5.1 Classification of EMC Problems   12
      1.5.2 EMC Problems Amenable to Modeling   13
      1.5.3 Types of Signals in EMC Models   16
            1.5.3.1 *Evaluation of the Frequency Spectra*   18
      1.5.4 Limits of Modeling   19

  1.6 Who is Using Modeling and to Whom Modeling is Useful   20

  References   22

  Problems   23

2. SYSTEM DECOMPOSITION FOR EMC MODELING 26

  2.1 Application of Analytical Methods in EMC   26
      2.1.1 System Design Phase   26
      2.1.2 System Construction Phase   28
      2.1.3 EMC Verification   28
            2.1.3.1 *Immunity Testing*   28
            2.1.3.2 *Emission Testing*   29

	2.1.4	Summary of the Use of Analytical Models	29
2.2	Topological Description of Systems		30
	2.2.1	Electromagnetic Topology	30
		2.2.1.1 Topological Diagram	31
		2.2.1.2 Electromagnetic Energy Points of Entry	32
		2.2.1.3 EMC Design	33
	2.2.2	Electromagnetic Interaction with the System	34
		2.2.2.1 Interaction Sequence Diagram	35
	2.2.3	Generalized EMC Design Principles Based on EM Topology	36
2.3	Modeling Accuracy		38
	2.3.1	Errors Inherent in the Analysis	38
	2.3.2	Balanced Accuracy of Analysis	38
References			39
Problems			39

PART II LOW-FREQUENCY CIRCUIT MODELS — 45

3. LUMPED-PARAMETER CIRCUIT MODELS — 47

3.1	Introduction		47
3.2	Conducted Disturbances in Circuits		48
	3.2.1	Thévenin and Norton Representations	48
	3.2.2	Models for Passive Two-Port Circuits	50
		3.2.2.1 Open-Circuit Impedance Parameters	51
		3.2.2.2 Short-Circuit Admittance Parameters	53
		3.2.2.3 Chain Parameters	56
		3.2.2.4 Two-Port Parameter Relationships	58
		3.2.2.5 Other Two-Port Representations	58
	3.2.3	Two-Port Models for Circuits with Sources	59
	3.2.4	Treatment of Multiport Circuits	62
	3.2.5	Example of Conducted Disturbances in Electrical Power Systems	63
		3.2.5.1 Generation of Harmonic Currents	64
		3.2.5.2 Determination of the Mains Impedance	67
		3.2.5.3 Estimation of the Harmonic Current Source	71
3.3	Disturbances in Circuits Induced by Electromagnetic Fields		72
	3.3.1	Magnetic Field Coupling	72
		3.3.1.1 Weak-Coupling Approximations	74
		3.3.1.2 Calculation of Mutual and Self-Inductances	77

		3.3.2	Electric Field Coupling	81

        3.3.2 Electric Field Coupling   81
            3.3.2.1 Weak-Coupling Approximations   82
            3.3.2.2 Calculation of Mutual and Self-Capacitances   83
        3.3.3 General Field Coupling at Low Frequencies   86
            3.3.3.1 Example: Crosstalk Between Two Parallel Traces on a Printed Circuit Board   86
        3.3.4 General Methods of Reducing Low-Frequency Interference   90
        3.3.5 Specific Measures to Reduce Capacitive Coupling   91
            3.3.5.1 Effect of a Cylindrical Shield Around a Conductor   91
            3.3.5.2 Discussion   94
        3.3.6 Specific Measures to Reduce Inductive Coupling   94
    3.4 Disturbances Caused by Common Ground Returns   95
    3.5 Extension of Circuit Modeling to High Frequencies   101
    References   103
    Problems   104

PART III HIGH-FREQUENCY AND BROADBAND COUPLING MODELS   111

4. RADIATION MODELS FOR WIRE ANTENNAS   113

    4.1 Introduction   113
    4.2 Radiation of Electromagnetic Fields in the Frequency Domain   114
        4.2.1 Overview   114
        4.2.2 Radiation from Elementary Sources   116
            4.2.2.1 Electric Dipole   117
            4.2.2.2 Magnetic Dipole   120
        4.2.3 Radiation from Extended Sources   122
            4.2.3.1 Center-Fed Wire Antenna   123
            4.2.3.2 Integral Equation for the Wire Antenna   126
                4.2.3.2.1 Solution by the Method of Moments   127
                4.2.3.2.2 Approximate Solution for the Antenna Problem   129
        4.2.4 Dipole Radiation in the Presence of Other Bodies   135
            4.2.4.1 Electric Dipoles over a Perfect Ground   135
            4.2.4.2 Electric Dipoles in a Parallel-Plate Region   138

viii CONTENTS

		4.2.4.3 Electric Dipoles in a Cavity	141
		4.2.4.3.1 Solution by the Method of Images	142
		4.2.4.3.2 Eigenmode Solution	144
		4.2.4.4 Electric Dipoles near a Sphere	147
		4.2.4.5 Electric Dipoles over an Imperfectly Conducting Earth	150
	4.2.5	Evaluation of Magnetic Field Components	153
4.3	Reception and Scattering of Electromagnetic Fields in the Frequency Domain		154
	4.3.1	General Considerations	154
	4.3.2	Approximate Solution for the Thin Wire	155
		4.3.2.1 Determination of the Induced Current	155
		4.3.2.2 Scattered Field	156
4.4	Electric-Field Integral Equations in the Time Domain		157
	4.4.1	Overview	157
	4.4.2	The Integrodifferential Equation	158
	4.4.3	Extension to a Wire over a Lossy Ground	159
	4.4.4	Numerical Solution for the EFIE for Thin Wires in the Time Domain	161
4.5	Singularity Expansion Method		161
	4.5.1	Background	161
	4.5.2	Mathematical Description of SEM	164
	4.5.3	SEM Representation of the Antenna Current	166
	4.5.4	SEM Representation of the Scattering Current	167
	4.5.5	SEM Representation of the Radiated Fields	168
	4.5.6	SEM Representation of Scattered Fields	168
	4.5.7	Example of SEM Applied to the Approximate Antenna Analysis	169
		4.5.7.1 Induced Current for the Antenna Problem	169
		4.5.7.2 Induced Current for the Scattering Problem	174
References			175
Problems			178

5. RADIATION, DIFFRACTION, AND SCATTERING MODELS FOR APERTURES   183

5.1	Introduction		183
5.2	EM Field Penetration Through Apertures		185
	5.2.1	Scalar Diffraction Theory	185
		5.2.1.1 Kirchhoff Approximation	187

	5.2.1.2	Dirichlet Solution	188
	5.2.1.3	Neumann Solution	188
	5.2.1.4	Discussion of the Scalar Solutions	189
		5.2.1.4.1 Rectangular Aperture	191
		5.2.1.4.2 Circular Aperture	191
5.2.2	General Vector Field Diffraction		192
	5.2.2.1	Fundamentals	192
	5.2.2.2	Application to the Aperture Penetration Problem	194
5.2.3	Far-Field Vector Field Diffraction		196
	5.2.3.1	Example for a Rectangular Aperture	198
5.2.4	Aperture Integral Equation		199
	5.2.4.1	Example of Aperture Field Calculation	203
5.2.5	Equivalent Area of an Aperture		203

5.3 Radiation from Extended Antennas — 207

5.4 Low-Frequency Approximation — 208
   5.4.1 Dipole Moments — 208
   5.4.2 Aperture Polarizabilities — 209
       5.4.2.1 Corrections of the Polarizabilities for Aperture Loading — 210

5.5 Wideband and Transient Responses of Apertures — 211
   5.5.1 Wideband Responses — 212
   5.5.2 Direct Time-Domain Calculations — 216

References — 216

Problems — 217

PART IV TRANSMISSION LINE MODELS — 221

6. TRANSMISSION LINE THEORY — 223

6.1 Overview of Transmission Line Models — 223
   6.1.1 Lumped and Distributed Circuit Parameters — 223
   6.1.2 Lumped and Distributed Excitations — 224
       6.1.2.1 Examples of System Excitation — 225
   6.1.3 Two-Conductor and Multiconductor Systems — 226
   6.1.4 Transmission Line and Antenna Mode Responses — 226
   6.1.5 Telegrapher's Equations for a Two-Conductor System — 228
       6.1.5.1 Evaluation of the Line Parameters — 229

6.2 Frequency-Domain Responses 231
   6.2.1 Solution of the Telegrapher's Equations for a Two-Conductor Line 232
      6.2.1.1 Chain Parameter Representation of a Two-Wire Line 233
      6.2.1.2 Other Two-Port Representations for the Two-Wire Line 234
      6.2.1.3 Applications of Two-Port Representations 235
      6.2.1.4 The Π and T Equivalent Circuits of the Two-Wire Line 235
   6.2.2 Lumped-Source Excitation of Transmission Lines 237
   6.2.3 Terminated Lines: Voltage Reflection Coefficient 239
   6.2.4 General Solution for a Terminated Line 240
      6.2.4.1 Example of Line Response for a Voltage Source Excitation 241
   6.2.5 Load Responses for a Finite Line 242
      6.2.5.1 BLT Equation 244
      6.2.5.2 Example of Frequency-Domain Voltage Response of a Line 244
      6.2.5.3 Validation of the Transmission Line Models 245
   6.2.6 Multiconductor Transmission Lines 247
      6.2.6.1 Impedance and Admittance Matrices 248
      6.2.6.2 Natural Propagation Modes 250
      6.2.6.3 Diagonalization of the [P] and [R] Matrices 251
      6.2.6.4 Modal Voltages and Currents 251
      6.2.6.5 Solution of the Modal Equations 252
      6.2.6.6 Calculation of the Propagation Matrix and Diagonalization Matrix Elements 253
      6.2.6.7 Open-Circuit Voltage of a Semi-infinite Multiconductor Line Excited by a Voltage Source 258
      6.2.6.8 Simplified Modeling by the Equal-Velocity Assumption 259
   6.2.7 BLT Equation for Multiconductor Lines 260
      6.2.7.1 Simplifications of the BLT Equation 266
   6.2.8 Chain Parameters for a Multiconductor Line 266
   6.2.9 Example of the Use of Multiconductor Line Models 267

6.3 Time-Domain Transmission Line Responses 268
   6.3.1 Time-Harmonic Excitation 268
   6.3.2 Nonsinusoidal Traveling Waves 268
   6.3.3 Analytical Transformation from the Frequency Domain to the Time Domain 269

	6.3.4	Numerical Transformation of the Solution from the Frequency to the Time Domain	271
	6.3.5	Numerical Solution of the Telegrapher's Equations in the Time Domain	272
	6.3.6	Inductive and Capacitive Terminations in the Time Domain	274
	6.3.7	Bergeron's Graphical Solution in the Time Domain	275
		6.3.7.1 Principle of Bergeron's Method	275
		6.3.7.2 Numerical Application of Bergeron's Method	276
		6.3.7.3 Solution of the Nodal Matrix Equation	279
		6.3.7.4 Example: Transient State of a Circuit After Closing Two Interrupters	280
	6.3.8	Electromagnetic Transients Program	282
		6.3.8.1 Transmission Line Response Using EMTP	282
		6.3.8.2 Modeling a Test Installation with Two Parallel Lines	282
6.4	Determination of Line Inductance Parameters		283
	6.4.1	Inductance Measurement	286
	6.4.2	Analytical Inductance Evaluation	287
		6.4.2.1 Geometrical Means Distance Between Two Circuits	289
		6.4.2.2 Mutual Inductance per Unit Length	289
		6.4.2.3 Self Inductance per Unit Length	291
		6.4.2.4 Mutual and Self-Inductances of Lines with the Earth as a Return Conductor	292
		6.4.2.4.1 Mutual and Self-Inductances of Lines over Perfectly Conducting Ground	292
6.5	Determination of Line Capacitance Parameters	293
	6.5.1	Measurement of the Capacitance Parameters	294
	6.5.2	Analytical Capacitance Evaluation	295
		6.5.2.1 Partial Capacitances	296
		6.5.2.2 Static Capacitances	297
		6.5.2.2.1 Sign of the Partial Capacitances	297
	6.5.3	Calculation of the Static Capacitances	298
		6.5.3.1 Conductors in a Homogeneous Medium over the Ground	298
		6.5.3.2 Transmission Lines in Nonhomogeneous Media	300
		6.5.3.3 Use of the Finite Element Method to Calculate Partial Capacitances	300

		6.5.3.4	*Integral Equation Evaluation of the Per-Unit-Length Capacitance Matrix*	301
		6.5.3.5	*Capacitance Calculation Using Inductance Values*	308
	References			310
	Problems			313

7. FIELD COUPLING USING TRANSMISSION LINE THEORY — 321

- 7.1 Introduction — 321
- 7.2 Two-Wire Transmission Line — 326
  - 7.2.1 Derivation of the Telegrapher's Equations with an External Excitation — 326
    - 7.2.1.1 *First Telegrapher's Equation* — 327
    - 7.2.1.2 *Second Telegrapher's Equation* — 329
    - 7.2.1.3 *Modification of the Telegrapher's Equations for a Finitely Conducting Wire* — 332
    - 7.2.1.4 *Modification for a Lossy Medium Surrounding the Line* — 333
  - 7.2.2 Alternative Forms of the Telegrapher's Equations — 333
    - 7.2.2.1 *Total Voltage Formulation* — 333
    - 7.2.2.2 *Scattered Voltage Formulation* — 334
    - 7.2.2.3 *Numerical Example of the Two Formulations* — 336
  - 7.2.3 Solution for the Line Current and Voltage — 338
  - 7.2.4 Solution for the Load Currents and Voltages: The BLT Equation — 339
  - 7.2.5 Load Responses for Plane-Wave Excitation — 340
  - 7.2.6 Examples of Line Responses — 342
    - 7.2.6.1 *Frequency-Domain Responses* — 342
    - 7.2.6.2 *Transient Response* — 343
- 7.3 Single Line Over a Perfectly Conducting Ground Plane — 345
  - 7.3.1 Derivation of the Telegrapher's Equations — 345
    - 7.3.1.1 *Total Voltage Formulation* — 346
    - 7.3.1.2 *Scattered Voltage Formulation* — 347
    - 7.3.1.3 *Comments on the Line Excitation from an EM Scattering Viewpoint* — 350
    - 7.3.1.4 *Modifications of the Telegrapher's Equations* — 352
  - 7.3.2 Solution to the Telegrapher's Equations for Load Responses — 353

		7.3.2.1 BLT Sources	353
	7.3.3	Load Responses for Plane-Wave Excitation	353
		7.3.3.1 Comparison with Two-Wire Results	355
	7.3.4	Validation of the Coupling	356
	7.3.5	Load Response for Non-Plane-Wave Excitation	357
7.4	Treatment of Highly Resonant Structures		364
	7.4.1	Single-Wire Line	364
		7.4.1.1 Numerical Example	364
		7.4.1.2 Expansion of the BLT Resonance Matrix	366
	7.4.2	Extension to Multiconductor Lines	367
7.5	Radiation from Transmission Lines		368
	7.5.1	Reciprocity Theorem	368
	7.5.2	Radiating Transmission Line	370
	7.5.3	Example of the Radiation from a Transmission Line	372
7.6	Transmission Networks		373
	7.6.1	Network Analysis by Thévenin Transformations	375
		7.6.1.1 Example of a Network Response Using Thévenin Transformations	379
	7.6.2	Development of the Network BLT Equation	380
7.7	Transmission Lines with Nonlinear Loads		382
	7.7.1	Volterra Integral Equation	382
	7.7.2	Example of a Single Transmission Line with a Nonlinear Load Impedance	384
References			388
Problems			390

8. EFFECTS OF A LOSSY GROUND ON TRANSMISSION LINES — 395

8.1	Introduction		395
8.2	Derivation of the Telegrapher Equations		396
	8.2.1	Total Voltage Formulation	397
		8.2.1.1 First Telegrapher Equation	397
		8.2.1.2 Second Telegrapher Equation	400
	8.2.2	Scattered Voltage Formulation	402
	8.2.3	Termination Conditions	403
		8.2.3.1 V–I Relationships	403
		8.2.3.2 Ground Impedance	404
	8.2.4	Solution of the Telegrapher Equations	404

8.3	Per-Unit-Length Line Parameters		405
	8.3.1	Equivalent Circuit for the Line	405
	8.3.2	Frequency-Domain Representations for the Ground Impedance	406
	8.3.3	Time-Domain Representation of the Ground Impedance	410
8.4	Reflected and Transmitted Plane-Wave Fields		411
	8.4.1	Plane-Wave Reflection and Transmission from the Earth	412
		8.4.1.1 *General Expressions for the Fields*	412
		8.4.1.2 *Excitation Fields for a Transmission Line*	415
	8.4.2	Transient Field Reflected from the Ground	417
		8.4.2.1 *Transient Fields Evaluated by the FFT*	417
		8.4.2.2 *Direct Evaluation of the Transient Reflected E-Field*	420
8.5	Examples of Aboveground Transmission Line Responses		422
	8.5.1	Variations in Earth Conductivity	424
	8.5.2	Variations with Angle of Incidence	424
	8.5.3	Variations with Line Height	428
8.6	Buried Cables		428
	8.6.1	Summary of Rigorous Solution	431
		8.6.1.1 *Integral Equation*	431
		8.6.1.2 *Soil Impedance*	432
		8.6.1.3 *Cable Impedance*	433
		8.6.1.4 *Solution for the Current*	434
	8.6.2	Transmission Line Approximation	435
	8.6.3	Additional Simplifications to the TL Solution	437
	8.6.4	Example of Current Responses on an Infinite Buried Cable	438
	8.6.5	Application to Buried Lines of Finite Length	439
References			443
Problems			444

PART V SHIELDING MODELS   449

9. SHIELDED CABLES   451

9.1	Introduction	451
9.2	Fundamentals of Cable Shield Coupling	452

	9.2.1	Definitions of Transfer Impedance and Transfer Admittance	453
	9.2.2	Relative Importance of Z'_t and Y'_t	455
9.3	EM Coupling Through a Solid Tubular Shield		455
	9.3.1	Transfer Impedance	455
	9.3.2	Transfer Admittance	457
9.4	Models for Braided Shields		457
	9.4.1	EM Field Penetration and Diffraction into Braided Shields	459
	9.4.2	Single Aperture Excitation	460
	9.4.3	Multiple Apertures	462
	9.4.4	Expressions for the Aperture Polarizabilities	464
	9.4.5	Shield Transfer Characteristics in Terms of Braid Weave Parameters	465
		9.4.5.1 Transfer Impedance	465
		9.4.5.2 Transfer Admittance	466
		9.4.5.3 Dielectric Filling in the Cable	467
		9.4.5.4 Comparison with Measurements	467
	9.4.6	Improved Expressions for the Transfer Impedance of Braided Shields	468
		9.4.6.1 Tyni's Model	469
		9.4.6.2 Demoulin's Model	472
		9.4.6.3 Kley's Model	473
		9.4.6.4 Comparison of Demoulin's and Kley's Models	477
	9.4.7	Effect of an Axial Magnetic Field Component	479
	9.4.8	Alternative Expression for the Transfer Admittance of Braided Shields	479
9.5	Calculated Responses of a Braided Cable		480
	9.5.1	External Transmission Line	480
	9.5.2	Internal Excitation Sources	483
	9.5.3	Internal Load Responses	483
	9.5.4	Numerical Example of a Shielded Cable System	484
9.6	Cables with Shield Interruptions		488
	9.6.1	Introduction	488
	9.6.2	Cable Connectors	493
	9.6.3	Pigtail Terminations	494
	9.6.4	Discontinuous Shields	495
	9.6.5	Example of an Interrupted Cable Shield	496
References			501
Problems			503

10. SHIELDING — 505

- 10.1 Introduction — 505
- 10.2 Generalized Shield Concept — 506
- 10.3 Shielding Mechanisms — 508
  - 10.3.1 Shielding of Static Fields — 508
    - *10.3.1.1 Electrical Shielding* — 508
    - *10.3.1.2 Magnetostatic Shielding* — 510
  - 10.3.2 Shielding of Time-Varying Fields: Eddy Current Shielding — 515
    - *10.3.2.1 General Concepts* — 515
    - *10.3.2.2 Skin Effect and Skin Depth* — 518
    - *10.3.2.3 Plane-Wave Shielding by an Infinite Metal Plate* — 519
    - *10.3.2.4 Plane-Wave Shielding by Two Infinite Parallel Plates* — 524
    - *10.3.2.5 Plane-Wave Shielding by a Conducting Mesh* — 524
  - 10.3.3 Summary of Shielding Dependence on Frequency — 527
- 10.4 Volumetric Shields — 528
  - 10.4.1 Closed, Homogeneous Metal Shield — 528
    - *10.4.1.1 Shielding of Time-Harmonic Fields* — 528
      - *10.4.1.1.1 Evaluation of the H-Field Shielding Effectiveness* — 528
      - *10.4.1.1.2 Limitations of the Shielding Expressions* — 532
      - *10.4.1.1.3 Examples of the H-Field Shielding Effectiveness* — 533
      - *10.4.1.1.4 Determination of the E-Field Shielding Effectiveness* — 534
    - *10.4.1.2 Shielding of Transient Electromagnetic Fields* — 534
  - 10.4.2 Closed Metallic Mesh Shield — 540
    - *10.4.2.1 Induced Responses Within a Mesh-Protected Shield* — 543
- 10.5 Shielding of Non-Plane-Wave Fields — 544
  - 10.5.1 Overview of Near-Field Shielding — 544
  - 10.5.2 Shielding Between Two Circular Loops — 544
- References — 547
- Problems — 548

APPENDICES

APPENDIX A: TABLES OF PHYSICAL CONSTANTS   550

APPENDIX B: VECTOR ANALYSIS AND FUNCTIONS   553

APPENDIX C: PER-UNIT-LENGTH LINE PARAMETERS   557

APPENDIX D: GROUNDING RESISTANCE PARAMETERS   563

APPENDIX E: COAXIAL CABLE AND CONNECTOR DATA   566

APPENDIX F: COMPUTER SOFTWARE   570

INDEX   595

PREFACE

The study of electromagnetic compatibility (EMC) is a young old science. It is relatively old because the problem of radio-frequency interference (RFI) arose nearly 100 years ago with the first use of radio waves as a communication medium. However, it is only in the last 20 to 25 years that progress in numerical computation has allowed scientists and engineers not only to propose models for the physical phenomena underlying this interference, but to use these models to better understand and visualize these phenomena and to mitigate the effects of interference.

The development and use of models has been the focus of much human activity. To quote the late Peter Johns[†], developer of the TLM model:

> Throughout history men have been making models of the physical events they observe. From cave pictures to surrealist art, from models of the atom to models of the universe, all have sought to focus attention on particular ideas for analysis or for communication to other people. As engineers we must be particularly skilled in the art of making or choosing the mathematical models of our engineering concepts. Desirable properties of a model, such as the enhancement of areas for which analysis is required and rejection of areas of no interest, seem obvious. However the computational power at the fingertips of the professional engineer is increasing enormously through the proliferation of calculators and computers. Thus the methods for analyzing models are changing, and this means that the models we have chosen in the past may not always be the best for the present and future.

Although numerical models can be very useful in our understanding of electrical phenomena, they are inherently limited in their ability to predict as much as we would like. As pointed out by Johns, sometimes the fundamental assumptions going into a mathematical foundation of a model are themselves approximations to reality. As a consequence, the model based on these approximations may have a limited range of validity. In other instances, the situation being modeled (i.e., the "real world") can be so complex that accurate modeling becomes difficult, if not impossible, to carry out. In such cases, one often resorts to measurement or experimental tests to understand the phenomenon.

Is it then a lack of modesty to present to the scientific and technical community a book on EMC modeling? We think not, for several reasons. The first is that incredible progress in electromagnetic field modeling has already been achieved, particularly in the last 15 years. We think that the

---

[†]Johns, P. B., "The Art of Modelling," *Electronics and Power*, August 1979, pp. 565–569. Reprinted by permission from the Institution of Electrical Engineers (IEE).

present state of the art in modeling represents a good basis for further progress in this field, and that such progress is likely to be very rapid in the future.

The second reason is that whereas present models may be far from perfect, they are very useful in aiding *understanding* of the fundamental principles of EM interference control. Much can be learned from an imperfect model if the user takes the effort to understand why the model does not work as it should. This understanding leads naturally to new, and hopefully improved, models.

In writing this book, we believe that graduate students, postdoctoral researchers, and senior researchers, as well as electrical engineers engaged in the research and development work for practical applications in the EMI area, will benefit in having at their disposal material which presently is only found dispersed throughout journals, technical reports, or conference papers. Much of the material in this book is in constant evolution, due to research work in progress at universities and laboratories throughout the world. During the development of this book, however, we were obliged to present the status of this subject of EMC modeling at a certain fixed point in time—when the book was being written. A typical example is the model of the transfer impedance of braided cables. This model, far from being perfect, does not yet reflect all the complexity of the real braid shield. It does represent, however, a big step in our understanding of what happens in a braided shield, and our effort has been aimed at giving the reader an adequate picture of the important work conducted in this area during the last few years.

The material in this book is divided into five parts. Part 1 presents the customary introduction to the subject matter treated in this book: *EMC analysis methods and computational models*. Chapter 1 reviews the overall concept of model development and discusses the effect that modeling has had in the area of EMC. Some of the various types of signals, both transient and continuous wave (CW), are reviewed, and the link between the transient domain and the frequency domain is explained by the Fourier transform.

A key aspect of model development is how to take an electrically complex system and break it up into smaller, more manageable pieces for which an EM model can be developed. This is done through the concept of *electromagnetic topology*, which is the topic of Chapter 2. As discussed in this chapter, the word *topology* as used in this book is not cast in a rigorous mathematical context, but rather, provides a conceptual tool in trying to view the system as if the observer were an electromagnetic wave impinging on an electrical system, such as an aircraft. Where are the global shielding surfaces that keep you from entering into the interior of the system? Where are the points of entry into the system that allow you to pass? How do you move from one region to another within the system? What effect do you have on components within the system? The answers to these questions are related to viewing the system in a topological manner.

Part 2 of this book consists of a chapter on lumped-parameter circuit models. At sufficiently low frequencies, interference between two electrical circuits can be described rather well by low-frequency circuit models. In this chapter, conducted interference models are introduced by a discussion of Thévenin and Norton equivalent circuits. This concept is then generalized to the case of active and passive two-port networks. In addition to the direct wire connections between circuits, capacitive and inductive field coupling can also be important in low-frequency models. These coupling mechanisms are discussed in this chapter, as is the case of galvanic coupling between two circuits connected to a common, lossy conductor.

In Part 3, models suitable for higher frequencies are discussed. As the dimensions of coupled circuits begin to approach the size of the wavelength of the EM fields, the circuit models are no longer adequate and the models must take into account the wave nature of the fields. Using Maxwell's equations, these models lead to the concept of radiation. Chapter 4 presents an overview of the radiation process: first from elementary electric and magnetic dipole sources and later from extended line sources (i.e., from wire antennas). Several examples of electric dipole radiation in the presence of perturbing bodies (a ground plane, a sphere, within parallel plates, and in a cavity) are given. A key aspect in the analysis of wire antennas is knowledge of the current distribution on the structure. This distribution can be estimated, approximated from the integral equation describing the current, or computed numerically using the method of moments. One topic of current interest is the singularity expansion method (SEM), which provides a link between the resonant behavior of *RLC* circuits and antennas.

Chapter 5 concludes Part 3 with a discussion of radiation, diffraction, and scattering models for apertures, which cannot be analyzed using the relatively simple one-dimensional wire models of Chapter 4. In this chapter, the scalar diffraction theory for apertures is first discussed, followed by the more rigorous vector field diffraction. This theory is also applied to radiating antennas. A case of special importance, especially in braid shield modeling, is when an aperture is small compared to the wavelength. In this case, the fields penetrating the aperture can be modeled as arising from equivalent electric and magnetic dipole moments located at the aperture, with the strengths of the dipoles being related to the size and shape of the aperture through aperture *polarizabilities*.

Part 4, dealing with transmission line models, contains three chapters. Chapter 6 lays the foundation for the discussions that follow. The concept of distributed (per-unit-length) line parameters is introduced, and their use in the telegrapher's equations for the line current and voltage is discussed. Using the two-conductor line as an example, the solution to these equations is developed, first for simple traveling waves on the line, and later, for lumped voltage or current sources. A particularly simple representation for the load responses of a transmission line is the BLT equation, which is examined in this chapter. In addition, the solution of transmission line problems directly in the time domain is discussed. The chapter concludes with a detailed discussion of the calculation of the per-unit-length line inductance and capacitances.

Chapter 7 continues the discussion of transmission line modeling by examining the issues of EM field excitation of lines. It is pointed out that transmission line coupling models provide only part of the complete solution—the differential mode response, with the common-mode response (i.e., the antenna mode) being neglected. Fortunately, for many practical cases, including lines over a conducting ground plane, the transmission line model provides an accurate way of computing the induced responses. Also in this chapter, models for highly resonant transmission lines, radiation from lines, transmission line networks, and lines with nonlinear loads are formulated and discussed.

The important case of transmission lines in the presence of a lossy earth is developed in Chapter 8. Derivation of the telegrapher's equations for this case is presented, and the determination of per-unit-length line parameters for aboveground and buried lines is presented. Furthermore, the behaviour of the incident, reflected, and transmitted EM fields from the lossy earth is discussed. Several examples of cable responses due to distributed EM field excitation are provided.

The last part of the book, Part 5, discusses models for shielding. Continuing with transmission line issues, Chapter 9 presents information on the protection of cables through the use of solid or braided coaxial shields. The classical shielding formula for a solid shield is reviewed, and the more recent work for modeling the behavior of braided cables is outlined. In addition, the issue of discrete breaks in a shield (due to connectors or pigtail connections) is discussed and appropriate computational models are suggested.

Chapter 10 investigates the more general aspects of EM shielding for enclosures, such as screen rooms. The usual infinite slab shielding model is briefly discussed, and it is pointed out that this provides an unrealistic view of the shielding provided by real shields. More appropriate models are those that have a finite shielded volume, bounded by a finite surface area—called the *volumetric shield*. In this chapter a number of useful formulas for shielding are presented and illustrated.

The book concludes with appendices that present information on appropriate physical constants useful for model development, data for various types of transmission lines, and pertinent vector identities. Also included is general information about the four computer programs that accompany this book. These programs are designed based on the models developed in the text and may be used to calculate various transmission line responses.

At the end of each chapter, exercises further an understanding of the basic theory behind the models and indicate how some of the models can be applied to practical situations of EMI prediction and control. Several of these exercises utilize the computer programs provided with the text.

Writing this book has been an interesting adventure for us. The material presented here results from many years of collective work, not only by ourselves, but by many of our colleagues. It is truly impressive to consider the past efforts on the part of researchers throughout the world that have gone into the formulation, development, and use of computational models for electromagnetics. We see that much has been accomplished; however, even more remains to be done. In this book we have only scratched the surface of this immense technical area.

F. M. TESCHE

Dallas, Texas

M. V. IANOZ

Lausanne, Switzerland

T. KARLSSON

Linköping, Sweden

ACKNOWLEDGMENTS

This work is the result of the author's activities in the field of electromagnetics over many years. During this period, they have collaborated with many people, and it is this common work, as well as the discussions with other researchers, that has contributed to their knowledge and to the various problems discussed here.

Michel Ianoz cites the collaboration over a period of more than 10 years with Professor C. Mazzetti of the University of Rome, Professor C. A. Nucci of the University of Bologna, and Dr. F. Rachidi of the Ecole Polytechnique Fédérale of Lausanne on lightning electromagnetic effects, which has permitted him to clarify various questions and to develop contributions in the field of EMC which are partially reflected in this book. Many other ideas pertaining to lightning problems have been developed during fruitful discussions with Professor M. Uman of the University of Florida, and Dr. M. Rubinstein, presently at the Swiss PTT. He wishes to acknowledge many helpful discussions with Dr. C. Baum, Dr. D. Giri, Dr. W. Radasky, and E. F. Vance on EM field coupling phenomena; with Professor B. Demoulin on transfer impedance modeling of cables and with Professor G. Costache on numerical methods applied to EMC. He also wants to acknowledge his pioneering collaboration with M. Aguet, with whom he initiated the first EMC research in Switzerland in the late 1970s, as well as over 15 years of technical interaction with F. M. Tesche.

Professor Ianoz also acknowledges that the three years of collaboration among eight European universities in the European Research Program on EMC has contributed to his portion of the book. He thanks his colleagues F. Canavero (Politècnico di Torino), M. D'Amore (Università di Roma), P. Degauque (Université de Lille), K. Feser (Universitaet Stuttgart), B. Jecko (Université de Limoges), J-C. Sabonnadiere (Institut Polytechnique de Grenoble), and J. L. ter Haseborg (Technische Universitaet Hamburg–Harburg) for intensive discussions and exchange of ideas.

Fred Tesche would like to recognize the tutelage and subsequent collaboration of Dr. C. E. Baum (Phillips Laboratory, U.S. Air Force) and Dr. K. S. H. Lee (Kaman Science Corp.) during his early years working on transient EM problems. His subsequent work with Drs. T. K. Liu and D. V. Giri at LuTech, Inc. and E. F. Vance (consultant) in the EMP area has helped to form the basis of many of his contributions to the book. Dr. Tesche also wishes to recognize helpful technical discussions with a number of colleagues over the past 25 years of EM investigation. These include Dr. R. L. Hutchens (BDM), Lt. Col. R. Vandre (U.S. Army), Dr. Al Bahr (SRI), Dr. Kendall Casey (SRI), Dr. Peter Mani (formerly of the Swiss NEMP Laboratory), Bruno Brändli (Swiss NEMP Laboratory), the late Drs. Ray

Latham (Northrop) and Lennart Marin (Kaman Sciences Corp.), and his co-authors, Professor Ianoz and Dr. Karlsson. The special assistance of Dr. F. Rachidi (EPFL) in providing a number of numerical results pertaining to line coupling is gratefully acknowledged.

Dr. Tesche also thanks P. R. Barnes (Oak Ridge National Laboratory) and Dr. Armin Kälin (Swiss NEMP Laboratory) for their support and interest in his research activities during the course of writing this book. Furthermore, the Swiss Federal Institute of Technology in Lausanne (EPFL) is recognized for offering Dr. Tesche, on two different occasions, the post of academic guest for the purpose of collaborating with his co-authors and writing portions of the text. Finally, recognition of the continued advice and spousal support from Sharon Wesson is offered, without which the writing of this book would not have been possible.

Torbjörn Karlsson considers 15 years of research on EMP as being the basis of his understanding of EMC. During that period he had the privilege to meet and discuss items related to Maxwell's equation with many EMP experts. To mention a few, the late Dr. Lennart Marin was a great source of inspiration for formulating and solving theoretical problems; Dr. Carl E. Baum has taken an encouraging role in, among other things, topological discussions; E. Vance has given much advice on grounding and shielded cables; and co-author Fred Tesche has been very helpful in the development of numerical analyses.

The early collaboration with Berndt Backlund at the Swedish Defense Research Institute laid the foundation for comprehension of the important experimental verification of theoretical work. In more recent years cooperation with Sven Garmland at Emicon has been a condition for the development of sophisticated EMC measurements.

A special acknowledgment from all three authors is due to Professor Clayton Paul of the University of Kentucky for his careful review of the manuscript and for the many interesting and helpful suggestions he has provided the authors to improve the quality of the book.

EMC ANALYSIS METHODS AND COMPUTATIONAL MODELS

PART I
PRELIMINARIES

CHAPTER 1

Introduction to Modeling and EMC

Modeling of an electrical system or a physical process is a very useful tool for an analyst. It permits the simulation of system behavior for a wide variety of initial conditions, excitations, and system configurations—often in a much shorter time than would be required to build and test a prototype. Because of the similarity of the simulation and experimental processes, the term *numerical experiment* is often used to describe the use of a model to predict the outcome of a particular set of conditions.

Modeling is very common in the area of electromagnetic compatibility (EMC). This chapter begins with a discussion of modeling in general and suggests how models might be used in EMC studies for the development of new systems, the retrofit hardening of old systems, and in the interpretation of experimental measurements. In short, modeling assists us in better understanding the basic mechanisms of EMC.

1.1 THE CONCEPT OF MODELING

The concepts of modeling and imaging have always been fundamental to human beings, who have long tried to build models of things that could not be seen or felt by the senses. Throughout history, the process of modeling has changed. In early times, modeling was the representation of an abstract *idea* into something concrete. More recently, physical and experimental *phenomena* have been represented by an abstraction.

The mythological Greek gods were often represented as having human forms. Egyptians have preferred animals, or human and animal combinations, to portray their gods. An inverse approach is found in the oldest book of humanity, where it is written: "And God said, Let us make man in our image, after our likeness...," and later, "so God created man in his *own* image..." (Genesis 1–26,27). Was this not an attempt to model the inconceivable, the immaterial, the abstract concept of the One whose figure was never represented and whose name (YHVH, unpronounceable) should not be pronounced?

The use of an image, portraying the role played by the god in mythology, can be considered as the first attempt in the development of the concept of a model. It is likely that without the early attempts in visualizing the abstract concept of a god, the much more difficult process of developing an abstract mathematical model of a natural phenomenon such as lightning would have been impossible.

The first abstract models appeared when the link between mathematics, which itself has developed as an abstract science, and experimental observation of natural phenomena was achieved. In an early attempt to add rigor and traceability to this process, the *scientific method* was formulated by Aristotle. The main features of this method are illustrated in Figure 1.1, in which experimental observations of a particular event are used to validate the correctness of a hypothesis about the relationship between an observed effect and its cause. This offered a picture of reality independent of the observer, and forms the basis of classical physics.

In the area of electrical engineering, one of the very first models was the representation of the magnetic effect produced by an electrical current using the lines of force of the magnetic field, which could be visualized using iron filings. Subsequently, enhancement of the magnetic lines of force was obtained by creating a multiloop circuit, and this led to the development of the concept of inductance. A conductor carrying an electric current is one of the simplest models used by the electrical engineer. It is represented by a resistor that accounts for the Joule effect (i.e., heat dissipation in the conductor), in series with an inductance.

As pointed out by Johns [1.1], successful development of a model for numerical calculations first requires a suitable theoretical model for relating a cause to an effect, and then a numerical implementation of the theory to form the numerical model. As a result, errors in either the theoretical formulation of the problem, or in the numerical implementation of the theory, can result in an incorrect numerical model. As an example, Johns

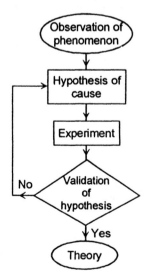

FIGURE 1.1. Basis of the scientific method.

cites a computer solution of the diffusion equation for solving heat-flow problems. Although this kind of model is widely used in science and industry, it is fundamentally incorrect, as it predicts instantaneous action at a distance, an effect known to be incorrect from Einstein's theory.†

As a consequence, the development of a numerical model is more complicated than is the application of the scientific method, as indicated in Figure 1.2. Note that like the scientific method, this method for model development relies on independent experiments or observations to validate the overall model, which consists of both the theory and the numerical implementation of the solution.

In the broad area of electromagnetics, models of observable phenomena are all based on Maxwell's equations, which constitute the most general theoretical models developed for electrical phenomena. For phenomena that are linear in nature, appropriate models can be developed directly from these equations. For nonlinear behavior, it is necessary to combine measurements of that nonlinear behavior together with the Maxwell equations to develop the model. As the number of measurements cannot be infinite, the development of nonlinear models often requires experience and intuition.

The accumulation of knowledge, together with a deeper understanding of the physics of nature, has given us the possibility of explaining more complex phenomena by building more sophisticated models. Mathematical models are often a function of several parameters, with one of the most

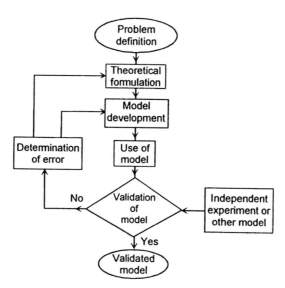

FIGURE 1.2. Development of a numerical model, based on the scientific method.

† The fact that the theory is basically incorrect, however, does not limit the usefulness of the model if it is able successfully to relate cause and effect in a predicted range of values. In this manner, the model is an approximate, yet useful, rendition of reality.

6 INTRODUCTION TO MODELING AND EMC

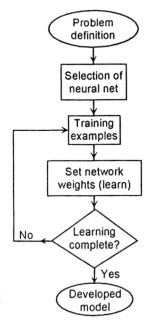

FIGURE 1.3. Characteristics of a model using a neural network.

complex models being that of weather forecasting. The challenge presented by such complex models is twofold: determining the necessary parameters to develop the many equations describing the phenomenon, and determining a way of solving the resulting system of equations in a reasonable time.

Today, the traditional concept of modeling is being challenged by alternative methods of fuzzy logic or decision processes using neural networks that are based on learning using real data. Figure 1.3 illustrates such a learning model. While such modeling techniques can produce meaningful results, they are not useful in providing a basic *understanding* of the physical process being modeled. They simply provide an output given an input—much as a "black box."

To teach such a neural network adequately, several hundred training examples might be required. Frequently, only a limited amount of experimental data is available, and this may not be enough. This lack of data can lead to "rules of thumb," which constitute a loose and informal collection of knowledge based on unstructured observations of a physical event.

1.2 VALIDATION OF MODELS

When a model is developed, the fundamental question to be answered is in regard to the relevance of the model for describing the particular phenomenon or process. This relevance depends both on the quality of the approximations made in representing reality, and on the convenience for the use of the model by the analyst. The quality and the degree of the approximation of the model can be determined only by a *validation* of the model results with experimental measurements. It is useless for an engineer to develop a model without checking it using real data, or at least with experiments that simulate the real situation.

Similarly, a motivation in developing a model is to have the capability of performing extensive parametric studies, in which independent parameters describing a model can be varied over a specified range in order to gain a global understanding of the response. A model that takes an inordinate amount of computer time to run one case, not to mention a complete parametric study, is not particularly desirable. For such purposes, a more relevant model concept might be one that provides results more rapidly, even if a degradation in solution accuracy results.

1.2.1 Example of Experimental Model Validation

A typical example showing the various ways in which a model can be validated is given by the modeling of lightning-induced voltages on an overhead conductor. This particular model can be constructed using three steps:

1. The spatial-temporal distribution of the current in the lightning channel discharge must be determined.
2. The electric and magnetic fields created by the lightning channel current are then evaluated.
3. The coupling of these fields to the overhead conductor is ultimately determined to provide a knowledge of the induced current and voltage.

To validate the first two steps in this model, full-scale measurements using natural or triggered lightning are necessary, as laboratory conditions simulating a lightning discharge are relevant only for the predischarge part of the phenomenon (i.e., for the linear portion of the response), not for the entire response. The third part of the model, involving the field-to-line coupling, can be validated in any of the electromagnetic pulse (EMP) simulators, which creates an impulsive electromagnetic plane wave provided that the proper excitation field is used in the simulator.

1.2.2 Model Validation Using Nonexperimental Methods

Although validation of a model by experimental results is the most convincing test of the validity of a model, at times it is unfeasible to conduct the necessary experiments due to cost, logistics, scale size, time duration, or a myriad of other obstacles. In such cases the important area of validation must be accomplished using alternative, indirect means.

8 INTRODUCTION TO MODELING AND EMC

One possible way is to use other validated models to validate a new model. For example, a newly developed antenna analysis computer model might be validated by comparing its results with those calculated for the same geometry using an older, perhaps slower computer code which has already been validated through extensive studies involving hundreds of users. Alternatively, a validated model that uses a different computational technique can be used to check on a model that employs a different method of solution. For example, a finite-difference solution of Maxwell's equations might be used to validate a solution obtained by using an integral equation's solution.

Alternatively, it is possible to validate a model partially using one or more "gedanken" experiments, in which the results of the model are compared with a priori knowledge of the behavior of the system. Note that this provides the necessary, but not sufficient, evidence as to the correctness of the solution. As an example, the following concepts can frequently be used to examine the validity of a model:

- Conservation of energy
- Causality (or turn-on time of a response)
- Times of arrival of waveform response components
- Low- or high-frequency asymptotic behavior of spectral responses
- Other known physical constraints of the solution (i.e., finite Q, etc.)

1.3 BUILDING MODELS IN ELECTROMAGNETICS

The modeling process in the field of electromagnetic compatibility (EMC) means the establishment of a relationship between a cause (such as a source of interference) and its effect (such as the response of a circuit, which can be part of an installation). This relationship can be established in several ways, depending on the type of problem, its complexity, and the degree of approximations with respect to an exact formulation. The possible methods involve:

- Using circuit theory for describing the conducted disturbances, such as overvoltages, voltage dips, voltage interruptions, harmonics, and common ground coupling.
- Using an equivalent circuit with either distributed or lumped parameters, such as in low-frequency electromagnetic (EM) field coupling expressed in terms of mutual inductances and stray capacitances, field-to-line coupling using the transmission line approximation, and cable crosstalk.
- Formulating the problem in terms of formal solutions to Maxwell's

equations and the appropriate field boundary conditions, as for example in problems involving antenna scattering and radiation.

The use of basic circuit theory concepts for electromagnetics permits the development of a system of linear equations to represent the mathematical formulation of a model. However, only a very limited number of physical processes can be described in such a simple way so as to give a closed-form solution to such a set of equations. An early step that permitted significant progress in electromagnetic modeling was the development of numerical methods for the inversion of a large system of equations. This development, coupled with the tremendous advances in computer technology, has resulted in the possibility of developing complex models that would have been impossible 20 years ago.

In discussing computational techniques for the case of electromagnetic field coupling to structures, Miller [1.2] refers to the relationship between the cause and its effect as a *field propagator*. He employs this concept to classify the various solution techniques for EM modeling, as summarized in Table 1.1. The development of a computer model, together with the analytical and numerical issues in realizing such models, are discussed in [1.2]. This paper presents an exhaustive survey of computational electromagnetics. Many of these computational methods can be useful in the area of EMC modeling.

1.4 EMC MODELING: A HISTORICAL OVERVIEW

As EMC is related to electrical noise suppression, its birth could be related to the use of very low currents in practical applications, such as in the use of Morse code transmission of information over long distances using telegraphic wires. Direct- or indirect-strike lightning effects could have been a cause of such interference, but no major equipment damage seems to have been recorded, probably due to the robustness of the telegraph devices.

In 1895, the first radio signals were transmitted by Marconi in Italy [1.3]. He also was the first to transmit over the Atlantic Ocean in 1902. At the same time, Popov in Russia transmitted radio signals over various distances. The use of electromagnetic waves to transmit information thus created a new source of possible interference to electrical systems.

For many years little modeling or mathematical estimation of the degree of EM interference was attempted, and only practical, rule-of-thumb protection measures were taken to eliminate electromagnetic interference (EMI). For instance, as recently as the 1950s, the rooftop antenna of a wireless set was supposed to capture the desired signals and the set was designed to discriminate the useful signal from other signals as well as the general atmospheric noise. However, wideband signals like those produced by atmospheric discharges due to cloud-to-cloud or cloud-to-ground lightn-

TABLE 1.1 Classification of Model Types in Computational Electromagnetics

Field Propagator	Description	Application	Requirements	Problem Type	Characterization
Integral operator	Green's function for infinite medium or special boundaries	Radiation	Determining the originating sources of a field	Solution domain	Time or frequency
Differential operator	Maxwell curl equations or their integral counterparts	Propagation	Obtaining the fields distant from a known source	Solution space	Configuration or wavenumber
Modal expansions	Solutions of Maxwell's equations in particular coordinate system and expansion	Scattering	Determining the perturbing effects of medium	Dimensionality	One, two, or three dimensions
Optical description	Rays and diffraction coefficients			Electrical properties of medium and/or boundary	Dielectric; lossy; perfect conducting; anisotropic; homogeneous; nonlinear
				Boundary geometry	Linear; curved; segmented; compound; arbitrary

Source: E. K. Miller, "A Selective Survey of Computational Electromagnetics," *IEEE Trans. Antennas Propag.*, Vol. AP-36, No. 9, September 1988. © 1988 Institute of Electrical and Electronics Engineers.

ing, were strong enough to disturb AM radio reception significantly on medium and particularly long waves (150 to 1600 kHz). With such communications systems, protection was rather primitive: At the approach of a thunderstorm it was recommended that the antenna wire be disconnected and grounded to avoid destruction of the radio set from a voltage surge induced by lightning.

In 1934, radio transmission became so important that a certain protection from industrial noise was found to be necessary. As a result, a regulatory commission to define rules and recommendations for avoiding radio interference was created: the Comité International Spécial des Perturbations Radioélectriques (CISPR).

It is interesting to note that as in other fields of science, EMC has its forerunners whose early theories found no large response or application at the time of their publication. In 1934, in the *Bell System Technical Journal*, S. A. Schelkunoff published a theory on electromagnetic field penetration through metallic shields [1.4]. This theory was applied much later in the early 1960s to the calculation of the transfer impedance of tubular sheaths, just at a time when the interest for such a subject was no longer simply theoretical but found important practical applications due to the development of EMC requirements for newly developed systems.

The development of electronic components that are increasingly sensitive to lower energies, together with their use for military applications, has created the need to find solutions to make these installations compatible with the electromagnetically disturbed environments [1.5, 1.6]. Furthermore, the decrease in normal signal levels, the increase in the operating frequency, and the use of digital electronics in modern systems have led to the need for heightened EMC considerations, and ultimately, better EMC models.

EMC modeling activities were spurred by the development of electronics for military applications. This was due to the requirement to design reliable installations hardened against a high level of electromagnetic interference. The most typical example of such military-created interference is the nuclear electromagnetic pulse (NEMP), in particular the high-altitude EMP (HEMP), which is a transient plane electromagnetic wave produced by a high-altitude nuclear explosion. This requirement has permitted the development of a wide variety of EMC models [1.7], including models for field coupling to aircraft and antennas, shielding models, field-to-transmission line coupling models, and EM propagation studies. Furthermore, the many available EMP simulators have permitted the validation of such models.

These coupling models developed for HEMP analysis are used today to predict a variety of effects, ranging from lightning-induced voltages on overhead power and distribution lines to interference in electronic circuits of airplanes. Furthermore, for extremely complex facilities in which measurements are the only realistic way of assessing the electrical behavior of the system, EMC coupling and interaction models can be useful in the design process as well as in interpreting the measured data.

The move away from the rule-of-thumb approach to EMC has been documented in a number of books on techniques and models for EMC purposes: [1.8] through [1.14] are representative of these texts.

1.5 CONSIDERATIONS FOR EMC MODELING

1.5.1 Classification of EMC Problems

The large variety of EMC problems stems from the many possible types of EMI sources. To develop a model for a particular EMC problem, the first step is to try to classify the type of "signal" path connecting the EMI source to the victim. The coupling between the source and the victim can occur through (see Section 2.3.2) electrical conduction or EM fields.

The sources of EMI are much more difficult to classify, due to their diversity. However, to develop a model, the first issue to be determined is the type of source causing the interference. A top-level classification puts sources into two separate categories, depending on the initial cause: natural noise, or man-made noise. For natural noise, the five categories of sources listed in Table 1.2 cover the principal natural phenomena that are a cause of disturbance for the electrical equipment. Man-made noise is so varied that it is difficult to be exhaustive in its discussion. A classification of sources due to different causes, with examples as discussed by Morgan [1.15], is presented in Table 1.3.

An alternative classification method that is useful for determining the modeling approach is to separate the sources into continuous wave (CW) and transient. Examples of EMI sources classified in this matter are shown in Table 1.4. From the modeling point of view, the frequency-domain approach is more suited for continuous EMI sources. Conversely, time-domain models are more appropriate for transient sources. After having determined the type of EMI source, several additional parameters are needed for the analysis, such as the amplitude of the disturbance signal and its frequency for the fixed-frequency emitters, or the rise and time to half-value for an impulse source.

Examples of characteristic parameters for some typical fixed-frequency or impulsive interference sources are given in Table 1.5. In addition, Table 1.6 shows the effective radiated power and the range of field strengths of various

TABLE 1.2 Classification of Natural EMI Sources

Static noise due to precipitation
Atmospherics (noise from lightning around the world)
Nearby and medium-distant lightning
Solar noise effects (whistlers, solar-disturbed and quiet radio noise)
Cosmic radio noise

TABLE 1.3 Classification of Man-Made EMI Sources

EMI sources due to the power network and its equipment:
  Switching operations   Static and rotary converters
  Power faults   Rectifiers
  Electric motors   Contractors

EMI sources due to industrial and commercial equipment:
  Arc furnaces   Fluorescent lamps
  Induction furnaces   Neon displays
  Air conditioning   Medical equipment
  Computer and switching circuits

EMI sources due to machines and tools:
  Workshop machines   Rotary saws
  Rolling mills   Compressors
  Cotton mills   Ultrasonic cleaners
  Welders

EMI sources due to vehicle systems:
  Automobile and vehicle ignition systems
  Electric locomotives

EMI sources due to communication systems:
  Radio broadcast stations   Citizens'-band
  TV stations   Mobile telephones
  Radar   Remote control door-opening transmitters

EMI sources due to consumer devices:
  Microwave ovens   Vacuum cleaners
  Refrigerators/freezers   Hair dryers
  Thermostats   Shavers
  Mixers   Light dimmers
  Washing machines   Personal computers

Source: D. Morgan, *A Handbook for EMC Testing and Measurement*, IEE Electrical Measurement Series 8, Peter Peregrinus, London, 1994. (Reprinted by permission.)

radio, TV, and radar emitters. The multiplicity of sources and the variety of situations in these tables shows that it is practically impossible to try to solve all EMC problems by modeling, or to predict the behavior of an installation in all possible situations. Modeling is therefore limited in its scope and capability in treating real-life problems.

1.5.2 EMC Problems Amenable to Modeling

Typical examples of EMC problems for which modeling can be applied successfully are the following:

1. Conducted disturbances

TABLE 1.4 Continuous and Transient EMI Sources

Sources of Continuous EMI (Sources with Fixed Frequency)	Sources of Transient EMI (Sources with a Large Frequency Spectrum)
Broadcast stations	Lightning
High-power radar	Nuclear EMP
Electric motor noise	Power line faults (sparking)
Fixed and mobile communications	Switches and relays
Computers, visual display units, printers	Electric welding equipment
High-repetition-rate ignition noise	Low repetition ignition noise
Ac/multiphase power rectifers	Electric train power pickup arcing
Solar and cosmic radio noise	Human electrostatic discharge

Source: D. Morgan, *A Handbook for EMC Testing and Measurement*, IEE Electrical Measurement Series 8, Peter Peregrinus, London, 1994. (Reprinted by permission.)

TABLE 1.5 Frequencies and Noise Levels from Typical Interference Sources

Source Type	Comments
Power mains disturbance	Double exponential transients with rise times of about 1 ms, fall times of tens of ms, and peak value of about 10 kV
	100-kHz ringing waveform with 0.5-ms rising edge
	Power dips up to 100 ms long
	Power frequency harmonics up to 2 kHz
Unintended radiators	
Switches and relays	Transients with rise times of a few ns and levels up to 3 kV producing frequencies into the VHF band
Commutator motors	Frequencies up to 300 MHz at repetition rates up to 10 kHz
Human electrostatic discharge	Rise times of 1 to 10 ns
Switching semiconductors	Rise times from 20 to 1000 ns at a repetition rate of 1 kHz to 10 MHz for voltages up to 300 V
Switched-mode power supplies	Continuous spectrum of noise from 1 kHz to 100 MHz
Digital logic	Continuous noise from kHz up to 500 MHz
Industrial and medical equipment	Metals heating in the range 1 to 199 kHz; medical equipment operates from 13 to 40 MHz at a high power of hundreds of watts
Intended radiators	
Broadcast stations	See Table 1.6
Other RF transmitters, including radar	See Table 1.6

Source: D. Morgan, *A Handbook for EMC Testing and Measurement*, IEE Electrical Measurement Series 8, Peter Peregrinus, London, 1994. (Reprinted by permission.)

TABLE 1.6 Effective Radiated Power and Field Strengths from Authorized Services

Service	Frequency Range (MHz)	Effective Radiated Power (dBW)	Usual Range of Separation Distances (km)	Estimated Range of Field Strengths (V/m)
Low frequency (LF) Communications and navigation aids	0.014–0.5	54	5–20	0.25–1.0
AM broadcast	0.5–1.6	47–50	0.5–2	0.2–6.0
HF amateur	18–30	30	20–100	3–15
HF communications	1.6–30	40	4–20	0.001–0.25
Citizens' band (CB)	27–27.5	11	10–100	0.3–3.0
VHF amateur	50–225	30–39	0.1–0.5	0.35–5.0
Fixed and mobile communications	30–470 900–1000	17–21	0.04–0.2	0.2–1.5
Television (VHF)	54–216	50–55	0.5–2.0	1.0–7.3
FM broadcast	88–108	50	0.25–1.0	2.0–8.3
Television (UHF)	470–890	67	0.5–3.0	1–3
Radar	1000–10,000	90–97	2–20	0.1–90

  a. Voltage interruptions and voltage dips
  b. Overvoltages due to direct lightning effects on the power or telecommunication network
  c. Harmonics
  d. Overvoltages due to switching operations (closing and opening operations, transformer taps, opening of nonbalanced loads in networks with insulated or impedance grounded neutral)
 2. Radiated disturbances
  a. Lightning EMP (LEMP) and nuclear EMP (NEMP) effects. Concerning lightning specifically, three specific models are necessary to evaluate lightning-induced voltages on lines and cables:
   • A model for the current propagation in the discharge.
   • A model for the electromagnetic field propagation over a real soil with finite conductivity.
   • A coupling model.
  This last model is applied also for NEMP and discussed in Chapters 6 and 7.
  b. Antenna radiation. This subject is developed in Chapter 4.
  c. Crosstalk between two circuits. An example of crosstalk solved by using a code based on Bergeron's method is given in Section 6.3.7. An important contribution to crosstalk modeling has been provided by Paul (see, e.g., [1.16] and [1.17]).
  d. Problems related to printed circuit boards (PCBs). EMC related to

16 INTRODUCTION TO MODELING AND EMC

PCBs have been discussed by Gravell and Wilson [1.18], who have listed three basic issues:
- Assuring voltage and current waveform integrity throughout the interconnecting paths.
- Suppression of conducted and radiated emission.
- Hardening against susceptibility failure.

They also provide an exhaustive literature review of the subject.

e. The interconnection of two computer boards by a multiwire cable. A common design involves ribbon cables due to cost considerations. Because of the possible presence of common-mode currents on the cables, this configuration can constitute an efficient radiating system and may cause EMI problems, as discussed in Chapter 7.

f. Shielding effectiveness and penetrations. These two items are discussed in Chapters 9 and 10.

1.5.3 Types of Signals in EMC Models

As mentioned previously, EMI can arise from unwanted electrical signals being injected directly into a victim by conducting wires or by the interaction of the victim with electromagnetic fields. In either case, the EMI can be characterized by either a time-domain *waveform* of some pertinent quantity, or by the corresponding Fourier transform of the quantity, which constitutes the frequency-domain *spectrum* of the disturbance. The physical quantities that are usually represented in EMI problems include voltages, currents, charge, electric fields, and magnetic fields. In developing EMC analysis models, it is important to define carefully the type of EMI that is being considered. Some models will be appropriate for limited classes of EMI environments, while others will be widely applicable to a variety of different signals.

Table 1.7 illustrates several general classes of disturbing signals that are pertinent for the models discussed in this book. The table illustrates both the transient waveform and the frequency-domain spectrum that is indicative of the bandwidth of the signal. Waveform 1 is an idealized, continuous sinusoidal signal, having an amplitude A and a frequency f_0. This waveform might be used to represent the EMI produced by a 50- or 60-Hz power main in the vicinity of a sensitive electronic device. Because there is no start or stop to this waveform, its frequency-domain spectrum consists only of a single frequency, f_0, as illustrated in the table.

The infinitely long waveform is an idealization, however, as there must always be a start and a stop to the waveform. An extreme example is waveform 2 in the table, which is a sinusoidal pulse function where the sinusoid is nonzero only over the interval t_0 to t_1. The corresponding spectrum still has a peak at the sinusoidal oscillation frequency f_0, but the energy in the waveform also exists at other frequencies. As the pulse width

TABLE 1.7 Different Types of Waveforms and Spectra Encountered in EMC Models

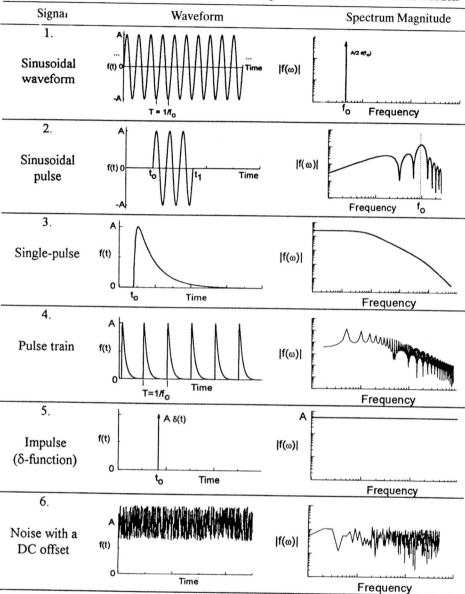

$(t_1 - t_0)$ increases, the waveform begins to look more like waveform 1 and the spectrum becomes more sharply peaked at f_0. Conversely, as the pulse width decreases, the waveform has significant energy away from the carrier frequency and a simple monochromatic analysis may not be accurate.

Waveform 3 is a single transient signal, starting at a time t_0 and slowly decaying away. With a more complex waveshape, it could represent the disturbing E-field produced by a lightning or electrostatic discharge (ESD). This is clearly a wideband signal, with a significant amount of energy at low frequencies. To understand the effects of this signal on an electrical system, it is necessary to compute the response in the entire frequency range for which the excitation is significant.

Waveform 4 is a periodic train of pulses such as waveform 3, occurring at time intervals T. This spectrum has many peaks occurring at frequencies $f_n = n/T$ with $n = 1, 2, 3, \ldots$, together with an overall shape similar to that of the single pulse. If a particular system happens to be particularly sensitive to signals at one of the harmonic frequencies f_n, it can be seen that the pulsing of the waveform can be much more damaging to the system than just the single pulse.

The impulse (δ-function) is shown as waveform 5. Although this is a mathematical artifice, since all physical waveforms have finite rise and fall times, it is particularly useful for analysis purposes. Note that for this distribution function, the frequency spectrum amplitude is a constant, implying that all frequencies are present in the signal with equal amplitude.

Finally, random transient noise is also encountered in EMC models. Waveform 6 portrays a random white noise signal superimposed on a dc signal. This class of waveforms might be encountered in the design of a low-noise, cryogenic frontend for a microwave receiver. Each of these transient disturbances can be represented symbolically by the function $f(t)$. As noted in the examples, this function need not be continuous but must be a single-value function of time. Since the squared function $f^2(t)$ is proportional to an instantaneous power quantity, the time integral of this quantity is proportional to the total energy contained in the signal, as

$$W \propto \int_{-\infty}^{\infty} |f(t)|^2 \, dt \qquad (1.1)$$

For any real system or process, this expression must be finite. The infinite sinusoidal waveform of Table 1.7 clearly violates this requirement unless it is assumed to start and stop at distant times.

1.5.3.1 Evaluation of the Frequency Spectra.
Each of the time-domain signals of Table 1.7 can also be described by their frequency spectra, which consists of a real and an imaginary part, or equivalently, a magnitude and phase. In this table, only the magnitude function is illustrated. Using the theoretical development of the Fourier transform [1.19], a reasonably well-behaved function of time $f(t)$ can be thought of as being comprised of a superposition of many different sinusoidal waveforms, each having a distinct amplitude and phase. There are several different forms for this expression

[1.20], one of which is the exponential form:

$$f(t) = \frac{1}{2\pi} \int_{-\infty}^{\infty} F(\omega) \exp(j\omega t)\, d\omega \qquad (1.2a)$$

The spectral density, $F(\omega)$, is a complex-valued function that represents the amplitude of a sinusoidal waveform component given by the phasor $\exp(j\omega t)$. It may be determined from a knowledge of the time function through the inverse relation,

$$F(\omega) = \int_{-\infty}^{\infty} f(t) \exp(-j\omega t)\, dt \qquad (1.2b)$$

As a consequence of these expressions, it is possible to reconstruct a transient waveform from its frequency spectrum. Examples of the various spectral magnitudes for the transient waveforms are also shown in Table 1.7. Note that for real-valued time functions, these spectra have the property that $F(\omega) = F^*(\omega)$, where * denotes the complex conjugate. Consequently, only the spectrum for positive frequencies is needed to compute the transient response.

The ability to construct a transient response from its frequency spectrum is very useful, as many analysis techniques for linear systems have been developed using the concept of a time-harmonic analysis [1.21]. This amounts to determining the response of a system to a single sinusoidal excitation waveform of infinite duration, with no defined turn-on time. Of course, this analysis must be performed for a large number of frequencies, which is usually referred to as wideband analysis.

From a practical standpoint, however, evaluation of the integrals in Eq. (1.2) can be time consuming, especially if there are several thousand points in the transient waveform or in the frequency spectrum. An efficient numerical method for performing Fourier transforms is the fast Fourier transform (FFT), which was popularized by Cooley and Tukey. This algorithm is a discrete Fourier transform that is based on powers of 2 and has been discussed by many authors, including [1.22] and [1.23].

Throughout this book, it is assumed that the reader is familiar with the concepts of Fourier transformation and linear system analysis. These will form the basis of all analytical and numerical models to be discussed later. Most of the models are developed in the frequency domain, yielding responses that are explicit functions of the frequency ω (or f). Transient results usually are obtained by numerically evaluating the Fourier integral in Eq. (1.2a).

1.5.4 Limits of Modeling

It can be said that all basic physical phenomena representing the interface between a disturbing EMI source and a victim circuit can be modeled to one degree or another. The basic question, therefore, is that of the accuracy of the model. Presently, the main limit to EMC modeling arises from the physical and electrical complexity of the real circuits or installations. As an example it is possible to model rather accurately the coupling between an electromagnetic field and several parallel transmission lines, or an inductive coupling between one high current line and several parallel low signal lines. However, when cables consisting of nonshielded conductors, coaxial lines, or bundles, each with several tens of conductors, are installed with a complicated routing in an automotive type of vehicle, it is virtually impossible to perform a calculation because:

- The electrical characteristics (inductances and capacitances) are too difficult to determine.
- The number of independent parameters describing the model become so large that the dimensions of the computer codes and the computer time become prohibitive.

Another difficulty can come from the multidimensional size of an EMI problem. It is obvious that several sources can produce disturbances that can affect a single victim and that several coupling paths can exist between one or several sources and the victim. A way of solving such a problem as proposed by Goedbloed [1.24] is to consider the sources and the coupling possibilities individually. This can be done if the sources and the coupling paths are independent of each other, but at times this assumption can be wrong. An example is a digital circuit of a computer submitted to the noise of an air conditioner [1.24]. No interference occurs until a radio signal of a nearby radio station gives a dc shift, which erodes part of the disturbance margin of the components, causing the air conditioner to become an EMI source.

1.6 WHO IS USING MODELING AND TO WHOM MODELING IS USEFUL

Up to the end of the 1980s, the approach to EMC was very different in different countries. In the United States, where the development of EMC had begun earlier, there was strong legislation against EM interference. In 1981 a federal law on computing devices (FCC part 15, subpart J) was adopted [1.25]. This strong U.S. law was a response to the explosive development of personal computer usage and with the continuing development of consumer electronics products. Germany also adapted EMC legislation, closely following the recommendations of the CISPR. In Switzerland, the Swiss PTT published recommendations, and in France, no culprit was considered as long as no victim was noted.

The situation changed radically in 1989, when the European Community (today the European Union) published the EMC Directive [1.26]. This directive states that in the 15 states of the Union, after January 1, 1996, "each apparatus used in an installation must comply individually to the relevant EMC requirements and tests." After this mandate, a remarkable effort has been made to produce the necessary standards and to define tests methods for homologation and certification.

In this context, modeling becomes useful, not only to retrofit an installation when a problem occurs, but also as a tool for designing new products. It has been estimated that EMC at the design level represents about 5 to 7% of product cost. If a prototype is already built, the introduction of EMC measures can make the product 50% more expensive, and the retrofit can double the costs.

The development of analytical and numerical modeling techniques has had a marked impact on the area of EMC. These techniques are used in the design, construction, test, and evaluation phases of:

- Defense electronics systems
- Communications and data transmission systems
- Power utilities
- Consumer electronics

In this context, analytical and numerical methods are used for a wide variety of purposes, including:

- Predicting system-level responses to external and internal EMI
- Evaluating the behavior of EMC protection measures
- Processing measured system test data

The application of analytical methods in these areas involves different modeling techniques, computational requirements, and accuracy expectations. For example, it is usually much more difficult to perform an accurate calculation of the EMI-induced voltage or current in an electrical component in a system than it is to process and draw conclusions from measurements of the same responses. Although internal response calculations may be inaccurate, calculations of relative changes of a response due to changes in the system design can be used to provide guidance to the designer for constructing a system that is hardened against EMI.

As discussed above, it is not possible to model everything. However, even for the most complex of systems, it is usually possible to obtain the order of magnitude of possible interference levels and thus provide guidance on the directions in which to look for improvements in the design. This can be of great benefit to the designer, and it is hoped that this book will provide the necessary material for this purpose.

REFERENCES

1.1. Johns, P. B., "The Art of Modeling," *Electron. Power*, August 1979, pp. 565–569.

1.2. Miller, E. K., "A Selective Survey of Computational Electromagnetics," *IEEE Trans. Antennas Propag.* Vol. AP-36, No. 9, September 1988.

1.3. Baker, W. J., *A History of the Marconi Company*, St. Martin's Press, New York, 1974.

1.4. Schelkunoff, S. A., "Theory of Lines and Shields," *Bell Syst. Tech. J.*, Vol. 13, No. 4 (1934), pp. 522–579.

1.5. Ryser, H., "Electromagnetic Compatibility," *Hasler Rev.*, Vol. 17, No. 2, 1984, pp. 33–40.

1.6. Joehl, W., "A General and Systematic Survey of NEMP Protection Measures," *Report FMB 78-1*, Research Institute for Protective Construction (FMB), Zurich, January 1978.

1.7. Lee, K. S. H., ed., *EMP Interaction: Principles, Techniques and Reference Data*, Hemisphere, New York, 1989.

1.8. Keiser, B. E., *Principles of Electromagnetic Compatibility*, Artech House, Dedham, MA, 1979.

1.9. Ianovici, (Ianoz) M., and J. J. Morf, eds., *Compatibilité Electromagnétique*, Presses Polytechniques Romandes, Lausanne, Switzerland, 1983.

1.10. Ott, H. W., *Noise Reduction Techniques in Electronic Systems*, Wiley, New York, 1988.

1.11. Paul, C. R., *Introduction to Electromagnetic Compatibility*, Wiley, New York, 1992.

1.12. Degauque, P., J. Hamelin, eds., *Electromagnetic Compatibility*, Oxford University Press, Oxford, 1993 (also in French).

1.13. Christopoulos, C., *Principles and Technique of Electromagnetic Compatibility*, CRC Press, Boca Raton, FL, 1995.

1.14. Perez, R., ed., *Handbook of Electromagnetic Compatibility*, Academic Press, New York, 1995.

1.15. Morgan, D., *A Handbook for EMC Testing and Measurement*, IEEE Electrical Measurement Series 8, Peter Peregrinus, London, 1994.

1.16. Paul, C. R., "Prediction of Crosstalk in Ribbon Cables: Comparison of Model Predictions and Experimental Results," *IEEE Trans.*, Vol. EMC-20, No. 3, August 1978, pp. 394–406.

1.17. Paul, C. R., "Transmission-Line Modeling of Shielded Wires for Crosstalk Prediction," *IEEE Trans. Electromagn. Compat.*, Vol. EMC-23, No. 4, November 1981, pp. 345–351.

1.18. Gravelle, L. B., and P. F. Wilson, "EMI/EMC in Printed Circuit Boards: A Literature Review," *IEEE Trans. Electromagn. Compat.*, Vol. EMC-34, No. 2, May 1992, pp. 109–116.

1.19. Papoulis, A., *The Fourier Integral and Its Applications*, McGraw-Hill, New York, 1962.

1.20. Hildebrand, F. B., *Advanced Calculus for Applications*, Prentice Hall, Englewood Cliffs, NJ, 1963.

1.21. Cheng, D. K., *Analysis of Linear Systems*, Addison-Wesley, Reading, MA, 1959.

1.22. Press, W. H., et al., *Numerical Recipes*, Cambridge University Press, Cambridge, 1986.

1.23. Bingham, E. O., *The Fast Fourier Transform*, Prentice Hall, Englewood Cliffs, NJ, 1974.

1.24. Goedbloed, J. J., *Electromagnetic Compatibility*, Prentice Hall, Englewood Cliffs, NJ, 1992.

1.25. Charoy, A., "Perturbations radioélectriques des appareils de traitement de l'information: la normalisation française s'aligne," *Rev. Gen. Electr.*, No. 10, November 1986, pp. 14–17.

1.26. Council directive of May 3, 1989 on the applicability of the laws of the member states relating to electromagnetic compatibility.

PROBLEMS

1.1 A good example of the development of a model is found in circuit theory, where complex phenomena are represented by simple approximations. Consider the resistor shown in Figure P1.1, which consists of a block of carbon material sandwiched between two conducting endplates that are connected to wire leads. The common view of this circuit element is that it can be represented by a simple resistance R and that the instantaneous voltage and current through the element are related by Ohm's law, $v = R I$. As shown in the figure, however, this simple model is valid only at low frequencies. At high frequencies, the capacitance between the plates and the inductance of the resistor leads becomes important. Furthermore, mutual capacitance between

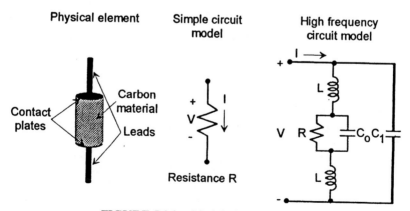

FIGURE P1.1. Models for a resistor.

the leads exists. This high-frequency behavior of the resistor can be modeled by a more complicated interconnection of elements, as shown in the figure.

(a) Develop a suitable high-frequency model for a capacitor.

(b) Develop a similar high-frequency model for an inductive coil. What happens to the model if the coil is wrapped around a material with high permeability μ? What happens to the model if μ is a nonlinear function of the magnetic field H?

1.2 Examine the following systems or system components and discuss at which frequencies they are electrically small: **(a)** a large commercial aircraft; **(b)** a fighter aircraft; **(c)** a table radio or TV set; **(d)** a hand-held calculator; **(e)** an electronic wristwatch; **(f)** a microchip circuit. Assume that they are located in free space, with $c = 3 \times 10^8$ m/s.

1.3 Discuss the spectral characteristics (i.e., frequency, wavelength, bandwidth, modulation, polarization, etc.) of the following radiated emissions: **(a)** AM radio signals; **(b)** FM signals; **(c)** television broadcasts; **(d)** digital telephony; **(e)** lightning EMI; **(f)** spread-spectrum communications; **(g)** NEMP; **(h)** 50- to 60-Hz "hum"; **(i)** electrostatic discharge.

1.4 For the Fourier transform pair $f(t) \Leftrightarrow F(\omega)$ given by Eq. (1.2), prove Parseval's theorem for expressing the energy contained in the signal:

$$W = \int_{-\infty}^{\infty} |f(t)|^2 \, dt = \frac{1}{2\pi} \int_{-\infty}^{\infty} |F(\omega)|^2 \, d\omega$$

1.5 Given a transient waveform $f(t) = A_0 e^{-\alpha t} U(t)$, where $U(t)$ is the unit step function defined as 0 for $t < 0$ and 1 for $t \geq 0$:

(a) Evaluate the Fourier transform $F(\omega)$, and

(b) Evaluate the spectral power density W.

1.6 Assume that $f(t)$ in Problem 1.5 is shifted in time such that

$$f(t) = A_0 e^{-\alpha(t-t_0)} U(t-t_0)$$

(a) Show that the new Fourier spectrum is given by $F(\omega) e^{-j\omega t_0}$.

(b) How does this time shift affect the energy contained in the new spectrum?

1.7 Given two Fourier transform pairs $f_1(t) \Leftrightarrow F_1(\omega)$ and $f_2(t) \Leftrightarrow F_2(\omega)$, show that the Fourier transform operation is linear, so that

$$\alpha f_1(t) + \beta f_2(t) \Leftrightarrow \alpha F_1(\omega) + \beta F_2(\omega)$$

where α and β are real-valued constants.

1.8 For the Fourier transform pairs $f_1(t) \Leftrightarrow F_1(\omega)$ and $f_2(t) \Leftrightarrow F_2(\omega)$, prove the convolution theorem:

$$F_1(\omega)F_2(\omega) \Leftrightarrow \int_{-\infty}^{\infty} f_1(\tau)f_2(t-\tau)\,d\tau = \int_{-\infty}^{\infty} f_1(t-\tau)f_2(\tau)\,d\tau \equiv f_1(t)*f_2(t)$$

1.9 Consider $f_1(t) = e^{-t}U(t)$ and $f_2(t) = e^{-2t}U(t)$, where $U(t)$ is the unit step function.
  (a) Plot the resulting convolution of the two functions $g(t) = f_1(t) * f_2(t)$.
  (b) Compute the Fourier transform of $g(t)$ by evaluating the integral of Eq. (1.2b).
  (c) Compute the Fourier transform of $g(t)$ by evaluating the transforms of f_1 and f_2 individually and using the convolution theorem.

1.10 Show that for $f(t)$ assumed to be a real-valued function, its Fourier transform has the property that $F(\omega) = F^*(-\omega)$, where * denotes the complex conjugate. This is a useful property when evaluating inverse Fourier transforms numerically.

1.11 Show that for the Fourier transform pair $f(t) \Leftrightarrow F(\omega)$, the following are true:

  (a) $F(0) = \int_{-\infty}^{\infty} f(t)\,dt$ **(b)** $|F(\omega)| \leq \int_{-\infty}^{\infty} |f(t)|\,dt$

CHAPTER 2

System Decomposition for EMC Modeling

In this chapter, the roles that analysis plays in the area of EMC are discussed. Key to the application of any mathematical model representing the behavior of a system is knowledge of how to decompose the system into simpler parts. This may be accomplished using the concept of EM shielding topology, which is introduced in this chapter.

2.1 THE APPLICATION OF ANALYTICAL METHODS IN EMC

Analytical modeling techniques can be used throughout the planning, design, and construction phases of an electrical system. As an example of the different areas requiring analytical techniques for a typical system, consider the diagram shown in Figure 2.1.

2.1.1 The System Design Phase

Given a specified set of system operational requirements and a specification of the EMI environment that the system is expected to encounter, a preliminary design of the system can be developed. Depending on the function and mission criticality of the system, a specific EMC requirement will be defined. Prior to developing a detailed design of the system, a set of first-order bounding calculations might be performed to assist in developing the EMC design concept and in providing protection allocations for the system. This involves:

- Use of the specified EM environment for the system
- Developing simple system-level coupling models
- Estimating the electrical levels to be experienced by internal equipment

Such stress levels, when compared with failure or upset levels of the equipment or components, provide the system designer with the needed

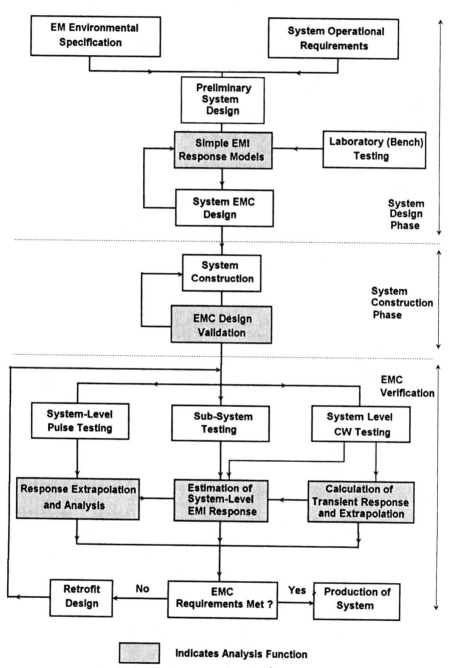

FIGURE 2.1 Overall role of analysis in system development.

information to perform trade-off studies of different conceptual EMC designs.

After a suitable EMC design concept has been established for the system and the detailed design process begun, EMC models continue to be used to provide the designer with detailed information on the behavior of EM environment control elements and techniques incorporated into the design. This is usually an iterative process, with elements of the design being refined with the help of a continuing analysis throughout the design phase. Such calculations are frequently applied to simplified, idealized geometries, which are referred to as *canonical* problems. The results of such analysis can provide an answer to an isolated technical question or concern and do not attempt to give information about the entire system-level response. The results of these studies allow the designer to begin obtaining information as to whether or not his design for the system is feasible, and if not, what steps should be taken to improve the design.

Often, in the design process, small-scale laboratory or "bench" tests are used to complement the analytical studies of canonical problems and to provide accurate technical data for evaluating the design procedure. Additional analytical studies of the system responses are usually performed at this stage to understand the results of the laboratory testing, and to help integrate the measured data into the overall system EMC design.

2.1.2 System Construction Phase

Once the system's design has been finalized and the construction phase begins, analytical models continue to be used for the purpose of providing support for the validation of the system EMC. Variations of the original design will arise, and these must be scrutinized as to their effect on the system EMC. In addition, analysis is frequently used to ensure that the original EMC design is being implemented correctly.

2.1.3 EMC Verification

2.1.3.1 Immunity Testing. After the system has been constructed, it is necessary to verify that the system meets the original EMC design goal of not being adversely affected by externally produced EM environments. System-level EMI simulations, preferably at the expected EMI levels, are usually the best approach for this verification. However, such simulations are usually imperfect and do not provide complete verification of the system EMC. In such cases, analytical modeling techniques have been developed to extrapolate measured system responses to the actual EM environment. For linear systems, such response extrapolations are straightforward, involving the manipulation of frequency-domain spectra, or equivalently, convolutions of time-domain functions. For nonlinear systems, however, extrapolation of a low-level system response to a higher level can be difficult and may require significant numerical effort.

At times, alternative test techniques are used for EMC immunity verification. Subsystem testing can be performed using a radiated EM field, or current injection excitation. The measured results can be combined analytically to provide an estimate of the total system response. It is to be stressed that in doing this, however, significant errors can arise in the computed system response. For cases of EMI involving transient excitations, full-scale testing can involve pulse testing that reproduces the actual transient excitation, or a nonpulse, broadband (CW) test, converting the measured spectral responses to the time domain and extrapolating to a threat-related response using analytical methods. In addition to these uses, analysis procedures and techniques have been developed for the acquisition, correction, interpretation, and presentation of measured data.

2.1.3.2 Emission Testing. Equally important in the design of a system is that its normal operation must not cause interference with nearby systems. To validate that this design requirement has been met, EMC emission testing is often required. Most of the same elements and procedures that arise for EMC immunity testing discussed above are found in emission testing. The main difference is that the system under consideration is now the source of the potential EMI and is no longer considered as the potential victim of the EMI. A key theoretical concept linking these two subject areas—immunity testing and emission testing—is electromagnetic reciprocity, which is discussed in Chapter 7.

2.1.4 Summary of the Use of Analytical Models

Despite the widespread use of modeling techniques in EMC, the main utility of such analysis is frequently misunderstood. Analytical methods are not useful in attempting to predict a detailed and accurate system-level response to EMI. They are not accurate enough, nor is sufficient system detail available to the analyst for this purpose. There is, however, a useful role for analytical methods. As noted in Figure 2.1, the overall application of analytical techniques can be broken down into the following general areas:

- Solution of simple canonical problems with the goals of providing guidance for EMC design and of understanding and interpreting measured data
- Data acquisition and processing
- Development of system-level EMC assessments using measured data

Relative to most experimental programs, performing analytical studies of the system electrical response is inexpensive. As described later in this book, there are a wide variety of existing computer models and analysis techniques that permit a rapid estimation of the electromagnetic behavior of simplified system problems. Such numerical studies require nothing more than a computer (together with the necessary software) and a knowledgeable analyst's time; yet they can yield engineering insight into the feasibility of various design options and provide crucial guidance for future test programs.

In terms of EMC test guidance, a pretest analysis program can be employed for obtaining information on levels of expected responses. Consequently, this can be used to select the appropriate measurement quantities and locations. These data are also useful to have during the testing itself to ensure that the measurement equipment is set up and functioning properly.

Finally, a very important role of analysis is in the post-test processing of measured data for the subsequent inference of system EMC. It is rare that conclusions can be made directly from the raw data obtained in a test program; such data must be processed, corrected, extrapolated, or perhaps displayed in a manner that EMC design conclusions can be extracted.

2.2 TOPOLOGICAL DESCRIPTION OF SYSTEMS

One of the main techniques for ensuring EMC in an electrical system is to create EM *barriers* [2.1, 2.2] between an EMI source and potentially sensitive equipment. These barriers can consist of electromagnetic shields, filters, surge suppressors, or nonlinear devices. The barrier serves to exclude externally generated EMI from sensitive equipment. Conversely, the barrier can contain harsh EM environments and prevent them from radiating from their sources. This concept is not new. It is essentially the same technique that is used in a Faraday shield. Thus the EMC design is simple to summarize: Surround the region to be protected by a closed metallic shield, and at the locations where there must be penetrations, install suitable protective devices.

2.2.1 Electromagnetic Topology

The term *electromagnetic topology* has been coined to describe the nature of the EM shields of a system, as discussed in [2.3]. It is generally not described in an abstract mathematical sense, although the subject has been examined in this manner in [2.4]. Here, we use this term as a description of the electrical nature of one or more shielding surfaces that form EM barriers within the system [2.5, 2.6]. Electromagnetic topology pertains to the description of the shielding surfaces comprising the system and to the various penetrations that occur in the shields. A system is said to be designed "in a topological manner" when consideration of the proper configuration of the shielding surfaces and the penetrations have been made.

Detailed discussions of EM topology and design guidelines for a system are available in [2.7] and [2.8].

2.2.1.1 Topological Diagram. To illustrate the concept of EM topology, Figure 2.2a illustrates an aircraft subjected to some sort of externally generated EMI, say from a distant lightning discharge. This external environment is described by a transient EM waveform, having a defined amplitude, polarization, and angle of incidence. It strikes the aircraft,

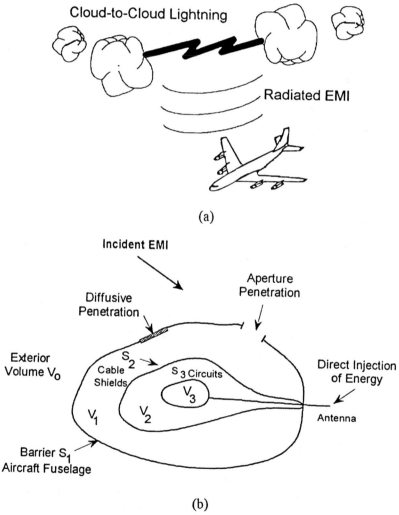

FIGURE 2.2 Example of shielding topology: (a) aircraft excited by lightning discharge; (b) electromagnetic shielding topology for the aircraft.

inducing currents and charges in the outer skin of the aircraft, as well as producing responses inside the aircraft.

Let us assume that we wish to try to estimate the induced voltage or current responses at a specified location inside the aircraft. Conceptually, this problem is made easier by considering the system to be comprised of one or more shielding surfaces, one inside the other, which exclude the incident EMI from the system interior. This is shown in Figure 2.2b. The exterior region outside the aircraft and containing the sources of the lightning EMI is denoted as V_0, and the shielding enclosure of the aircraft fuselage is denoted as S_1. Other shielded volumes within the aircraft are denoted as V_1, V_2, and V_3, with the corresponding surfaces bounding these volumes being denoted by S_2 and S_3. As discussed in [2.5], each system will have a different topological shielding diagram, depending on its own unique construction.

2.2.1.2 Electromagnetic Energy Points of Entry.
In practice, each shielding surface, or barrier, will have one or more locations where EM energy can enter. Referred to as *points of entry* (POE), these are usually in the form of apertures, penetrating wires, or EM field diffusion. These penetrations result in a shield that is not perfect. For a well-designed and compatible system, the number of energy penetration points in the shield will be small. It is usually possible to identify several simplified EM penetration models to describe the electrical behavior of these penetrations The solutions to these canonical problems are then used to infer the shielding behavior of the larger system.

In the simple aircraft example of Figure 2.2b, the different POEs are identified. Each one poses a simple, canonical problem that may be solved to gain an understanding of the overall response of the system. For example, there can be a communications antenna on the aircraft, and this will provide a direct injection of lightning-induced energy into the interior. This field-to-antenna coupling can be calculated using the simple antenna coupling models described in Chapter 4 and propagation to internal electronics of the energy collected can be treated using the transmission line models of Chapter 6. The effects of the induced voltages or currents within the electronic loads connected to the antenna can then be studied using a circuit modeling code, such as SPICE [2.9].

The aperture penetrations in this example are due primarily to the presence of windows or composite material panels used in the airframe construction. They can be treated using the models developed in Chapter 5. If the entire aircraft is metallic, the effects of diffusion can be practically neglected. However, in some cases calculation of the levels of diffusive penetration may be desired. In these cases, the aircraft can be modeled as a cylindrical shell and the diffusive penetration models of Chapter 10 applied. The EM field inside volume V_1, which results from superposition of the aperture and diffusive penetrations of the surface S_1, will represent the exciting field for shielded cables running within the aircraft. The shield surfaces are represented by S_2, and the volume inside the cables is V_2.

This topological shielding concept can be applied to a wide variety of systems. It is important to keep in mind that the frequency spectrum of EMC responses scale inversely with the size of the system being considered. Thus, decreasing a system's size by a factor of 100 will increase the range of major spectral responses by a factor of 100. What is important in the use of the topological decomposition of a system is that it provides the concept as to how to partition the system into components that can be analyzed separately.

Consider, for example, the case of a ground-based communications facility that is located near a power substation. Occasional switching of the power system mains may create switching surges that propagate along the mains to the facility and are injected into the building. In addition, radiative EMI propagates from the mains and couples to the site's antennas and communications lines. In this case, the main POEs consist of the power line and the other wire penetrations into the facility. Other POEs, such as the building windows or doors (apertures), conductive buried water or gas lines (conductive penetrations), or perhaps direct field penetration through the building walls (diffusion), may exist in the facility. However, as these penetrations do not pertain directly to operation of the internal electrical systems, they are usually not as important as those that are connected directly to the electronics.

2.2.1.3 EMC Design.
For a system having one or more well-defined shielding surfaces or barriers against the EMI, the protection concept is clear: All important POEs within the barrier must be provided with some sort of a protection device to limit passage of the EMI into the system. These protection measures can be in the form of power line filters, filters on signal lines, communication wires or other low-voltage sensor lines, voltage-limiting diodes or other switches, wire meshes, EMI gaskets, and so on. Each class of POE may have several different protection options, depending on its particular configuration.

For a complex system not designed around the concept of a well-controlled EM topology, it can be difficult to identify smaller subproblems to analyze independently. Most of the system parts may be interconnected in a complex manner, with the consequence that the entire system must be analyzed together. For this reason, the errors involved in analyzing an uncontrolled system can be very large. In addition, developing an EMC plan for such a system is usually difficult. By structuring the design of a system around its EM topology, we have not only a clear way of designing a system for EMC, but a way of analyzing, validating, and verifying the system's EMC. For a system designed with a well-defined shielding topology in mind, it is possible to analyze and perform tests on each of the penetrations and associated protection devices, instead of conducting an analysis or test on the entire system.

2.2.2 Electromagnetic Interaction with the System

For performing an approximate analysis of a system response, the topological concept is very useful. This permits dividing a complicated chain of events into a number of simpler parts, starting from the knowledge of the incident field and ending with the internal component response. Each of these simple problems can be analyzed independently from each other, and the results combined to provide an estimate of the response for the entire system [2.5].

As an example of this procedure, the EMI effects on the system in Figure 2.2 can be thought of as first arising from the surface current and charge densities induced on the conducting exterior of the aircraft by the incident lightning fields. Such a process is referred to as *external coupling* and can be described by a transfer function from the external sources to these surface response quantities. In doing this, the presence of apertures, antennas, and internal electronics are ignored.

Figure 2.3 illustrates this process. The external EMI source (the lightning) can create a charge and current response on the exterior structure of the aircraft, either by a direct conduction of energy if the lightning strikes the aircraft, or by EM field coupling if the lightning strike is distant. Each of these possibilities is shown in the figure. Note that the term *exterior structure* refers not only to the aircraft skin, but to other exposed conductors, such as the blades of antennas, long trailing wires, and so on. Depending on the frequency of the EMI, various models describing the field coupling are used, ranging from low-frequency capacitive and inductive coupling to the higher-frequency radiative coupling. Each of these concepts is developed in more detail later in the book. Of course, in the present example of Figure 2.2, there is no direct conducting path from the source to the aircraft. However, this path can exist in some instances, as in the case of a direct lightning strike to the aircraft.

Once these surface response functions are determined, additional calculations of the EM energy penetration into the shielded region are made. This process referred to as *electromagnetic penetration*, involves one of the mechanisms discussed previously. As in the coupling problem, simplifications are made to permit decoupling of the internal region from the external region. That is, penetration through an aperture is assumed to be independent of the backing behind it. This is an implication of an assumption of good shielding.

EMI energy inside the system can then be redistributed from the electrical point of entry (POE) to an internal location by a mechanism called *electromagnetic propagation*. At low frequencies, the redistribution of internal energy is not described precisely by electromagnetic wave propaga-

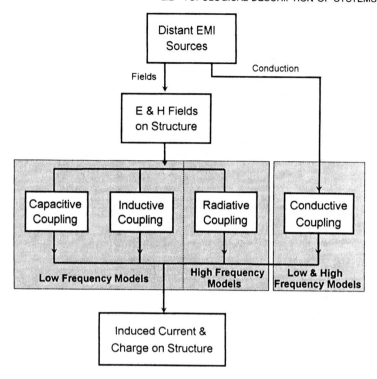

FIGURE 2.3 EM coupling process.

tion, as the fields are quasistatic in nature. Nevertheless, the term *propagation* is used in this context. At higher frequencies, the energy redistribution mechanism typically involves true transmission line propagation of signals from one region to another, or direct EM field radiation. As in previous cases, relatively simple models for estimating the propagation effects are used here, requiring knowledge of the excitation by the POE and the details of the internal system geometry.

2.2.2.1 Interaction Sequence Diagram.
This entire process is referred to as *electromagnetic interaction*. In complicated systems having more than one shielding surface or *level*, the process of coupling, penetration and propagation will be repeated until the component level is reached. This structured approach to viewing and analyzing complex systems permits division of the system into a number of simpler problems. It results in a sequence of analysis steps shown in an *interaction sequence diagram*, illustrating the flow of analysis and the models required for treating any specific system. An example of such a diagram for external EMI sources acting on a system having a single shielding layer is indicated in Figure 2.4. Examples of other

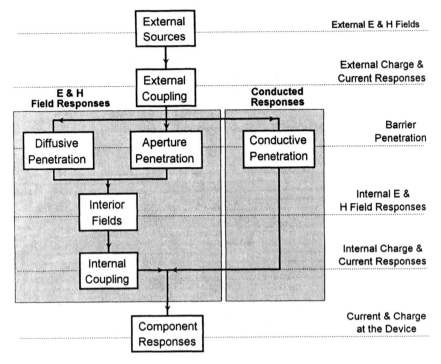

FIGURE 2.4 Interaction sequence diagram for a system with external excitation.

interaction sequence diagrams for systems having additional layers of shielding are presented in [2.10].

Not all EMC problems involve external sources of EMI. It is possible to have interference sources located within the system. For these cases the same topological concepts may be used, considering shielding layers located farther within the system as the primary barriers against the EMI. Figure 2.5 illustrates an interaction sequence diagram suitable for internal sources. Note that in this figure, the possibility of having a direct, or conducting, connection between the source and the barrier is included.

The interaction sequence diagram of Figure 2.5 is similar in form to that of Figure 2.4 but with the aperture penetration replaced by braid penetration. The latter penetration mechanism refers to EM fields leaking into the interior of shielded cables through small holes in a braided cable, a topic discussed in Chapter 9.

2.2.3 Generalized EMC Design Principles Based on EM Topology

As mentioned earlier, electromagnetic topology forms the basic EMC design with introduction of the concept of an EM barrier between the source of interference and the victim. The barrier is usually thought of as a physical

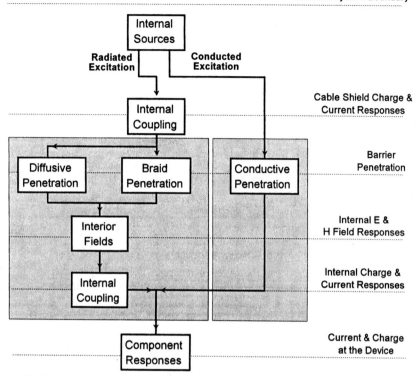

FIGURE 2.5 Interaction sequence diagram for internal EMI sources.

shield. However, this concept can be extended to include alternative methods of field reduction. A *generalized* EM barrier for protecting a victim circuit can be created by three fundamental mechanisms: EM field attenuation, source–victim separation, and source–victim orthogonalization.

The field attenuation mechanism is the common protection solution obtained by inserting attenuating or reflecting material between the source and victim. Often, this is viewed as a shielding problem (discussed in more detail in Chapters 9 and 10). However, another common and cost-effective solution for the reduction of the collected EM energy is to consider the use of a ground plane near the radiating source or receiving circuit. Examples of this type of generalized shield or barrier are found in printed circuit board design, which use the ground plane to reduce the radiation from strip lines. On a larger scale, the use of continuous cable trays in communication facilities have proven to be an efficient way of reducing EMI coupling and crosstalk between cable bundles.

Another method of reducing the EMI is by physically (or electrically) separating the source and victim circuitry. This separation can be viewed as being part of a generalized barrier (see Chapter 10) within the system topology. Furthermore, the relative orientations of the source and emitter can sometimes be adjusted to reduce, or possibly eliminate, the interference. This also forms a protection approach. In much of what follows, we concentrate on the first protection, or hardening, concept: that of reducing the EMI strength by some type of physical shield. The other approaches to EMI reduction should not be forgotten, however.

2.3 MODELING ACCURACY

2.3.1 Errors Inherent in the Analysis

Because analysis methods can be applied to system problems with a much lower cost than the experimental methods, and because analytical methods occur in all phases of the design evaluation process, a common misconception is that EMC for a complex system can be determined by analytical methods alone. It is then assumed that this can provide a cost-effective alternative to some form of system testing. Experience in using modeling in EMC has shown that present-day analytical techniques are incapable of performing system-level estimates if the models are not combined with test or measurement results. This limitation is due, in part, to the following:

- The high degree of system complexity
- Errors and uncertainties in the system description
- System nonlinearities that cannot be predicted
- The difficulty of predicting upsets (too many possible system states)

Simply put, most systems are too complex, and analytical techniques are too primitive to allow an EMC assessment by analysis alone. Analysis and experimental techniques complement each other. They must be used together.

2.3.2 Balanced Accuracy of Analysis

An important aspect in the analysis of a topologically controlled system is the specification of the accuracy of the individual calculations that comprise the overall solution of the system response. The best possibility is when there is no error at all in each of the individual models and calculations. This is obviously unrealistic, however, due to the many sources of error described above. In a practical case, it is desired to obtain the best possible system-level response for a specified analytical effort.

One way of approaching this goal is to require that each subproblem contribute roughly the same amount of uncertainty to the overall response [2.5]. This requirement is difficult to quantify exactly, but the guiding principle is clear: Do not attempt to develop a "perfect" external interaction calculation if all of the other penetration and internal propagation models have large errors. Depending on the circumstances, therefore, detailed calculations of a subproblem response may be required in one case but a simple back-of-the-envelope calculation warranted in another.

REFERENCES

2.1. Baum, C. E., "Electromagnetic Topology: A Formal Approach to the Analysis and Design of Complex Electronic Systems," *Proceedings of the 4th Symposium and Technical Exhibition on EMC, Zurich*, March 10–12, 1981, pp. 209-214.

2.2. Vance, E. F., and W. Graf, "The Role of Shielding in Interference Control," *IEEE Trans. Electromagn. Compat.*, Vol. EMC-30, No. 3, August 1988.

2.3. Baum, C. E., "The Role of Scattering Theory in Electromagnetic Interference Problems," in *Electromagnetic Scattering*, P. L. E. Uslenghi, ed., Academic Press, New York, 1978.

2.4. Baum, C. E., "Electromagnetic Topology for the Analysis and Design of Complex Electromagnetic Systems," pp. 467–547 in *Fast Electrical and Optical Measurements*, Vol. I, I. E. Thompson and L. H. Luessem, eds., Martinus Nijhoff, Dordrecht, the Netherlands, 1986.

2.5. Tesche, F. M., "Topological Concepts for Internal EMP Interaction," *IEEE Trans. Electromagn. Compat.*, Vol. EMC-20, No. 1, February 1978.

2.6. Tesche, F. M., "Introduction to Concepts of Electromagnetic Topology as Applied to EMP Interaction with Systems," in *Interaction Between EMP, Lightning and Static Electricity with Aircraft and Missile Avionics Systems*, NATO/AGARD Lecture Series Publication 144, May 1986.

2.7. Vance, E. F., "EMP Hardening of Systems," *Proceeding of the 4th Symposium and Technical Exhibition on EMC*, Zurich, March 10–12, 1981.

2.8. Tesche, F. M., "Design Guidelines for EMP Hardening of the MX Weapons System," *Report AFWL-TR-82-124*, Air Force Weapons Laboratory, Kirtland AFB, NM, March 1983.

2.9. Paul, C. R., *Introduction to Electromagnetic Compatibility*, Wiley, New York, 1992.

2.10. Lee, K. S. H., ed., *EMP Interaction: Principles, Techniques and Reference Data*, Hemisphere, New York, 1989.

PROBLEMS

2.1 Compare and contrast the meanings of *electromagnetic interaction* and *electromagnetic coupling*.

2.2 Identify the various ways that EM energy can penetrate a shielded enclosure and the various protection techniques that can be used to reduce these penetrations.

2.3 In the generic system illustrated in Figure 2.4, which coupling path is usually most significant in determining the internal responses? Why?

2.4 The general EM interaction and coupling problem has the possibility of a scattered or reflected field from a body affecting the radiation from the primary source through the mechanism of mutual coupling (or multiple scattering). In the EM topology concept discussed in this chapter, however, this effect is neglected. For example, in Figure 2.4, the behavior of the external sources is assumed to be independent of the external coupling, the external coupling is assumed to be independent of the penetrations, and so on. Justify this assumption and discuss cases when this one-way signal flow concept should not be used.

2.5 (Adapted from [2.9] with thanks to C. Paul.) For each of the following cases, identify the signal source, the signal propagation path, the signal receptor, and the possible interference source that could disrupt the desired reception of the signal.

(a) AM radio transmission to the human ear; **(b)** TV transmission to the human eye; **(c)** radar target identification; **(d)** transfer of 60-Hz power to an air conditioner; **(e)** transfer of digital computer data to a computer; **(f)** operation of an AM radio in an automobile.

2.6* A ground-based communication facility consists of a radio tower 100 m high, two computers, two equipment vans, and a power generator. A top view is shown in Figure P2.6. The design is supposed to protect against a lightning electromagnetic pulse (LEMP) caused by a lightning stroke hitting the ground 50 m from the radio tower, as indicated on the figure. For this system, discuss the design and the needed protection measures. What additional considerations may be appropriate?

2.7* Propose a solution for hardening the fourth floor of a six-floor building containing critical telephone switching equipment (Figure P2.7). Identify all the coupling paths for different kinds of possible disturbances [e.g., direct or indirect lightning effects, citizens' band (CB) radio]. Discuss qualitatively the protection measures that could be used. (Problem proposed by Mr. M. Wik.)

2.8 A network analyzer is used to make a measurement of the emissions from a computer system when it is radiated simultaneously by a high-powered continuous-wave source. To eliminate the effects of

---

*Problems denoted by an asterisk were used in EMP short courses conducted by the Summa Foundation in 1983 through 1989. These contributions by C. Baum and the other lecturers are gratefully acknowledged.

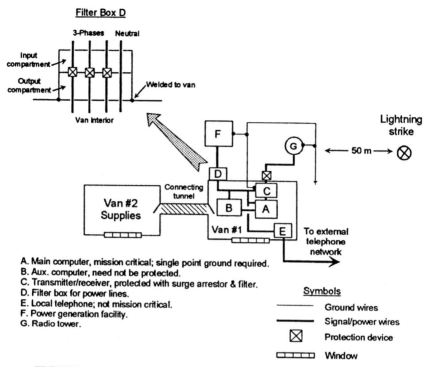

FIGURE P2.6 Ground-based communication facility (top view).

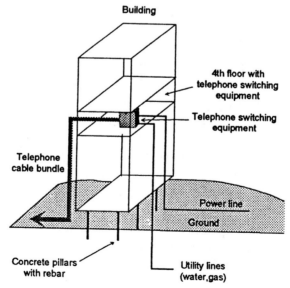

FIGURE P2.7 Six-floor building requiring hardening.

42 SYSTEM DECOMPOSITION FOR EMC MODELING

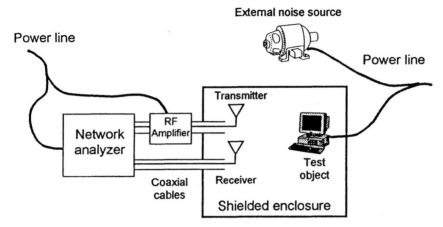

FIGURE P2.8 Equipment configuration for a radiated emission test of a computer system.

nearby noise sources, this measurement is conducted in a shielded enclosure, as shown in Figure P2.8. Focusing your attention on the receiver port of the network analyzer, develop a shielding topology diagram for this system together with an interaction sequence diagram indicating the paths of energy flow from the various sources to the receiver port.

2.9 For the measurement equipment setup in Figure P2.9, the receiver is located within a shielded enclosure and the transmitter is located outside. Consequently, point A is located in an unshielded region, and point B is in a shielded region. Following both of these points along the interior of the coaxial line, we see that the interior of the network analyzer can be considered either as a shielded or unshielded region. Reconcile this ambiguity and draw a topological shielding diagram and interaction sequence diagram for this example.

2.10 It is clear that the object in Figure P2.10a can be considered to be a shielded enclosure having a single small imperfection in the form of an

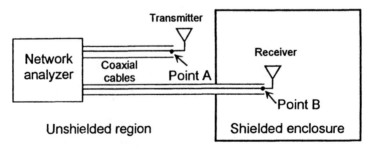

FIGURE P2.9 Measurement equipment location.

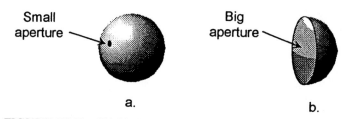

FIGURE P2.10 Shielded enclosure with small and big apertures.

aperture. However, as the aperture becomes bigger, as shown in Figure 2.10b, at some point the enclosure can no longer be considered as a good shield. At this point, the remaining part of the enclosure no longer can be thought of as comprising part of the shielding topology. Discuss when this occurs and how the frequency of the EM environment enters this problem. What happens if the EM environment is a transient waveform?

PART II
LOW-FREQUENCY CIRCUIT MODELS

ERRATUM
EMC Analysis Methods and Computational Models,
by Frederick M. Tesche, Michel V. Ianoz, and Torbjörn Karlsson
© 1997 by John Wiley & Sons, Inc.   ISBN 0-471-15573-X
The following page is missing from some copies of this book:

■ CHAPTER 3

Lumped-Parameter Circuit Models

Low-frequency circuit theory is a powerful tool for developing EMC models. It is applicable when the physical size of the system is much smaller than the wavelength of the disturbance. This chapter begins with a discussion of the two-port representation of circuits, which forms a basic foundation for developing circuit models for EMC predictions. Later, the subjects of field coupling by capacitive and inductive mechanisms are discussed. Finally, the coupling of two circuits through a common resistive path is discussed.

3.1 INTRODUCTION

For cases where the frequency of EMI signals is low, with the wavelength of the interference much larger than the system under consideration, circuit models are useful for predicting system responses to EMI. These models can be applied when the interfering sources are connected directly to the victim circuit or when the victim circuit is located near the source and is excited by the electromagnetic fields produced by the source.

There are many examples of the occurrence of these types of interference:

- Conducted interference flowing into devices connected to the power mains
- Conducted or field-induced interference conducted between two electronic or electric circuits within the same system
- Field-induced interference coupled between power transmission lines and nearby telecommunications lines
- Field-induced interference between power and data transmission lines within a power substation
- Field-induced interference between the ac power mains and nearby sensitive electronic equipment

For the purpose of this discussion, we define the interference between circuits to be either a *conducted disturbance* or a *radiated disturbance* originating from a source circuit. In the first part of this chapter we discuss several circuit modeling approaches for computing the effects of EMI sources connected directly to load equipment. These models involve the use of conventional ac circuit analysis techniques, which are discussed in [3.1] to [3.4]. The intent of this chapter is not to review the basics of circuit modeling, but to illustrate how these modeling techniques may be applied in a broad sense to the area of EMC. It is assumed that the reader has a solid background in circuit analysis and its applications.

Later in this chapter, low-frequency models for computing source–victim interaction by EM field coupling are described. These include coupling by inductive, capacitive, and resistive mechanisms. The important issues of high-frequency radiated coupling are discussed in Chapters 4 and 5.

3.2 CONDUCTED DISTURBANCES IN CIRCUITS

3.2.1 Thévenin and Norton Representations

Figure 3.1 illustrates a common source–victim configuration. An electrical circuit containing one or more active sources of EMI is connected directly to another circuit by conducting wires or a cable. Often referred to as a *hard-wired* connection, this can provide a direct injection of the EMI into the victim. An example of this configuration is that of an electrical motor (a potentially noisy source of interference due to the armature-brush contacts) interfering with a poorly designed radio (the victim) by producing electrical noise which is conducted through ac power lines.

Analysis of this configuration of EMI source and victim is facilitated by the use of a Thévenin or Norton equivalent circuit [3.2], as shown in Figure 3.2. This permits representation of the source at its external connection ports by a simple combination of a passive impedance element together with either an active voltage or active current source. Note that by the term *port* we refer to a pair of wires exiting from the source circuit for which a voltage and current can be defined. In representing the source in this way, all details of the internal responses within the source are lost; only the external response of the source is maintained. However, this model is applicable regardless of the complexity of the internal source circuitry, as long as the internal EMI source circuit can be considered to be linear.

The equivalent source parameters, $V_s(\omega)$ and $Z_s(\omega)$ for the case of the Thévenin circuit and $I_s(\omega)$ and $Y_s(\omega)$ for the Norton circuit, are complex-valued functions of the angular frequency $\omega = 2\pi f$ and are described by a

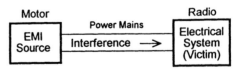

FIGURE 3.1 Victim circuit directly connected to an EMI source.

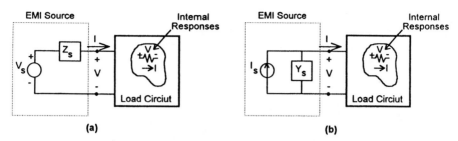

FIGURE 3.2 Equivalent-circuit representations for the EMI source: (*a*) Thévenin equivalent circuit; (*b*) Norton equivalent circuit.

real and an imaginary part (or magnitude and phase). In the discussions to follow, dependence on the frequency ω is not explicitly shown.

Although the representations of the source in Figure 3.2*a* and *b* appear to be different, they must provide the same voltage and current responses to an arbitrary load. Consequently, the following relationship exists between the open-circuit parameters of the Thévenin circuit and the short-circuit parameters of the Norton circuit:

$$Y_s = \frac{1}{Z_s} \qquad (3.1a)$$

$$I_s = \frac{V_s}{Z_s}. \qquad (3.1b)$$

The effect of the EMI source on the victim circuit can be determined if the impedance of the victim circuit is known. With the impedance of the load denoted as Z_L, the interfering current and voltage at the connected terminal of the load are expressed as

$$I = \frac{V_s}{Z_s + Z_L} \qquad (3.2a)$$

$$V = \frac{V_s Z_L}{Z_s + Z_L} \qquad (3.2b)$$

for the EMI source represented by the Thévenin equivalent circuit of Figure 3.2*a*. A similar expression may be derived for the source having the Norton representation of Figure 3.2*b*.

Given knowledge of the electrical configuration of both the source and the victim circuitry, conventional ac circuit modeling techniques may be used to determine specific expressions for the parameters in the equations above for the voltage and current at the victim's input port. Furthermore, an internal response within the victim is frequently desired, and this also may be determined using conventional circuit analysis. These more detailed circuit analysis topics are not discussed further here, and the reader is referred to standard texts for this subject [3.1, 3.2].

Using the equivalent-circuit concept, the source is characterized in a manner that is independent of the load. The parameters V_s and Z_s (or I_s and Y_s) depend only on the nature of the source, and consequently, the source model will be appropriate for determining the response of an arbitrary load. If a detailed circuit model of the entire source–victim circuit were developed, the response calculation would be specific to the problem analyzed. It would be difficult to generalize the solution to other loads. Thus the use of the equivalent-circuit concept permits a more general analysis.

In some cases the source circuit can be complicated and a detailed analysis is difficult to conduct. In this event it is possible to consider *measuring* the Norton or Thévenin circuit parameters. This can provide accurate estimates of load responses if the measurements are conducted properly. It must be remembered that both V_s and Z_s (as well as I_s and Y_s) are complex-valued phasors, so that both the magnitude and phase functions are required. Furthermore, both the source strength *and* the source impedance (or admittance) parameters are needed. It is not sufficient to measure or compute only one.

3.2.2 Models for Passive Two-Port Circuits

An extension of the Thévenin and Norton representation of a single-port circuit is two-port circuit theory [3.2]. Here, an arbitrarily complex (but again linear) circuit is assumed to have two pairs of terminals identified. At each of these ports, a voltage and a current can be defined, and several relationships between these observables can be developed. For the moment it will be assumed that there are no sources located inside the two-port circuit. The presence of voltage and current at the ports is due to the existence of external excitation sources only. An extension of this case to that involving internal sources is discussed in the next section.

An example of a such a configuration is shown in Figure 3.3, where an external EMI source is connected to a victim circuit through a two-port network. This could be the configuration if a passive interference filter were connected to the ac power mains feeding the noisy motor discussed previously. Such a two-port circuit can affect the transmission of the EMI produced by the motor and reduce (and hopefully eliminate) the interfer-

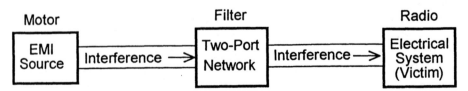

FIGURE 3.3 EMI source connected to a load through a two-port circuit.

ence at the victim circuit. Furthermore, the internal response location within the load of Figure 3.2 could be defined as a second port, and the load circuit of this figure would be considered as a two-port network.

Given a knowledge of the electrical composition of a two-port filter, it is possible to perform a detailed circuit analysis of the entire source–filter–load circuit. However, as discussed previously, this would be an analysis specific to the particular problem. If the load or the source were changed, it would be necessary to perform the analysis over again with the new circuit configuration. A better approach for the analysis is to represent the intermediate two-port (the filter) in a manner that is *independent* of the source or load. Figure 3.4 illustrates a general two-port circuit, where two voltages and currents are defined at each port. By convention, a positive current flows into the network, as indicated in the figure.

3.2.2.1 Open-Circuit Impedance Parameters.

For a one-port network without sources, the v–i relationship at the pair of terminal wires is simply expressed as $V = ZI$. For the corresponding two-port network, this relationship is in the form of a matrix equation,

$$\begin{bmatrix} V_1 \\ V_2 \end{bmatrix} = \begin{bmatrix} z_{11} & z_{12} \\ z_{21} & z_{22} \end{bmatrix} \begin{bmatrix} I_1 \\ I_2 \end{bmatrix} \quad (3.3)$$

where the elements z_{ij} are called the *open-circuit impedance parameters*. The matrix is referred to as the *Z-matrix* of the two-port. The values of the individual impedance elements in this equation can be determined by computing (or measuring) selected port voltages and currents with the other port open-circuited, as

$$z_{11} = \left. \frac{V_1}{I_1} \right|_{I_2=0} \qquad z_{12} = \left. \frac{V_1}{I_2} \right|_{I_1=0} \qquad \text{etc.} \quad (3.4)$$

As in the case of the Thévenin or Norton equivalent circuits of a one-port circuit, the representation of the two-port in Eq. (3.3) is valid *only* at the two terminals of the network. All information about the responses inside the two-port is lost.

Passive linear circuits are bilateral in nature, leading directly to the requirement of reciprocity [3.2]. In terms of the elements of the Z-matrix, reciprocity requires that

$$z_{12} = z_{21} \quad (3.5)$$

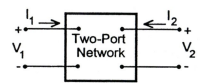

FIGURE 3.4 Voltage and current relationships for a general two-port network.

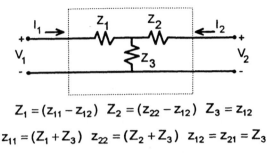

$Z_1 = (z_{11} - z_{12}) \quad Z_2 = (z_{22} - z_{12}) \quad Z_3 = z_{12}$

$z_{11} = (Z_1 + Z_3) \quad z_{22} = (Z_2 + Z_3) \quad z_{12} = z_{21} = Z_3$

FIGURE 3.5 T-network representation for the Z-matrix.

Thus only *three* independent parameters are needed to represent this type of two-port network. For this common class of networks, the port v–i relationship of Eq. (3.3) can be obtained from the equivalent T-network shown in Figure 3.5. Of course, the internal electrical configuration of the actual circuit may be much more complex, but the T-network is sufficient for modeling its behavior at the ports. The values of the individual circuit elements are expressed in terms of the Z-parameters and are indicated in the figure.

Two-port networks are usually connected to a source circuit and a load circuit, as shown in Figure 3.3. The source may be represented by its general Thévenin circuit, with input impedance Z_s and voltage source V_s, and the load is a passive impedance element, Z_L. The resulting interconnected network is shown in Figure 3.6. A simple circuit analysis yields the following expression for the load current in terms of the Z-parameters:

$$I_L = \frac{z_{12}}{(z_{11} + Z_S)(z_{22} + Z_L) - z_{12}^2} V_S \qquad (3.6a)$$

The load voltage is given by $V_L = Z_L I_L$. Similar expressions can be de-

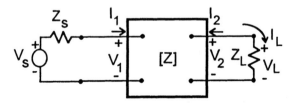

FIGURE 3.6 Loaded two-port excited by a Thévenin circuit source.

veloped for the input (or driving) current and voltage of the network:

$$I_1 = \frac{Z_L + z_{22}}{(z_{11} + Z_S)(z_{22} + Z_L) - z_{12}^2} V_S \tag{3.6b}$$

$$V_1 = \frac{(Z_L + z_{22})z_{11} - z_{12}^2}{(z_{11} + Z_S)(z_{22} + Z_L) - z_{12}^2} V_S \tag{3.6c}$$

The open-circuit impedance representation of a two-port network is also useful when two or more two-port circuits are connected together in series, as illustrated in Figure 3.7. Denoting the 2×2 Z-matrix of Eq. (3.3) by $[Z]$ and the two-element voltage and current vectors as $[V]$ and $[I]$, respectively, this equation can be written compactly as $[V] = [Z][I]$. Using this notation for the configuration shown in Figure 3.7, with $[Z_A]$ referring to the impedance parameters of network A and $[Z_B]$ for network B, the overall Z-matrix for the composite circuit becomes

$$[V_T] = ([Z_A] + [Z_B])[I_T] \tag{3.7}$$

3.2.2.2 Short-Circuit Admittance Parameters.
The inverse of the Z-matrix equation, when it exists, provides an expression for the two-port currents in terms of the port voltages, in a manner similar to that of the Norton equivalent circuit. This dual representation is given by an admittance matrix, or Y-matrix, of the form

$$\begin{bmatrix} I_1 \\ I_2 \end{bmatrix} = \begin{bmatrix} z_{11} & z_{12} \\ z_{21} & z_{22} \end{bmatrix}^{-1} \begin{bmatrix} V_1 \\ V_2 \end{bmatrix} = \begin{bmatrix} y_{11} & y_{12} \\ y_{21} & y_{22} \end{bmatrix} \begin{bmatrix} V_1 \\ V_2 \end{bmatrix} \tag{3.8}$$

where $y_{12} = y_{21}$ for reciprocal networks. As in the case of the Z-matrix, the individual Y-matrix elements can be found by calculations or measurements of individual port currents and voltages, but this time with the ports

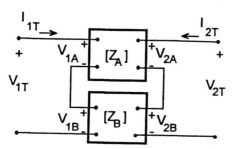

FIGURE 3.7 Series combination of two two-port networks.

short-circuited. The elements are given by

$$y_{11} = \left.\frac{I_1}{V_1}\right|_{V_2=0} \qquad y_{12} = \left.\frac{I_1}{V_2}\right|_{V_1=0} \qquad \text{etc.} \tag{3.9}$$

A lumped-element circuit representing the two-port network in terms of its Y-parameters is the dual network of that for the Z-parameters shown in Figure 3.5. This is illustrated in Figure 3.8 as a Π-network, with the values of the three independent admittance elements Y_a, Y_b, and Y_c indicated in the figure.

The Y-matrix representation is useful for analyzing a two-port network that is excited by a Norton equivalent circuit as shown in Figure 3.9. In this case the voltage response at the load, denoted by V_L, is given by an equation similar to Eq. (3.6) as

$$V_L = \frac{-y_{12}}{(y_{11} + Y_S)(y_{22} + Y_L) - y_{12}^2} I_S \tag{3.10a}$$

and the load current is $I_L = Y_L V_L$. The driving-point voltage and current are given as

$$V_1 = \frac{Y_L + y_{22}}{(y_{11} + Y_S)(y_{22} + Y_L) - y_{12}^2} I_S \tag{3.10b}$$

$$I_1 = \frac{(Y_L + y_{22})y_{11} - y_{12}^2}{(y_{11} + Y_S)(y_{22} + Y_L) - y_{12}^2} I_S \tag{3.10c}$$

The Y-parameter representation for two-ports is useful for combining the effects of circuits connected in parallel, as shown in Figure 3.10. Here two networks, A and B, have 2×2 Y-matrices $[Y_A]$ and $[Y_B]$, respectively, and

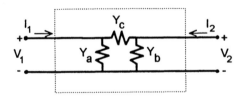

$Y_a = (y_{11} + y_{12}) \quad Y_b = (y_{22} + y_{12}) \quad Y_c = -y_{12}$

$y_{11} = (Y_a + Y_c) \quad y_{22} = (Y_b + Y_c) \quad y_{12} = y_{21} = -Y_c$

FIGURE 3.8 Π-network representation for the Y-matrix.

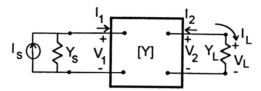

FIGURE 3.9 Loaded two-port excited by a Norton circuit source.

the v–i relationship for the total combination is expressed as

$$[I_T] = [[Y_A] + [Y_B]][V_T] \tag{3.11}$$

Because the port v–i relationships for the circuits in Figures 3.5 and 3.8 are the same, there must be a relationship between the Z and Y parameters of these two circuits. These are known as the T–Π transformation [3.2]. For converting from the T- to Π-circuit representation, these are expressed as

$$Y_a = \frac{Y_1 Y_3}{Y_1 + Y_2 + Y_3} = \frac{Z_2}{Z_1 Z_2 + Z_2 Z_3 + Z_1 Z_3} \tag{3.12a}$$

$$Y_b = \frac{Y_2 Y_3}{Y_1 + Y_2 + Y_3} = \frac{Z_1}{Z_1 Z_2 + Z_2 Z_3 + Z_1 Z_3} \tag{3.12b}$$

$$Y_c = \frac{Y_1 Y_2}{Y_1 + Y_2 + Y_3} = \frac{Z_3}{Z_1 Z_2 + Z_2 Z_3 + Z_1 Z_3} \tag{3.12c}$$

where $Y_1 = 1/Z_1$, and so on. The corresponding expressions for the Π-to-T

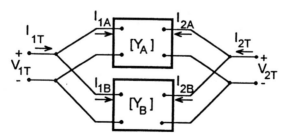

FIGURE 3.10 Parallel combination of two-port networks.

transformation are

$$Z_1 = \frac{Z_a Z_c}{Z_a + Z_b + Z_c} = \frac{Y_b}{Y_a Y_b + Y_b Y_c + Y_a Y_c} \quad (3.13a)$$

$$Z_2 = \frac{Z_b Z_c}{Z_a + Z_b + Z_c} = \frac{Y_a}{Y_a Y_b + Y_b Y_c + Y_a Y_c} \quad (3.13b)$$

$$Z_3 = \frac{Z_a Z_b}{Z_a + Z_b + Z_c} = \frac{Y_c}{Y_a Y_b + Y_b Y_c + Y_a Y_c} \quad (3.13c)$$

where $Y_a = 1/Z_a$, and so on.

Not all circuits can be represented by Z or Y parameters. Consider the circuit in Figure 3.5 with $Z_1 = Z_2 = 0$, (i.e., a circuit with just a shunt impedance element Z_3). In this case, the Z-matrix is

$$[Z] = \begin{bmatrix} Z_3 & Z_3 \\ Z_3 & Z_3 \end{bmatrix} \quad (3.14)$$

which is singular and cannot be inverted to provide the Y-matrix representation. Similarly, if $Y_a = Y_b = 0$ in Figure 3.8, the resulting Y-matrix is singular. The Z-matrix is undefined for the circuit of a simple series admittance in a circuit. Both of these circuits are physically realizable, however, and there is a way to describe their two-port behavior mathematically, as discussed next.

3.2.2.3 Chain Parameters.
Because the two-port network is assumed to be linear, other linear combinations of the port voltages and currents can also be used to describe its behavior. A useful set of parameters for circuits that are cascaded together are the *chain parameters*. Also known as the *ABCD* or *transmission parameters*, these are defined by the matrix expression

$$\begin{bmatrix} V_1 \\ I_1 \end{bmatrix} = \begin{bmatrix} A & B \\ C & D \end{bmatrix} \begin{bmatrix} V_2 \\ I_2' \end{bmatrix} \quad (3.15)$$

The location and orientation of the port voltages and currents for the chain parameters are shown in Figure 3.11. Observe that the direction of the current at port 2 is reversed from that in the definitions of the Z and Y parameters, and this current is denoted as I_2'. As a consequence of the assumed reciprocal nature of the two-port, the chain parameters satisfy the

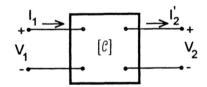

FIGURE 3.11 Voltage and current definitions for the chain parameters.

relation $AD - BC = 1$. Thus, as in the case of the Z and Y parameters, only three independent chain parameters are needed to describe the two-port.

It is frequently desired to have an expression for the voltage and current at port 2 rather than that provided in Eq. (3.15). This can be obtained by inverting the equation to provide the following expression:

$$\begin{bmatrix} V_2 \\ I'_2 \end{bmatrix} = \begin{bmatrix} D & -B \\ -C & A \end{bmatrix} \begin{bmatrix} V_1 \\ I_1 \end{bmatrix} \tag{3.16}$$

For the loaded two-port of Figure 3.6 excited by a Thévenin circuit, the load current can be expressed in terms of the chain parameters as

$$I_L = \frac{V_S}{(A + CZ_S)Z_L + DZ_S + B} \tag{3.17}$$

The most useful feature of the chain-parameter representation is in the treatment of cascaded two-ports, as shown in Figure 3.12. With $[\mathscr{C}_A]$ and $[\mathscr{C}_B]$ representing the chain parameter matrices for networks A and B in the figure, the overall chain parameter matrix for the combined circuit is given as

$$[\mathscr{C}_T] = [\mathscr{C}_A][\mathscr{C}_B] \tag{3.18}$$

Table 3.1 summarizes the chain parameters for several simple two-ports. With these parameters, a complex circuit can be constructed as simple series and shunt elements, and the overall chain parameters determined by a series of matrix multiplications. For example, the circuit in Figure 3.13 contains two series elements and one shunt element. The overall chain parameter matrix for this circuit is expressed as

$$\begin{bmatrix} A & B \\ C & D \end{bmatrix} = \begin{bmatrix} 1 & R \\ 0 & 1 \end{bmatrix} \begin{bmatrix} 1 & 0 \\ j\omega C & 1 \end{bmatrix} \begin{bmatrix} 1 & j\omega L \\ 0 & 1 \end{bmatrix} \tag{3.19}$$

These parameters are also useful for treating ladder-type networks having a large number of identical sections. If the chain parameters, $[\mathscr{C}]$, of a single section can be determined, the overall chain parameter matrix for n cascaded sections is given as $[\mathscr{C}_T] = [\mathscr{C}_A]^n$.

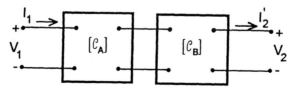

FIGURE 3.12 Cascaded two-port networks.

TABLE 3.1 Chain Parameters for Simple Circuit Elements[a]

Circuit Element Description	Circuit Diagram	Chain Matrix
Series impedance	Z (series resistor)	$\begin{bmatrix} 1 & Z \\ 0 & 1 \end{bmatrix}$
Shunt admittance	Y (shunt)	$\begin{bmatrix} 1 & 0 \\ Y & 1 \end{bmatrix}$
Ideal transformer, 1:n turns ratio	1:n transformer	$\begin{bmatrix} 1/n & 0 \\ 0 & n \end{bmatrix}$
Inductive coupling	L_1, L_2, M	$\begin{bmatrix} \dfrac{L_1}{M} & \dfrac{j\omega(L_1 L_2 - M^2)}{M} \\ \dfrac{1}{j\omega M} & \dfrac{L}{M_2} \end{bmatrix}$

[a] For reciprocal networks, $AD - BC = 1$.

3.2.2.4 Two-Port Parameter Relationships. The Z, Y, and chain parameters of a two-port cannot be specified independently. As they all describe the same circuit, various relationships can be derived to express one set of parameters from another. Table 3.2 presents these expressions for determining one set of parameters from another.

3.2.2.5 Other Two-Port Representations. Other linear combinations of port voltages and currents are possible, and this leads to the *hybrid*, *inverse hybrid*, and the *scattering* parameters as means of representing the external behavior of two-port networks [3.5]. The hybrid parameters are used frequently in modeling individual devices such as transistors. The scattering parameters are useful for describing transmission lines or other problems involving forward and backward traveling waves. For the present circuit

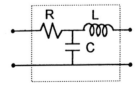

FIGURE 3.13 Simple two-port network modeled by chain parameters.

TABLE 3.2 Relationships Between Two-Port Parameters[a]

To: \ From:	$[Z]$	$[Y]$	Chain
$[Z]$	$\begin{bmatrix} z_{11} & z_{12} \\ z_{21} & z_{22} \end{bmatrix}$	$\begin{bmatrix} \dfrac{y_{22}}{|y|} & -\dfrac{y_{12}}{|y|} \\ -\dfrac{y_{21}}{|y|} & \dfrac{y_{11}}{|y|} \end{bmatrix}$	$\begin{bmatrix} \dfrac{A}{C} & \dfrac{\Delta}{C} \\ \dfrac{1}{C} & \dfrac{D}{C} \end{bmatrix}$
$[Y]$	$\begin{bmatrix} \dfrac{z_{22}}{|z|} & -\dfrac{z_{12}}{|z|} \\ -\dfrac{z_{21}}{|z|} & \dfrac{z_{11}}{|z|} \end{bmatrix}$	$\begin{bmatrix} y_{11} & y_{12} \\ y_{21} & y_{22} \end{bmatrix}$	$\begin{bmatrix} \dfrac{D}{B} & -\dfrac{\Delta}{B} \\ -\dfrac{1}{B} & \dfrac{A}{B} \end{bmatrix}$
Chain	$\begin{bmatrix} \dfrac{z_{11}}{z_{21}} & \dfrac{|z|}{z_{21}} \\ \dfrac{1}{z_{21}} & \dfrac{z_{22}}{z_{21}} \end{bmatrix}$	$\begin{bmatrix} -\dfrac{y_{22}}{y_{21}} & -\dfrac{1}{y_{21}} \\ -\dfrac{|y|}{y_{21}} & -\dfrac{y_{11}}{y_{21}} \end{bmatrix}$	$\begin{bmatrix} A & B \\ C & D \end{bmatrix}$

[a] $|z| = z_{11}z_{22} - z_{12}z_{21}$; $|y| = y_{11}y_{22} - y_{12}y_{21}$; $\Delta = AD - BC$; $z_{12} = z_{21}$, $y_{12} = y_{21}$; $AD - BC = 1$ for reciprocal networks.

application, however, the latter parameters are not as useful as are the Z, Y, and chain parameters. Consequently, they are not discussed further in this chapter.

3.2.3 Two-Port Models for Circuits with Sources

The preceding discussion has considered passive two-port networks. For two-port networks containing internal EMI sources, we can develop a generalized Z-, Y-, or chain-parameter representation for the port voltage and current. As for the case of the single-port circuit represented by a Thévenin equivalent circuit in Figure 3.14a, an active two-port can be represented by the Z-parameter circuit. This is shown in Figure 3.14b.

The Z-matrix representation for an active two-port consists of the previously discussed passive Z-matrix, plus two voltage sources located in series with the terminals. With the choice of the source polarities shown in the figure, the v–i expression for the active network becomes

$$\begin{bmatrix} V_1 \\ V_2 \end{bmatrix} = \begin{bmatrix} z_{11} & z_{12} \\ z_{21} & z_{22} \end{bmatrix} \begin{bmatrix} I_1 \\ I_2 \end{bmatrix} + \begin{bmatrix} V_{s1} \\ V_{s2} \end{bmatrix} \qquad (3.20)$$

In representing the circuit in this way, the effects of all of the internal sources in the network are accounted for by the two sources at the ports.

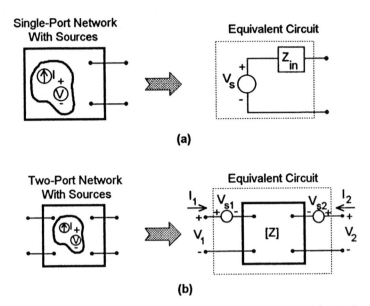

FIGURE 3.14 Z-matrix representation for an active circuit: (a) single-port Thévenin circuit; (b) two-port Thévenin circuit.

The dual network for the active two-port is analogous to the Norton equivalent circuit of the active source in Figure 3.2b. As shown in Figure 3.15, this consists of the Y-matrix circuit with two current sources shunted across both terminals.

The two-port $v-i$ relationships in this case are given as

$$\begin{bmatrix} I_1 \\ I_2 \end{bmatrix} = \begin{bmatrix} y_{11} & y_{12} \\ y_{21} & y_{22} \end{bmatrix} \begin{bmatrix} V_1 \\ V_2 \end{bmatrix} + \begin{bmatrix} I_{s1} \\ I_{s2} \end{bmatrix} \qquad (3.21)$$

Similarly, the chain parameter representation may be used for active two-ports. Figure 3.16 illustrates the fact that this case requires both a voltage and a current source at port 1 of the circuit. These are denoted by V_s and I_s,

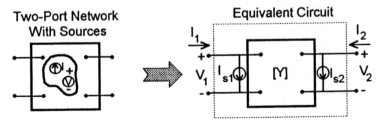

FIGURE 3.15 Y-matrix representation for an active two-port.

3.2 CONDUCTED DISTURBANCES IN CIRCUITS

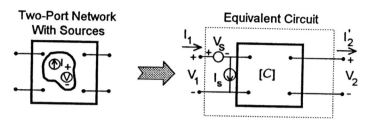

FIGURE 3.16 Chain parameter representation for an active two-port.

respectively. Note that, as before, the direction of the current at port 2 has been changed and this current is denoted as I'_2.

With this representation, the v–i relationships at the ports are given as

$$\begin{bmatrix} V_1 \\ I_1 \end{bmatrix} = \begin{bmatrix} A & B \\ C & D \end{bmatrix} \begin{bmatrix} V_2 \\ I'_2 \end{bmatrix} + \begin{bmatrix} V_s \\ I_s \end{bmatrix} \qquad (3.22)$$

As was the case with the passive two-port parameters, the two-port sources for the Z, Y, and chain parameters are all related. Table 3.3 summarizes these relationships.

TABLE 3.3 Relationships Between Active Two-Port Sources

To: \ From:	Impedance Matrix Sources $\begin{bmatrix} V_{s1} \\ V_{s2} \end{bmatrix}$	Admittance Matrix Sources $\begin{bmatrix} I_{s1} \\ I_{s2} \end{bmatrix}$	Chain Matrix Sources $\begin{bmatrix} V_s \\ I_s \end{bmatrix}$
Impedance matrix sources $\begin{bmatrix} V_{s1} \\ V_{s2} \end{bmatrix}$	$\begin{bmatrix} V_{s1} \\ V_{s2} \end{bmatrix}$	$\begin{bmatrix} -\dfrac{y_{22}}{\|y\|} I_{s1} + \dfrac{y_{12}}{\|y\|} I_{s2} \\ \dfrac{y_{21}}{\|y\|} I_{s1} - \dfrac{y_{11}}{\|y\|} I_{s2} \end{bmatrix}$	$\begin{bmatrix} V_s - \dfrac{A}{C} I_s \\ -\dfrac{I_s}{C} \end{bmatrix}$
Admittance matrix sources $\begin{bmatrix} I_{s1} \\ I_{s2} \end{bmatrix}$	$\begin{bmatrix} -\dfrac{z_{22}}{\|z\|} V_{s1} + \dfrac{z_{12}}{\|z\|} V_{s2} \\ \dfrac{z_{21}}{\|z\|} V_{s1} - \dfrac{z_{11}}{\|z\|} V_{s2} \end{bmatrix}$	$\begin{bmatrix} I_{s1} \\ I_{s2} \end{bmatrix}$	$\begin{bmatrix} I_s - \dfrac{D}{B} V_s \\ \dfrac{I_s}{B} \end{bmatrix}$
Chain matrix sources $\begin{bmatrix} V_s \\ I_s \end{bmatrix}$	$\begin{bmatrix} V_{s1} - \dfrac{z_{11}}{z_{21}} V_{s2} \\ -\dfrac{V_{s2}}{z_{21}} \end{bmatrix}$	$\begin{bmatrix} -\dfrac{I_{s2}}{y_{21}} \\ I_{s1} - \dfrac{y_{11}}{y_{21}} I_{s2} \end{bmatrix}$	$\begin{bmatrix} V_s \\ I_s \end{bmatrix}$

62 LUMPED-PARAMETER CIRCUIT MODELS

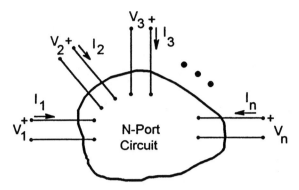

FIGURE 3.17 Generalization of a passive N-port circuit.

3.2.4 Treatment of Multiport Circuits

The concepts of one- and two-port circuits can be extended to the case of an N-port circuit without internal sources, as illustrated in Figure 3.17. In doing this the open-circuit impedance and short-circuit admittance relationships of Eqs. (3.3) and (3.8) become generalized matrix equations of the form $[V] = [Z][I]$ and $[I] = [Y][V]$, respectively, where $[V]$ and $[I]$ are N-vectors and $[Z]$ and $[Y]$ are $N \times N$ matrices. For an N-port circuit with internal sources, Eqs. (3.20) and (3.21) can be generalized in a similar manner, with the effects of the internal sources being represented by the appropriate voltage or current sources at each of the ports in Figure 3.17.

The chain parameter circuit representation can also be generalized to circuits having two multiport terminals, as shown in Figure 3.18. Letting the vectors $[V_1]$ and $[I_1]$ represent the m voltages and currents at the multiport terminal 1, and $[V_2]$ and $[I_2]$ represent the similar quantities at the n ports of terminal 2, the general chain parameter representation for this circuit is a

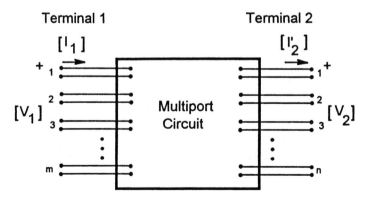

FIGURE 3.18 Multiport network represented by the chain parameters.

matrix equation whose elements are themselves matrices. This class of equation, known as a *supermatrix equation*, has the form

$$\begin{bmatrix} [V_1] \\ [I_1] \end{bmatrix} = \begin{bmatrix} [A] & [B] \\ [C] & [D] \end{bmatrix} \begin{bmatrix} [V_2] \\ [I'_2] \end{bmatrix} \quad (3.23)$$

where $[A]$, $[B]$, $[C]$, and $[D]$ are $m \times n$ matrices.

An alternative form of the multiport circuit is one in which there are an equal number of ports in the terminals 1 and 2, together with a common voltage reference conductor for each of the two-wire ports. In this manner, the multiport network appears as in Figure 3.19. This is frequently used to represent a section of multiconductor transmission line, as discussed in Chapter 6.

3.2.5 Example of Conducted Disturbances in Electrical Power Systems

The electrical power system provides a conducting link between a large number of customers and is one of the most important ways that low-frequency disturbances can propagate from a source of EMI to a victim. The various circuit modeling concepts discussed in preceding sections can be used to analyze such problems.

The most frequently encountered sources of EMI conducted by the power mains are:

- Current or voltage components within the power system at harmonics of the supply frequency
- Fluctuations of the power supply voltage

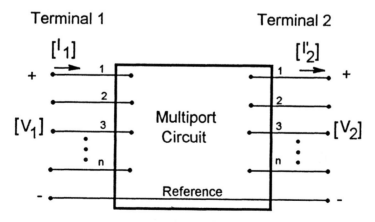

FIGURE 3.19 Multiport network having a common voltage reference.

- Voltage drops due to microswitches
- Overvoltages due to switching operations or lightning discharges

Modeling of the power system and these EMI sources is useful in understanding how to mitigate their effects. In what follows we illustrate the analysis of the effects of power system harmonics generated and conducted in the mains.

3.2.5.1 Generation of Harmonic Currents.

Normally, the electrical power system operates with a sinusoidal voltage waveform at a constant frequency of $f_0 = 50$ or 60 Hz. If all power system elements and user loads are linear, the resulting current flow within the system is also a sinusoidal function at the same operating frequency. However, if the system contains elements having a nonlinear voltage–current relationship, the current flowing in the mains can have a highly distorted waveform. A Fourier analysis of such waveforms shows that in addition to the principal component at the power frequency f_0, there are waveform components at the harmonic frequencies $\pm 2f_0$, $\pm 3f_0$, $\cdots$. Typical examples of equipment generating such harmonics include:

- Electronic circuits (i.e., loads) containing diodes and thyristors
- Rectifiers or inverters
- Switching power supplies†
- Power transformers or other circuits containing saturable magnetic cores
- Frequency converters, which are very difficult to filter.

The determination of the level of harmonic currents in loads and other parts of the power system is important. In many types of common classes of equipment, there are specified immunity levels for such interference. If these levels are not met, the following events may occur:

- Delay of supply voltage zero crossing, which can disturb the operation of electronic devices
- Increased VAR demand from power generation
- Additional losses in capacitances, rotating machines, and transformers
- Disturbances in control circuits
- Disturbances in signal transmission using the mains
- An increase in the current flowing in the neutral conductor of a

---

† Switching power supplies and frequency converters also generate harmonics other than $n \times f_0$. These are related to the switching frequency f_s.

three-phase grounded-wye power transmission system, resulting in transformer heating
- Undesired torques in rotating machines and increased noise in motors

Figure 3.20 shows a partial schematic of a three-phase electrical power system containing generation, transformers, lines, and consumer loads. We are interested in computing the behavior of the current (and voltage) flowing into one of the consumer loads at the low-voltage side of the system, say load 1.

For consumer load 1, Figure 3.21 shows an equivalent circuit for one of the three phases of the power mains, together with the various EMI sources occurring within the system. The high- and medium-voltage transformers are represented by impedance elements Z_{THV} and Z_{TMV}, and the connecting transmission line is represented by impedance Z_{line}. The power generator is assumed to have a zero source impedance. Consequently, its effect on load 1 is represented by the Norton current source I_1, shunting the transformer and line elements. This source operates as a constant current source at a frequency f_0.

In parallel with the primary current source is another source I'_1, representing the harmonic current components generated within the power transformers. This can be thought of as a parallel combination of several

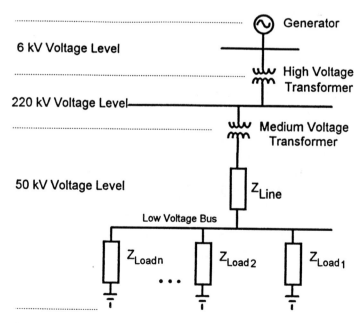

FIGURE 3.20 Circuit diagram of the electrical power system.

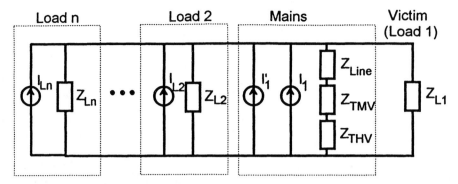

FIGURE 3.21 Equivalent network of one phase of the power system as seen by load 1.

current generators at frequencies $2f_0$, $3f_0$, and so on, or as a single generator providing a periodic but nonsinusoidal waveform.

The other power system loads (loads 2 to n) appear as additional shunt Norton equivalent circuits across load 1. Each is modeled by an impedance element characteristic of the service load, together with a current source representing the EMI-generated current within the load. Note that in this example any EMI generated within load 1 is neglected.

The rigorous analysis of the circuit of Figure 3.21 is difficult, if not impossible, because of the nonlinearities within the system. Each of the current sources, with the exception of I_1, will have values that are dependent on each other in a nonlinear manner. Unfortunately, their nonlinear characteristics are not always well known. Furthermore, the power system loads are always fluctuating, and this will also change the nonlinear behavior.

Notwithstanding these difficulties, it is often desired to obtain a rough indication of the harmonic content of the current flowing into the victim load 1. This may be desired for the purpose of designing a robust electrical circuit or for verifying that the interference does not exceed a specified level. To do this, the several harmonic current sources in Figure 3.21 can be approximated as behaving independently, and combined into a single current source representing the overall effect of the harmonically generated currents. Combining all the impedance elements together into a single equivalent mains impedance results in the circuit of Figure 3.22. To use this circuit for determining the response of load 1, it is necessary to know the mains impedance at the fundamental frequency f_0 and at the harmonic frequencies (up to some practical maximum number, say 11). Additionally, a specification of the levels of the nonlinear current source at each of the harmonic frequencies, $2f_0$, $3f_0$, and so on, is needed. With this information, the nth harmonic current flowing through load 1 can be approximated as

$$I_n \approx \frac{(Z_{L1})^{-1}}{(Z_{L1})^{-1} + (Z_{mains})^{-1}} I_h(f_n) \qquad (3.24)$$

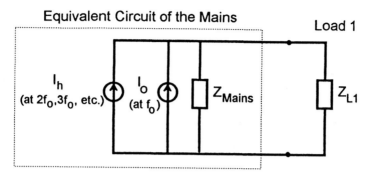

FIGURE 3.22 Single-phase equivalent circuit for harmonic interference within the power mains.

where $I_h(f_n)$ is the strength of the combined harmonic current source at frequency f_n.

3.2.5.2 Determination of the Mains Impedance. As seen from load 1, the mains impedance at frequencies f_0, $2f_0$, $3f_0$, and so on, can be either measured or calculated. At the fundamental power system frequency, it is usually possible to neglect the parallel combinations of the load impedances, $Z_{L2}, Z_{L3}, \ldots, Z_{Ln}$, which are typically much larger than the series impedance elements of the transformers and the transmission lines. However, at the higher harmonic frequencies, this may not be possible, due to the capacitive reactance occurring in the loads. Thus the impedance of the mains is given as

$$Z_{\text{mains}} = \left[(Z_{\text{line}} + Z_{\text{THV}} + Z_{\text{TMV}})^{-1} + \sum_{2}^{n} (Z_{Li})^{-1} \right]^{-1} \qquad (3.25)$$

Clearly, the values of these impedances will depend on the construction details of the power system under consideration. The impedance of a line or busbar is described by a per-unit-length resistance R' and reactance X' (usually given in Ω/km at a reference frequency of 50 or 60 Hz). HV transmission is usually achieved through aerial lines such as that shown in Figure 3.23a, except for special cases when short sections of cables are used (e.g., for connections within cities or submersed power links). For power distribution systems, the line configurations can be either aerial or buried, depending on design practice and the geographical and geological peculiarities (earthquake danger) of each country. The resistance and reactance values for such lines can be very different, depending on the line details.

Conductors for three-phase cables for medium- and low-voltage distribution networks (voltage < 110 kV) are usually symmetrically spaced, with the return (neutral) conductor being a cylindrical sheath around the three phases, as shown, in Figure 3.23b. Denoting the self-inductances of the three

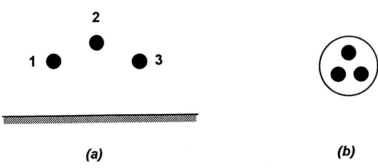

FIGURE 3.23 Cross sections of three-phase cables: (*a*) overhead line for high voltages; (*b*) low-voltage cable with sheath.

phases by M_{11}, M_{22}, M_{33}, respectively, and the mutual inductances as M_{12}, M_{23}, M_{31}, symmetry requires that

$$M_{11} = M_{22} = M_{33} \tag{3.26a}$$

$$M_{12} = M_{23} = M_{31} \tag{3.26b}$$

and the sum of the currents in the three phases is equal to zero

$$I_1 + I_2 + I_3 = 0 \tag{3.27}$$

The voltage drop on phase 1 can be written as

$$V_1 = R'I_1 + R_n'^1 I_1 + j\omega M_{11}' I_1 + R_n' I_2 + j\omega M_{12}' I_2 + R_n' I_3 + j\omega M_{31}' I_3 \tag{3.28}$$

Taking Eqs. (3.27) and (3.28) into account, relation (3.28) becomes

$$V_1 = R_1' I_1 + j\omega(M_{r11} - M_{12}')I_1 = R_1' I_1 + j\omega L_1' I_1 \tag{3.29}$$

in which

$$L_1' = M_{11}' - M_{12}' \tag{3.30}$$

is called the *cyclic inductance*. Calculations pertaining to the impedance of the mains when the distribution is made by cables can be performed using relation (3.29) and the cyclic inductance as seen from the three-phase matrix relation

$$\begin{bmatrix} V_1 \\ V_2 \\ V_3 \end{bmatrix} = \begin{bmatrix} R'^1 + j\omega L_1' & 0 & 0 \\ 0 & R_2' + j\omega L_2' & 0 \\ 0 & 0 & R_3' + j\omega L_3' \end{bmatrix} \begin{bmatrix} I_1 \\ I_2 \\ I_3 \end{bmatrix} \tag{3.31}$$

3.2 CONDUCTED DISTURBANCES IN CIRCUITS

It is the cyclic inductance value that is usually given by cable manufacturers in their product data sheets.

For aerial lines such as that shown in Figure 3.23a, the three-phase configuration is nonsymmetric, because the distances between the three phases are not equal. In this case, Eqs. (3.26a) and (3.26b) are not valid. As developed in [3.6], the consequence of such a phase imbalance is that the phase inductances are given as

$$L'_1 = L'_3 = L'_2 + \frac{\mu_0}{2\pi} \ln 2 \qquad (3.32)$$

and additional coupling terms $(\mu_0/2\pi) \ln 2$ are present in two off-diagonal terms in the matrix of Eq. (3.31). At a frequency of 50 Hz, the term

$$\omega \left(\frac{\mu_0}{2\pi} \right) \ln 2 \approx 0.044 \, \Omega/\text{km} \qquad (3.33)$$

which is not important with respect to typical reactance values for aerial lines (see Table 3.4). Consequently, aerial lines can be approximated as being symmetrical and the matrix relation in Eq. (3.31) used for the power mains impedance calculations.

Table 3.4 shows typical resistance and reactance (at 50 Hz) values for aerial lines. For HV lines, no typical value can be chosen due, to the wide variation of line geometries. Consequently, a range of values has been given as a function of cross section and distance between conductors. For medium- and low-voltage lines, the 7-mm diameter and 150-mm^2 cross section, respectively, can be considered as typical values [3.7], and these have been chosen for the table.

Table 3.5 shows resistance and reactance values for cables which are used only for medium- and low-voltage distribution [3.7]. The cross section of 150 mm^2 is here considered as being the most typical. Additional data on

TABLE 3.4 Typical Per-Unit-Length Impedance Components of Aerial Power Lines

Line Class	R' (Ω/km)	X' (Ω/km)
HV power transmission 110–500 kV	0.02–1.0 (function of cross section)	0.15: small phase separation 0.65: large phase separation
Medium-voltage distribution (conductor diameter $\approx$ 7 mm)	0.555	0.361
Low-voltage distribution (copper, 150 mm^2)	0.152	0.287

TABLE 3.5 Typical Per-Unit-Length Impedance Components of Power Cables

Line Class	R' (Ω/km)	X' (Ω/km)
Medium-voltage (20 kV) with plastic sheath, copper, 3×150 mm^2	0.158	0.107
Low-voltage plastic sheath, copper, 4×150 mm^2	0.153	0.072

typical line parameters are provided in [3.8] at a frequency of $f_{ref} = 60$ Hz. To calculate the resulting impedance of a line of length ℓ at the nth harmonic of a power frequency of $f_0 = 50$ Hz, the following expression can be used:

$$Z_{line}(nf_0) \approx \left(R' + j\frac{nf_0}{f_{ref}} X' \right)\ell \qquad (3.34)$$

The values of R_T and X_T, the resistive and reactive components of the impedance of a typical 16/0.4-kV power transformer at a reference frequency of 50 Hz, as seen from the low-voltage side, are shown in Table 3.6. These values also include the influence of the impedance of a 5-km line supplying the high-voltage side of the transformer, for which $R_L = 1.3$ mΩ and $X_L = 1.1$ mΩ at the low-voltage side.

In addition to the inductive and resistive components of the mains, any compensating capacitors or capacitive components of the service loads can be very important in determining the mains impedance. Figure 3.24 illustrates an example of the mains impedance calculated for a particular configuration of the power system. Note that there is a considerable variation of the impedance with frequency, a fact that illustrates the

TABLE 3.6 Resistance and Reactance of Different 16/0.4-kV Power Transformers at 50 Hz, as Seen from the Low-Voltage Side

Transformer Rating (kVA)	R_L (mΩ)	X_L (mΩ)	Transformer Rating (kVA)	R_L (mΩ)	X_L (mΩ)
63	42.1	104	400	5.4	20.0
75	34.7	89.0	500	4.4	16.0
100	24.8	69.0	630	3.5	13.0
160	13.7	45.0	800	3.0	11.0
200	10.9	37.0	1000	2.5	9.10
250	8.6	30.0	1250	2.3	7.60
300	7.1	26.0	1600	1.8	6.5

Source: [3.7].

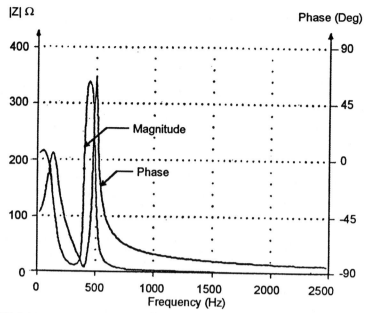

FIGURE 3.24 Example of the power mains impedance as a function of frequency.

importance of taking into account both the inductive and capacitive elements within the system.

3.2.5.3 Estimation of the Harmonic Current Source.

The values of the harmonic currents in the circuit of Figure 3.22 can be measured directly using a network analyzer or calculated using the Fourier transform of a measurement of the nonsinusoidal current waveform. As an example of harmonic currents generated in a metallurgical plant, Table 3.7 shows measured harmonics at various locations within the facility.

Various standards also give the amplitudes of harmonics for different types of devices using empirical formulas validated by measurements. For example, measurements of the harmonic components of currents flowing in power lines feeding rectifiers have provided the following approximate expression for the amplitude of the nth harmonic current for $5 \leq n \leq 31$ [3.9]:

$$I_n = \frac{I_1}{1.2(n - 5/n)} \qquad (3.35)$$

where I_1 is the current at the fundamental frequency f_0. The phase of the harmonics can be approximated by

$$\phi_n = n\alpha + (n \pm 1)\beta \qquad (3.36)$$

TABLE 3.7 Percentage of Harmonic Currents Measured at Two Locations in a Metallurgical Plant

Harmonic Number	220-kV Mains	15-kV Bus at Load
1	99.97	97.8
3	0.6	0.9
5	0.7	17.3
7	0.7	10.0
11	1.6	4.1
13	1.3	3.2
17	0.4	2.0
19	0.4	1.4

where α is the angle representing the delay in the switching time in the rectifier and β is the phase shift between the primary and secondary voltages of the transformer supplying the rectifier.

3.3 DISTURBANCES IN CIRCUITS INDUCED BY ELECTROMAGNETIC FIELDS

Another way in which EMI can affect a victim circuit is by electromagnetic field coupling. In contrast to the conducting-wire injection of EMI into the victim, this type of coupling involves the electromagnetic fields emanating from the EMI source and affecting the victim. For the low-frequency regime under discussion in this section, the type of coupling usually depends on either the E-field or the H-field, depending on the nature of the source and victim circuitry.

3.3.1 Magnetic Field Coupling

As an example of coupling that is due predominately to the magnetic field, Figure 3.25a shows an EMI source and victim equipment, both represented by single-port equivalent circuits. The portions of the circuits responsible for the coupling appear as two loops. It is assumed that both the source and victim equivalent networks within the enclosures are well shielded, with no EM coupling between the internal components. In the discussion to follow, we continue to assume that the frequency of the EMI disturbance is low. This implies that the wavelength of the disturbing signal, $\lambda = c/f$, is such that $\lambda \gg d$, where d is a typical dimension of the overall source–victim ensemble.

The sources within the EMI enclosure induce a current I_1 in the exposed circuit loop shown in Figure 3.25a. This current creates a magnetic flux density B which links the pickup loop in the victim circuit. If there is a time

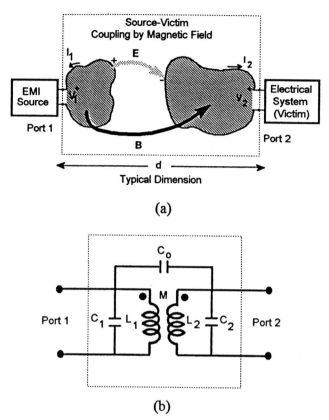

FIGURE 3.25 Low-frequency magnetic field coupling of source and victim circuits; (a) circuit geometry; (b) two-port representation of field coupling.

variation of this B-field, by Lenz's law a resulting voltage is induced in the pickup loop. This coupling between the source and victim circuits is known as *inductive coupling*, and leads to induced voltages within the victim's circuitry.

At the same time, the EMI source induces a voltage V_1 at the terminals of the exposed source circuit. This leads to a charge distribution on the loop. The charge creates an electric field E which extends through space, and a portion of this field may terminate on other charges induced in the pickup loop of the victim circuit. Since the E-field varies with time, the resulting time variation of the induced charge creates a current flow in the victim circuit. This is referred to as *capacitive coupling*.

Inductive and capacitive coupling exist together, and in a general high-frequency EMI problem, they must both be included in the calculation of the victim response. However, for the assumed loop geometries of the coupling elements shown in Figure 3.25a, the magnetic field coupling is the primary coupling mechanism at low frequencies. This coupling can be represented by a two-port network of exposed loops as shown in Figure 3.25b. The exposed circuit at port 1 (the EMI source) has an inherent inductance L_1 and a parasitic capacitance C_1. The exposed circuit at the victim has similar parameters, L_2 and C_2. The magnetic flux linking the two circuits is represented by the mutual inductance M and the coupling capacitance between the two circuits is C_0. All of these parameters are dependent on the geometry and environment surrounding the circuits and are typically found by measurements or by computational models. For low frequencies, the effects of the parasitic capacitances C_1 and C_2 are usually negligible.

The induced response in the victim circuit may be computed using the loaded two-port circuits of Figure 3.6 or 3.9, with the two-port impedance or admittance parameters corresponding to those for the two-port coupling network of Figure 3.25b. Neglecting the parasitic capacitances C_1 and C_2, the coupling circuit can be represented by an equivalent π-circuit as shown in Figure 3.8.† For this representation, the admittance elements Y_a, Y_b, and Y_c are

$$Y_a = \frac{L_2 - M}{j\omega(L_1 L_2 - M^2)} \qquad (3.37a)$$

$$Y_a = \frac{L_1 - M}{j\omega(L_1 L_2 - M^2)} \qquad (3.37b)$$

$$Y_c = j\omega C_0 + \frac{M}{j\omega(L_1 L_2 - M^2)} \qquad (3.37c)$$

These lumped parameters can be used directly in the expressions in Figure 3.8 to provide the y_{ij} parameters for the Y-matrix describing the coupling. Equation (3.10) can then be used to determine the resulting voltage across the input port of the victim circuit. Using the T-to-Π transformation, the Z_1, Z_2, and Z_3 element values for the T-network of Figure 3.5 can also be determined, and the corresponding values of z_{ij} determined. For this last case, Eq. (3.6) provides the current flowing into the victim circuit. Either method provides the proper response; however, the expressions for Z_1, Z_2, and Z_3 are much more complicated than are those above for Y_a, Y_b, and Y_c.

3.3.1.1 Weak-Coupling Approximations.
The preceding development for field coupling between the source and victim is general, in that the responses in the victim circuit are able to affect the behavior of the source circuit. This is a direct result of the two-port representation of field coupling.

---

† Note that this development neglects the effects of conductor resistance, which for this circuit configuration is usually more important than the capacitive effects.

In many cases, however, the source and victim are sufficiently distant that the reaction of the victim responses back at the source is negligible. In this case, the two circuits are said to be weakly coupled, and simplifications to the coupling equations can be made.

Referring to Figure 3.25b, weak coupling implies that $M^2 \ll L_1 L_2$ and $1/j\omega C_0 \gg j\omega L_1$ or $j\omega L_2$. Under these conditions, and assuming that the effects of the parasitic capacitances C_1 and C_2 are still negligible compared with the circuit inductances, the admittance elements for the Π equivalent circuit for coupling of Figure 3.25b become

$$Y_a \approx (j\omega L_1)^{-1} \tag{3.38a}$$

$$Y_b \approx (j\omega L_2)^{-1} \tag{3.38b}$$

$$Y_c \approx j\omega C_0 + M(j\omega L_1 L_2)^{-1} \tag{3.38c}$$

For this weak-coupling case, the y-parameters of Figure 3.8 are given approximately as $y_{11} \approx Y_a$, $y_{22} \approx Y_b$, and $y_{12} = y_{21} \approx -Y_c$. Representing the EMI source in Figure 3.25a by its Norton equivalent circuit (see Figure 3.2b), and assuming that the victim circuit presents a load admittance Y_L to the coupling loop at port 2 of Figure 3.25a, the induced load voltage may be approximated using Eq. (3.10a) as

$$V_L \approx \frac{j\omega C_0 + M(j\omega L_1 L_2)^{-1}}{[(j\omega L_1)^{-1} + Y_s][(j\omega L_2)^{-1} + Y_L]} I_s \tag{3.39}$$

This last expression can be thought of as having two separate excitation terms corresponding to current sources operating on the victim circuit. The first term arises from the capacitive coupling between the source and victim circuits and is given by

$$I' = \frac{j\omega C_0}{(j\omega L_1)^{-1} + Y_s} I_s = j\omega C_0 V_1 \tag{3.40a}$$

where V_1 is the voltage induced by the EMI source across port 1. The second term arises from the magnetic field coupling of the two circuits and is

$$I'' = \frac{M(j\omega L_1 L_2)^{-1}}{(j\omega L_1)^{-1} + Y_s} I_s \approx \frac{M}{L_2} I_1 \tag{3.40b}$$

For the loop coupling of the circuit configuration in Figure 3.25a, the major contribution to the response is by the latter excitation source arising from the magnetic field excitation. In many discussions of loop coupling, the electric field coupling term of Eq. (3.40a) is neglected. As will be illustrated shortly, this term can be important under certain circumstances.

With both of these sources defined, the voltage induced in the victim load is now expressed simply as

$$V_L \approx \frac{I' + I''}{(j\omega L_2)^{-1} + Y_L} \tag{3.41}$$

and the load current is $I_L = Y_L V_L$.

Figure 3.26a shows a circuit model representing this case of weak coupling, with the behavior of the EMI source being independent of that of the victim circuit. It is possible to transform the current source arising from the inductive coupling I'' into an alternative, and more common form, involving a voltage source. This is shown in Figure 3.26b. Here this current source is transformed by a Norton-to-Thévenin transformation, and the resulting voltage source is located in series with the inductance L_2. This source is given by

$$V' = j\omega M I_1 \tag{3.42}$$

Note that at low frequencies, the loop current I_1 in the EMI source circuit

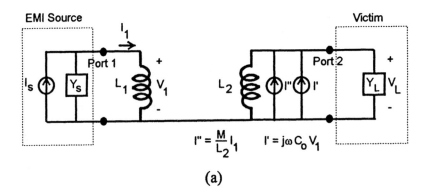

(a)

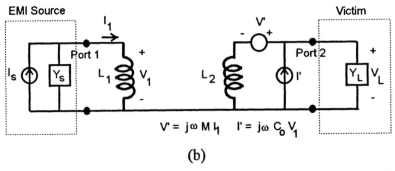

(b)

FIGURE 3.26 Coupled source–victim circuits for weak magnetic field coupling of the circuits of Figure 3.20: (a) two current sources; (b) current and voltage source.

can be large, while the loop voltage V_1 is practically shorted out by the loop. Thus the current source in the victim circuit is $I' \approx 0$, and the main excitation is due to the source V' arising from the time rate of change of the magnetic field linking the victim circuit. These circuits are referred to as being magnetically coupled.

3.3.1.2 Calculation of Mutual and Self-Inductances.

The mutual and self-inductances of realistic circuit configurations are usually difficult to know precisely, due to the irregular geometries of the circuits as well as the presence of magnetic material in the surrounding environment. However, rough estimates can be made of these quantities by making several simplifying assumptions. Assuming that two loop circuits are located in free space, the mutual inductance between the two is given as

$$M = \frac{\int_{S_2} \mathbf{B}_1 \cdot d\mathbf{S}_2}{I_1} \tag{3.43}$$

where I_1 is the current flowing in the driven circuit (loop 1) and $\mathbf{B}_1$ is the magnetic flux produced by this current. This flux is integrated over the surface S_2 formed by loop 2. As long as there is no anisotropic magnetic material present, the reciprocity principle implies that the same mutual inductance results if loop 2 is driven with a current I_2 and the flux through areas S_1 of loop 1 is computed.

An alternative form for the mutual inductance may be obtained from vector potential considerations. This provides the Neumann form of the mutual inductance of two loops as a double integral around the contours of both circuit loops:

$$M = \frac{\mu_0}{4\pi} \oint_{C1} \oint_{C2} \frac{d\mathbf{l}_1 \cdot d\mathbf{l}_2}{r} \tag{3.44}$$

As shown in Figure 3.27a, r is the distance from each of the integration points $d\mathbf{l}_1$ and $d\mathbf{l}_2$.

This equation can be applied to the special case of two coaxially oriented loops of radii a and b, separated by a distance d, as shown in Figure 3.27b. In this case, M is given by the equation [3.10]

$$M = \mu_0 \sqrt{ab} \left[\left(\frac{2}{\kappa} - \kappa\right) K(\kappa) - \frac{2}{\kappa} E(\kappa) \right] = \mu_0 \sqrt{ab}\, F(\kappa) \tag{3.45}$$

where

$$\kappa^2 = \frac{4ab}{d^2 + (a+b)^2} \tag{3.46}$$

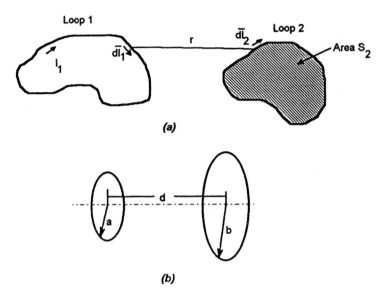

FIGURE 3.27 Mutual coupling configurations: (*a*) two arbitrary loops; (*b*) two concentric, circular loops.

and $K(\cdot)$ and $E(\cdot)$ are the complete elliptic integrals, defined as

$$K(\kappa) = \int_0^{\pi/2} \frac{d\phi}{\sqrt{1 - \kappa^2 \sin^2\phi}}$$

$$E(\kappa) = \int_0^{\pi/2} \sqrt{1 - \kappa^2 \sin^2\phi}\, d\phi \qquad (3.47)$$

A significant amount of effort has been spent in both tabulating these integrals and in developing suitable approximating forms for the mutual inductance of the circular loops [3.11]. However, with the advent of readily available computers and software for special function evaluation [3.12], direct evaluation of Eq. (3.45) is practical. Figure 3.28 presents a plot of the quantity $F(\kappa) \equiv [(2/\kappa) - \kappa]K(\kappa) - (2/\kappa)E(\kappa)]$ as a function of the parameter κ, which may be used in Eq. (3.45) to evaluate the mutual inductance of the loops.

For the case when $a \ll d$, the parameter κ is small and $F(\kappa)$ can be expanded in a small-argument approximation, giving the following expression for M [3.10]:

$$M = \frac{\mu_0 \pi a^2 b^2}{2(b^2 + d^2)^{3/2}} \qquad (3.48)$$

A useful expression is available for the mutual inductance between two

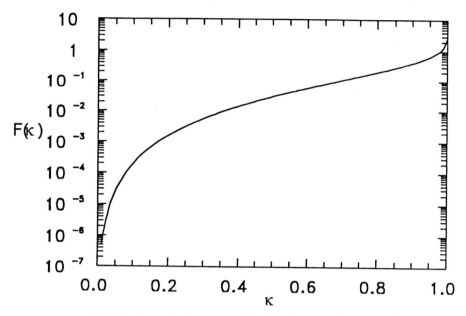

FIGURE 3.28 Plot of normalized inductance function $F(\kappa)$.

parallel current elements displaced from each other as shown in Figure 3.29a. Reference [3.10] provides the relation

$$M = \frac{\mu_0}{4\pi}\left[\ln\frac{(A+a)^a(B+b)^b}{(C+c)^c(D+d)^d} + (C+D) - (A+B)\right] \quad (3.49)$$

for the mutual inductance, where the distances A through D are indicated in the figure. This expression is useful when computing the mutual inductance of square-loop structures such as that shown in Figure 3.29b. Here, all

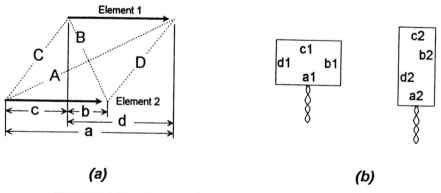

FIGURE 3.29 Geometry for two parallel conductor segments.

mutually perpendicular elements (like a_1 and c_2) will have a zero contribution to the total mutual coupling of the loops, due to the dot product in Eq. (3.44). The remaining terms can be evaluated using Eq. (3.49).

The self-inductance of an *isolated* circular current loop also can be found using Eq. (3.45). In doing this, loops 1 and 2 are considered to be overlapping, with the difference in loop radii being equal to the radius of the wire comprising the loop. In this manner the integrand is not singular and the self-inductance is given as

$$L = \mu_0 \sqrt{a(a-r_0)} \left[\left(\frac{2}{\kappa} - \kappa \right) K(\kappa) - \frac{2}{\kappa} E(\kappa) \right] \equiv \mu_0 \sqrt{a(a-r_0)} F(\kappa) \tag{3.50}$$

Here, a is the loop radius and r_0 is the wire radius, and the parameter κ is given by

$$\kappa^2 = \frac{4a(a-r_0)}{(2a-r_0)^2} \tag{3.51}$$

This expression may also be evaluated using the function $F(\kappa)$ in Figure 3.28.

For cases where $r_0 \ll a$, the elliptic integrals can be evaluated asymptotically, and the self-inductance of the loop is approximated by

$$L \approx \mu_0 a \left(\ln \frac{8a}{r_0} - 2 \right) \tag{3.52}$$

The self-inductance of a square loop of radius r_0 and side length a may be evaluated using Eq. (3.44), as illustrated in [3.10]. Letting $p = a/r_0$, this inductance is expressed as

$$L = \frac{2\mu_0 a}{\pi} \left(\ln \frac{p + \sqrt{1+p^2}}{1+\sqrt{2}} + \frac{1}{p} - 1 + \sqrt{2} - \frac{1}{p}\sqrt{1+p^2} \right) \tag{3.53a}$$

and for loops where $r_0 \ll a$, this expression is approximated by

$$L = \frac{2\mu_0 a}{\pi} \left(\ln \frac{2p}{1+\sqrt{2}} - 2 + \sqrt{2} \right) \tag{3.53b}$$

Note that in these calculations of the self-inductance given in Eqs. (3.50) and (3.53), the contribution by the internal flux within the conductor has been neglected. This term, which is most important at low frequencies when the current is uniformly distributed across the cross section of the conductor, is a per-unit-length quantity given as

$$L'_{int} = \frac{\mu_0}{8\pi} \tag{3.54}$$

The additional contribution of the internal flux is then calculated as L'_{int} times the loop length.

3.3.2 Electric Field Coupling

An alternative source–victim coupling configuration is shown in Figure 3.30a, where the field coupling is due primarily to the E-field. The EMC source voltage V_1 can induce large charge densities on the coupling elements. As there is no closed path from one conductor to another at port 1, the current I_1 and its resulting magnetic field will be small. However, as in the previous case, both E and H fields exist, and both contribute to the field coupling.

The equivalent-circuit model for this coupling is shown in Figure 3.30b. The terms C_1 and C_2 represent the self-capacitances of the conductors responsible for coupling at ports 1 and 2, and C_0 represents the mutual

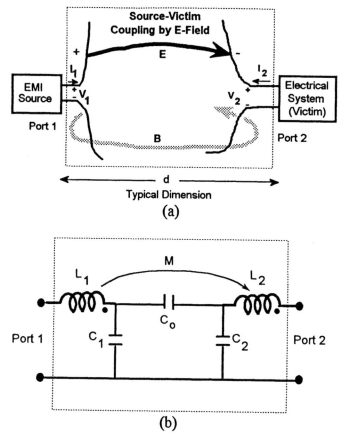

FIGURE 3.30 Low-frequency electric field coupling of source and victim circuits: (a) circuit geometry; (b) two-port representation of field coupling.

capacitance between the two. L_1 and L_2 are the stray inductances of the wires, and M is the mutual inductance. In most practical cases involving wire coupling as shown in the figure, the effects of these inductances are negligible compared with the capacitances.

The equivalent circuit of Figure 3.30b can be put into a form more suitable for analysis by transforming the Π network of capacitors into a T network using Eq. (3.13). Neglecting the effects of the stray inductance of the wires, L_1 and L_2, but retaining the mutual coupling inductance M, a T network in the form of Figure 3.5 results. This network has the following parameters:

$$Z_1 = \frac{C_2}{j\omega[C_1C_2 + C_0(C_1 + C_2)]} \quad (3.55a)$$

$$Z_2 = \frac{C_1}{j\omega[C_1C_2 + C_0(C_1 + C_2)]} \quad (3.55b)$$

$$Z_3 = j\omega M + \frac{C_0}{j\omega[C_1C_2 + C_0(C_1 + C_2)]} \quad (3.55c)$$

3.3.2.1 Weak-Coupling Approximations.
As in the case of the inductive coupling discussed previously, these impedance expressions may be used in the definitions of the open-circuit impedance parameters in Figure 3.5 to evaluate the victim load current response given in Eq. (3.6). Doing this assumes a Thévenin equivalent-circuit representation of the EMI source.

The practical case of weak coupling simplifies the solution. For the case when $C_0 \ll C_1$ and C_2, the T-circuit impedance elements are

$$Z_1 \approx (j\omega C_1)^{-1} \quad (3.56a)$$

$$Z_2 \approx (j\omega C_2)^{-1} \quad (3.56b)$$

$$Z_3 \approx j\omega M + C_0(j\omega C_1 C_2)^{-1} \quad (3.56c)$$

and Eq. (3.6a) becomes

$$I_L \approx \frac{j\omega M + C_0(j\omega C_1 C_2)^{-1}}{[(j\omega C_1)^{-1} + Z_s][(j\omega C_2)^{-1} + Z_L]} V_s \quad (3.57)$$

Observe that this equation is identical in form to the response in Eq. (3.39) for the inductive coupling. Thus we can again identify two types of source terms: a voltage source defined as $V'_1 = j\omega M I_1$ and a current source $I' = j\omega C_0 V_1$. This results in the circuit in Figure 3.31 for capacitive coupling. Note that at low frequencies, the current I_1 flowing into the capacitance at the source is small, so that $V' \approx 0$, and the coupling is dominated by the capacitive coupling source $I' = j\omega C_0 V_1$.

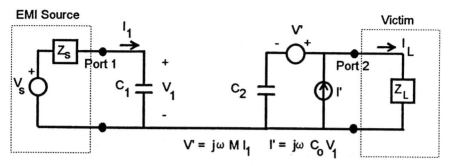

FIGURE 3.31 Equivalent circuit for weak electric field coupling in Figure 3.30.

3.3.2.2 Calculation of Mutual and Self-Capacitances.
The estimation of the self- and mutual coupling capacitances of wirelike structures is more difficult than for the inductances discussed previously. This is because the capacitances depend on the charge distributions on the coupling elements, and these distributions can change, depending on the details of the local geometry of the wires. In the case of the inductances, the current flow in the wires was assumed to be a constant, and independent of the loop geometry.

To illustrate the calculation of the capacitances for use in the coupling model, consider the idealized geometry of two parallel, thin cylindrical wires of total length $2\mathcal{L}_1$ and $2\mathcal{L}_2$, radii a_1 and a_2, and separation d, as illustrated in Figure 3.32. Each wire has a pair of terminals at the center of the conductor, and voltages V_1 and V_2 can be either impressed or measured at these locations.

The capacitance of this two-wire structure is defined by a matrix relationship between V_1 and V_2 defined at the two terminals and the electric

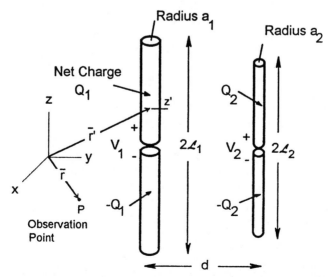

FIGURE 3.32 Geometry of two thin cylinders for capacitance calculations.

charge residing on the wires. Letting Q_1 be the charge on the upper half of wire 1 and Q_2 be the charge on wire 2 (with $-Q_1$ and $-Q_2$ on the lower half of the wires), the charge–voltage relationship is written as

$$\begin{bmatrix} Q_1 \\ Q_2 \end{bmatrix} = \begin{bmatrix} C_{11} & C_{12} \\ C_{21} & C_{22} \end{bmatrix} \begin{bmatrix} V_1 \\ V_2 \end{bmatrix} \tag{3.58}$$

where the terms C_{ij} are referred to as the capacitive coefficients of the structure and must be determined numerically for each specific wire geometry.

To determine the C_{ij} terms, it is necessary to determine the charge distributions on each wire. For thin wires oriented in the z direction, these are described by linear charge densities $\rho_1(z)$ and $\rho_2(z)$, (in C/m). These charge distributions are used to compute the electric potential at any observation point denoted by $\mathbf{r}$ by integrating over the wires, as

$$\Phi(\mathbf{r}) = \frac{1}{4\pi\epsilon_0} \left[\int_{\text{wire 1}} \frac{\rho_1(\mathbf{r}')\, dz'}{|\mathbf{r} - \mathbf{r}'|} + \int_{\text{wire 2}} \frac{\rho_2(\mathbf{r}')\, dz'}{|\mathbf{r} - \mathbf{r}'|} \right] \tag{3.59}$$

By imposing voltages V_1 and V_2 at the terminals of the wires, the top halves of the conductors are maintained at constant potentials of $+V_1/2$ and $+V_2/2$, and the bottom halves are at $-V_1/2$ and $-V_2/2$. Letting the observation point $\mathbf{r}$ approach each of the conductors results in a set of coupled integral equations for the charges [3.13].

These integral equations can be solved numerically by the moment method [3.14]. This involves representing the charge densities by n basis (or expansion) functions on each wire and then matching Eq. (3.59) at n locations on each wire. This results in a $2n \times 2n$ matrix equation that must be inverted numerically. If the charges on the conductors are represented by a sum of pulse functions of width Δ, integrals of the form

$$\phi(z_i, z_j) = \int_{z_i - \Delta/2}^{z_i + \Delta/2} \frac{d\xi}{\sqrt{(\xi - z_j)^2 + K^2}} \tag{3.60}$$

must be evaluated. The terms z_i and z_j denote the source and observation location along the axis of the cylinders. When the source and observation points are located on different wires, the term K in Eq. (3.60) is given by $K = d$. For z_i and z_j on the same wire, $K = a_i$. The integral of Eq. (3.60) is well defined and is approximated by

$$\phi \approx \frac{\Delta}{\sqrt{(z_i - z_j)^2 + K^2}} \tag{3.61}$$

when the source and observation point are far apart [i.e., when $(z_i - z_j)^2 \gg$

a_1^2 or a_2^2]. For cases where the source and observation points are close, Eq. (3.60) can be integrated analytically as

$$\phi = \ln \frac{\sqrt{(2(z_i - z_j) + \Delta)^2 + 4a^2} + 2(z_i - z_j) + \Delta}{\sqrt{(2(z_i - z_j) - \Delta)^2 + 4a^2} + 2(z_i - z_j) - \Delta} \quad (3.62)$$

where $a = a_1$ or a_2, depending on the conductor being considered.

Figure 3.33 shows an example of the numerically calculated mutual capacitance between the two wires, as a function of the distance d between the wires. This quantity is defined as

$$C_0 \equiv c_{12} = \left.\frac{Q_1}{V_2}\right|_{V_1=0} = c_{21} = \left.\frac{Q_2}{V_1}\right|_{V_2=0} \quad (3.63)$$

where Q_1 (or Q_2) is the total charge on the top half of the wire 1 (or 2) and is given by $Q_1 = \int_0^{\mathcal{L}_1/2} \rho_1(z)\, dz$. In this figure the aspect ratios of each wire is the same, with $\mathcal{L}/a = 100$. This capacitance is a negative quantity and is presented in normalized form as $-C_0/(4\pi\epsilon_0 \mathcal{L}_1)$.

Figure 3.34 shows the corresponding normalized self-capacitance of an

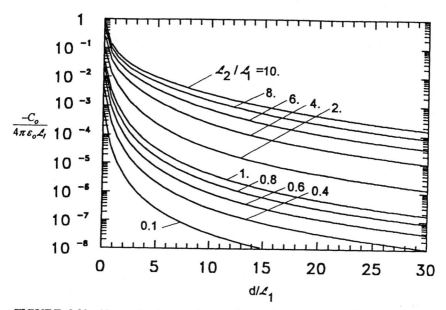

FIGURE 3.33 Normalized mutual capacitance between two thin wires with $a_1/\mathcal{L}_1 = a_2/\mathcal{L}_2 = 0.01$.

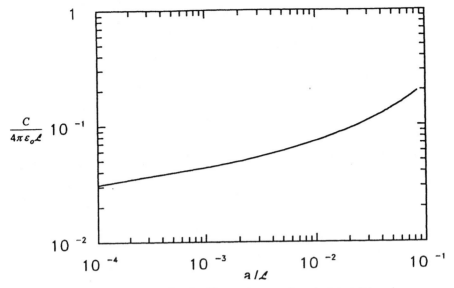

FIGURE 3.34 Normalized self-capacitance of an isolated thin wire.

isolated wire, defined as

$$C \equiv c_{11} = \frac{Q_1}{V_1} \quad \text{for } d \to \infty \qquad (3.64)$$

This is shown as a function of the aspect ratio $a/\mathscr{L}$.

3.3.3 General Field Coupling at Low Frequencies

The two different types of coupling just discussed are special cases of a more general type of low-frequency coupling in which *both* the electric and magnetic fields exist together and contribute to the source terms V' and I' in the victim circuitry [3.15]. For weak coupling, the inductances in Figure 3.26b or the capacitances in Figure 3.31 are replaced by more general impedance elements as shown in Figure 3.35. The EMI source induces responses V_1 and I_1 in the generalized load in the input circuit, Z_L, and both of these contribute to the victim excitation. The following example illustrates this more general type of field coupling.

3.3.3.1 Example: Crosstalk Between Two Parallel Traces on a Printed Circuit Board.
As an example of the practical use of field coupling models, let us estimate the crosstalk between two weakly coupled printed circuit board (PCB) traces (lands) which are located parallel to each other (Figure 3.36). A typical size of a PCB is such that the lengths of the lines $\mathscr{L}$ are on

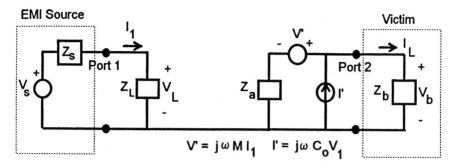

FIGURE 3.35 Circuit diagram for generalized weak coupling by both E and H fields.

the order of 5 cm. Based on the criteria that the wavelength of the EMI must be such that $\lambda \geq 10\ell$ for the low-frequency modeling approach to be valid, we note that the maximum frequency for this analysis will be $f_{max} = c/\lambda = c/10\ell \approx 600$ MHz. For pulsed signals within the PCB, this bandwidth limitation implies that the low-frequency model is valid for pulses having a 10 to 90% rise time of $t_r \approx 0.44/f_{max}$, or about 0.7 ns [3.16]. Data transmission can occur on lines having much larger dimensions (e.g., $\ell \approx 5$ m) [3.17]. In this case the corresponding maximum frequency is equal to 6 MHz, with a minimum rise time of 70 ns.

In this example, the primary line is excited by a lumped voltage source located at one end, and this induces a current I_1 to flow in the line. Thus this configuration is similar to that of Figure 3.25a, in which there is a significant magnetic flux coupling to the receptor. However, the voltage source V_s also produces a charge density on the excitation line, and this gives rise to capacitive coupling between the two lines. Both of these coupling mechanisms must be considered.

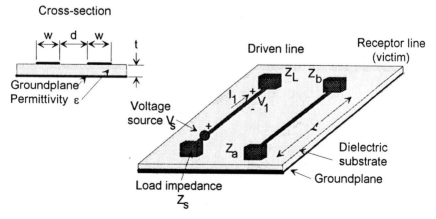

FIGURE 3.36 Geometry of two coupled lines for crosstalk estimation.

For long, straight, circular wire conductors located over a perfect ground with no dielectric layer, the mutual inductance and capacitance elements M and C_0 discussed in previous sections can be calculated from expressions given in [3.18] for the per-unit-length inductance and capacitance values for multiconductor lines. This information is summarized in Appendix C. For the present case, however, the PCB traces are not circular but appear as strip conductors having a width w and a separation d. Furthermore, the presence of the dielectric material affects the value of the capacitance. As discussed in [3.19], the capacitance and inductance of these traces can differ significantly from those of circular wires, and a numerical calculation is generally needed to determine these parameters. Several computer programs are provided in [3.19] for calculating these parameters, given the details of the conductor geometries. The interested reader should consult this reference for further information.

Under the assumption of weak coupling, the current and voltage responses in the driving line, I_1 and V_1, can be expressed in terms of the source excitation voltage V_s as

$$I_1 = \frac{V_s}{Z_s + Z_L} \tag{3.65}$$

$$V_1 = \frac{Z_L V_s}{Z_s + Z_L} \tag{3.66}$$

To determine the voltage or current responses in either of the load elements of the victim line, the circuit model shown in Figure 3.37 can be used. Notice that in this case, both the voltage and current sources are needed to describe the field coupling. Performing a simple analysis of this circuit provides the following expressions for the load voltages, V_a and V_b:

$$V_a = \frac{Z_a}{Z_a + Z_b}(j\omega C_0 Z_b V_1 + j\omega M I_1) \tag{3.67}$$

$$V_b = \frac{Z_b}{Z_a + Z_b}(j\omega C_0 Z_a V_1 - j\omega M I_1) \tag{3.68}$$

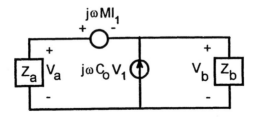

FIGURE 3.37 Equivalent circuit of victim conductor and termination impedances.

3.3 DISTURBANCES IN CIRCUITS INDUCED BY ELECTROMAGNETIC FIELDS

and the load currents are given by $I_a = V_a/Z_a$ and $I_b = V_b/Z_b$.

Substituting the values of V_1 and I_1 into these last equations gives expressions for the load voltages in terms of the source voltage V_s:

$$V_a = \frac{Z_a}{(Z_a + Z_b)(Z_s + Z_L)} (j\omega C_0 Z_b Z_L + j\omega M)V_s \quad (3.69a)$$

$$V_b = \frac{Z_b}{(Z_a + Z_b)(Z_s + Z_L)} (j\omega C_0 Z_a Z_L - j\omega M)V_s \quad (3.69b)$$

To assist in interpreting the behavior of the induced responses in the receptor circuit loads, it is convenient to define two response coefficients. Note that these coefficients are referred to as *response coefficients*, not *coupling coefficients*, since they depend on both the field coupling parameters and on the impedances of the receptor circuit. From the capacitive coupling terms in Eqs. (3.67) and (3.68), we define the capacitive response term k_C as

$$k_C = \frac{C_0 Z_a Z_b Z_L}{(Z_a + Z_b)(Z_s + Z_L)} \quad (3.70a)$$

Similarly, the inductive coupling term provides a response coefficient k_M as

$$k_M = \frac{M}{(Z_a + Z_b)(Z_s + Z_L)} \quad (3.70b)$$

With these response coefficients, the voltages at either end of the receptor line are expressed simply as

$$V_a = (j\omega k_C + j\omega k_M Z_a)V_s \quad (3.71a)$$

$$V_b = (j\omega k_C - j\omega k_M Z_b)V_s \quad (3.71b)$$

As a numerical example, assume that the overall PCB trace lengths are $\mathcal{L} = 5$ cm, the edge-to-edge conductor separation is $d = 0.5$ cm, the conductor widths are $w = 0.2$ cm, and the dielectric substrate thickness is $t = 0.1$ cm. The relative permittivity $\epsilon = 3.6$, which is typical of a plastic material. Using the MSTRP.FOR program of [3.19], the per-unit-length line parameters may be determined, and the corresponding total mutual inductance and capacitance for the 5-cm line is given by $M = 4.131 \times 10^{-10}$ H and $C_0 = -3.607 \times 10^{-14}$ F. Assuming that all of the load impedances are matched to the characteristic impedance of a single trace (about 53.5 Ω), we have $Z_s = Z_L = $

$Z_a = Z_b = Z_c = 53.5 \, \Omega$. The load voltages of Eq. (3.69) become

$$V_a = j\omega(-4.825 \times 10^{-13} + 1.930 \times 10^{-12})V_s = j\omega 1.448 \times 10^{-12}V_s \qquad (3.72a)$$

$$V_b = j\omega(-4.825 \times 10^{-13} - 1.930 \times 10^{-12})V_s = -j\omega 2.413 \times 10^{-12}V_s \qquad (3.72b)$$

Observe that in these expressions the influence of the capacitive coupling is just as important as is the inductive coupling. Neglecting the capacitive coupling is a frequent mistake made in coupling analysis. However, there are special cases where the capacitive coupling term can be very small compared with the inductive term. Consider the case where the load impedance on the excitation line Z_L approaches zero. In this case, $k_c = 0$ and the inductive response term becomes

$$k_M = \frac{M}{Z_s(Z_a + Z_b)} \qquad (3.73)$$

Similarly, there can be instances where the capacitive coupling dominates. If the load on the excitation line is open-circuited, $Z_L = \infty$, and this results in the inductive response term $k_M = 0$. The capacitive response term is

$$k_C = \frac{C_0 Z_a Z_b}{Z_a + Z_b} \qquad (3.74)$$

Similar variations of the type of coupling also occur for different values of the termination impedances of the receptor conductor.

3.3.4 General Methods of Reducing Low-Frequency Interference

The generalized equivalent circuit for calculating low-frequency interference of a victim circuit in Figure 3.35 suggests several methods for reducing the effects of the interference on the victim circuitry [3.20]. These are as follows:

- Reduce the EMI signal strength V_s.
- Reduce the E or H field strengths at the victim circuit (i.e., minimize M and C_0).
- Design a mismatch between the primary fields produced by the EMI source circuit and the type of fields to which the victim circuit responds
- Design the load Z_L to not cause the interference.

As an example of this last item, consider the fact that if Z_L is primarily capacitive (see Figure 3.35), the induced response I_1 in the source circuit will be small, and the main coupling will be by the E-field. Thus the main induced source in the victim circuit will be I' arising from the E-field. If the victim reception element Z_a is designed to be primarily inductive, this current source will be shunted by the inductance at low frequencies, and the current flowing into the load impedance Z_b will be small.

Measures taken to reduce the EM field strengths in order to minimize M and C_0 can include:

- Increasing the separation distance between the source and victim circuits.
- Inserting conducting or magnetic material between the source and victim to reduce the EM field strengths, either by attenuation or by reflection (i.e., shielding).
- Changing the geometry of the EMI-producing element by twisting wires or minimizing loop areas to reduce the resulting EM fields.

3.3.5 Specific Measures To Reduce Capacitive Coupling

As the capacitive coupling is the consequence of the presence of an electric field, a metallic barrier (metallic sheath) of any kind of metal constitutes an efficient way of reducing the interference between the source and the victim. This barrier can be:

- A metallic cylindrical shield around a cable or a bundle of cables.
- A metallic box in which sensitive circuits and the connections between these circuits are placed.
- A metallized plastic box using metallic spray painting.
- A continuous cable tray (distributed ground plane).
- A simple wire short-circuited at the two ends, parallel to the sensitive cable.

In this discussion we treat the cable as an unprotected conductor that carries a desired signal or information. The metal forming the barrier can be any conducting diamagnetic material such as copper, aluminum, or silver painting, or a conducting ferromagnetic material such as iron.

3.3.5.1 Effect of a Cylindrical Shield Around a Conductor. Consider the two-wire conductor system shown in Figure 3.38. Let conductor 1 be the source of a low-frequency disturbance at a potential V_1 and conductor 2 be the victim conductor which has a load resistance R attached. The voltage across this load, V_R, which is induced by the victim source, is to be minimized. For this example, three distinct cases should be considered:

1. The conductor remains unshielded.
2. The conductor is covered by a conducting shield, leaving the load resistance unshielded.

92 LUMPED-PARAMETER CIRCUIT MODELS

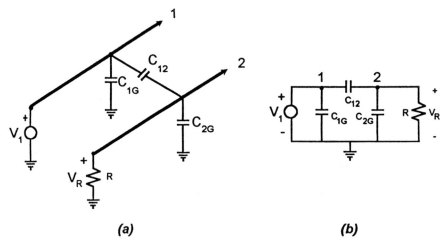

FIGURE 3.38 Capacitively coupled circuits without shielding: (*a*) physical configuration; (*b*) equivalent circuit.

3. The conductor and any termination impedances (loads) are completely enclosed by the shield.

The first case was discussed in Section 3.3.3, where it was shown that the coupling of the source to the victim is due to the capacitance C_{12}, as illustrated in Figure 3.38.

The second case is illustrated in Figure 3.39, where a conducting shield surrounds the victim wire 2 but does not enclose the load completely. In this case there is still a wire-to-wire capacitance C_{12}, permitting a direct coupling from the source to the victim wire. This coupling, however, is reduced considerably due to the fact that C_{12} acts only over the unshielded sections of wire 2. Other capacitances, such as the shield-to-ground capacitance C_{SG}, the wire 2-to-shield capacitance C_{2S}, and the wire 1-to-shield C_{1S}, are introduced into the problem, and these influence the response at the load. Figure 3.39*b* shows the equivalent circuit for this configuration. Frequently, the shield is connected directly to the ground, so that point S in the figure is on the ground. This eliminates the effect of C_{SG}. Figure 3.39*c* illustrates the resulting circuit for this configuration. In both circuits in parts (b) and (c) of the figure, it is evident that the capacitance C_{12} plays a key role in determining the response at the element R. Since this capacitance has not been reduced completely by the shield, the source and victim circuits are still coupled together, and a response will be induced in R.

Although the partially shielded configuration in Figure 3.39 can result in a reduction of the induced noise in the victim circuit due to the reduction of the coupling capacitance, C_{12}, it is not the best that can be obtained, because there is still the capacitive coupling at the unshielded ends of the line. A

3.3 DISTURBANCES IN CIRCUITS INDUCED BY ELECTROMAGNETIC FIELDS

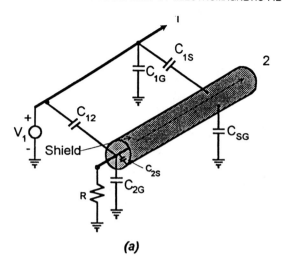

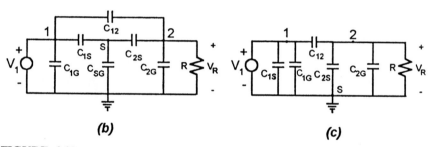

FIGURE 3.39 Capacitively coupled circuits with partial shielding: (*a*) physical configuration; (*b*) equivalent circuit; (*c*) equivalent circuit with the shield grounded.

better solution is to enclose the entire victim line *and the load resistance* in the shield. This is consistent with the EM topological shielding concepts described in Chapter 2. Figure 3.40 shows this configuration and the equivalent circuit. Note that in this case the internal resistance R is completely isolated from the exterior source circuit, with the exception of a voltage-controlled current source

$$I = -j\omega C_t V_s \tag{3.75}$$

where C_t is called the transfer capacitance of the enclosure and cable shields. The product $C_t V_s$ can be thought of as a charge on the inner conductor induced by the E-field that is able to penetrate the shield imperfections (such as the holes in a braided cable). For good shields, this E-field is very small, and consequently, C_t is usually many orders of magnitude smaller than C_{12}. In this case the internal response V_R is also very small.

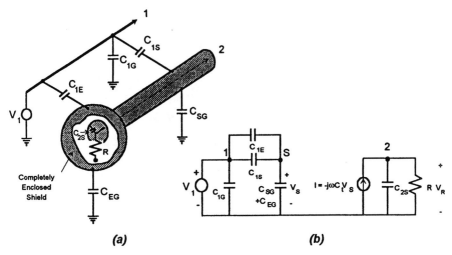

FIGURE 3.40 Capacitively coupled circuits with complete shielding; (*a*) physical configuration; (*b*) equivalent circuit.

3.3.5.2 Discussion. Due to progress in EMC knowledge in the last few years, it is now less common to encounter situations in which the ends of a shielded cable are open or when the shield is connected to the grounded case of a measurement device, for instance, by a pigtail.† This implies that, in general, the capacitive coupling of conductors that extend beyond the shield, as in Figure 3.38, should be an exception. However, real cable connectors mounted at the ends of coaxial cables are not always of perfect quality, and the stray capacitance C_{12} can exist in some cases, even if its value is very small. This subject of coupling through cable shields is discussed in more depth in Chapter 9.

3.3.6 Specific Measures to Reduce Inductive Coupling

Protecting a circuit against the effects of magnetic coupling is more difficult than for electric coupling. The electrostatic shield shown in Figure 3.40 is not at all effective in reducing magnetic coupling if a diamagnetic or paramagnetic material having $\mu_m \approx \mu_0 = 1$ is used. It is possible, however, to create a barrier against low-frequency magnetic fields by introducing a shield of ferromagnetic material with a high value of μ_r (mu-metal or permalloy) between the interference source and the victim. The effect of such screens with high magnetic permeability will be to concentrate the flux lines into the

---

† Although it is known that pigtails are not desirable, they are used in many systems that can tolerate the inductive coupling through the pigtail connection. In the process industry and in nuclear power plants, several decimeters of pigtail are allowed. In small systems, pigtails on the orders of centimeters in length are sometimes found to eliminate feed-through capacitors, which may be impractical in some cases.

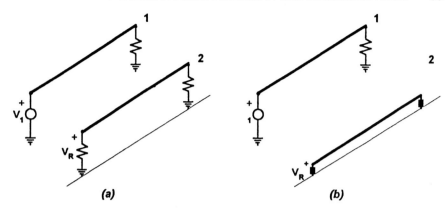

FIGURE 3.41 Method of reducing the magnetic field coupling of two circuits: (a) original circuit; (b) modified circuit with smaller mutual coupling.

shield, as in this case $B_{mat} \gg B_{air}$. These types of screens are used to shield monitors for PCs in which the electronic beam of the display can be turned by the presence of a strong low-frequency magnetic field. Such fields can arise due to the presence of large current-carrying electrical conductors (e.g., the traction mains of an electrified railway) or the presence in the same building as the computer of a medium- or low-voltage transformer station.

The use of mu-metal shielding can be very expensive, however. Another efficient way of reducing inductive interference is by minimizing the mutual inductance of the source and victim circuits. For the two-wire line configuration discussed previously, the victim wire can be repositioned close to the conducting ground, as shown in Figure 3.41. Doing this reduces the area through which the magnetic field from the victim circuit links the receptor circuit, thereby reducing the induced voltage. Alternatively, it is possible to position the source circuit close to the ground plane to reduce the B-field produced by this circuit. Either or both possibilities can reduce the magnetic field coupling in a cost-effective manner.

An alternative method for reducing the magnetic field coupling when the victim circuit consists of a two-wire line rather than a single wire with a ground return) is simply to twist the two conductors together (see, e.g., [3.21]). The effect of the twisting causes the external magnetic flux to induce alternative positive and negative voltages in the receptor wire pair, as demonstrated in Figure 3.42. This compensation process repeats itself along the cable, thereby providing a smaller overall induced signal in the victim.

3.4 DISTURBANCES CAUSED BY COMMON GROUND RETURNS

Another way that EMI from a disturbing source can affect a victim circuit is by pickup in a common ground path. Grounding is often implemented with

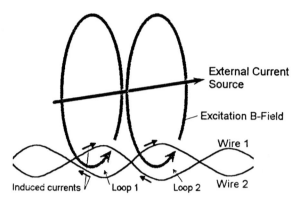

FIGURE 3.42 Effect of twisting two conductors in the victim circuit.

the goal of protecting personnel from electrical hazards and of eliminating EMI. If done properly, both goals can be achieved. However, if done improperly, grounding can make the EMI problem even worse. In this section we will discuss issues pertaining to low-frequency coupling of circuits by a common ground system [3.20].

Consider the case of a dc current being injected into a conducting half-space by an electrode located at point A (for $x = -d/2$) on the surface and then removed from an electrode at point B (for $x = d/2$), as shown in Figure 3.43. The conductivity of the half-space is denoted by σ. For this configuration we desire to compute the resistance presented by the ground between terminals A–B, as well as the electric potential on the surface of the ground, due to the current flowing through the earth. The latter quantity is useful because if there is a second (victim) circuit connected to terminals A'–B', it will provide the voltage exciting this circuit. Thus the mutual coupling between these two circuits by resistive coupling may be determined.

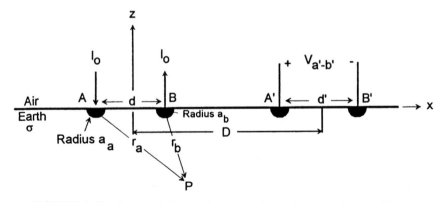

FIGURE 3.43 Current injection into two electrodes on a lossy half-space.

For an assumed spherical current injection electrode of radius a located in an infinite conducting medium, the electrical potential Φ away from the electrode is described by the equation $\nabla^2 \Phi = 0$ and the current density in the material is given by $\mathbf{J} = \sigma \mathbf{E} = -\sigma \nabla \Phi$. The well-known solution (see [3.22]) for the potential due to a point electrode is $\Phi = k/r$, and the corresponding expression for the current density is $\mathbf{J} = \hat{r} k\sigma/r^2$. The coefficient k may be evaluated by integrating the current density over the electrode surface as

$$I_0 = \int_{\text{electrode}} J_r \, ds = \int_0^{2\pi} \int_0^{\pi} \frac{k\sigma}{a^2} a^2 \sin\theta \, d\theta \, d\phi = 4\pi k\sigma \tag{3.76}$$

so that the potential within the infinite medium may be written as

$$\Phi = \frac{I_0}{4\pi\sigma} \frac{1}{r} \tag{3.77}$$

For the same current I_0 injected into a half-electrode at location A on the surface of the half-space as shown in Figure 3.43, the potential in the earth produced by this electrode is twice that of Eq. (3.77), as

$$\Phi_a = \frac{I_0}{2\pi\sigma} \frac{1}{r} \tag{3.78}$$

Dividing the potential difference between one electrode at $r = a$ and a reference point at infinity by the injected current gives the resistance of the electrode to the reference point as

$$R_{a-\infty} = \frac{1}{2\pi\sigma} \left(\frac{1}{a} - \frac{1}{\infty} \right) = \frac{1}{2\pi\sigma a} \quad \Omega \tag{3.79}$$

As shown in Figure 3.43, electrode A injects current into the earth and electrode B removes it, with the resulting potential at point P being given by

$$\Phi(P) = \Phi_a + \Phi_b = \frac{I_0}{2\pi\sigma} \left(\frac{1}{r_a} - \frac{1}{r_b} \right) \tag{3.80}$$

where r_a and r_b are the distances from each electrode to the observation point P. The resistance between these two electrodes may be found by evaluating the potential difference between the two and dividing by the current as

$$R_{ab} = \frac{\Phi(A) - \Phi(B)}{I_0} = \frac{1}{2\pi\sigma} \left(\frac{1}{a_a} + \frac{1}{a_b} - \frac{1}{d - a_a} - \frac{1}{d - a_b} \right) \tag{3.81}$$

Here a_a is the radius of electrode A, a_b is the corresponding radius for electrode B, and d is the separation between the two. For the case when

$a_a = a_b = a$ and $d \gg a$, the resistance becomes

$$R_{ab} = \frac{1}{\pi a \sigma} \quad \Omega \tag{3.82}$$

The voltage sensed by the victim circuit connected to the earth at locations $A'-B'$ can be calculated from the difference of the potentials of Eq. (3.77) at these points:

$$V_{a'b'} = \Phi(A') - \Phi(B') = \frac{I_0}{2\pi\sigma}\left[\left(\frac{1}{r_{aa'}} - \frac{1}{r_{ba'}}\right) - \left(\frac{1}{r_{ab'}} - \frac{1}{r_{bb'}}\right)\right] \tag{3.83}$$

where $r_{aa'}$ denotes the distance from electrode A to electrode A', and so on. From this expression it is possible to define the transfer resistance between the source current and the voltage $V_{a'b'}$ as

$$R_t = \frac{V_{a'b'}}{I_0} \quad \Omega \tag{3.84}$$

Figure 3.44 shows a perspective view of the current injection and receptor electrodes, the lines of constant potential on the earth's surface around the electrodes, and the victim circuit connected at $A'-B'$.

For a victim circuit located parallel with the excitation circuit, as shown in Figure 3.44, the transfer resistance can be expressed as

$$\begin{aligned} R_t &\approx \frac{1}{2\pi\sigma} \frac{32 dd' D}{16 D^4 - 8D^2(d^2 + d'^2) - 2d^2 d'^2 + (d^4 + d'^4)} \\ &\approx \frac{1}{\pi\sigma} \frac{dd'}{D^3} \quad \Omega \quad (\text{for } D \gg d \text{ and } d') \end{aligned} \tag{3.85}$$

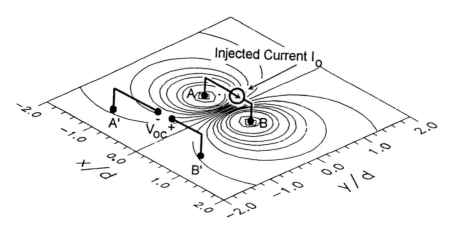

FIGURE 3.44 Source and victim circuit coupled by a common earth-return conductor.

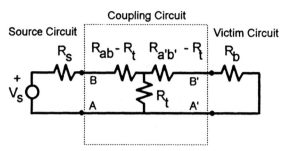

FIGURE 3.45 Equivalent circuit for common earth-return coupling.

where D is the separation of the midpoints of the two circuits and d' is the separation of the electrodes of the victim circuit.

The equivalent T-circuit used to represent the source–victim coupling in Figure 3.5 also can be used for the present problem. Figure 3.45 shows the resulting T-circuit driven by a Thévenin source generator at the terminals A–B and loaded by a resistance R_b across terminals A'–B'. The resistance element R_{ab} is given by Eq. (3.81) for the source circuit, $R_{a'b'}$ is given by the same equation with a different electrode radius, if necessary, and the transfer resistance R_t is given by the evaluation of Eq. (3.84).

The preceding discussion of coupling on a conductive surface can be extended to the case of a finitely conducting plate of thickness t. With a two-sided plate, the victim circuit can be located on the same side as the source circuit as before, or it can be positioned on the opposite side and the plate will serve as more of a shield to reduce the coupling. Such a configuration is shown in Figure 3.46. If the electrical conductivity of the plate is sufficiently high, or if the plate is very thick, the mutual inductive and capacitive coupling between the two circuits is negligible. However, if there is finite conductivity in the wall material, the current I_1 flowing in the excitation circuit will still induce a voltage $V_{a'b'}$ on the back side in the lossy conductor, and this voltage will excite currents in the victim circuit.

The thickness of the plate conductor will affect both the transfer impedance between the source and victim circuits and the contact resistances

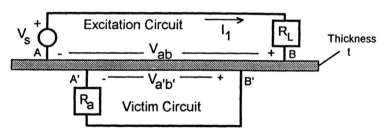

FIGURE 3.46 Source–victim circuit coupling on opposite sides of a common ground return.

of the source and victim circuits with the plate. These effects may be analyzed using image theory, which places successive images of the electrodes in the top and bottom surfaces of the plate. For each electrode, this produces a doubly infinite set of images that contribute terms to the potential. For the identical hemispherical electrodes of radii a, separated a distance d, and located along the x-axis as shown in Figure 3.43, the resistance between the two electrodes R_{ab} is given by

$$R_{ab} = \frac{1}{\pi\sigma}\left[\frac{1}{a} + 2\sum_{n=1}^{\infty}\frac{1}{\sqrt{a^2 + (2nt)^2}} - \frac{1}{\sqrt{d^2 + (2nt)^2}}\right] \quad \Omega \quad (3.86)$$

The first term $(1/a)$ in the brackets is the same as for the electrodes in the semi-infinite half-space as in Eq. (3.82). Consequently, the summation term accounts for the finite thickness of the plate. Note that each term in the sum behaves asymptotically as $1/n$ and consequently, is not strictly summable independently. However, the two terms in the sum, when taken together, fall off faster than $1/n$, and the infinite series converges. Unfortunately, there is no closed-form expression available for the resulting summation, so a numerical treatment must be used to obtain specific numerical values.

For the victim circuit located on the top of the plate, a useful quantity is the potential distribution Φ on the plate. This is similar to the potential distribution that is shown in Figure 3.44. With this quantity defined, the voltage between any two points on the surface can readily be determined by a simple subtraction of the values of Φ at the contact points. The transfer resistance can then be determined using Eq. (3.84). On the top surface, this potential has the form

$$\Phi_{top}(x, y) = \frac{I_0}{2\pi\sigma}\left[\left(\frac{1}{r_a} - \frac{1}{r_b}\right) + 2\sum_{n=1}^{\infty}\left(\frac{1}{\sqrt{r_a^2 + (2nt)^2}} - \frac{1}{\sqrt{r_b^2 + (2nt)^2}}\right)\right]$$

(3.87)

where $r_a = \sqrt{(x + d/2)^2 + y^2}$ and $r_b = \sqrt{(x - d/2)^2 + y^2}$. For an infinitely thick plate (i.e., the half-space) the summation term vanishes.

For victim circuits located on the bottom of the plate, the potential on the plate surface is given by

$$\Phi_{bot}(x, y) = \frac{I_0}{\pi\sigma}\sum_{n=1}^{\infty}\left\{\frac{1}{\sqrt{r_a^2 + [(2n-1)t]^2}} - \frac{1}{\sqrt{r_b^2 + [(2n-1)t]^2}}\right\} \quad (3.88)$$

As the thickness of the plate becomes large, this term may be seen to approach zero, as would be expected for a very thick conducting shield.

3.5 EXTENSION OF CIRCUIT MODELING TO HIGH FREQUENCIES

When the frequency of an EMI disturbance increases and the wavelength starts to become comparable to the dimensions of the circuits, the low-frequency circuit models discussed in previous sections become inaccurate. The effects of stray inductance and parasitic capacitance of the circuit elements can become important. Figure 3.47 shows an example of low- and high-frequency models for simple R, L, and C circuit elements.

At low frequencies the primary R, L, and C values of circuit elements can be reasonably well known, due to a knowledge of the design of the circuit elements. At high frequencies, however, the parasitic elements are usually not well controlled, and they may not be available to the analyst. If a high-frequency analysis is desired, it is usually necessary that these parameters be measured. In addition, these parameters are typically highly geometry dependent, so that the actual values may vary from case to case, depending on how they are installed in a circuit.

The high-frequency models of real devices can become quite complex [3.23]. Figure 3.48a shows the usual low-frequency model for a simple

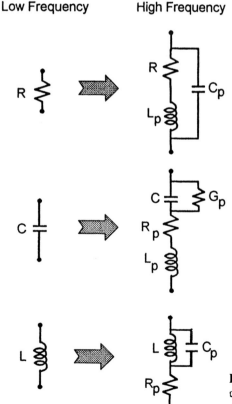

FIGURE 3.47 Low- and high-frequency models for simple circuit elements.

102 LUMPED-PARAMETER CIRCUIT MODELS

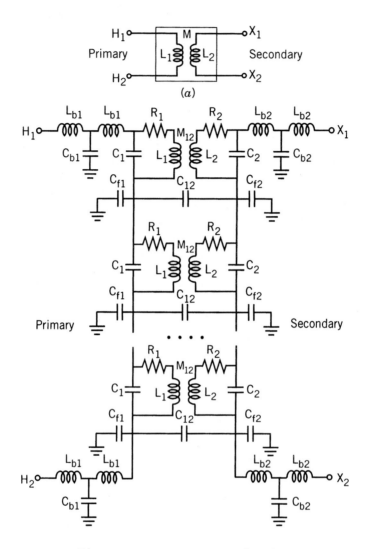

FIGURE 3.48 Transformer circuit models: (a) low frequency; (b) high frequency. (From [3.23].)

transformer. At high frequencies, the turn-to-turn parasitic capacitances, C'_h, the winding-to-frame capacitances C'_{11} and C'_{22}, and the capacitance between the two windings cannot be neglected. These elements, plus the distributed inductance and resistance within the coils, are shown in Figure 3.48b. In addition, the lumped capacitances of the transformer bushings, C_{b1} and C_{b2}, are included. These various elements cannot be calculated, nor can they be estimated easily from two-port measurements made at the terminals of the transformer. Consequently, the high-frequency equivalent circuit is not useful for calculations of the device's response. Its main use is in assisting in understanding observed experimental data of the circuit responses.

Along with the problems in representing circuit elements at high frequencies, there are difficulties in computing the EM field coupling to circuits. Consider the loop coupling illustrated in Figure 3.24. If the loop diameter is comparable to a wavelength, there will be a spatial variation of the magnetic flux density over the area of the loop. At higher frequencies, the magnetic flux can even change sign over the area. This causes the simple quasistatic coupling models to be incorrect as the frequency increases. Coupling models at high frequency are a very important aspect of EMC studies, however, and these are discussed in Chapter 4.

REFERENCES

3.1. Paul, C. R., *Analysis of Linear Circuits*, McGraw-Hill, New York, 1989.

3.2. Balabanian, N., *Fundamentals of Circuit Theory*, Allyn and Bacon, Boston, 1961.

3.3. Chang, D. C., *Analysis of Linear Systems*, Addison-Wesley, Reading, MA, 1962.

3.4. Scott, R. E., *Linear Circuits*, Parts 1 and 2, Addison-Wesley, Reading, MA, 1964.

3.5. Fink, D. G., and D. Christiansen, *Electronics Engineers' Handbook*, 2nd edition, McGraw-Hill, New York, 1982.

3.6. Aguet, M. and J. J. Morf, *Energie Electrique*, Traité d'Electricité de l'Ecole Polytechnique Fédérale de Lausanne, Vol. 12, Giorgi, St. Saphorin, Switzerland, 1981.

3.7. Recommendations de l'Association Suisse des Electriciens, limitation des perturbations électriques dans les réseaux publics de distribution, *ASE 3600-2.1897*, Norme Suisse SN 413600-2.

3.8. Stevenson, W. D., Jr., *Elements of Power System Analysis*, McGraw-Hill, New York, 1962.

3.9. CIGRE Working Group 36-05, "Harmonics, Characteristic Parameters, Methods of Study, Estimates of Existing Values in the Network, *Electra*, No. 77, 1981.

3.10. Ramo, S., J. R. Whinnery, and T. Van Duzer, *Fields and Waves in Communication Electronics*, 2nd edition, Wiley, New York, 1965.

3.11. Rosa, E. B., and F. W. Grover, "Formulas and Tables for the Calculation of Mutual and Self Inductance," *Bulletin of the National Bureau of Standards*, Vol. 8, U.S. Department of Commerce and Labor, Washington, DC, 1912.

3.12. Press, W. H., et al., *Numerical Recipes*, Cambridge University Press, Cambridge, 1986.

3.13. Sylvester, P., *Modern Electromagnetic Fields*, Prentice Hall, Englewood Cliffs, NJ, 1967.

3.14. Harrington, R. F., *Field Computation by Moment Methods*, reprinted by the author, Syracuse University, Syracuse, NY, 1968.

3.15. Paul, C. R., "On the Superposition of Inductive and Capacitive Coupling in Crosstalk Prediction Models," *IEEE Trans. Electromagn. Compat.*, Vol. EMC-24, No. 3, August 1982, pp. 335–343.

3.16. Stremler, F. G., *Introduction to Communication Systems*, 2nd edition, Addison-Wesley, Reading, MA, 1982.

3.17. Paul, C. R., "Prediction of Crosstalk in Ribbon Cables: Comparison of Model Predictions and Experimental Results," *IEEE Trans. Electromagn. Compat.*, Vol. EMC-20, No. 3, August 1978, pp. 394–406.

3.18. Lee, K. S. H., ed., *EMP Interaction: Principles, Techniques and Reference Data*, Hemisphere, New York, 1990.

3.19. Paul, C. R., *Analysis of Multiconductor Transmission Lines*, Wiley, New York, 1994.

3.20. Ott, H. W., *Noise Reduction Techniques in Electronic Systems*, 2nd ed., Wiley-Interscience, New York, 1988.

3.21. Paul, C. R., and J. A. McKnight, "Prediction of Crosstalk Involving Twisted Pairs of Wires, Part I, A Transmission Line Model for Twisted Wire Pairs, and Part II, A Simplified Low-Frequency Prediction Model," *IEEE Trans. Electromagn. Compat.*, Vol. EMC-21, No. 2, May 1979, pp. 92–114.

3.22. Sunde, E. D., *Earth Conduction Effects in Transmission Systems*, Van Nostand, New York, 1949.

3.23. Vance, E. F., "Electromagnetic-Pulse Handbook for Electric Power Systems," *Report DNA 3466F*, Defense Nuclear Agency, Washington, DC, February 4, 1975.

PROBLEMS

3.1 A device is represented by a Thévenin equivalent circuit with parameters V_{oc} and Z_{in} and is loaded by a generalized impedance Z_L. Derive the conditions for maximum power transfer to the load impedance.

3.2 The Thévenin equivalent circuit of a physical circuit is desired at a frequency f by making measurements at the terminals of the circuit. The open-circuit voltage V_{oc} is measured to be 20 V, and the short-circuit current is determined to be 2.2 A. Subsequently, a 10-Ω resistor is connected across the terminals, and the resulting voltage across it is

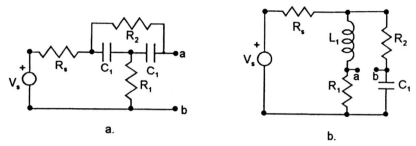

FIGURE P3.3 Sample electrical circuits.

found to be 11.1 V. Finally, when a variable capacitor is connected at the terminals, it is found that the resulting current reaches a maximum at a particular value, given by C_0. Determine the voltage and impedance of the Thévenin equivalent for the circuit. What is the maximum power that the circuit can supply?

3.3 Evaluate the Thévenin equivalent circuits at terminals a–b for the networks shown in Figure P3.3.

3.4 Derive Eqs. (3.6a), (3.6b) and (3.6c).

3.5 Derive the T-to-Π transformation given in Eq. (3.12).

3.6 Compute the [Z], [Y], and chain matrices for the circuits shown in Figure P3.6.

3.7 In Figure P3.7, the details of the network inside the box are unknown. However, the short-circuit admittance parameters y_{ij} are assumed to be determined from external port measurements. Derive the expression for the transfer admittance with the resistive load in place, $Y_{21} = I_2/V_s$: (a) by first using the equations defining the y-parameters

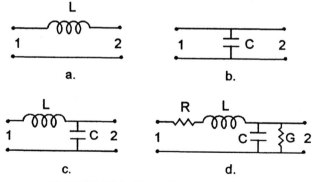

FIGURE P3.6 Several two-port circuits.

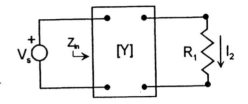

FIGURE P3.7 Loaded two-port network described by y-parameters.

in Eq. (3.8), (b) by using the Π equivalent of the network, and then (c) by applying Thévenin's theorem at the output terminals.

3.8 Let $R_1 = 10\,\Omega$ in Figure P3.7. Use the results of Problem 3.7 to evaluate the voltage across the load resistance for networks described by the following parameters (note that $s = j\omega$):

(a) $[Y] = \begin{bmatrix} 2e^{j30°} & 5e^{-j135°} \\ 5e^{-j135°} & 2e^{j30°} \end{bmatrix}$
(b) $[Y] = \begin{bmatrix} 3s+1 & -2s \\ -2s & 5s+2 \end{bmatrix}$

(c) $[Y] = \begin{bmatrix} \dfrac{10(s+1)}{s} & -10 \\ -10 & \dfrac{10(s+1)}{s} \end{bmatrix}$
(d) $[Z] = \begin{bmatrix} \dfrac{10(s+1)}{s} & -10 \\ -10 & \dfrac{10(s+1)}{s} \end{bmatrix}$

3.9 Compute the input impedance of the two-port network of Figure P3.6 using (a) the chain ($ABCD$) parameters of the network, and (b) the z-parameters.

3.10 Write four expressions for the chain parameters A, B, C, and D in terms of just two terminal quantities as in Eqs. (3.9) and (3.4).

3.11 Derive Eq. (3.17).

3.12 There are many sign conventions in electromagnetics (e.g., the right-hand rule relating current flow and the resultant magnetic flux direction). The actual response of a system, however, should be independent of such sign conventions. To illustrate this, consider two coupled magnetic circuits, as shown in Figure P3.12. Take a coin and by flipping heads or tails, define arbitrarily the senses of the currents I, voltages V and fluxes Φ in the table below, using the following convention: for the head of the coin, select a positive or (upward) direction ↑ or ⇑ for the indicated quantity, or for the tail of the coin, select a negative or (downward) direction ↓ or ⇓ for the quantity.
(a) Complete the diagram in Figure P3.12 by indicating the directions of the voltages, currents, and fluxes, based on the chosen directions.
(b) Write the relations between the terminal voltages V_1 and V_2, the flux-induced voltage contributions, V_{i1} and V_{i2}, the loop resistances R_1 and R_2, and the loop current I_1 and I_2 for the two circuits.

I_1	V_1	V_{i1}	Φ_{L1}	Φ_1	I_2	V_2	I_{i2}	Φ_{L2}

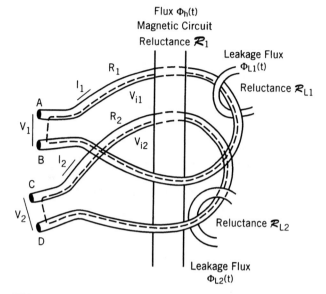

FIGURE P3.12 Two magnetically coupled loops.

- (c) Write the relations between voltages V_{i1} and V_{i2} and the fluxes Φ_h, Φ_{L1}, and Φ_{L2} for the two circuits.
- (d) Write the three expressions for the fluxes Φ_h, Φ_{L1}, and Φ_{L2} in terms of the currents I_1 and I_2, using the magnetic circuit reluctances $\mathcal{R}_1$, $\mathcal{R}_{L1}$, and $\mathcal{R}_{L2}$. (The reluctance of a magnetic circuit is the analog of the resistance of an electrical circuit. See P. Lorrain and D. P. Corson, *Electromagnetic Fields and Waves*, W.H. Freeman, New York, 1988, p. 381 for more details on magnetic circuits.)
- (e) Eliminate among the equations V_{i1}, V_{i2}, Φ_h, Φ_{L1} and Φ_{L2} to obtain two equations in terms of R, L and M.

3.13 For the two coaxial coils shown in Figure 3.27b, compute the self-inductances L_a and L_b, assuming that $a = 1$ cm, $b = 2$ cm, and the wire radius $r = 0.05$ mm. Compute and plot the mutual inductance of the loops for $0.1 \leq d \leq 10$ cm.

3.14 Investigate the nature of the crosstalk coupling between the two lines in Figure 3.36 as the load impedances vary from a short circuit ($\approx 0\,\Omega$) to an open circuit ($\approx \infty$).

3.15 Discuss qualitatively the E-field and H-field coupling to the shielded

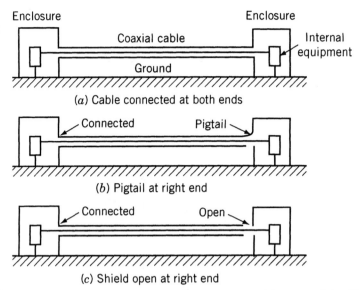

(a) Cable connected at both ends

(b) Pigtail at right end

(c) Shield open at right end

FIGURE P3.15 Two equipment enclosures connected by a coaxial cable.

enclosures and the coaxial cables in the cases shown in Figure P3.15. In which case is the magnetic field coupling to the cable shield most significant? In which case is the electric field coupling to the shield the greatest? Which case will be most troublesome for the internal electronics?

3.16 Consider the case of source–victim coupling through a common ground return, as shown in Figure 3.46. Assuming that the circuit connections to the lossy material are hemispherical with radii a, use Eq. (3.86) to develop a three-dimensional surface plot (see Figure 8.6 for an example) of the normalized resistance of the plate $(\pi\sigma a)R_{ab}$ as a function of the two normalized variables d/a and t/a.

(a) For what plate thicknesses can the thickness of the plate be neglected?

(b) Assuming that the plate is made of copper ($\sigma = 5.76 \times 10^7$ S/m), compute the plate resistance.

(c) Repeat this calculation for a plate made of carbon ($\sigma \sim 3.0 \times 10^4$ S/m).

(d) Suppose that the source and load impedances are $R_L \approx R_a \approx 50\,\Omega$. How do these plate resistances compare with these loads?

3.17 Consider the R, C and L circuit elements shown in Figure 3.47. Assume that the principal circuit values are $R = 50\,\Omega$, $C = 1\,\mu F$, and $L = 1$ mH.

(a) For each of the high-frequency models, choose your own parasitic inductance, capacitance, and resistance values, based on experience, your own measurements, or from circuit handbooks. Discuss and justify these values.

(b) Compute and plot, as a function of frequency, the terminal impedance *magnitudes* for both the low- and high-frequency models.

(c) Indicate the frequencies at which the low-frequency models become inadequate.

(d) Determine which parasitic circuit element is most critical in determining the high-frequency deviation of the model.

3.18 The metallic box of an apparatus is grounded with a conductor of 1 m length and a cross section of 1 mm^2. The floor of reinforced concrete connected to the grounding structure of the building is considered to be an equipotential plane. Assuming that the inductance of a wire is about 1 µH/m and that the capacitance between the metallic box and the floor is equal to 30 pF, calculate the variation of the grounding wire versus frequency. Using the impedance locus in the complex plane, show that this impedance has a maximum for a certain frequency.

PART III
HIGH-FREQUENCY AND BROADBAND COUPLING MODELS

CHAPTER 4
Radiation Models for Wire Antennas

As the frequency of the EM fields increases to the point where the wavelength becomes comparable to the dimensions of the source and victim equipment, the EMC models developed in earlier chapters become inaccurate. This is because these models were based on a quasistatic assumption for the fields. At higher frequencies, the fields become wavelike and an alternative modeling approach is required. This more general approach is based on solutions to Maxwell's equations, which describe the EM field behavior at both high and low frequencies. In this chapter we discuss models for computing the radiation and reception from wire antennas.

4.1 INTRODUCTION

Two basic types of problems are of interest in this chapter: the radiation or antenna problem, and the scattering or reception problem. The radiation problem involves an active source connected to an electrically conducting structure (i.e., an antenna). This source injects currents onto the antenna, and these currents produce an EM field propagating away from the radiator. Near the antenna the EM fields can be rather complicated. However, as the distance from the antenna increases, the field behavior becomes much simpler, with an amplitude that falls off as $1/r$ with distance, and with mutually orthogonal H and E fields lying in the plane transverse to the propagation direction. Far from the antenna, the radiated field appears as a local plane wave. The goal of the antenna analysis problem is to calculate the EM fields produced at a specified observation location, given a specification of the excitation source and a knowledge of the shape and composition of the antenna structure.

Often, the radiation from such a source is desired, as in the case of a radio transmitter. Such an antenna is referred to as a *deliberate antenna,* and consideration must be given to increasing the efficiency of the radiation process. At other times, radiation is not desired, as in the case of EM field emission from harmonics produced in a switching power supply connected to a long electrical cable. In this case, the electrical cable is said to be an *inadvertent antenna*, and steps must be taken to reduce the radiation efficiency.

The reception problem is the opposite of the radiation problem, where an EM field from a distant source propagates to a collection of conductors (the receiving antenna) and induces currents to flow. These currents, in turn, induce a voltage in an electrical load connected to the antenna, and this can be viewed as a signal received from the distant source. Depending on the nature of the equipment, the receiving antenna can again be considered to be either a deliberate or an inadvertent antenna.

The currents induced in the receiving antenna structure by the incident EM field also will create EM fields that propagate away from the antenna. These fields are known as *scattered fields*, and in some instances, they are the desired observable quantities. An example is a radar system, in which the return signal from a distant target is received and processed to provide information as to the target's distance and velocity.

In either the transmission or the reception case, the key to understanding the behavior of EM fields is the knowledge of the *current distribution* on the conductors. If the conductors are electrically short (i.e., the length is small compared with the wavelength), the induced current can appear as a constant or a linear function. However, as the conductors become longer (or as the frequency increases), the current distribution takes on a traveling-wave behavior. This results in a sinusoidally varying spatial variation in the current that is different for each type of antenna and which changes with frequency.

There are several possible approaches for determining the current distribution on an antenna. The easiest is to use a simple approximation to the current distribution. Although it is not precise, it can provide useful results, especially for the case of thin-wire conductors. A more accurate approach is to try to solve a specific boundary value problem involving Maxwell's equations for the antenna. This yields an integral equation for the current which must be solved numerically for the current distribution [4.1]. Alternatively, the region surrounding the antenna can be gridded and a numerical solution to the differential forms of Maxwell's equations can be obtained [4.2].

In this chapter we discuss more details of the radiation and scattering of EM energy from wire antennas at frequencies where the static models are no longer valid. In this development we concentrate on the models that permit rapid calculations of antenna responses, as opposed to the more accurate solutions to the current distributions discussed in [4.1] and [4.2] and which involve considerable computer resources.

4.2 RADIATION OF ELECTROMAGNETIC FIELDS IN THE FREQUENCY DOMAIN

4.2.1 Overview

A time-varying current, produced by the motion of an electric charge, will create a time-varying magnetic field at a distance away from the current

4.2 RADIATION OF ELECTROMAGNETIC FIELDS IN THE FREQUENCY DOMAIN

source. Similarly, the moving charge produces a time-varying electric field. For the behavior of these fields and charges to be self-consistent, they must be all related to each other in a certain manner. For time-harmonic sources varying with a fixed angular frequency ω and varying as $e^{j\omega t}$, the relationships describing the electromagnetic fields at a point P in free space are Maxwell's equations [4.3]:

$$\nabla \times \mathbf{E} = -j\omega \mathbf{B} \qquad \nabla \times \mathbf{H} = \mathbf{J} + j\omega \mathbf{D}$$
$$\nabla \cdot \mathbf{D} = \rho \qquad \nabla \cdot \mathbf{B} = 0 \qquad (4.1)$$

These fundamental equations are supplemented by the constitutive relations $\mathbf{B} = \mu_0 \mathbf{H}$ and $\mathbf{D} = \epsilon_0 \mathbf{E}$, where $\epsilon_0 = 8.85 \times 10^{-12}$ F/m is the permittivity of the surrounding medium and $\mu_0 = 4\pi \times 10^{-7}$ H/m is the permeability. In Eq. (4.1) the volume current density $\mathbf{J}(\text{A}/\text{m}^2)$ and the volume charge density ρ (C/m^3) can be viewed as the primary sources of the fields, $\mathbf{E}$ and $\mathbf{H}$. Although it may not be readily apparent, Eq. (4.1) also implies a relationship between $\mathbf{J}$ and ρ, which is known as the continuity equation:

$$\nabla \cdot \mathbf{J} = -j\omega \rho \qquad (4.2)$$

Although it is possible to derive general vector wave equations for the $\mathbf{E}$ and $\mathbf{H}$ fields [4.4], a simpler solution is to introduce a vector and scalar potential function, $\mathbf{A}$ and Φ, from which the fields may be evaluated as

$$\mathbf{B} = \nabla \times \mathbf{A} \qquad (4.3)$$

$$\mathbf{E} = -\nabla \Phi - j\omega \mathbf{A} \qquad (4.4)$$

The potentials $\mathbf{A}$ and Φ are related by the Lorentz gauge condition $\nabla \cdot \mathbf{A} = -j\omega \mu_0 \epsilon_0 \Phi$, so the expression for the E-field above may be expressed entirely in terms of the vector potential as

$$\mathbf{E} = \frac{1}{j\omega \mu_0 \epsilon_0} \nabla(\nabla \cdot \mathbf{A}) - j\omega \mathbf{A} \qquad (4.5)$$

The reason for using these potential functions instead of calculating the fields directly is that they depend on the sources $\mathbf{J}$ and ρ in a simple manner.

It can be demonstrated that both $\mathbf{A}$ and Φ are solutions to the Helmholtz wave equation

$$\nabla^2 \Phi + k^2 \Phi = -\frac{\rho}{\epsilon_0} \qquad (4.6)$$

$$\nabla \mathbf{A} + k^2 \mathbf{A} = -\mu_0 \mathbf{J} \qquad (4.7)$$

where $k = \omega\sqrt{\mu_0 \epsilon_0} = \omega/c$ and $c = 3.0 \times 10^8$ m/s is the speed of light in vacuum. It should be noted that Eq. (4.7) is to be interpreted *only* in a rectangular coordinate system as three independent equations, one for each

116 RADIATION MODELS FOR WIRE ANTENNAS

rectangular component of **A**:

$$\nabla^2 A_x + k^2 A_x = -\mu_0 J_x \tag{4.8a}$$

$$\nabla^2 A_y + k^2 A_y = -\mu_0 J_y \tag{4.8b}$$

$$\nabla^2 A_z + k^2 A_z = -\mu_0 J_z \tag{4.8c}$$

For a distribution of isolated charges and currents located in free space at $\mathbf{r}_s$ within the volume V_s shown in Figure 4.1, the solution to Eqs. (4.6) and (4.7) for Φ and **A** at an observation location defined by $\mathbf{r}_0$ may be expressed as integrals over the sources in the volume as [4.4]

$$\Phi(\mathbf{r}_0) = \frac{1}{4\pi\epsilon_0} \int_{\text{vol}} \rho(\mathbf{r}_s) \frac{e^{-jkR}}{R} dV_s \tag{4.9}$$

$$\mathbf{A}(\mathbf{r}_0) = \frac{\mu_0}{4\pi} \int_{\text{vol}} \mathbf{J}(\mathbf{r}_s) \frac{e^{-jkR}}{R} dV_s \tag{4.10}$$

where $R = |\mathbf{r}_0 - \mathbf{r}_s|$ is the distance between the field observation location at P and the source point. Once these potentials are evaluated for a given collection of sources, the E and B fields can then be calculated using Eqs. (4.3) and (4.4).

4.2.2 Radiation from Elementary Sources

Although the expressions in the preceding section can be used to compute the radiation from an arbitrary distribution of current and charge, it is

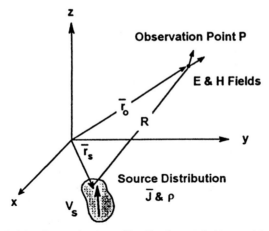

FIGURE 4.1 General source distribution and the resulting EM fields.

4.2 RADIATION OF ELECTROMAGNETIC FIELDS IN THE FREQUENCY DOMAIN

instructive to examine the behavior of the fields radiated by very simple sources, the electric and magnetic dipole.

4.2.2.1 Electric Dipole. As an example of the EM fields produced by a very simple source distribution, consider the case of a z-directed current element of infinitesimal length dl, located at the origin of the coordinate system $(x,y,z) = (0,0,0)$. This is defined by a volume current density of $\mathbf{J}(\mathbf{r}_s) = I\delta(x)\delta(y)\hat{z}$ and provides an electric current dipole moment $\mathbf{p} = (1/j\omega)I\,dl\,\hat{z}$, as shown in Figure 4.2a. This source is frequently referred to as a *Hertzian dipole*, and can be thought of as arising from an oscillation of a point charge along the z axis.

The vector potential produced by this current element is also in the $\hat{z}$ direction and is given by Eq. (4.10) as

$$\mathbf{A}(\mathbf{r}_0) = \frac{\mu_0}{4\pi} I\,dl\, \frac{e^{-jkr_0}}{r_0} \hat{z} \qquad (4.11)$$

To calculate the differential EM fields produced by this current element using Eq. (4.5), it is convenient to express all quantities in a spherical (r,θ,ϕ) coordinate system. The vector potential has components $A_r = A_z \cos\theta$ and $A_\theta = -A_z \sin\theta$, and the fields can be found by performing the required vector operations. This yields the following expression for the E and H field components in the spherical coordinate system:

$$E_r = \frac{I\,dl\,\cos\theta}{2\pi} k^2 Z_0 \left[\frac{1}{(kr_0)^2} + \frac{j}{(kr_0)^3} \right] e^{-jkr_0} \qquad (4.12a)$$

$$E_\theta = \frac{I\,dl\,\sin\theta}{4\pi} k^2 Z_0 \left[\frac{j}{kr_0} + \frac{1}{(kr_0)^2} - \frac{j}{(kr_0)^3} \right] e^{-jkr_0} \qquad (4.12b)$$

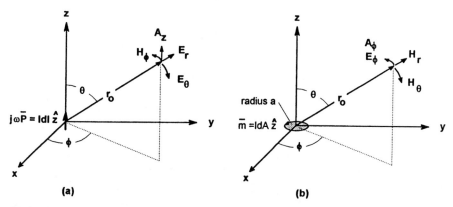

FIGURE 4.2 Fields produced by electric and magnetic dipole moments: (*a*) electric dipole; (*b*) magnetic dipole.

$$H_\phi = \frac{I\,dl\,\sin\theta}{4\pi} k^2 \left[\frac{j}{kr_0} + \frac{1}{(kr_0)^2}\right] e^{-jkr_0} \qquad (4.12c)$$

$$E_\phi = 0 \qquad H_r = 0 \qquad H_\theta = 0 \qquad (4.12d)$$

where $Z_0 = \sqrt{\mu/\epsilon} \approx 377\,\Omega$ is the free space wave impedance.

Note that in these expressions the term $kr_0 = \omega r_0/c = 2\pi r_0/\lambda$ is a measure of the distance r_0 in wavelengths. Close to the source where $kr_0 \ll 1$, the $1/(kr_0)^3$ terms dominate in the expressions for the E-field components, the $1/(kr_0)^2$ term dominates in the H-field, and the exponential term $e^{-jkr_0} \approx 1$. Observation locations in this region are said to be in the *near-zone* of the source, and the vector field components are approximated by

$$\mathbf{E} \approx \frac{1}{j\omega} \frac{I\,dl}{4\pi\epsilon r_0^3} (\hat{r}\cdot 2\cos\theta + \hat{\theta}\sin\theta) \qquad (4.13a)$$

$$\mathbf{H} \approx \frac{I\,dl}{4\pi r_0^2} \hat{\phi}\sin\theta \qquad (4.13b)$$

where $\hat{r}$, $\hat{\theta}$, and $\hat{\phi}$ are the unit vectors.

Farther away from the source (or at a higher frequency) is the *intermediate zone*, where $kr_0 \approx 1$. Here the fields must be described by Eqs. (4.12), since each of the terms with kr_0 in the denominator can have comparable contributions. At distances where $kr_0 \gg 1$, the $1/kr_0$ terms dominate and the fields simplify to planar electromagnetic waves. In this *radiation zone* or *far field*, the fields have the simple form

$$\mathbf{E} = jkZ_0 I\,dl\,\sin\theta\,\frac{e^{-jkr_0}}{4\pi r_0}\,\hat{\theta} \qquad (4.14a)$$

$$\mathbf{H} = jkI\,dl\,\sin\theta\,\frac{e^{-jkr_0}}{4\pi r_0}\,\hat{\phi} \qquad (4.14b)$$

Note that in the far field, $\mathbf{E}$ and $\mathbf{H}$ are mutually orthogonal, and $|\mathbf{E}|/|\mathbf{H}| = Z_0$. Furthermore, the term e^{-jkr_0}/r_0 represents a wave propagating away from the source with a propagation constant k and a $1/r_0$ amplitude falloff.

It is interesting to examine the spatial variation of the fields produced by the infinitesimal current element at various distances from the source. From Eq. (4.12) we note that the E-fields depend on the parameter kr_0. Figure 4.3 illustrates the rotationally symmetric E-field magnitudes for the several different values of kr_0. These spatial field patterns are normalized to have a unit amplitude. In reality, the fields are significantly stronger for smaller values of kr_0—a fact that is not evident in this figure.

4.2 RADIATION OF ELECTROMAGNETIC FIELDS IN THE FREQUENCY DOMAIN

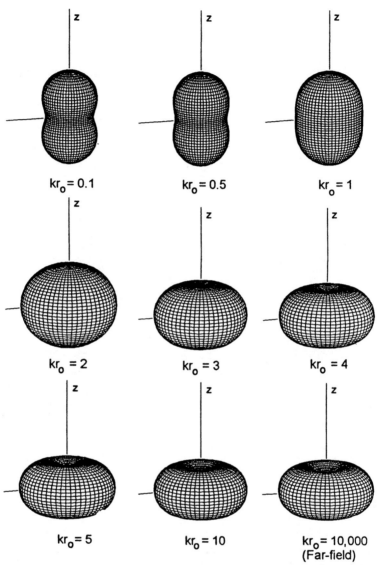

FIGURE 4.3 Spatial plots of the magnitude of the E-field produced by a z-directed current element for different values of kr_0.

Usually, when thinking about the radiation from a dipole source, we have in mind the far-field pattern described by Eq. (4.14a). This provides a radiation pattern with zero E-fields along the axis of the source (i.e., along the z-axis), and a maximum of the radiated field broadside to the source. This pattern is the last one in Figure 4.3, indicated by $kr_0 = 10,000$. Closer to the source, however, the field behavior is considerably different, with the null in the field along the axis disappearing. In fact, for values of $kr_0 < 1$, *the field is largest along the z-axis*. It is important to keep in mind these changes in the field patterns, especially when making measurements with radiating antennas in the near field.

The Poynting vector **P**, a measure of the complex power density flowing in electromagnetic fields, is defined as

$$\mathbf{P} = \mathbf{E} \times \mathbf{H}^* \quad \text{W/m}^2 \tag{4.15}$$

where * represents the complex conjugate of the field. The real part of this vector gives the time-averaged real power density as

$$\mathbf{P}_{avg} = \frac{1}{2} \text{Re}(\mathbf{E} \times \mathbf{H}^*) \quad \text{W/m}^2 \tag{4.16}$$

with the factor of $\frac{1}{2}$ resulting from the averaging of the sinusoidal waveforms. In the farzone of the Hertzian dipole source, $\mathbf{P}_{avg}$ is in the radial direction and is expressed as

$$\mathbf{P}_{avg} = \frac{1}{2}(E_\theta H'_\phi) = (I\, dl)^2 \frac{k^2 Z_0}{32\pi^2 r_0^2} \sin^2\theta \hat{r} \quad \text{W/m}^2 \tag{4.17}$$

The total power radiated by this source is given by an integral of the power density over a closed sphere of radius r surrounding the source:

$$W = \oint_S \mathbf{P} \cdot d\mathbf{S} = \int_0^\pi \int_0^{2\pi} \mathbf{P} \cdot \hat{r} r_0^2 \sin\theta\, d\theta\, d\phi$$

$$= \frac{\pi Z_0}{3} I^2 \left(\frac{dl}{\lambda}\right)^2 = \frac{Z_0 k^2}{12\pi}(I\, dl)^2 \quad \text{W} \tag{4.18}$$

4.2.2.2 Magnetic Dipole.
Another useful elementary radiating source is the infinitesimal *magnetic dipole*, as shown in Figure 4.2b. This results from the circulation of a current I around a loop having an infinitesimal area dA. For the case shown in the figure where the current loop is in the x–y plane, the strength of this source is defined by the magnetic dipole moment $\mathbf{m} = I\, dA\, \hat{z}$. As in the case of the electric dipole, it is required that the size of this source be much smaller than the observation distance r_0, as well as being much smaller than a wavelength.

For the magnetic dipole, the radiated fields may be evaluated by determining the vector potential, as done for the electric dipole [4.5]. In doing this it can be noted that there is a duality between the electric and magnetic dipole. Letting $\mathbf{H}_e$ and $\mathbf{E}_e$ represent the fields for the electric dipole source given in Eqs. (4.12), and $\mathbf{H}_m$ and $\mathbf{E}_m$ represent the fields arising from the magnetic dipole, it is noted in ref.[4.6] that dual relationships exist between the electric and magnetic dipole sources; see Table 4.1.

An important consequence of these relations is that by simply interchang-

4.2 RADIATION OF ELECTROMAGNETIC FIELDS IN THE FREQUENCY DOMAIN

TABLE 4.1 Dual Relationships Between Electric and Magnetic Sources

Electric Source	Converts to:	Magnetic Source
E-field	$\Rightarrow$	H-field
H-field	$\Rightarrow$	$-E$-field
$\mathbf{p}$	$\Rightarrow$	$j\omega\mu\mathbf{m}$
Electric current $\mathbf{J}$	$\Rightarrow$	Magnetic current $\mathbf{J}_m$
μ	$\Rightarrow$	ϵ
ϵ	$\Rightarrow$	μ
$c = 1/\sqrt{\epsilon\mu}$	$\Rightarrow$	$c = 1/\sqrt{\mu\epsilon}$
$Z_0 = \sqrt{\mu/\epsilon}$	$\Rightarrow$	$1/Z_0 = \sqrt{\epsilon/\mu}$

ing the E- and H-fields (along with a change in sign) and by replacing $I\,dl$ by $j\omega\mu I\,dA$ and $1/Z_0$ for Z_0 in Eq. (4.12), the fields produced by an infinitesimal z-directed magnetic dipole source can be determined. These fields are given as

$$H_r = \frac{(jkI\,dA)\cos\theta}{2\pi} k^2 \left[\frac{1}{(kr_0)^2} - \frac{j}{(kr_0)^3}\right] e^{-jkr_0} \quad (4.19a)$$

$$H_\theta = \frac{(jkI\,dA)\sin\theta}{4\pi} k^2 \left[\frac{j}{kr_0} + \frac{1}{(kr_0)^2} - \frac{j}{(kr_0)^3}\right] e^{-jkr_0} \quad (4.19b)$$

$$E_\phi = -\frac{(jkI\,dA)Z_0 \sin\theta}{4\pi} k^2 \left[\frac{j}{kr_0} + \frac{1}{(kr_0)^2}\right] e^{-jkr_0} \quad (4.19c)$$

$$H_\phi = 0 \quad E_r = 0 \quad E_\theta = 0 \quad (4.19d)$$

Figure 4.4 provides an example of the E- and H-field lines around a finite electric and magnetic dipole source. At observation locations far from the sources ($r_0 \gg dl$ or $r_0 \gg a$, where a is the radius of the loop), the field shapes are seen to be identical, as may be noted in Eqs. (4.12) and (4.19). Near the sources, however, the fields are different, with the E-field lines

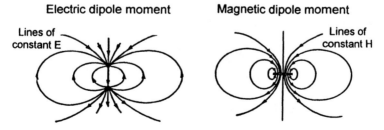

FIGURE 4.4 Examples of the E and H fields produced by electric and magnetic dipole sources.

terminating on the charges of the electric dipole and the H-field lines linking the circulating loop of current. This illustrates the fundamental difference between the E- and H-fields. It is important to remember that the equivalence between the electric and magnetic dipoles in Table 4.1 is valid only for distances far from the sources.

4.2.3 Radiation from Extended Sources

Practical antennas are seldom represented by a single infinitesimal dipole source. However, the expressions for the fields from the infinitesimal sources can be used to calculate the radiated fields by dividing the antenna into a number of smaller parts, each of which appears more like the point dipole. Figure 3.25a has illustrated the fact that at low frequencies, an EMI source connected to a loop radiator produces primarily a magnetic field that can affect a nearby victim circuit. Similarly, Figure 3.30a shows the source connected to two open wires, resulting in a strong electric field environment. As the frequency of the EMI source increases in these examples, significant E and H field components are produced by both of these antennas. To compute these fields, the antennas can be segmented as shown in Figure 4.5, with the resulting field at point P being determined as a sum over all of the individual dipole contributions.

As noted in Figure 4.5a, the open-wire antenna is represented by a collection of electric dipole sources, varying in strength depending on the current distribution along the wire. This results in the field at P being determined as a sum of terms along the antenna, or more precisely, by a line integral along the wire. Similarly, the loop antenna is represented by a collection of magnetic dipoles that are located over the area of the large loop. Thus, the field at P is found from a sum (or surface integration) of terms over the loop area, as illustrated in Figure 4.5b. Typically, surface integrations are more time consuming than are line integrals, so an alternative approach for computing the fields radiated from the loop antenna is to consider segmenting the loop current into electric dipole sources as in

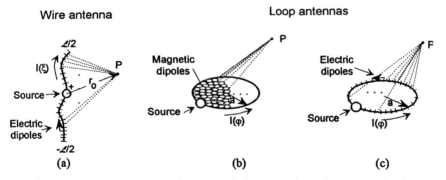

FIGURE 4.5 Segmentation of an extended antenna into elementary dipole sources.

Figure 4.5c. As long as the correct current distribution is used for the loop antenna current, either approach will give the same radiated field. Because the electric dipole can be used for both the wire and the loop antennas and because only a line integral must be used to find the fields, we concentrate on the use of the electric dipole in the remainder of this chapter.

4.2.3.1 Center-Fed Wire Antenna.
As an example of the use of Eq. (4.12) for the calculation of the fields radiated form an extended wire antenna, consider the case of an EMI source connected to a wire radiator, as shown in Figure 4.6. At very low frequencies we have seen that this antenna produces mainly electric fields in the vicinity of the source. We wish to extend this analysis to higher frequencies and for distances far from the source. A simple model for computing the EM fields produced by this radiator is a straight-wire, center-fed antenna of total length $\mathcal{L}$. The voltage source, V_0, induces an antenna current that varies in amplitude and phase along the wire and vanishes at each end of the conductor.

As mentioned above, the accurate determination of the antenna current distribution requires the use of extensive numerical calculations, often requiring the solution to an integral equation for the current. This is discussed in Section 4.2.3.2. An alternative technique, which is often used to approximate the current by a simple functional form, is based on an a priori knowledge of the actual distribution. This knowledge is gained from either experimental observation or from previous experience with antennas, possibly involving numerical calculations for different antenna structures.

At sufficiently low frequencies where $\lambda > \mathcal{L}/2$, a reasonable guess for the current distribution along the wire is given by the function

$$I(z) = I_0 \left(1 - \frac{2|z|}{\mathcal{L}}\right) \tag{4.20a}$$

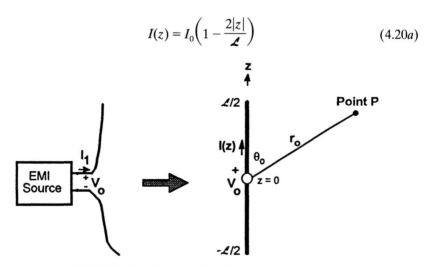

FIGURE 4.6 Center-fed radiating wire antenna.

which vanishes at each end of the antenna. At higher frequencies, the current distribution along the line can vary sinusoidally and an approximation to the current distribution is [4.7]

$$I(z) = I_0 \sin\left[k\left(\frac{\mathcal{L}}{2} - |z|\right)\right] \qquad (4.20b)$$

Of course, the current amplitude I_0 must also be specified in these equations, and this is difficult to do without actually solving for the current in a rigorous manner or by using some other approximation.

With the wire current specified and a knowledge of the radiation from a point source, the radiated field at a point P due to the extended wire radiator can be determined by an integral over the wire. The location of this point is defined by the angle θ_0 and distance r_0, as indicated in Figure 4.6a. Frequently, we are interested in the fields in the far field, where $kr_0 = 2\pi r_0/\lambda \gg 1$. In this case, only the $1/r_0$ terms in Eq. (4.12) are significant in determining the radiated field, and only E_θ and H_ϕ field components exist. The radiated E-field is then expressed as

$$E_\theta = \int_{\text{wire}} dE_\theta = \frac{jkZ_0}{4\pi} \int_{-\mathcal{L}/2}^{\mathcal{L}/2} \frac{I(z) \sin\theta\, e^{-jkr}}{r} dz \qquad (4.21)$$

where the angle θ and the distance r both vary with the position z along the antenna. This expression can be evaluated by noting that in the far field the following approximations can be made:

$$\frac{1}{r(z)} \approx \frac{1}{r_0} \qquad \sin\theta(z) \approx \sin\theta_0 \quad \text{and} \quad e^{-jkr(z)} \approx e^{-jkr_0} e^{jkz\cos\theta_0}$$

In evaluating Eq. (4.21), it is customary to drop the e^{-jkr_0} phase factor and remove the r_0 term from the denominator. Substituting in the functional form for the current in Eq. (4.20b) gives the radiated E-field as

$$r_0 E_\theta = \frac{jkZ_0 \sin\theta_0}{4\pi} \int_{-\mathcal{L}/2}^{\mathcal{L}/2} I_0 \sin\left[k\left(\frac{\mathcal{L}}{2} - |z|\right)\right] e^{jkz\cos\theta_0} dz \qquad (4.22)$$

The integration in Eq. (4.22) can be performed analytically, providing the final expression

$$r_0 E_\theta \approx j60 I_0 \left[\frac{\cos[(k\mathcal{L}/2)\cos\theta_0] - \cos(k\mathcal{L}/2)}{\sin\theta_0}\right] \qquad (4.23)$$

The normalized companion H-field component is simply $r_0 H_\phi = r_0 E_\theta/Z_0$.

The total radiated power for this antenna is similar to Eq. (4.18) and is given by

$$W = \oint_S \mathbf{P} \cdot d\mathbf{S} = \int_0^\pi \int_0^{2\pi} \mathbf{P} \cdot \hat{r} r^2 \sin\theta_0 \, d\theta_0 \, d\phi$$

$$= \frac{Z_0 I_0^2}{4\pi} \int_0^\pi \frac{\{\cos[(k\mathcal{L}/2)\cos\theta_0] - \cos(k\mathcal{L}/2)\}}{\sin\theta_0} \, d\theta_0. \quad (4.24)$$

This integral cannot be evaluated directly, but it can be expressed in terms of the Si and Ci integrals, or integrated numerically.

A common antenna is the half-wave dipole in which the total length $\mathcal{L} = \lambda/2$. In this case, $k\mathcal{L}/2 = \pi/2$ and the expressions for the radiated field and total power become

$$r_0 E_\theta \approx j60 I_0 \frac{\cos[(\pi/2)\cos\theta_0]}{\sin\theta_0} \quad (4.25)$$

$$W = \frac{Z_0 I_0^2}{4\pi} \int_0^\pi \frac{\{\cos[(\pi/2)\cos\theta_0]\}^2}{\sin\theta_0} \, d\theta_0 \approx 36.55 I_0^2 \quad (4.26)$$

where the integral for the total power has been evaluated numerically.

In some circumstances it is useful to define a radiation resistance R_r to relate the total radiated power W to the input current through the relationship $W = \frac{1}{2} I_0^2 R_r$. In this manner the radiation resistance for the center-fed antenna can be calculated from Eq. (4.24) as

$$R_r = \frac{2W}{I_0^2} = \frac{Z_0}{2\pi} \int_0^\pi \frac{\{\cos[(k\mathcal{L}/2)\cos\theta_0] - \cos(k\mathcal{L}/2)\}^2}{\sin\theta_0} \, d\theta_0 \quad (4.27)$$

and for the half-wave dipole, the resistance is determined from Eq. (4.26) to be

$$R_r = \frac{2W}{I_0^2} \approx 73.09 \, \Omega \quad (4.28)$$

It should be remembered that the radiation resistance defined by Eq. (4.27) is only approximate, as it depends on the assumption of the current distribution on the antenna. In reality, the antenna's input impedance Z_{in} is a complex-valued quantity, consisting of a real part and an imaginary part. For an assumed perfectly conducting antenna, the real part of Z_{in} corresponds to the radiation resistance, and it must be a positive quantity for all frequencies. The imaginary part is related to the reactive energy stored in the near field of the antenna. The latter susceptance cannot be evaluated from the far field expressions and must rely on a near-field analysis, as discussed in the next section.

4.2.3.2 Integral Equation for the Wire Antenna.

The example of the radiating antenna shown in Figure 4.6 is a rather special one, in that the source is located at a symmetry point of the structure. The more general (and more interesting) case is for an arbitrary source location along the antenna. As stressed in earlier sections, the prime requirement is calculating the radiated field or power is the knowledge of the current on the antenna (or at least, an approximation to the current). For the center-fed antenna, the current distribution is easily guessed. For a more general antenna, this current distribution is not so evident.

Consider the case of the same antenna of Figure 4.6 of length $\mathcal{L}$ but with an arbitrary source location. For convenience, we will shift the $z = 0$ origin to be at the bottom of the antenna so that it extends from $z = 0$ to $z = \mathcal{L}$. The voltage source V_0 is assumed to be at a location z_s such that $0 < z_s < \mathcal{L}$. The geometry of this antenna is shown in Figure 4.7b. In the development that follows we assume that the antenna structure is a thin wire so that the radius of the wire $a \ll \mathcal{L}$.

As mentioned earlier, one way of determining the current distribution is to solve an integral equation for the wire current. Consider for the moment the excitation of the antenna by a general incident E-field as shown in Figure 4.7a. The incident electric field $\mathbf{E}^{\text{inc}}$ induces a current I in the wire, which, in turn, produces a scattered field $\mathbf{E}^{\text{sca}}$ so that the total tangential E-field $\mathbf{E}_z^{\text{tot}}$ is zero along the surface of the wire (i.e., $\mathbf{E}_z^{\text{inc}} + \mathbf{E}_z^{\text{sca}} = \mathbf{E}_z^{\text{tot}} = 0$).

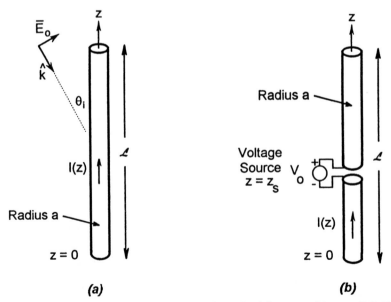

FIGURE 4.7 Thin-wire linear antenna: (*a*) excited by an arbitrary E-field; (*b*) excited by a localized voltage source at $z = z_s$.

4.2 RADIATION OF ELECTROMAGNETIC FIELDS IN THE FREQUENCY DOMAIN

The expression for the scattered field produced by a volume current density **J** is given by Eqs. (4.5) and (4.10) as

$$\mathbf{E}^{sca}(\mathbf{r}_0) = \frac{1}{j\omega\epsilon_0} [\nabla(\nabla \cdot) + k^2] \int_{vol} \mathbf{J}(\mathbf{r}_s) \frac{e^{-jkR}}{4\pi R} dV_s \qquad (4.29)$$

Specializing this equation for only a z-directed current, extracting the z-component of the E-field, and then equating it to $-E_z^{inc}$ for observation points $\mathbf{r}_0$ located on the wire results in the Pocklington integrodifferential equation [4.8] for the wire current,

$$-j\omega\epsilon_0 E_z^{inc}(z) = \left(\frac{d^2}{dz^2} + k^2\right) \int_{wire} I(z') \frac{e^{-jkR}}{4\pi R} dz' \qquad (4.30)$$

where the distance R is given by

$$R = \sqrt{(z - z')^2 + a^2}$$

The derivation of this integral equation has used the thin-wire approximation [4.9], in which the wire current is assumed to be a line source located along the z-axis and the field observation point $\mathbf{r}_0$ is located on the wire surface.

4.2.3.2.1 Solution by the Method of Moments.
To discuss the solution of the integral equation (4.30) for the unknown current $I(z')$, we see that the equation has the general form

$$\int_a^b K(z; z') f(z') dz' = g(z) \qquad (4.31)$$

where the kernel $K(z;z')$ and the forcing function $g(z)$ are both known functions and the function $f(z')$ is to be determined. To determine $f(z')$, a number of techniques are available. The point–matching technique is widely employed and usually gives results which are reasonably accurate. A more accurate method is Galerkin's method [4.10], which is related to the Rayleigh–Ritz variational approach. In addition, it is possible to obtain a solution by using a least squares technique [4.11]. In the now-classic text by Harrington, [4.1], it is shown that each of these approaches is a special case of a general technique called the *method of moments*.

To illustrate use of the moment method, consider representing the integral equation of Eq. (4.31) in operator form as

$$\mathcal{K} f = g \qquad (4.32)$$

and expanding the unknown function f by a series of normalized basis functions f_n in the form

$$f = \sum_n i_n f_n \qquad (4.33)$$

where i_n are the amplitudes of each mode. Equation (4.32) can be written as

$$\sum_n i_n \mathcal{K} f_n = g \qquad (4.34)$$

Here $\mathcal{K}$ is assumed to be a linear operator, so that it can be moved inside the summation sign. The summation over the index n is infinite in general, and the functions f_n should form a complete set. However, for an approximate solution, this summation can be truncated at some value N.

By defining a finite set of M testing functions w_m, which are defined over the same range as f_n, it is possible to take M moments of Eq. (4.34), giving

$$\sum_N i_n \langle w_m, \mathcal{K} f_n \rangle = \langle w_m, g \rangle \qquad m = 1, 2, \ldots, M \qquad (4.35)$$

Here $\langle w_m, g \rangle$ represents a suitable inner product between the two functions. Denoting $\langle w_m, g \rangle$ by the vector element v_m and $\langle w_m, \mathcal{K} f_n \rangle$ by the matrix element $z_{m,n}$, Eq. (4.35) can be written in a matrix form as:

$$[z_{m,n}][i_n] = [v_m] \qquad (4.36a)$$

Note that with a slight rearrangement of terms, the integral equation (4.30) takes the form of Eq. (4.36a), where the elements $z_{m,n}$ represent impedance quantities and v_m represent voltages. Elsewhere in the literature, this system of linear equations for the antenna current takes on different forms. One common representation of this equation is

$$[Z][I] = [V] \qquad (4.36b)$$

If $M = N$ and $[z_{m,n}]$ is nonsingular, it is possible to invert this matrix equation and obtain the approximation to f, since

$$[i_n] = [z_{m,n}]^{-1}[v_m] \qquad (4.37)$$

and f is related to i_n through Eq. (4.33). If $M > N$, it is possible to obtain a least-squares solution for the function f [4.1].

The choices of the basis functions f_n are arbitrary, and it is important to try to find the best type of functions for a particular problem. The simplest type of basis function is one having subrange support, in which f_n is defined as unity within a small cell or zone in the interval $[a,b]$, and zero elsewhere. Other types of basis function may be defined, such as ones that approximate the current within a particular cell by a linear or quadratic-topped pulse instead of a rectangular pulse. Of course, it is possible to define the basis functions to have a range over the entire interval $[a,b]$, such as a family of sine functions. However, it has been shown that under certain circumstances this type of basis function leads to relatively unstable solutions [4.12]. Moreover, if this choice of basis functions is made, longer computation times in evaluating the matrix elements are required.

The choice of the testing functions w_m is also arbitrary, and its form must be decided upon before proceeding with the solution. By letting w_m be a delta function within the mth cell, it can be seen that the resulting matrix equation is that obtained by the point-matching method. The results obtained by using this choice might not be as accurate as those found by using a more complicated testing function, but shorter computational times result. Letting w_m have the same form as the basis function f_n results in Galerkin's method. It has been shown that this implies automatic conservation of energy in scattering problems, so care must be used in checking this kind of solution by using energy balance comparisons [4.13.]

The evaluation of the matrix elements $[z_{m,n}]$ can be difficult for two reasons. The $1/R$ term in Eq. (4.30) is nearly singular, and care must be used in performing the numerical integrals required for the inner product. In addition, the differential operator $[d^2/dz^2 + k^2]$ requires a finite-difference representation to be able to put the equation in a discrete form. Details of this implementation of the moment method are beyond the scope of this book and are discussed thoroughly in refs. [4.1] and [4.6].

Using the moment method, several computer programs have been developed for treating wire antenna and scattering problems. One common program is the Numerical Electromagnetics Code (NEC) [4.14], which has become a standard for such analysis. Many of the integral equation results for wire antennas and transmission lines in this book have been obtained using the NEC code.

4.2.3.2.2 Approximate Solution for the Antenna Problem.

Although the use of a standard computer program such as NEC will permit the solution of a wide variety of antenna problems in the frequency domain, the computer time and memory requirements can be quite large, especially if a time-domain solution is ultimately desired. As a result, it is instructive to explore approximate methods for analyzing antenna and scattering problems involving wire structures.

An alternative solution for determining the wire current begins by viewing Eq. (4.30) as a differential equation for a function $\Lambda(z)$:

$$\Lambda(z) \equiv \int I(z') \frac{e^{-jkR}}{4\pi R} dz' \qquad (4.38)$$

Note that $\Lambda(z) = A_z(z)/\mu_0$, where A_z is the vector potential defined in Eq. (4.10). The solution to the differential equation (4.30) consists of a homogeneous and a particular solution, with two constants of integration chosen to satisfy the boundary conditions that $I = 0$ at each end of the wire [4.15]. This solution for $\Lambda(z)$ may be written as

$$\Lambda(z) = C_1 \sin kz + C_2 \cos kz - \frac{j\omega\epsilon_0}{k} \int_0^z E_z^{\text{inc}}(z') \sin k(z - z') \, dz' \quad (4.39)$$

which is the Hallén integral equation for the thin wire. This equation can be solved numerically using the moment method [4.15].

An approximate method for solving this equation for the current I has been discussed by Schelkunoff [4.16]. This involves writing Eq. (4.38) as

$$\Lambda(z) = \frac{I(z)}{4\pi} \int_0^{\mathcal{L}} \frac{dz'}{R} + \frac{1}{4\pi} \int_0^{\mathcal{L}} \frac{I(z')e^{-jkR} - I(z)}{R} \, dz' \quad (4.40)$$

It is convenient to define the parameter $\Omega(z)$ as

$$\Omega(z) = \int_0^{\mathcal{L}} \frac{dz'}{R} \approx \ln \frac{4(h^2 - a^2)}{a^2} + \frac{1}{2}\left[\frac{a^2}{(h+z)^2} + \frac{a^2}{(h-z)^2}\right] + \cdots \quad (4.41)$$

where $h = \mathcal{L}/2$. For cases where the radius a is small compared with $\mathcal{L}$, this term may be approximated as

$$\Omega(z) \approx 2 \ln \frac{2h}{a} = 2 \ln \frac{\mathcal{L}}{a} \equiv \Omega_0 \quad (4.42)$$

Writing $\Omega(z)$ as $\Omega_0 + [\Omega(z) - \Omega_0]$, substituting this into Eqs. (4.40) and the result into (4.39) yields the following equation for the current:

$$I(z) = \frac{4\pi}{\Omega_0}\left[C_1 \sin kz + C_2 \cos kz - \frac{j\omega\epsilon_0}{k} \int_0^z E_z^{\text{inc}}(z') \sin k(z - z') \, dz'\right]$$

$$+ \left[1 - \frac{\Omega(z)}{\Omega_0}\right]I(z) - \frac{1}{\Omega_0}\int_0^{\mathcal{L}} \frac{I(z')e^{-jkR} - I(z)}{R} \, dz' \quad (4.43)$$

Neglecting the last two terms, which are small, the approximate current on the wire now can be written explicitly as

$$I(z) \approx C_1' \sin kz + C_2' \cos kz - \frac{j4\pi}{Z_0\Omega_0} \int_0^z E_z^{\text{inc}}(z') \sin k(z - z') \, dz' \quad (4.44)$$

where C_1' and C_2' are to be determined from the boundary conditions requiring that the current vanish at the ends of the wire at $z = 0$ and $z = \mathcal{L}$.

EVALUATION OF THE ANTENNA CURRENT. For the antenna driven by a voltage generator of V_0 volts applied at position z_s on the conductor, the excitation can be modeled as a small ring of magnetic current proportional to the tangential E-field produced by the voltage source [4.6]. The magnetic current "frill" source radiates as if it were in free space, producing the incident tangential field $E_z^{\text{inc}}(z)$ along the antenna which is used in Eq. (4.44). An investigation of this incident E-field shows that a very good approximation is to assume that the excitation E-field exists only in the vicinity of the source, and that the source is infinitesimally small:

$$E_z^{\text{inc}}(z) = V_0 \delta(z - z_s) \tag{4.45}$$

This type of source is known as a *slice generator*.

Inserting this incident field into Eq. (4.44), performing the required integration, and evaluating the constants C_1 and C_2 so that the solution for the current vanishes at the ends of the antenna gives the following expression for the antenna current distribution:

$$I(z) = I_a \left\{ \sin kz \cos kz_s [1 - U(z - z_s)] \right.$$
$$\left. + \sin kz_s \cos kz \, U(z - z_s) - \cos k\mathcal{L} \sin kz \, \frac{\sin kz_s}{\sin k\mathcal{L}} \right\} \tag{4.46a}$$

$U(z - z_s)$ is the Heaviside function defined as 0 for $z - z_s < 0$ and 1 for $z - z_s \geq 0$. A more compact form for the current is obtained after some algebraic manipulation:

$$I(z) = I_a \frac{\sin kz_< \sin k(\mathcal{L} - z_>)}{\sin k\mathcal{L}} \tag{4.46b}$$

where $z_<$ denotes the smaller of z and z_s, and $z_>$ signifies the larger of z and z_s. In these expressions, the leading term I_a is defined as

$$I_a = \frac{j4\pi V_0}{Z_0 \Omega_0} \tag{4.47}$$

To check the validity of this solution, it can be verified from Eq. (4.46) that this current vanishes at each end of the antenna as required and has singularities occurring at frequencies where $\sin k\mathcal{L} = \sin(2\pi f \mathcal{L}/c) = 0$, or for $f_\alpha = \alpha c/2\mathcal{L}$, $\alpha = 1, 2, \ldots$. These correspond to the usual resonance frequencies of an antenna of length $\mathcal{L}$. Also note that the expression for the current is continuous through the source at z_s and has a slope discontinuity representing the change in potential along the antenna due to the voltage source.

As an example of the behavior of the induced current, and for an indication of the expected accuracy of this approximate solution, Figure 4.8 plots the magnitude of antenna current as a function of position along the antenna. For this illustration, the antenna consisted of a very thin, straight

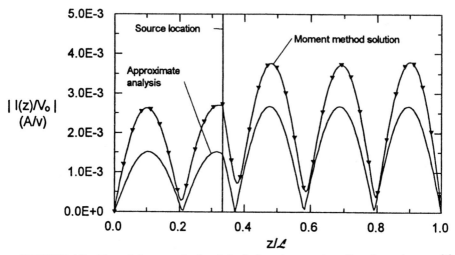

FIGURE 4.8 Plot of the magnitude of the induced current on the wire antenna with source at $z_s = \mathcal{L}/2$ and $a/\mathcal{L} = 10^{-4}$, for $k\mathcal{L} = 15$.

wire with $a/\mathcal{L} = 10^{-4}$. The voltage source V_0 was located one-third of the distance from one end (at $z_s = \mathcal{L}/3$) and the frequency was such that $k\mathcal{L} = 2\pi f \mathcal{L}/c = 15$. This plot illustrates the approximate current distribution as computed using Eq. (4.46), along with the "correct" solution obtained by a numerical solution of the integral equation (4.30) using the method of moments.

The approximate analysis is seen to be within about a factor of 2 of the numerical solution, but it is obtained at a fraction of the computer cost required for the moment method solution. If such an approximate analysis is to be used for a practical engineering study or simulation, one should carefully evaluate the other uncertainties in the problem (i.e., lack of precise definition of the EM environment, unknown material parameters, nonideal geometries, etc.) to put this modeling error into proper perspective.

We see from Eq. (4.47) that by developing this approximate solution from the integral equation for the current, the strength of the current magnitude I_a is determined in terms of the excitation voltage source. This is in contrast to the solution in Eq. (4.20b), in which the current amplitude I_0 remains unspecified. We can define the input admittance of the antenna at the terminal at z_s as

$$Y_{in}\big|_{z=z_0} = \frac{I(z_s)}{V_0} = j\frac{4\pi}{Z_0\Omega_0}\frac{\sin kz_s(\sin k\mathcal{L}\cos kz_s - \cos k\mathcal{L}\sin kz_s)}{\sin k\mathcal{L}} \quad (4.48)$$

which is purely reactive. Similarly, the input impedance of the antenna is $Z_{in} = 1/Y_{in}$.

Because this expression is derived from the first term in an expansion for the current, it is seen that the radiation damping of the current is not properly taken into account, and this results in the purely reactive input admittance and an infinite current response at the resonant frequencies. The correct solution for the antenna current will contain the effects of radiation damping, and this will serve to limit the amplitude of the current at the resonances. As discussed in [4.17], it is possible to include this damping in the approximate solution, and this will be investigated later in the chapter.

As an example of the wideband behavior of the antenna input current calculated by this approximate approach, Figure 4.9 plots the magnitude of the normalized current $|I/V_0|$ (i.e., the input admittance) for the thin antenna. These results are plotted as a function of $k\mathcal{L}$. For these calculations, the source was located at the midpoint of the antenna. The curve labeled "approximate analysis" comes from an evaluation of the input admittance using Eq. (4.48). The curve labeled "moment method solution" results from a numerical solution of Eq. (4.30) using the method of moments. The latter solution contains the effects of radiation loss, and consequently, it provides responses with lower quality factors (Q's) at the resonant frequencies. At the resonant frequencies, the approximate response is formally infinite, but this is not apparent in the plot, due to the fact that the frequency sample points never coincide exactly with the resonances. Notice that there are periodic resonances in the current at frequencies where $k\mathcal{L} = \pi, 3\pi, 5\pi$, and so on. Although Eq. (4.48) suggests that there are also resonances at $k\mathcal{L} = 2\pi, 4\pi, 6\pi$, and so on, we see that these are not present

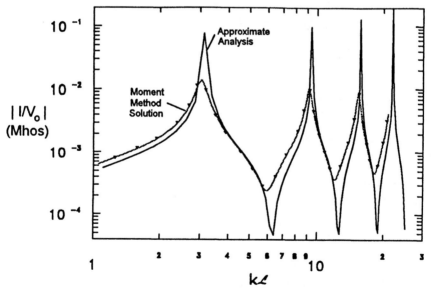

FIGURE 4.9 Spectrum of the normalized input current magnitude $|I/V_0|$ for a thin center-fed linear antenna ($a/\mathcal{L} = 10^{-4}$).

134 RADIATION MODELS FOR WIRE ANTENNAS

in this example, as they are canceled by zeros in the numerator. This is a direct consequence of having the antenna source at the midpoint of the antenna. If the source were to be moved slightly, these resonances would appear in the response spectrum.

EVALUATION OF THE RADIATED FIELD. The evaluation of the radiated E-field for this antenna in a direction given by the observation angle θ_0 can be determined by performing the integration indicated in Eq. (4.22), but using the current distribution of Eq. (4.46). In doing this, it is expedient to consider the phase reference to be at the source location at $z = z_s$. Thus, the radiated field can be expressed as the integral

$$r_0 E_\theta = \frac{jkZ_0 \sin \theta_0}{4\pi} \int_0^{\mathcal{L}} I(z) e^{jk(z-z_s)\cos\theta_0} \, dz$$

$$= -\frac{kV_1}{\Omega_0} \frac{\sin\theta}{\sin k\mathcal{L}} \left[\sin k(\mathcal{L} - z_s) \int_0^{z_0} \sin kz \, e^{jk(z-z_s)\cos\theta_0} \, dz \right.$$

$$\left. + \sin k(z_s) \int_{z_0}^{\mathcal{L}} \sin k(\mathcal{L} - z) e^{jk(z-z_s)\cos\theta_0} \, dz \right] \quad (4.49)$$

The integrals in Eq. (4.49) can be integrated analytically, resulting in the following expression for the radiated field:

$$r_0 E_\theta = \frac{V_1}{\Omega_0 \sin\theta_0} \frac{1}{\sin k\mathcal{L}} [\sin k(z_s - \mathcal{L})][\cos(kz_0 \cos\theta_0) - \cos kz_s]$$

$$- \sin kz_s \{\cos[k(z_s - \mathcal{L}) \cos\theta_0] - \cos k(z_s - \mathcal{L})\})$$

$$+ j(\sin k(z_s - \mathcal{L})[\cos\theta_0 \sin kz_s - \sin(kz_s \cos\theta_0)]$$

$$+ \sin kz_s \{\sin[k(z_s - \mathcal{L})\cos\theta_0] - \cos\theta_0 \sin k(z_s - \mathcal{L})\})] \quad (4.50)$$

Alternative forms of this solution are available by expressing the sine and cosine terms into complex exponentials. As a check of these equations, note that if the source location z_s is positioned at either end of the antenna at 0 or $\mathcal{L}$, the radiated field is identically zero. Furthermore, for the special case of a center-fed antenna, $z_s = \mathcal{L}/2$, and this expression reduces to

$$r_0 E_\theta = \frac{-2V_1}{\Omega_0} \frac{\sin(k\mathcal{L}/2)}{\sin k\mathcal{L}} \frac{\cos(k\mathcal{L}\cos\theta_0/2) - \cos(k\mathcal{L}/2)}{\sin\theta_0} \quad (4.51)$$

which has the same functional dependence on the angle of observation θ_0 as given in Eq. (4.23), but now, the leading term is defined in terms of the excitation voltage and the antenna parameters.

4.2.4 Dipole Radiation in the Presence of Other Bodies

The expressions for the EM fields produced by a point electric dipole in Eq. (4.12) are for the case of an isolated source in an unbounded region. Unfortunately, such a source configuration is not encountered frequently in practice. Antennas are usually located near the earth, on a vehicle, or next to other conducting bodies. The primary EM fields produced by such a source will interact with the nearby object, inducing secondary currents in the body, and these currents will modify the original fields radiated by the source.

The treatment of an arbitrary-shaped object near the source can be difficult and usually requires a numerical solution to an integral equation for the unknown currents induced in the object. There are, however, certain cases in which the interaction between the dipole source and the body can be treated rather simply. Some of these cases are discussed in this section. In this discussion we concentrate on the behavior of the electric dipole source. The corresponding case of the magnetic dipole can be treated in the same manner; the details are available in the literature [4.6, 4.10, 4.18].

4.2.4.1 Electric Dipoles over a Perfect Ground. The simplest case of a source located near a conducting object is that of an arbitrarily orientated current element located over an infinite, perfectly conducting ground plane, as illustrated in Figure 4.10. In this figure, the ground plane lies in the x–y

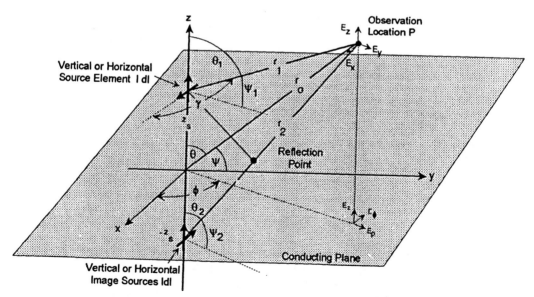

FIGURE 4.10 Geometry for a vertical or x-directed horizontal current element over a perfect ground plane.

plane at $z = 0$. For analysis purposes, the current element can be divided into a vertical dipole and a horizontal dipole, and the fields from both may be superimposed to provide the fields produced by the arbitrarily oriented source. Generally, the vertical current element will radiate more efficiently than the horizontal element in the presence of the ground.

Consider first the vertical current element. The fields produced by the elementary electric dipole moment $I\,dl$ located at a height z_s over the ground can be determined by formally solving Eq. (4.5) in the region $z \geq 0$, subject to the perfectly conducting boundary conditions on the E and H fields at $z = 0$: namely $E_x = E_y = 0$; and $H_z = 0$.

An alternative way of constructing the fields above the ground is to use image theory [4.6]. This permits calculation of the field very simply by locating an image dipole moment at the location $z = -z_s$ and then removing the plane. The resulting field at a point P above the $z = 0$ plane can be constructed from two contributions of terms such as Eq. (4.12) but with different values of r_0 and polar angle θ. As illustrated in Figure 4.10, the physical source contributes a direct contribution to the response at P and another contribution that is reflected from the plate but which appears as if it arrives from the fictitious image source.

All of the terms in Eq. (4.12) contribute to the fields at any location P. In many instances, however, we are interested in the *far fields* from the source and we retain only the $1/r$ terms in the equations. The E-field components for this geometry are most suitably expressed *in cylindrical coordinates* (ρ,ϕ,z) and they are given in [4.19] as

$$E_z \approx -j30kI\,dl\left(\cos^2\psi_1 \frac{e^{-jkr_1}}{r_1} + \cos^2\psi_2 \frac{e^{-jkr_2}}{r_2}\right) \qquad (4.52a)$$

$$E_\rho \approx j30kI\,dl\left(\cos\psi_1 \sin\psi_1 \frac{e^{-jkr_1}}{r_1} + \cos\psi_2 \sin\psi_2 \frac{e^{-jkr_2}}{r_2}\right) \qquad (4.52b)$$

$$E_\phi = 0 \qquad (4.52c)$$

where r_1 and ψ_1 are the distance and angle from the primary source, and r_2 and ψ_2 are the same for the image source. As noted in Figure 4.10, the angles ψ are measured from the horizontal direction. For the observation point at polar coordinates $(\rho_0, 0, z_0)$, these distances and angles are

$$r_1 = \sqrt{\rho_0^2 + (z_0 - z_s)^2} \qquad r_2 = \sqrt{\rho_0^2 + (z_0 + z_s)^2} \qquad (4.53a)$$

$$\psi_1 = \arctan\frac{z_0 - z_s}{\rho_0} \qquad \psi_2 = \arctan\frac{z_0 + z_s}{\rho_0} \qquad (4.53b)$$

Notice that for the observation point P on the plate, $z_0 = 0$ and $E_\rho = 0$, as required by the boundary condition that $E_{\text{tan}} = 0$.

A similar expression can be derived for the ϕ-component of the H-field by summing two terms involving Eq. (4.12c). This gives the expression

$$H_\phi = \frac{jk}{4\pi} I\, dl \left(\cos \psi_1 \frac{e^{-jkr_1}}{r_1} + \cos \psi_2 \frac{e^{-jkr_2}}{r_2} \right) \qquad H_\rho = H_z = 0 \quad (4.54)$$

The fields from the horizontal current element in Figure 4.10 can also be evaluated by image theory. The fields produced by this source are slightly more complex than for the vertical element, due to the absence of symmetry about the z-axis. These fields will depend on the cylindrical angle ϕ. For this horizontal source, the E-field components are conveniently expressed in *Cartesian coordinates*. For the current element assumed to be directed along the x-axis, the three components of the direct and image radiation-zone E-fields are given as

$$E_z \approx j30kI\, dl \cos\phi \left(\sin\psi_1 \cos\psi_1 \frac{e^{-jkr_1}}{r_1} - \sin\psi_2 \cos\psi_2 \frac{e^{-jkr_2}}{r_2} \right) \quad (4.55a)$$

$$E_x \approx -j30kI\, dl \left[\cos^2\phi \left(\sin^2\psi_1 \frac{e^{-jkr_1}}{r_1} - \sin^2\psi_2 \frac{e^{-jkr_2}}{r_2} \right) \right.$$
$$\left. + \sin^2\phi \left(\frac{e^{-jkr_1}}{r_1} - \frac{e^{-jkr_2}}{r_2} \right) \right] \quad (4.55b)$$

$$E_y \approx -j30kI\, dl \sin\phi \cos\phi \left[\left(\sin^2\psi_1 \frac{e^{-jkr_1}}{r_1} - \sin^2\psi \frac{e^{-jkr_2}}{r_2} \right) \right.$$
$$\left. + \left(\frac{e^{-jkr_1}}{r_1} - \frac{e^{-jkr_2}}{r_2} \right) \right]$$

$$(4.55c)$$

A similar set of equations can be derived for the H-field radiated from this source. This is left as an exercise for the reader.

For an example of the fields produced by the electric dipole source located over the perfect ground plane, consider a vertically directed current element at heights $z_s = 0$, 0.5, 1.0, and 2.0 wavelengths. The radiated field is the E_θ component, which is given in terms of the cylindrical components E_z and E_r in Eq. (4.52) as $E_\theta = -E_z \sin\theta + E_\rho \cos\theta$, where θ is the polar angle shown in the figure. We assume that the observation distance $r_0 \gg z_s$, so that $1/r_1 \approx 1/r_2 \approx 1/r_0$, and $\theta_1 \approx \theta_2 \approx \theta$. With these approximations, the E_θ

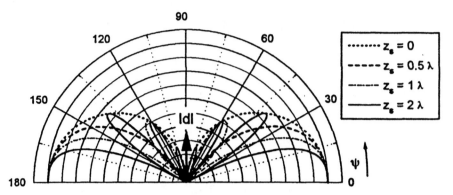

FIGURE 4.11 Plots of the normalized E-field radiated from a point electric dipole source located at a height z_s over a perfectly conducting plane.

component of the field in the far zone may be expressed as

$$r_0 E_\theta \approx (j60kI\ dl\ e^{-jkr_0}) \sin\theta \cos(kz_s \sin\theta) \tag{4.56}$$

Figure 4.11 shows the normalized radiated E-fields from these sources as a function of the angle $\psi = 90° - \theta$. For the source on the ground plane, the field pattern is a simple $\sin\theta$ pattern which is the same as for the dipole in free space (although the actual amplitude of the radiated field in Eq. (4.56) is a factor of 2 larger than for the free-space case). This pattern provides the largest radiation in the direction perpendicular to the source and no radiation along the polar axis. As the frequency increases, the E-field pattern in the broadside direction becomes more narrow and eventually, sidelobes appear in the pattern.

4.2.4.2 Electric Dipoles in a Parallel-Plate Region.

Image theory can also be used to develop expressions for the fields from a current element located within a parallel-plate region [4.20, 4.21]. As shown in Figure 4.12, two perfectly conducting plates lie in the $x-y$ plane at $z = 0$ and $z = d$. The primary sources, either vertical or horizontal in the x-direction, are at the point $(x,y,z) = (0,0,z_s)$ and can be imaged in the upper and lower plates. Subsequent images are also imaged, and this gives rise to a doubly infinite set of images: one set at the locations $z = 2nd + z_s$ and another at $z = 2nd - z_s$ (for $n = -\infty, \ldots, 0, \ldots, \infty$). Notice that for the vertical current element, all of the images are oriented in the same direction. For the horizontal dipole, however, the images alternate in direction.

The expressions for the fields produced by the point current element in Eq. (4.12) can be summed over each of the sources. As in the case of the single plate, if we keep only the $1/r$ terms, the E-fields radiated by the

4.2 RADIATION OF ELECTROMAGNETIC FIELDS IN THE FREQUENCY DOMAIN

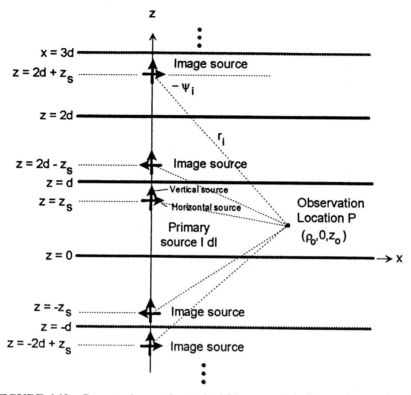

FIGURE 4.12 Current element located within a parallel-plate region, analyzed by the image method.

vertical source are given by

$$E_z \approx -j30kI\,dl\left(\sum_{n=-\infty}^{\infty}\cos^2\psi_{1n}\frac{e^{-jkr_{1n}}}{r_{1n}} + \sum_{n=-\infty}^{\infty}\cos^2\psi_{2n}\frac{e^{-jkr_{2n}}}{r_{2n}}\right) \quad (4.57a)$$

$$E_\rho \approx j30kI\,dl\left(\sum_{n=-\infty}^{\infty}\cos^2\psi_{1n}\sin\psi_{1n}\frac{e^{-jkr_{1n}}}{r_{1n}} + \sum_{n=-\infty}^{\infty}\cos\psi_{2n}\sin\psi_{2n}\frac{e^{-jkr_{2n}}}{r_{2n}}\right)$$
$$(4.57b)$$

$$E_\phi = 0 \quad (4.57c)$$

where the distances and angles r and ψ are similar to those for the source over the single plate and are given by

$$r_{1n} = \sqrt{\rho_0^2 + [z_0 - (2nd + z_s)]^2} \qquad r_{2n} = \sqrt{\rho_0^2 + [z_0 - (2nd - z_s)]^2}$$

(4.58a)

$$\psi_{1n} = \arctan \frac{z_0 - (2nd + z_s)}{\rho_0} \qquad \psi_2 = \arctan \frac{z_0 - (2nd - z_s)}{\rho_0}$$

(4.58b)

A similar development is possible for the E-fields radiated by the horizontal current source. The fields are

$$E_z = j30kI\,dl\cos\phi \left(\sum_{n=-\infty}^{\infty} \sin\psi_{1n}\cos\psi_{1n}\frac{e^{-jkr_{1n}}}{r_{1n}} - \sum_{n=-\infty}^{\infty} \sin\psi_{2n}\cos\psi_{2n}\frac{e^{-jkr_{2n}}}{r_{2n}} \right)$$

(4.59a)

$$E_x = -j30kI\,dl\left[\sum_{n=-\infty}^{\infty} (\cos^2\phi\sin^2\psi_{1n} + \sin^2\phi)\frac{e^{-jkr_{1n}}}{r_{1n}} \right.$$

$$\left. - \sum_{n=-\infty}^{\infty} (\cos^2\phi\sin^2\psi_{2n} + \sin^2\phi)\frac{e^{-jkr_{2n}}}{r_{2n}} \right]$$

(4.59b)

$$E_y = -j30kI\,dl\sin\phi\cos\phi\left[\sum_{n=-\infty}^{\infty} (1 + \sin^2\psi_{1n})\frac{e^{-jkr_{1n}}}{r_{1n}} \right.$$

$$\left. - \sum_{n=-\infty}^{\infty} (1 + \sin^2\psi_{2n})\frac{e^{-jkr_{2n}}}{r_{2n}} \right]$$

(4.59c)

As discussed in [4.20], a difficulty in evaluating the fields in the parallel-plate region arises in summing the image terms. Because each term is proportional to e^{-jk2dn}/n for large values of n, the series are very slowly converging, and for $kd = \pi$, 2π, 3π, and so on, the series do not converge. These frequencies correspond to the resonances of the parallel-plate region and are where one of the waveguide modes change from evanescent to propagating.

One way to speedup the convergence in Eqs. (4.57) and (4.59) is to use the relationship [4.22]

$$\sum_{n=1}^{\infty} \frac{e^{-jk2dn}}{n} = -\ln(1 - e^{-jk2d})$$

(4.60)

By adding and subtracting this expression in the summation terms, the resulting modified series converges more rapidly and permits a numerically efficient evaluation of the fields.

As an example of the behavior of an antenna located within a parallel-plate region, [4.20] describes the use of an integral equation solution for a thin wire, using expressions similar to Eqs. (4.57) and (4.59) to construct the system impedance matrix. A numerical solution then provides information about the current distribution on the antenna and the radiated fields. One useful parameter in assessing the antenna performance is the real part of the input admittance—the input conductance G. Figure 4.13 plots computed values for G for a center-fed dipole antenna of length $\mathcal{L} = \lambda/2$ and $\Omega = 2\ln(L/a) = 10$, which is located with its midpoint at $z_s = d/2$.

4.2.4.3 Electric Dipoles in a Cavity.
An extension of the parallel-plate geometry is obtained by adding four more perfectly conducting plates to enclose the dipole source completely. The determination of the EM fields within such a rectangular cavity has been discussed by several authors [4.23–4.25] in terms of electric or magnetic dyadic Green's functions $\overleftrightarrow{G}_e$ and $\overleftrightarrow{G}_m$ that are solutions to the vector wave equation.

The dyadic Green's function for the E-field within the cavity provides a knowledge of all three vector components for an arbitrarily oriented electric

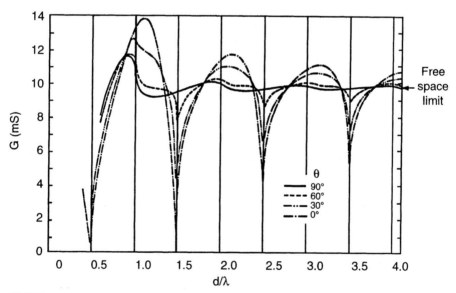

FIGURE 4.13 Input conductance of a $\mathcal{L} = \lambda/2$ center-fed dipole antenna with $\Omega = 2\ln(\mathcal{L}/a) = 10$ located in the middle of a parallel-plate waveguide, shown as a function of the plate spacing, d. ($\theta = 0°$ corresponds to the antenna parallel to the plates, $\theta = 90°$ corresponds to the antenna perpendicular to the plates.) (From [4.20] © 1972 Institute of Electrical and Electronics Engineers.)

142 RADIATION MODELS FOR WIRE ANTENNAS

dipole. Thus it is a dyad containing nine elements. In our discussion here, we limit the elementary dipole source to be only in the direction of the z-axis. In this way, we will need to compute only three components of $\overleftrightarrow{G}_e$. Other orientations of the dipole source, if needed, can then be determined by a rotation of the coordinate system.

Figure 4.14 illustrates the geometry of such an enclosure. The z-directed electric dipole source of moment $I\,dl$ is located at location (x_s, y_s, z_s) within the cavity, and the E-field is observed at location (x_0, y_0, z_0). The x, y, and z dimensions of the box cavity are x_e, y_e, and z_e, respectively.

4.2.4.3.1 Solution by the Method of Images.
One approach for solving for these Green's functions is to use image theory, as was done for the parallel-plate enclosure. This results in a triply infinite set of images in the x, y, and z directions, with the fields at the observation point obtained by summing the individual contributions of terms like Eq. (4.14). As noted in [4.25], summing the contributions from the images can be very time consuming, due to the slow convergence of the series.

Fortunately, in cases where a transient response of the cavity fields is desired, it is possible to limit the number of required terms in the frequency-domain series in a very simple way. First, an estimate of the maximum time of interest in the solution is needed. This will permit the definition of an imaginary sphere around the enclosure, in which all sources within the sphere will contribute to the transient response. Those sources located outside the sphere will contribute responses at a later time, and thus they

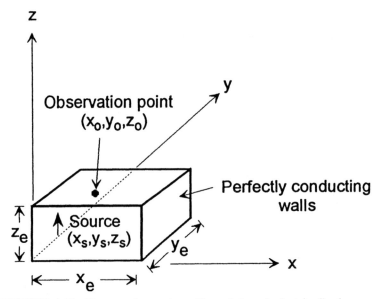

FIGURE 4.14 Rectangular cavity with an internal electric dipole source.

may be neglected in the sum. Of course, the frequency-domain response as calculated by the truncated image series will be in error, as the series has not converged sufficiently. However, the transient response will be accurate up to the time of arrival of the contribution from the furthest image source contained within the imaginary sphere.

As an example of an internal E-field response computed using the image Green's function, consider a cavity with overall dimensions $(x_e, y_e, z_e) =$

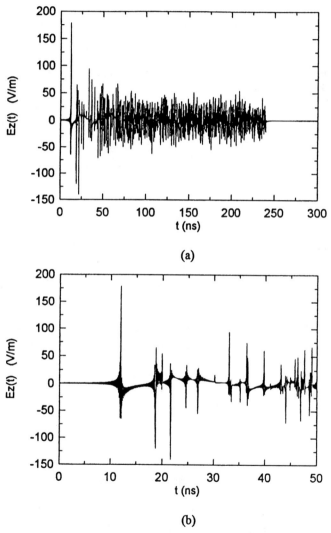

FIGURE 4.15 Transient E_z-field at $(x_0, y_0, z_0) = (5, 3, 0)$ m due to a pulsed dipole source at $(x_s, y_s, z_s) = (8.1, 2.45, 1.7)$ m: (*a*) late-time response; (*b*) early-time response.

(16.2, 4.9, 3.4) meters. Figure 4.15 shows the transient E_z field at a location $(x_0, y_0, z_0) = (5, 3, 0)$ due to a pulsed z-directed electric current element positioned at the center of a rectangular cavity at $(x_s, y_s, z_s) = (8.1, 2.45, 1.7)$ meters. For this calculation, a normalized transient electric dipole moment $I\,dl$ of the form

$$I(t)\,dl = I_0\,dl(e^{-\alpha t} - e^{-\beta t}) \tag{4.61}$$

was used, with $I_0\,dl = 1$, $\alpha = 4.0 \times 10^6\,\text{s}^{-1}$ and $\beta = 4.76 \times 10^8\,\text{s}^{-1}$. This transient waveform was transformed into the frequency domain using a fast Fourier transform (FFT), and the relevant E-field spectrum was computed using a sum of terms such as Eq. (4.14) at each frequency in the spectrum for the primary source and its images. The transient response was then computed using an inverse FFT.

Note that the jk term in the numerator of Eq. (4.14) amounts to a time derivative of the current element $I(t)\,dl$ in computing the radiated E-field. Thus the resulting E-field appears as the derivative of the current element. Also, observe in Figure 4.15a that at about $t = 230$ ns, the E-field response goes to zero, due to the fact that no further images in the enclosure walls have been considered. Beyond this cutoff time, the numerical solution is obviously in error.

Figure 4.15b illustrates the early-time behavior of the E-field. The initial spike in the response is due to the direct contribution from the dipole source. The subsequent responses arise from each of the images, each coming with a distinct time and amplitude, depending on the source, observation point, and enclosure geometry. A similar behavior for the tangential components of the H-field may be noted.

This computational approach involving images is very good for the early-time or high-frequency responses. Considering only the first 100 ns of a response, there will be only a small number of image sources contributing to the transient response. However, if the response at times on the order of micro- or milliseconds is desired, hundreds of thousands of images must be taken. The resulting response is very slowly converging and time consuming to evaluate. An alternative way of representing the fields is required in this case.

4.2.4.3.2 Eigenmode Solution. An alternative to the image method is to expand the cavity fields in terms of the eigenmodes of the cavity. This leads again to a triply infinite series that must be evaluated. Not only does this solution converge better at low frequencies, but it can be analytically transformed to a doubly infinite series and this leads to an even more rapid solution. Reference [4.24] presents a detailed derivation of the Green's functions for the rectangular cavity and presents results in a form useful for a direct numerical calculation. For a cavity having dimensions (x_e, y_e, z_e), the electric field produced by the *z-directed* dipole is given by

$$E(\mathbf{r}_0) = \frac{j\omega\mu}{k^2} I\,dl\,\delta(\mathbf{r}_0 - \mathbf{r}_s)\hat{z} - \frac{j\omega\mu}{k^2} I\,dl \sum_{n=0}^{\infty}\sum_{m=0}^{\infty}\sum_{l=0}^{\infty}$$

$$\frac{\epsilon_{0n}\epsilon_{0m}\epsilon_{0l}}{x_e y_e z_e [k^2 - (n\pi/x_e)^2 - (m\pi/y_e)^2 - (l\pi/z_e)^2]} \quad (4.62)$$

$$\times \left\{ \left[\left(\frac{n\pi}{x_e}\right)^2 + \left(\frac{m\pi}{y_e}\right)^2\right] \sin\frac{n\pi x_0}{x_e} \sin\frac{n\pi x_s}{x_e} \sin\frac{m\pi y_0}{y_e} \sin\frac{m\pi y_s}{y_e} \cos\frac{l\pi z_0}{z_e} \cos\frac{l\pi z_s}{z_e} \hat{z} \right.$$

$$- \frac{m\pi}{y_e}\left(\frac{l\pi}{z_e}\right) \sin\frac{n\pi x_0}{x_e} \sin\frac{n\pi x_s}{x_e} \cos\frac{m\pi y_0}{y_e} \sin\frac{m\pi y_s}{y_e} \sin\frac{l\pi z_0}{z_e} \cos\frac{l\pi z_s}{z_e} \hat{y}$$

$$\left. - \frac{n\pi}{x_e}\left(\frac{l\pi}{z_e}\right) \cos\frac{n\pi x_0}{x_e} \sin\frac{n\pi x_s}{x_e} \sin\frac{m\pi y_0}{y_e} \sin\frac{m\pi y_s}{y_e} \sin\frac{l\pi z_0}{z_e} \cos\frac{l\pi z_s}{z_e} \hat{x} \right\}$$

In this expression, the term $\epsilon_{0n} = 1$ for $n = 0$, $= 2$ for $n \neq 0$, and so on. The first term containing the δ-function is required only when the field is evaluated within the source region. For our purposes, we will be evaluating the fields away from the source, so only the triple-sum term will be required. A similar expression is reported in [4.24] for the Green's tensor for the magnetic field.

Although the expression in Eq. (4.62) gives an explicit equation for the *E*-field, it is computationally unwieldy, due to the very slow convergence of the sums. Reference [4.25] has studied the convergence of this series and reports needing 90,000 terms in the series in some cases. An alternative to a brute-force summation of the entire series is to use the summation formulas [4.22]

$$\sum_1^{\infty} \frac{\cos nx}{n^2 - a^2} = \frac{1}{2a^2} - \frac{\pi}{2a}\frac{\cos(x-\pi)a}{\sin \pi a} \quad \text{for } 0 \leq x \leq 2\pi \quad (4.63a)$$

$$\sum_1^{\infty} \frac{n \sin nx}{n^2 - a^2} = \frac{\pi}{2}\frac{\sin(\pi - x)a}{\sin \pi a} \quad \text{for } 0 \leq x \leq 2\pi \quad (4.63b)$$

to reduce the sums to two-dimensional series. Equation (4.63a) may be applied to the $\hat{z}$ component, and Eq. (4.63b) may be applied to the $\hat{y}$ and $\hat{x}$ terms. In doing this the expressions for the *E*-field components become

$$E_z(x_0, y_0, z_0) = \frac{-j\omega\mu I\,dl}{k^2(x_e y_e z_e)} \sum_{n=0}^{\infty}\sum_{m=0}^{\infty} \Gamma_{nm}\left(\frac{z_e}{2k'}\right)$$

$$\times \frac{\cos k'(z_0 + z_s - z_e) + \cos k'(|z_0 - z_s| - z_e)}{\sin k' z_e} \quad (4.64a)$$

with

$$\Gamma_{nm} = \epsilon_{0n}\epsilon_{0m}\left[\left(\frac{n\pi}{x_e}\right)^2 + \left(\frac{m\pi}{y_e}\right)^2\right]\sin\frac{n\pi x_0}{x_e}\sin\frac{n\pi x_s}{x_e}\sin\frac{m\pi y_0}{y_e}\sin\frac{m\pi y_s}{y_e}$$

(4.64b)

$$E_y(x_0, y_0, z_0) = \frac{-j\omega\mu I\,dl}{k^2(x_e y_e z_e)} \sum_{n=0}^{\infty}\sum_{m=0}^{\infty} \Gamma'_{nm}\left(\frac{z_e}{2}\right)$$

$$\times \frac{\sin k'[z_e - (z_0 + z_s)] + \text{sgn}(z_0 - z_s)\sin k'(z_e - |z_0 - z_s|)}{\sin k' z_e}$$

(4.64c)

with

$$\Gamma'_{nm} = \epsilon_{0n}\epsilon_{0m}\left(\frac{m\pi}{y_e}\right)\sin\frac{n\pi x_0}{x_e}\sin\frac{n\pi x_s}{x_e}\cos\frac{m\pi y_0}{y_e}\sin\frac{m\pi y_s}{y_e} \quad (4.64d)$$

$$E_x(x_0, y_0, z_0) = \frac{-j\omega\mu I\,dl}{k^2(x_e y_e z_e)} \sum_{n=0}^{\infty}\sum_{m=0}^{\infty} \Gamma''_{nm}\left(\frac{z_e}{2}\right)$$

$$\times \frac{\sin k'[z_e - (z_0 + z_s)] + \text{sgn}(z_0 - z_s)\sin k'(z_e - |z_0 - z_s|)}{\sin k' z_e}$$

(4.64e)

with

$$\Gamma''_{nm} = \epsilon_{0n}\epsilon_{0m}\left(\frac{n\pi}{x_e}\right)\cos\frac{n\pi x_0}{x_e}\sin\frac{n\pi x_s}{x_e}\sin\frac{m\pi y_0}{y_e}\sin\frac{m\pi y_s}{y_e} \quad (4.64f)$$

In these expressions, the term k' is defined as

$$k' = \sqrt{k^2 - \left(\frac{n\pi}{x_e}\right)^2 - \left(\frac{m\pi}{y_e}\right)^2} \quad (4.65)$$

with $k = \omega/c$. Note that for small values of the indices n and m the value of k' is real if $k^2 > (n\pi/x_e)^2 + (m\pi/y_e)^2$. However, as n and m increase, the value of k' eventually becomes imaginary, the sine and cosine terms in Eqs. (4.64) become exponentially damped, and the series begins to converge very rapidly. Consequently, for low frequencies, the series in Eqs. (4.64) are very efficient for evaluating the cavity fields.

To illustrate the calculation of the cavity fields using Eqs. (4.64), consider again the cavity with the dimensions $(x_e, y_e, z_e) = (16.2, 4.9, 3.4)$ meters with a time-harmonic, z-directed electric current element at the center $(x_s, y_s, z_s) = (8.1, 2.45, 1.7)$ meters. Figure 4.16 shows the spectrum of the E_z-field component at the observation point $(x_0, y_0, z_0) = (5, 3, 0)$ meters. Notice that there are resonances in this response occurring at frequencies corresponding

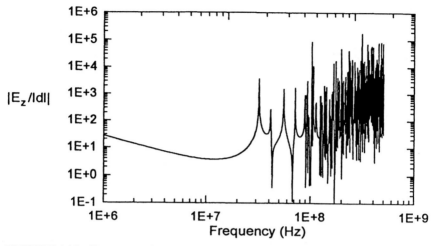

FIGURE 4.16 Frequency-domain spectrum of the E_z-field at $(x_0, y_0, z_0) = (5, 3, 0)$ m produced by a z-directed current element at $(x_s, y_s, z_s) = (8.1, 2.45, 1.7)$ m in a rectangular box enclosure.

to the zeros of the denominator term in Eq. (4.62):

$$f = \frac{c}{2}\sqrt{\left(\frac{n}{x_e}\right)^2 + \left(\frac{m}{y_e}\right)^2 + \left(\frac{l}{z_e}\right)^2} \qquad n, m, l = 0, 1, 2, \ldots. \qquad (4.66)$$

Because there is no loss in the problem, the responses at the resonances are infinite.

4.2.4.4 Electric Dipoles near a Sphere. Another source and object configuration that is useful for EMC modeling is that of a dipole source near a conducting sphere. This can be used for determining the perturbation of the radiation pattern of an emitter located near a conducting body. Figure 4.17 illustrates the geometry for this example. The dipole source is assumed to be radially directed and located along the z-axis, at a distance b from the sphere center. The sphere has a radius a and is assumed to be perfectly conducting. For this geometry it is desired to determine the E-field components at the observation location P given by the coordinates (r_0, θ_0, ϕ_0). Once this field is determined, a simple shift of the coordinate system will permit the determination of the fields produced by a radially directed dipole source at coordinates (b, θ_s, ϕ_s) [4.26].

The analysis of this problem is discussed in detail by Jones [4.10] for both perfectly conducting and lossy spheres. Moreover, this reference treats dipole sources having components in all three spatial directions. In the present discussion, however, we consider only the radially directed current source. The reader is referred to [4.10] for the expressions for the current sources in other orientations.

148 RADIATION MODELS FOR WIRE ANTENNAS

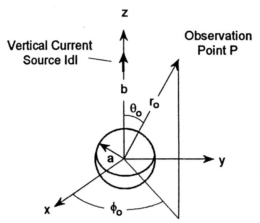

FIGURE 4.17 Radial electric dipole located near a perfectly conducting sphere.

The solution for this problem involves the introduction of spherical Bessel and Hankel functions, $j_n(kr)$ and $h_n^{(2)}(kr)$ and the associated Legendre polynomials $P_n^m(\cos\theta)$, all of which are described in [4.27]. For the radially directed dipole source at the pole of the sphere, there is axial symmetry around the z-axis, so the fields do not depend on the angle ϕ. Assuming the case of a perfectly conducting sphere, the E-field components in spherical coordinates are

$$E_r(r_0, \theta_0, \phi_0) = -\frac{Z_0}{4\pi} I\,dl \sum_{n=1}^{\infty} n(n+1)(2n+1) P_n(\cos\theta_0) \frac{1}{br_0} \quad (4.67a)$$

$$\times \left\{ h_n^{(2)}(kr_>) j_n(kr_<) - \frac{\dfrac{d}{d(ka)}[ka j_n(ka)]}{\dfrac{d}{d(ka)}[ka h_n^{(2)}(ka)]} h_n^{(2)}(kb) h_n^{(2)}(kr_0) \right\}$$

$$E_\theta(r_0, \theta_0, \phi_0) = -\frac{Z_0}{4\pi} I\,dl \sum_{n=1}^{\infty} (2n+1) \frac{dP_n(\cos\theta_0)}{d\theta_0} \frac{1}{br_0}$$

$$(4.67b)$$

$$\times \left\{ j_n(kr_<) \frac{d}{d(kr_>)}[kr_> h_n^{(2)}(kr_>)] \right.$$

$$\left. - h_n^{(2)}(kb) \frac{\dfrac{d}{d(ka)}[ka j_n(ka)]}{\dfrac{d}{d(ka)}[ka h_n^{(2)}(ka)]} \frac{d}{d(kr_0)}[kr_0 h_n^{(2)}(kr_0)] \right\}$$

$$E_\phi(r_0, \theta_0, \phi_0) = 0 \quad (4.67c)$$

where $r_>$ denotes the larger of r_0 and b and $r_<$ is the smaller of r_0 and b.

Frequently, we are interested in the far fields produced by the source and sphere. For this case the large-argument expression for the Hankel function can be used:

$$h_n^{(2)}(kr) \underset{kr \to \infty}{=} j^{n+1} \frac{e^{-jkr}}{kr} \quad (4.68)$$

Substituting this expression into Eq. (4.67), retaining only the $1/r_0$ terms for the radiated field, suppressing the phase term $\exp(-jkr_0)$, and using the fact that $dP_n(\cos\theta)/d\theta = P_n^1(\cos\theta)$ yields the following radiated field components:

$$r_0 E_\theta(r_0, \theta_0, \phi_0) = -\frac{Z_0}{4\pi b} I\, dl \sum_{n=1}^{\infty} j^n (2n+1) P_n^1(\cos\theta_0)$$

$$\times \left\{ j_n(kb) - \frac{\frac{d}{d(ka)}[kaj_n(ka)]}{\frac{d}{d(ka)}[kah_n^{(2)}(ka)]} h_n^{(2)}(kb) \right\} \quad (4.69a)$$

$$E_r(r_0, \theta_0, \phi_0) = E_\phi(r_0, \theta_0, \phi_0) = 0 \quad (4.69b)$$

For the special case of the dipole source located on the spherical surface, $b = a$ and Eq. (4.69a) reduces to

$$r_0 E_\theta(r_0, \theta_0, \phi_0) = \frac{Z_0}{4\pi ka^2} I\, dl \sum_{n=1}^{\infty} j^{n+1}(2n+1) \frac{P_n^1(\cos\theta_0)}{\frac{d}{d(ka)}[kah_n^{(2)}(ka)]}$$

$$(4.70)$$

Reference [4.28] discusses the analysis of a wire antenna located radially on a sphere and excited by a voltage source located between the sphere and the antenna. This analysis uses the expression for the E-field in Eq. (4.67a) to develop the integral equation for the unknown current distribution $I(r_s)$, which is solved using the moment method. Because Eq. (4.67) already takes into account the currents flowing on the sphere, the range of integration for the integral equation is only over the wire antenna.

Once the current is determined, the radiated field is calculated by integrating Eq. (4.69a) over the variable b along the antenna. Figure 4.18 shows the spatial dependence of the normalized radiated power (i.e., a plot of $|E_\theta|^2$, normalized to a peak value of unity) for this sphere–antenna combination for a quarter-wave antenna attached to the sphere at the $\theta = 0°$ axis. Values of the sphere radius a ranging from $a = 0.1\lambda$ to 0.6λ are considered. Note that because of the axial symmetry of the geometry, these

150 RADIATION MODELS FOR WIRE ANTENNAS

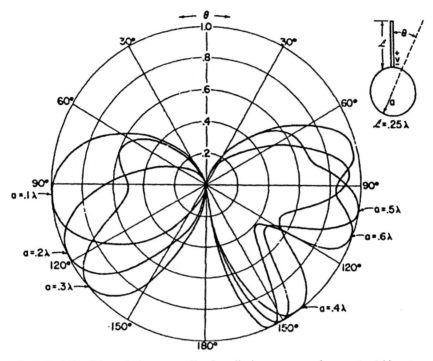

FIGURE 4.18 Plot of the normalized radiation pattern for a $\mathscr{L} = \lambda/4$ antenna mounted on a perfectly conducting sphere of radius a. (From [4.28] © 1976 Institute of Electrical and Electronics Engineers.)

radiation patterns are also symmetric around the polar axis. In the figure, only one-half of each radiation pattern is shown.

4.2.4.5 Electric Dipoles over an Imperfectly Conducting Earth.

Figure 4.19 illustrates another useful configuration for practical problems involving a lossy earth. Here an arbitrarily orientated electric dipole is located at a height z_s over the earth having an electrical conductivity σ and a relative dielectric constant ϵ_r. The theoretical development for determining the fields produced by such a current element has been described in [4.29–4.31]. This theory is well established and will not be rederived here; only the results are summarized.

As discussed by Norton [4.32], the rigorous formulation for the E-fields produced by this current element is described by the formulation of Sommerfeld. This solution can be approximated by considering wave contributions directly from the current element, from a fictitious image source in the ground, and from a lateral or "surface" wave. The direct contribution from the dipole source and the image contribution are the same as for the dipole over the perfect ground, except for the fact that the image contribution is reduced slightly by a multiplicative function that depends on

4.2 RADIATION OF ELECTROMAGNETIC FIELDS IN THE FREQUENCY DOMAIN

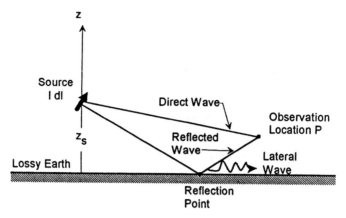

FIGURE 4.19 Dipole source over a lossy earth.

the earth properties, the frequency, and the geometry. It is known that as the earth becomes perfectly conducting, the lateral wave contribution vanishes, and the multiplicative function becomes unity, so that the image term becomes the usual image representing the reflection of the fields in the perfect ground, as discussed in Section 4.2.4.1.

As for the case of the dipole over the perfectly conducting plate, the arbitrarily oriented current element is decomposed into a vertical and a horizontal component, as shown in Figure 4.10. For the vertical dipole source, the E-field is again rotationally symmetric about the z-axis. In the far field, where the static and quasistatic (induction) fields may be neglected, the radiated E-field components (in cylindrical coordinates) produced at P are given in [4.19] as

$$E_z \approx -j30kI\,dl \left\{ \cos^2\psi_1 \frac{e^{-jkr_1}}{r_1} + R_v \cos^2\psi_2 \frac{e^{-jkr_2}}{r_2} \right. $$
$$\left. + \left[(1 - R_v)(1 - u^2 + u^4 \cos^2\psi_2)F(\chi) \frac{e^{-jkr_2}}{r_2} \right] \right\} \quad (4.71a)$$

$$E_\rho \approx j30kI\,dl \left\{ \cos\psi_1 \sin\psi_1 \frac{e^{-jkr_1}}{r_1} + R_v \cos\psi_2 \sin\psi_2 \frac{e^{-jkr_2}}{r_2} \right.$$
$$\left. - \left[(1 - R_v)u\sqrt{1 - u^2 \cos^2\psi_2}\left(1 + \frac{\sin^2\psi_2}{2}\right)F(\chi) \frac{e^{-jkr_2}}{r_2} \right] \right\} \quad (4.71b)$$

In these expressions, the first term corresponds to the direct contribution from the source, the second term is from the image, and the third term corresponds to the lateral wave. The term R_v is the Fresnel plane-wave reflection coefficient for a vertically polarized wave, which is discussed in standard texts [4.19,4.33]. This coefficient is expressed as

$$R_v = \frac{\epsilon_r\left(1 + \frac{\sigma}{j\omega\epsilon_r\epsilon_0}\right)\sin\psi_2 - \sqrt{\epsilon_r\left(1 + \frac{\sigma}{j\omega\epsilon_r\epsilon_0}\right) - \cos^2\psi_2}}{\epsilon_r\left(1 + \frac{\sigma}{j\omega\epsilon_r\epsilon_0}\right)\sin\psi_2 + \sqrt{\epsilon_r\left(1 + \frac{\sigma}{j\omega\epsilon_r\epsilon_0}\right) - \cos^2\psi_2}} \qquad (4.72)$$

More will be said about this reflection coefficient in Chapter 7, where plane-wave reflection from a lossy earth is discussed.

In Eq. (4.71), the terms r and ψ are the same as those defined in Eq. (4.53), and the following definitions are used: $u^2 = 1/[\epsilon_r - j(\sigma/\omega\epsilon_0)]$, $F(\chi) = 1 - j\sqrt{\pi\chi}\,e^{-\chi}\,\text{erfc}(j\sqrt{\chi})$, and the parameter χ is

$$\chi = -\frac{1}{2}jkr_2 u^2(1 - u^2\cos^2\psi_2)\left(1 + \frac{\sin\psi_2}{u\sqrt{1 - u^2\cos^2\psi_2}}\right) \qquad (4.73)$$

The E-fields radiated by a horizontal current element are slightly more complicated, again due to the absence of symmetry about the z-axis. In Cartesian coordinates, the E-fields in the radiation zone due *only to the primary and image sources* are

$$E_z \approx j30kI\,dl\cos\phi\left[\sin\psi_1\cos\psi_1\frac{e^{-jkr_1}}{r_1} - R_v\sin\psi_2\frac{e^{-jkr_2}}{r_2}\right] \qquad (4.74a)$$

$$E_x \approx -j30kI\,dl\left[\cos^2\phi\left(\sin^2\psi_1\frac{e^{-jkr_1}}{r_1} - R_v\sin^2\psi_2\frac{e^{-jkr_2}}{r_2}\right)\right.$$
$$\left. + \sin^2\phi\left(\frac{e^{-jkr_1}}{r_1} + R_h\frac{e^{-jkr_2}}{r_2}\right)\right] \qquad (4.74b)$$

$$E_y \approx -j30kI\,dl\sin\phi\cos\phi\left[\left(\sin^2\psi_1\frac{e^{-jkr_1}}{r_1} - R_v\sin^2\psi_2\frac{e^{-jkr_2}}{r_2}\right)\right.$$
$$\left. + \left(\frac{e^{-jkr_1}}{r_1} + R_h\frac{e^{-jkr_2}}{r_2}\right)\right] \qquad (4.74c)$$

In addition to the vertical reflection coefficient R_v introduced in Eq. (4.72), these expressions use a horizontally polarized reflection coefficient R_h, which is also given in [4.19] and [4.33] as

$$R_h = \frac{\sin\psi_2 - \sqrt{\epsilon_r\left(1 + \frac{\sigma}{j\omega\epsilon_r\epsilon_0}\right) - \cos^2\psi_2}}{\sin\psi_2 + \sqrt{\epsilon_r\left(1 + \frac{\sigma}{j\omega\epsilon_r\epsilon_0}\right) - \cos^2\psi_2}} \qquad (4.75)$$

In addition to the fields given by Eq. (4.74), the lateral-wave contribution must be included. This wave can be expressed in cylindrical coordinates as [4.19]

$$E_z \approx j30kI\,dl\left[\cos\phi\, u\sqrt{1 - u^2\cos^2\psi_2}(1 - R_v)(1 - u^2 + u^4\cos^2\psi_2)F(\chi)\frac{e^{-jkr_2}}{r_2}\right]$$

$$\times \cos\psi_2\left(1 + \frac{\sin^2\psi_2}{2}\right) \qquad (4.76a)$$

$$E_\rho \approx j30kI\,dl\left[\cos\phi\, u\sqrt{1 - u^2\cos^2\psi_2}(1 - R_v)(1 - u^2 + u^4\cos^2\psi_2)F(\chi)\frac{e^{-jkr_2}}{r_2}\right]$$

$$\times u\sqrt{1 - u^2\cos^2\psi_2}\,\frac{1 - \sin^2\psi_2 - [u^{-2}(1 - R_h)/(1 - R_v)][F(\chi')/F(\chi)]}{1 - u^2\cos^2\psi_2}$$

$$(4.76b)$$

$$E_\phi \approx j30kI\,dl\left[\cos\phi\, u\sqrt{1 - u^2\cos^2\psi_2}(1 - R_v)(1 - u^2 + u^4\cos^2\psi_2)F(\chi)\frac{e^{-jkr_2}}{r_2}\right]$$

$$\cdot \sin\phi(1 - R_h)F(\chi') \qquad (4.76c)$$

The terms k, u, ξ, $F(\chi)$, and R_v have been defined previously for the surface wave in Eq. (4.71). In Eq. (4.76), the parameter χ' is given by

$$\chi' = \frac{-jkr_2(1 - u^2\cos^2\psi_2)}{2u^2}\left(1 + \frac{\sin\psi_2}{u\sqrt{1 - u^2\cos^2\psi_2}}\right)^2 \qquad (4.77)$$

4.2.5 Evaluation of Magnetic Field Components

In earlier sections, the E-field was taken as the principal observation quantity. The magnetic field is also of interest, especially since in experimental programs, measurements of the H-field are often more accurate than for the E-field. The evaluation of explicit expressions for the H-field from a radiating current element in the presence of the various conducting bodies treated in the preceding section is a relatively easy task. However, an alternative approach is to evaluate the total H-field from a knowledge of the E-field and the Maxwell equation (4.1a).

In Cartesian coordinates, the individual H-field components are

$$H_x = \frac{j}{\omega\mu}\left(\frac{\partial E_z}{\partial y} - \frac{\partial E_y}{\partial z}\right) \tag{4.78a}$$

$$H_y = \frac{j}{\omega\mu}\left(\frac{\partial E_x}{\partial z} - \frac{\partial E_z}{\partial x}\right) \tag{4.78b}$$

$$H_z = \frac{j}{\omega\mu}\left(\frac{\partial E_y}{\partial x} - \frac{\partial E_x}{\partial y}\right) \tag{4.78c}$$

Thus the H-field can be determined from the spatial derivatives of the E-field. From a numerical standpoint, this involves representing the derivatives by a finite-difference operator, which requires separate evaluations of the E-field at different spatial locations to compute the H-field. Note that as $\omega \to 0$ the expressions in Eq. (4.78) become undefined. This suggests that at low frequencies the evaluation of the H-fields will contain errors, due to the numerical evaluation process.

4.3 RECEPTION AND SCATTERING OF ELECTROMAGNETIC FIELDS IN THE FREQUENCY DOMAIN

4.3.1 General Considerations

The description of the electromagnetic reception, or scattering, problem involves solution of the same integral equation (4.30) as for the antenna radiation problem, but with a different incident E-field exciting the antenna [4.1]. For the radiating antenna, the incident E-field exists only over a small region on the antenna, corresponding to the location of a driving voltage source. In the case of the scatterer or receiving antenna, the incident E-field covers the entire antenna and has a spatial dependence characteristic of the type of incident field being considered (i.e., a plane wave, cylindrical wave, etc.).

Because the integral equation is the same in both the radiation and scattering cases, the resulting system impedance matrix of Eq. (4.36) obtained from applying the method of moments to the integral equation is also the same. The only difference between the two problems, therefore, is a change in the excitation vector, or forcing function $[V]$. As most of the computational time needed for a moment method solution of the integral equation is spent in filling and inverting the system impedance matrix, once the radiation problem has been solved, the scattering problem for the same structure can be solved with only a minimal additional effort.

4.3.2 Approximate Solution for the Thin Wire

In this section we examine the reception and scattering of the thin wire antenna, using the same approximate techniques used previously for the antenna problem. This problem can be solved more accurately using the moment method. As we will see, however, the approximate solution is reasonably accurate for engineering applications, takes only a fraction of the time needed for the complete solution, and provides insight into the characteristics of scattered fields.

4.3.2.1 Determination of the Induced Current.
For the case of a thin wire illuminated by a plane-wave E-field with an angle of incidence θ_i, as shown in Figure 4.7a, the current induced along the wire is again described approximately by Eq. (4.44), but now with the incident field being given by the expression

$$E_z^{inc} = E_0 \sin\theta_i e^{jkz\cos\theta_i} \tag{4.79}$$

Thus, the integral term in Eq. (4.44) becomes

$$\int_0^z E_z^{inc}(z')\sin k(z-z')\,dz' = \frac{E_0 e^{jkz\cos\theta_i}}{k\sin\theta_i} \tag{4.80}$$

and the current induced on the wire can be written as

$$I(z) = C_1 \sin kz + C_2 \cos kz - I_s e^{-jkz} \tag{4.81a}$$

where the term I_s is

$$I_s = \frac{j4\pi E_0}{Z_0 \Omega_0 k \sin\theta_i} \tag{4.81b}$$

Again requiring that C_1 and C_2 be determined such that the current vanishes at $z = 0$ and $z = \mathcal{L}$, the final expression for the current on the scatterer becomes

$$I(z) = I_s\left[(\cos kz - e^{jkz\cos\theta_i}) + (e^{jk\mathcal{L}\cos\theta_i} - \cos k\mathcal{L})\frac{\sin kz}{\sin k\mathcal{L}}\right] \tag{4.82}$$

As in the antenna case, there are periodic resonances in this current at the zeros of $\sin k\mathcal{L}$.

As an example of the use of these expressions, consider equipment connected asymmetrically to a wire antenna at $z = z_0$, as in Figure 4.20. Using the expression from Eq. (4.82) to calculate a short-circuit current source, I_{sc}, and Eq. (4.48) to calculate the input admittance of the antenna Y_{in}, the excited antenna can be represented by a Norton equivalent circuit,

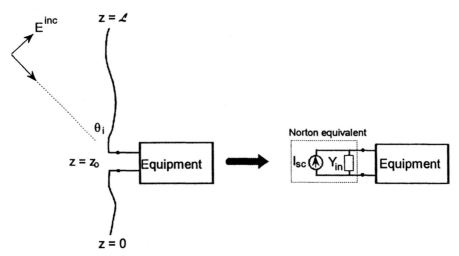

FIGURE 4.20 Example of EMI reception by equipment connected to a thin-wire antenna.

as indicated in the figure. This equivalent then can be used in an analysis of the equipment behavior due to the incident EM-field excitation.

4.3.2.2 Scattered Field.

The determination of the scattered field from the wire scatterer proceeds in exactly the same manner as for calculating the radiated field from the antenna. Since the scattered field produced by the incident field at angle θ_i can occur at an arbitrary observation angle θ_0, we have two distinct angles that define the problem. For cases when $\theta_i \neq \theta_0$, the process is referred to as *bistatic* scattering. For the special case when $\theta_i = \theta_0$, the scattering is called *monostatic*.

The scattered field is given by an integral similar to that in Eq. (4.22) as

$$r_0 E_\theta^{sca} = \frac{jkZ_0 \sin \theta_0}{4\pi} \int_0^{\mathscr{L}} I(z) e^{jkz \cos \theta_0} \, dz$$

$$= -\frac{E_0}{\Omega_0} \frac{\sin \theta_0}{\sin \theta_i} \frac{1}{\sin k\mathscr{L}} \int_0^{\mathscr{L}} [(\cos kz - e^{jkz \cos \theta_i}) \sin k\mathscr{L} \quad (4.83)$$

$$- (\cos k\mathscr{L} - e^{jk\mathscr{L} \cos \theta_i}) \sin kz] e^{jkz \cos \theta_0} \, dz$$

This integral can be evaluated analytically; however, the resulting expression is rather lengthy and will not be reproduced here. It is frequently desired to express the scattering in terms of the *radar cross section*, which is the ratio of

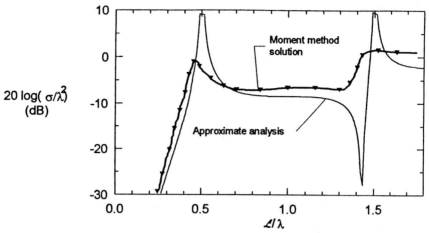

FIGURE 4.21 Plot of the normalized monostatic radar cross section of a thin wire with $\ell/a = 148.4$ for $\theta_i = \theta_0 = 90°$.

the scattered to incident power densities in the fields, as

$$\sigma(\theta_i, \theta_0) = \lim_{r_0 \to \infty} \left(4\pi r_0^2 \frac{P^{sca}}{P^{inc}} \right) = \lim_{r_0 \to \infty} \left(4\pi r_0^2 \frac{|E^{sca}|^2}{|E^{inc}|^2} \right) \qquad (4.84)$$

As an example of the scattered field calculated using this approximate method, and a comparison with the more accurate numerical results from [4.1] using the moment method, consider the case of a wire scatterer with a length/radius ratio $\ell/a = 148.4$ (see Figure 4.7a). For this calculation the incident field is broadside ($\theta_i = 90°$), and the backscatter field is calculated ($\theta_0 = 90°$). Figure 4.21 plots the normalized scattering cross section σ/λ^2 in dB for both methods. As can be noted, the agreement between the two solutions is not too bad away from the resonant frequencies of the wire. Near these frequencies, however, the lack of radiation resistance in the approximate model causes rather large errors. Nevertheless, the general trends in the data remain consistent for a wide range of frequencies.

4.4 ELECTRIC-FIELD INTEGRAL EQUATIONS IN THE TIME DOMAIN

4.4.1 Overview

Earlier we have examined the radiation and scattering process from the frequency-domain viewpoint. It is possible, however, to consider these phenomena directly in the time domain. As in the frequency domain, it is possible to develop an integral equation in the time domain that describes

4.4.2 The Integrodifferential Equation

For the thin-wire antenna or scatterer of arbitrary shape shown in Figure 4.22, the time-domain electric field integral equation (EFIE) for the induced current has been discussed in [4.34]. This equation is derived from the expression of the transient E-field tangent to the wire at an observation point P, arising from the wire current I as [4.35]

$$\mathbf{E}_{\tan}^{sca}(P, t) = \int_C L\{I(s', t')\hat{s}\} * g(r, r', t)\hat{s}' \, ds' \tag{4.85}$$

where the symbol * represents the convolution operator in the time domain. This E-field is denoted as E^{sca} because it is considered to be the field *scattered* by the wire. In this equation, the operator L is an integrodifferential operator, defined as

$$L = \mu_0 \frac{\partial}{\partial t} - \frac{1}{\epsilon_0} \nabla(\nabla \cdot) \int_0^{t'} d\tau \tag{4.86}$$

In Eq. (4.85) C is the curved axis of the wire, s and s' denote the coordinates of points P on the wire surface and P' on the wire axis, and $\hat{s}$ and $\hat{s}'$ are the unit vectors on the wire surface and on the wire axis, respectively.

The term $g(r,r',t)$ in Eq. (4.85) is an auxiliary Green's function [4.34], which is the solution of the equation

$$\nabla^2 g - \mu_0 \epsilon_0 \frac{\partial^2 g}{\partial t^2} = \delta(r - r', t - t') \tag{4.87}$$

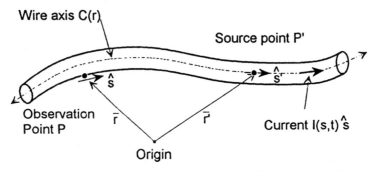

FIGURE 4.22 Thin wire showing unit vectors $\hat{s}$ and $\hat{s}'$ and the coordinate system.

and it is given by

$$g(r, t, r', t') = -\frac{1}{4\pi R} \delta\left(t - t' - \frac{R}{c}\right) \qquad (4.88)$$

where $R = |\mathbf{R}| = |\mathbf{r} - \mathbf{r}'|$, c is the velocity of light, and $\delta(x)$ is the Dirac delta function.

In the free space the scattered field of Eq. (4.85) can be expressed as

$$\mathbf{E}_{\tan}^{sca}(P, t) = \int_C \left[\mu_0 \frac{\partial I(s', t')}{\partial t} \hat{s} - \frac{1}{\epsilon_0} \nabla(\nabla \cdot) \int_0^{t'} I(s', \tau) \hat{s} \, d\tau \right] \frac{\delta(t - t' - R/c)}{4\pi R} \hat{s}' \, ds' \qquad (4.89)$$

Letting the observation point P approach the wire surface and using the boundary condition that $\hat{s} \cdot \mathbf{E}^{inc}(P, t) = -\hat{s} \cdot \mathbf{E}^{sca}(P, t)$ on the perfectly conducting wire results in the EFIE for the current in the thin wire. This equation is the time-domain analog of the frequency-domain equation of Eq. (4.30). This equation can be put into an alternative form which shows explicitly the terms of $1/R$ by expanding the differential operators. As discussed in [4.34], this EFIE becomes

$$\hat{s} \cdot \mathbf{E}^{inc}(r, t) = \frac{\mu_0}{4\pi} \int_C \left\{ \frac{\hat{s} \cdot \hat{s}'}{R} \frac{\partial}{\partial t} I(s', t') + c \frac{\hat{s} \cdot \mathbf{R}}{R^2} \frac{\partial}{\partial s'} I(s', t') \right.$$

$$\left. + c^2 \frac{\hat{s} \cdot \mathbf{R}}{R^3} \int_{-\infty}^{t'} \frac{\partial}{\partial s'} I(s', \tau) \, d\tau \right\} ds' \qquad (4.90)$$

4.4.3 Extension to a Wire over a Lossy Ground

The above time-domain expressions, valid for a wire in the free space, have been extended by Dafif and Jecko [4.36] for treating the case of a wire over a lossy ground. This approximate analysis is the time-domain counterpart to the time-harmonic method introduced by Miller et al. [4.37]. The geometry of this problem is shown in Figure 4.23. The analysis uses the fact that the contribution due to the reflection of a wave emitted by the dipole situated in the point P' on the wire axis follows the path $P'AP$, as if emitted by an image source situated at the point P^*. The contribution of the wave reflected on the soil at point A is given by the incident wave starting from point P^* convolved with a time-dependent reflection coefficient

$$d\mathbf{E}^{sca}(P, t) = L\left\{ \frac{sI(s', t' - R/c) \, ds'}{4\pi R} \right\} + [r(t)] * L\left\{ \frac{s^* I(s'^*, t'^* - R^*/c) \, ds'^*}{4\pi R^*} \right\} \qquad (4.91)$$

where $[r(t)]$ is the matrix of the reflection coefficients of the field radiated by

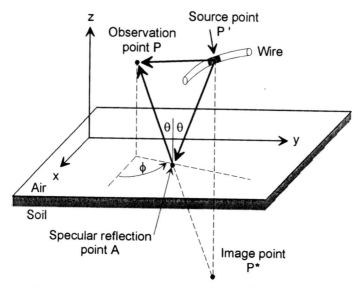

FIGURE 4.23 Incident and reflected waves for a thin wire over a lossy ground plane.

the thin wire [4.36]:

$$[r(t)] = \begin{bmatrix} r_v(t)\cos\phi' & -r_h(t)\sin\phi' & 0 \\ -r_v(t)\sin\phi' & r_h(t)\cos\phi' & 0 \\ 0 & 0 & r_v(t) \end{bmatrix} \quad (4.92)$$

with $r_v(t) = \mathcal{F}^{-1}[R_v(\omega)]$ and $r_h(t) = \mathcal{F}^{-1}[R_h(\omega)]$, the inverse Fourier transforms of the reflection coefficients for a vertically and horizontally polarized waves, respectively, which are given in Eqs. (4.72 and 4.75). Approximate expressions for both $r_v(t)$ and $r_h(t)$ are discussed in Section 8.4.2.2.

Applying the reflection coefficient to Eq. (4.90), the EFIE for a wire over a finite conducting ground becomes

$$\hat{s} \cdot \mathbf{E}^{\text{inc}}(r',t') = \frac{\mu_0}{4\pi} \int \left\{ \frac{\hat{s}\cdot\hat{s}'}{R} \frac{\partial}{\partial t} I(s',t_0') + c \frac{\hat{s}\cdot\mathbf{R}}{R^2} \frac{\partial}{\partial s} I(s',t_0') \right.$$

$$+ c^2 \frac{\hat{s}\cdot\mathbf{R}}{R^3} \int_0^{t_0'} \frac{\partial}{\partial s} I(s',\tau)\,d\tau$$

$$+ \frac{\hat{s}\cdot\hat{s}'*}{R} [r(t*)] \frac{\partial}{\partial t*} I(s'*,t_0'*) + c \frac{\hat{s}\cdot\mathbf{R}*}{R*^2} [r(t*)] \frac{\partial}{\partial s*} I(s'*,t_0'*)$$

$$\left. + c^2 \frac{\hat{s}\cdot\mathbf{R}*}{R*^3} [r(t*)] \int_0^{t'*} \frac{\partial}{\partial s} I(s'*,\tau)\,d\tau \right\} ds' \quad (4.93)$$

with $t'_0 = t' - R/c$, $t'^*_0 = t' - R^*/c$, $|\mathbf{R}| = |\mathbf{r} - \mathbf{r}'|$, and $|\mathbf{R}^*| = |\mathbf{r}^* - \mathbf{r}'^*|$.

The general form of the integral equation of a thin wire over a ground of finite conductivity can take different particular forms, depending on the nature of the soil. In the free space the reflection coefficients are equal to zero and Eq. (4.93) becomes equal to Eq. (4.90). For a perfect conducting soil, $R_v(\omega) = -1$ and $R_h(\omega) = 1$ (see Section 8.4.2.2), which gives $r_v(t) = -\delta(t)$ and $r_h(t) = \delta(t)$.

4.4.4 Numerical Solution for the EFIE for Thin Wires in the Time Domain

The EFIE in the time domain can be solved using the moment method, which has been introduced for the solution of the integral equation in the frequency domain (see Section 4.2.3.2.1). A specific computer code for time-domain solutions has been developed for thin wires in the free space by Van Blaricum [4.38] and by Landt et al. [4.39]. The case of the wire over a finite conducting ground has been considered by the University of Limoges group, where the Van Blaricum code has been substantially modified to take into account the presence of the soil [4.40].

The time-domain analysis approach has been applied to the study of different EMP simulators [4.41, 4.42], coaxial cables with nonlinear protection illuminated by an EMP [4.43], as well as to the coupling of electromagnetic waves to aircraft [4.44] and the analysis of the EM field radiated by distribution power lines [4.45].

4.5 SINGULARITY EXPANSION METHOD

4.5.1 Background

The singularity expansion method (SEM) is a technique developed by Baum [4.46] and other investigators [4.47] and is an extension of the integral equation solutions to complex frequencies: $j\omega \to s$. It is useful because it permits the representation of antenna and scatterer behavior by a few parameters and provides analytical representations for frequency- and time-domain behavior. Moreover, it provides insight into fundamentals of the radiation process.

SEM resulted from observations by Baum, and earlier by Stratton [4.4] and Schelkunoff [4.48], that the frequency-domain responses of antennas and scatterers have resonances associated with characteristic frequencies of the antenna structure (see Figure 4.9 for an example). Experimental evidence shows that these resonant frequencies are independent of the excitation, depending only on the size and shape of the antenna. From this observation came the idea that an antenna, like a lumped RLC circuit, can be analyzed in terms of its pole–zero pattern in the complex frequency plane.

To illustrate the concepts of SEM, it is useful to consider an example of a resonant circuit from conventional circuit theory, as shown in Figure 4.24. This is a series RLC circuit with a step-function excitation of V_0 volts occurring at $t = 0$. It is assumed that the initial conditions at $t = 0^-$ are all zero, and we wish to calculate the transient current through the circuit for $t > 0$.

The phasor expression for the current flowing in the circuit is given by

$$I(j\omega) = \frac{V(j\omega)}{R + j\omega L + 1/j\omega C} \tag{4.94}$$

If the voltage spectrum $V(j\omega)$ is specified, the expression for $I(j\omega)$ can be Fourier transformed numerically to find $i(t)$, an approach that is frequently used for circuit as well as antenna problems.

An alternative approach to the direct numerical evaluation is to use Laplace transform techniques [4.49] by continuing the response expressions analytically into the complex frequency plane $s = \sigma + j\omega$. For the step-function excitation, the voltage spectrum is

$$V(j\omega) \underset{j\omega \to s}{=} V(s) = \frac{V_0}{s} \tag{4.95}$$

and the current $I(s)$ becomes

$$I(s) = \frac{V(s)}{R + sL + 1/sC} = \frac{s(V_0/s)}{Ls^2 + Rs + 1/C} = \frac{V_0}{(s - s_1)(s - s_2)}$$

$$= \frac{V_0}{s_1 - s_2}\left(\frac{1}{s - s_1} - \frac{1}{s - s_2}\right) \tag{4.96}$$

where s_1 and s_2 are the natural resonant frequencies of the circuit. These are expressed in terms of the circuit parameters as

$$s_1, s_2 = -\frac{R}{2L} \pm \sqrt{\left(\frac{R}{2L}\right)^2 - \frac{1}{LC}} \tag{4.97}$$

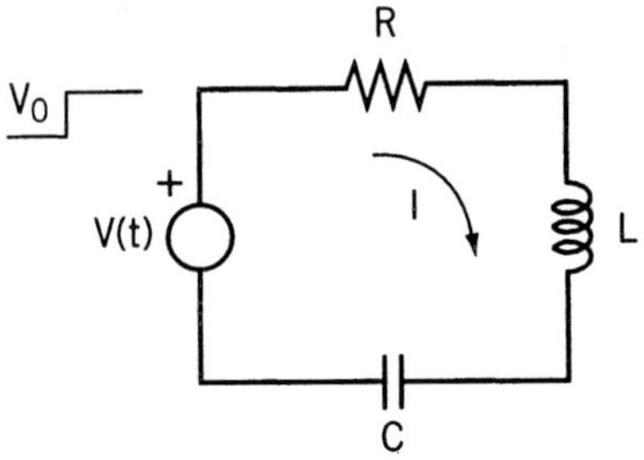

FIGURE 4.24 RLC circuit with a step-function excitation.

4.5 SINGULARITY EXPANSION METHOD

Noting that the inverse Laplace transform of a single-pole term may be computed analytically:

$$\mathcal{L}^{-1}\left(\frac{1}{(s-s_\alpha)}\right) = e^{s_\alpha t} \tag{4.98}$$

the step-function response for the current is determined as

$$i(t) = \frac{V_0}{s_1 - s_2}(e^{s_1 t} - e^{s_2 t})U(t) \tag{4.99}$$

where $U(t)$ is the Heaviside function: $U(t) = 0$ for $t \leq 0$, $=1$ for $t > 0$. This current response consists of two pole terms located in the frequency plane, with the residue of each pole providing the "strength" of the corresponding exponential waveform.

Consider a numerical example with the following parameters: $L = 1$, $R = 2$, $C = \frac{1}{5}$, which provide the frequencies $s_1, s_2 = -1 \pm j2$. For this circuit, the normalized frequency-domain current spectrum $|I(j\omega)/V_0|$ is shown in Figure 4.25. This response is obtained by evaluating Eq. (4.96) along the $j\omega$-axis. Note the resonance in the response spectrum near the frequency $\omega = 2$. Because the singularity in the response lies in the complex plane, off the $j\omega$-axis, the response is never truly singular.

From Eq. (4.99) the transient response can be evaluated as

$$i(t) = \frac{V_0 e^{-t}}{j4}(e^{j2t} - e^{-j2t})U(t) = \frac{V_0 e^{-t}}{2}\sin 2t \quad \text{for } t > 0 \tag{4.100}$$

and this is plotted in Figure 4.26. In this example we see that as long as the functional forms of the responses are known, the complete frequency-domain and time-domain responses can be determined by only a few numbers: the two complex frequencies, s_1 and s_2, and the voltage excitation

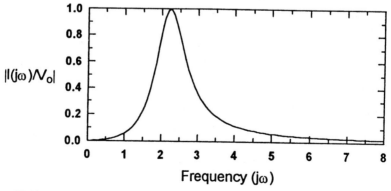

FIGURE 4.25 Spectral response for the current in the *RLC* circuit.

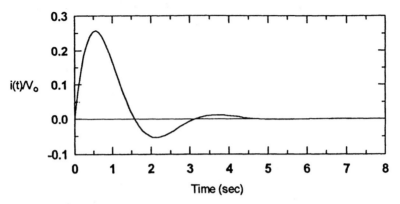

FIGURE 4.26 Step-function response of the RLC circuit.

term V_0. In a more general linear system, there might be additional multiplicative constants for each pole term, giving rise to two additional constants. This pole representation of the circuit response is very efficient. To represent the curves in Figure 4.25 or 4.26 directly, many data points would be needed.

SEM represents a successful attempt to extend this pole-residue analysis technique to the case of distributed circuits. In this manner, the much more complicated process of electromagnetic radiation and scattering from conducting bodies can be described in terms of complex poles in the s-plane and associated residues that are related to how the body is excited.

4.5.2 Mathematical Description of SEM

Although SEM can be developed for a general, three-dimensional conductor having a two-dimensional surface current, it is simpler to illustrate the use of SEM by treating a simple wire antenna excited by a voltage source, as shown in Figure 4.7b. As discussed in Section 4.2.3.2.1, the application of the moment method to the integral equation (4.30) allows us to write the equation in a matrix form as

$$[Z][I] = [V] \Rightarrow [I] = [Z]^{-1}[V] \qquad (4.101)$$

An investigation of the analytic properties of the impedance matrix $[Z]$ by Marin [4.50] showed that the *natural frequencies* (or resonant frequencies) s_α of the antenna are those frequencies where the determinant of $[Z]$ vanishes:

$$\det[Z(s_\alpha)] = 0 \qquad (4.102)$$

One way of physically thinking of this resonance is that at these frequencies a solution for the antenna current can exist with no excitation. Alternatively, if a source were somehow to excite the antenna at this complex frequency, the response would be infinite.

At each natural frequency the *natural current mode* $[M_\alpha(z)]$ is the current distribution existing on the wire and is obtained by solving the homogeneous equation

$$[Z]_{s=s_\alpha}[M_\alpha] = 0 \tag{4.103}$$

The *coupling vector* $[C_\alpha(z)]$ is the left-hand natural mode of the matrix $[Z]$ at the frequency s_α and satisfies the equation

$$[C_\alpha]^t[Z(s_\alpha)] = 0 \tag{4.104}$$

where the symbol t denotes the transpose. When the choice of basis and testing functions for the moment method are such that the matrix $[Z]$ is symmetric, it can be shown that $[C_\alpha] = [M_\alpha]$.

Under the assumption that the antenna has only first-order poles,[†] the solution for the current on the antenna $[I]$ as developed by Baum [4.46] can be written as singularity expansion of the natural resonance terms of the antenna, together with possible pole terms arising from the excitation function. In addition, because the antenna is a *distributed* system, there is the possibility of an additional entire function, denoted by $F_e(s)$, in the representation of the response.[‡] The possible requirement for such a function is evident from the Mittag–Leffler theorem [4.51], and this serves to distinguish the simple low-frequency circuit problems from the distributed antenna problems. A simple example of an entire function having no poles in the finite s-plane is the exponential function.

Using SEM, Baum shows how the current on a linear antenna of length ℓ can be written as an expansion over the singularities in the form

$$[I] = [Z]^{-1}[V]$$

$$= \sum_{\substack{\text{antenna}\\ \text{poles}}} \left[\left(\beta_\alpha \frac{[C_a]^t[V(s_\alpha)]}{s - s_\alpha} \right)[M_\alpha] \right]$$

$$+ \text{entire function } F_e(s) + \text{waveform pole terms } F_w(s) \tag{4.105a}$$

where β_α is a normalizing constant given by

$$(\beta_\alpha)^{-1} = [C_a]^t \left. \frac{\partial [Z]}{\partial s} \right|_{s=s_\alpha} [M_a] \tag{4.105b}$$

[†] Only first-order poles have been observed for isolated, loss-free bodies in free space.

[‡] A distributed system is one in which the propagation time across the system becomes important, as opposed to a lumped circuit, in which the propagation effects are negligible.

For the special case of the impulse response, in which the spectrum of the excitation function is a constant, there are no waveform pole terms.

In this expression it is convenient to define the *coupling coefficient* as the term

$$\eta_\alpha = [C_a]^t[V(s_\alpha)] \tag{4.106}$$

which contains all the information about the excitation of the antenna. Realizing that the discrete notation in Eqs. (4.105) and (4.106) represents continuous functions along the antenna, the SEM expression for the antenna current can be written as a continuous function as

$$I(s; z) = \sum_{\text{poles } \alpha} \beta_\alpha \frac{\eta_\alpha M_\alpha(z)}{s - s_\alpha} + F_e(s, z) + F_w(s, z) \tag{4.107a}$$

and the coupling coefficient becomes

$$\eta_\alpha = \int_0^L C_a(z')V(z'; s_\alpha)\, dz' \tag{4.107b}$$

As in the case of the circuit analogy, the transient response contribution from the pole terms can be calculated using Eq. (4.98). For a delta-function excitation (E^{inc} is a constant), the inverse Laplace transform of the current gives

$$i(t; z) = \sum_{\text{poles } \alpha} \beta_\alpha \eta_\alpha M_\alpha(z) e^{s_\alpha t} + \text{possible entire function contribution}$$

$$\tag{4.108}$$

Numerical examples of the calculation of the poles, modes, coupling coefficients, and resulting responses have been documented in [4.52].

4.5.3 SEM Representation of the Antenna Current

For the driven antenna with an impulsive voltage source at $z = z_s$, the excitation field is $E^{\text{inc}}(z) = V_0 \delta(z - z_s)$ and the coupling coefficient in Eq. (4.107b) becomes

$$\eta_\alpha = V_0 C_a(z_s) \tag{4.109}$$

The behavior of $F_e(s)$ has been difficult to determine, either from theoretical developments or from numerical studies. However, for the antenna excited at one location by an impulse it is suspected that $F_e(s) = 0$. Under this condition the current response is given by only the summation term:

$$I(s; z) = \sum_{\text{poles } \alpha} \beta_\alpha \frac{\eta_\alpha M_\alpha(z)}{s - s_\alpha} \quad (4.110)$$

and the resulting transient response for the antenna current is given by Eq. (4.108) with no entire function contribution.

4.5.4 SEM Representation of the Scattering Current

As suggested in Section 4.3.1, the scattering problem can be thought of as a superposition of driven antenna problems with the source moving along the length of the antenna and its strength varying according to the behavior of the incident field. Accordingly, for the E-field incident at an angle θ_i as shown in Figure 4.7a, the behavior of the field is described by Eq. (4.79). Integrating Eq. (4.105b) over this source distribution provides a coupling coefficient for the scattering case as

$$\eta_\alpha(s) = E_0 \sin \theta_i \int_0^L C_\alpha(z') e^{(sz'/c)\cos \theta_i} dz' \quad (4.111)$$

which is now a function of frequency, due to the exponential variation of the excitation field. As a result, the expression for the induced current in the wire is given by Eq. (4.110) with a numerator term depending on s:

$$I(s; z) = \sum_{\text{poles } \alpha} \beta_\alpha \frac{\eta_\alpha(s) M_\alpha(z)}{s - s_\alpha} \quad (4.112)$$

The frequency-dependent coupling coefficient in Eq. (4.112) is the *class 2* coefficient defined by Baum [4.46]. Because the original expansion term in Eq. (4.110) does not contain an entire function, the expansion for the scattering current lacks an additional term $F_e(s)$. However, it does contain a frequency-dependent numerator for each of the pole terms, and as a result, it cannot be immediately inverted into the time domain using residue calculus.

An alternative form for the current is possible by using the *class 1* coupling coefficients, which are the actual residues of the poles at s_α, together with an additive entire function to the series in Eq. (4.112). Consider writing the coupling coefficient as

$$\eta_\alpha(s) = \eta_\alpha(s_\alpha) + [\eta_\alpha(s) - \eta_\alpha(s_\alpha)] \quad (4.113)$$

where $\eta_\alpha(s_\alpha)$ is the frequency independent coupling coefficient. The current

168 RADIATION MODELS FOR WIRE ANTENNAS

then is given as a pole series with an additional entire function as

$$I(s, z) = \sum_{\text{poles }\alpha} \beta_\alpha \frac{\eta_\alpha(s_\alpha) M_\alpha(z)}{s - s_\alpha} + F_e(s, z) \quad (4.114)$$

where the entire function is expressed as

$$F_e(s, z) = \sum_{\text{poles }\alpha} \beta_\alpha \frac{[\eta_\alpha(s) - \eta_\alpha(s_\alpha)] M_\alpha(z)}{s - s_\alpha}$$

$$= E_0 \sin \theta_i \sum_{\text{poles }\alpha} \beta_\alpha M_\alpha(z) \frac{\int_0^L C_\alpha(z') (e^{(sz'/c)\cos \theta_i} - e^{(s_\alpha z'/c)\cos \theta_i}) \, dz'}{s - s_\alpha}$$

(4.115)

Its time-domain contribution to the current can be evaluated numerically.

4.5.5 SEM Representation of the Radiated Fields

To calculate the field radiated at an angle θ_0 by the impulse-excited antenna, the SEM representation of the current distribution in Eq. (4.110) can be used in the integral of Eq. (4.21) to yield

$$r_0 E_\theta = \frac{Z_0 \sin \theta_0}{4\pi} \frac{s}{c} \sum_{\text{poles }\alpha} \beta_\alpha \frac{\eta_\alpha}{s - s_\alpha} \int_0^L M_\alpha(z') e^{(sz'/c)\cos \theta_0} \, dz' \quad (4.116)$$

which again does not have an entire function in the expansion but does have a frequency-dependent numerator in each pole term. As in the case of the scattering current, the expression for the radiated field can be partitioned into a set of pole-residue terms and an entire function. Denoting the natural *far-field modes* [4.53] at each resonant frequency as $e_\alpha(\theta_0)$, where

$$e_\alpha(\theta_0) = \sin \theta_0 \int_0^L M_\alpha(z') e^{(s_\alpha z'/c)\cos \theta_0} \, dz' \quad (4.117)$$

the SEM representation for the radiated field becomes

$$r_0 E_\theta(s; \theta_0) = \frac{Z_0}{4\pi} \left[\sum_{\text{poles }\alpha} \frac{\beta_\alpha s_\alpha}{c} \frac{\eta_\alpha e_\alpha(\theta_0)}{s - s_\alpha} + F_e(s) \right] \quad (4.118)$$

4.5.6 SEM Representation of Scattered Fields

The treatment of the scattered fields proceeds in a similar manner. Replacing η_α in Eq. (4.116) by $\eta_\alpha(s)$ of Eq. (4.111) to account for a distributed excitation of the scatterer due to an incident E-field at an angle θ_i provides the scattered field at an angle θ_0. This expression can be put into a pole-residue expansion in exactly the same form as Eq. (4.117) but with a different entire function. Noting that $M_\alpha(z)$ and $C_\alpha(z)$ are the same for the wire antenna, we see that the integrals in Eqs. (4.111) and (4.117) are the same, except for the angle variable. Thus the coupling coefficient can be expressed in terms of the natural mode, a consequence of reciprocity. This SEM expression for the scattered field can be written as

$$r_0 E_\theta(s; \theta_0) = \frac{E_0 Z_0}{4\pi} \left[\sum_{\text{poles } \alpha} \frac{\beta_\alpha s_\alpha}{c} \frac{e_\alpha(\theta_i) e_\alpha(\theta_0)}{s - s_\alpha} + F_e(s) \right] \quad (4.119)$$

4.5.7 Example of SEM Applied to the Approximate Antenna Analysis

As a simple example of the use of SEM, let us reconsider the approximate solution for the antenna problem developed in Section 4.2.3.2.2. and discussed in [4.54]. Because this solution for the antenna current and radiated field can be expressed analytically, we will be able to obtain analytical expressions for the SEM terms. This is in contrast with the numerical SEM solution using the integral equation [4.52], in which only a numerical representation of the parameters is possible.

4.5.7.1 Induced Current for the Antenna Problem.
For the wire antenna of length $\mathcal{L}$, radius a, and a voltage source excitation V_0 at an arbitrary location $z = z_s$, the induced antenna current at an observation point $z = z_0$ is given in Eq. (4.46b) as

$$I(z_0) = j \frac{4\pi}{Z_0 \Omega_0} V_0 \frac{\sin kz_< \sin k(\mathcal{L} - z_>)}{\sin k\mathcal{L}} \quad (4.120)$$

where $z_<$ is the smaller of z_0 and z_s, $z_>$ is the larger of z_0 and z_s, and $\Omega_0 = 2 \ln(\mathcal{L}/a)$. To express this antenna current using SEM, we introduce the complex frequency variable $s = \sigma + j\omega$ and continue the expression from the $j\omega$-axis analytically onto the complex frequency plane, resulting in the following expression:

$$I(z_0) = \frac{4\pi}{Z_0 \Omega_0} V_0 \frac{\sinh(sz_</c) \sinh[s(\mathcal{L} - z_>)/c]}{\sinh(s\mathcal{L}/c)} \quad (4.121)$$

This response has simple poles at the complex frequency locations given by

$$s_\alpha = j \frac{\alpha \pi c}{\mathcal{L}} \quad \alpha = 0, \pm 1, \pm 2, \ldots \quad (4.122)$$

which correspond to the zeros of the $\sinh(s\mathcal{L}/c)$ term. Assuming that V_0 is a constant, Eq. (4.122) can be expressed as a sum of pole-residue terms [4.11] as

$$I(z_0) = \frac{4\pi V_0}{Z_0 \Omega_0} \sum_{\alpha=-\infty}^{\infty} \frac{R_\alpha}{s - s_\alpha} \tag{4.123}$$

where the residue is evaluated as

$$R_\alpha = \frac{\sinh(s_\alpha z_</c) \sinh[s_\alpha(\mathcal{L} - z_>)/c]}{\dfrac{d}{ds} \sinh(s\mathcal{L}/c)\Big|_{s=s_\alpha}} \tag{4.123b}$$

This yields the following SEM representation for the antenna current:

$$I(z_0) = \frac{4\pi V_0}{Z_0 \Omega_0} \frac{c}{\mathcal{L}} \sum_{\alpha=-\infty}^{\infty} \frac{\sin\left(\dfrac{\alpha\pi}{\mathcal{L}} z_s\right) \sin\left(\dfrac{\alpha\pi}{\mathcal{L}} z_0\right)}{s - s_\alpha} \tag{4.124}$$

This solution corresponds to Eq. (4.108), with the following definitions:

Natural frequencies: $\quad s_\alpha = j\dfrac{\alpha\pi c}{\mathcal{L}} \quad \alpha = 0, \pm 1, \pm 2, \ldots \tag{4.125a}$

Natural mode: $\quad M_\alpha(z_0) = \sin(\alpha\pi z_0/\mathcal{L}) \tag{4.125b}$

Coupling coefficient: $\quad \eta_\alpha = \displaystyle\int_0^{\mathcal{L}} V_0 \delta(z' - z_s) M_\alpha(z')\, dz' = V_0 \sin(\alpha\pi z_s/\mathcal{L}) \tag{4.125c}$

Normalization constant: $\quad \beta_\alpha = \dfrac{4\pi}{Z_0 \Omega_0} \dfrac{c}{\mathcal{L}} \quad$ (actually independent of α) $\tag{4.125d}$

Note that this expansion for the antenna current has a frequency-independent coupling coefficient with no entire function, as postulated for Eq. (4.110).

The expression in Eq. (4.124) yields the response spectrum of the antenna current for an impulsive voltage excitation $v(t) = V_0 \delta(t)$. Again using Eq. (4.80), together with the fact that the pole terms s_α occur in complex conjugate pairs, the inverse transform of Eq. (4.124) can be evaluated analytically to yield the unit impulse response (i.e., the response for $V_0 = 1$) as

$$i_{\text{impulse}}(z_0; t) = \frac{8\pi}{Z_0 \Omega_0} \frac{c}{\mathcal{L}} \sum_{\alpha=1}^{\infty} \sin\left(\frac{\alpha\pi}{\mathcal{L}} z_s\right) \sin\left(\frac{\alpha\pi}{\mathcal{L}} z_0\right) \cos\left(\frac{\alpha\pi c}{\mathcal{L}} t\right) \tag{4.126}$$

If the voltage source is described by a specified function of time, $v(t)$, the resulting antenna current can be obtained by a convolution with Eq. (4.126) as

$$i(z_0; t) = v(t) * i_{\text{impulse}}(z_0; t) = \int_{-\infty}^{t} v(\tau) i_{\text{impulse}}(z_0; t - \tau) \, d\tau \quad (4.127)$$

An alternative to this numerical approach is to perform the convolution in the s-domain analytically. Denote the Laplace transform of the transient excitation function $v(t)$ as $V(s)$. Assuming that the s-plane representation of $V(s)$ has only poles, the expansion of Eq. (4.121) in pole terms must include the waveform poles as well as the poles corresponding to the antenna resonances. Representing the poles of $V(s)$ as s_w, the resulting expansion for the step-function current response is

$$I(z_0) = \frac{4\pi}{Z_0 \Omega_0} \frac{c}{\mathcal{L}} \sum_{\alpha=-\infty}^{\infty} V(s_\alpha) \frac{\sin\left(\frac{\alpha\pi}{\mathcal{L}} z_s\right) \sin\left(\frac{\alpha\pi}{\mathcal{L}} z_0\right)}{s - s_\alpha}$$

$$+ \frac{4\pi}{Z_0 \Omega_0} \frac{c}{\mathcal{L}} \sum_{\substack{\text{waveform} \\ \text{poles } s_w}} \left[\sum_{\alpha=-\infty}^{\infty} \frac{\text{Res}(V(s))|_{s=s_w}}{s - s_w} \frac{\sin\left(\frac{\alpha\pi}{\mathcal{L}} z_s\right) \sin\left(\frac{\alpha\pi}{\mathcal{L}} z_0\right)}{s_w - s_\alpha} \right]$$

(4.128)

Thus, given the specific nature of the voltage excitation spectrum $V(s)$, each of the pole terms in Eq. (4.128) can be evaluated as an exponential in the time domain and the total response constructed as a sum over these functions.

As an example, consider the antenna of Figure 4.7b with $\mathcal{L} = 1$ m, $a = 0.5$ cm [$\Omega_0 = 2\ln(\mathcal{L}/a) = 10.6$], and with a step-function voltage source located at the midpoint of the antenna, $z_s = \mathcal{L}/2$. The step-function source $v(t) = V_0 U(t)$ has a Laplace transforms of $V(s) = V_0/s$, which contains a single waveform pole at $s_w = 0$. Inserting $V(s)$ into Eq. (4.128) and noting that since $s_w = 0$, the $+\alpha$ and $-\alpha$ terms in the waveform summation cancel identically, the step-excited antenna current spectrum is

$$I_{\text{step}}(z_0) = \frac{4\pi}{Z_0 \Omega_0} \frac{c}{\mathcal{L}} \sum_{\alpha=-\infty}^{\infty} \frac{V_0}{s_\alpha} \frac{\sin\left(\frac{\alpha\pi}{\mathcal{L}} z_s\right) \sin\left(\frac{\alpha\pi}{\mathcal{L}} z_0\right)}{s - s_\alpha}$$

(4.129a)

and the transient response is calculated directly to be

$$i_{\text{step}}(z_0; t) = \frac{8V_0}{Z_0 \Omega_0} \sum_{\alpha=1}^{\infty} \frac{1}{\alpha} \sin\left(\frac{\alpha\pi}{\mathcal{L}} z_s\right) \sin\left(\frac{\alpha\pi}{\mathcal{L}} z_0\right) \sin\left(\frac{\alpha\pi c}{\mathcal{L}} t\right) \quad (4.129b)$$

Figure 4.27 plots the transient behavior of this step-function response of

172 RADIATION MODELS FOR WIRE ANTENNAS

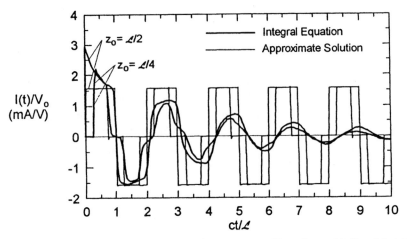

FIGURE 4.27 Current on thin center-fed antenna of length $\mathcal{L}$ with $\Omega = 2\ln(\mathcal{L}/a) = 10.6$ for a step-function excitation.

the antenna current for two observation locations: at the input of the antenna at $z_s = \mathcal{L}/2$ and at the location $z_0 = \mathcal{L}/4$. Although it is not readily apparent from the form of Eq. (4.129b), it is important to note that causality is enforced in this equation. The current at $z_0 = \mathcal{L}/4$ does not turn on prior to the normalized time of arrival $ct/\mathcal{L} = 0.25$. However, as mentioned previously, there is no accounting for radiation loss in the approximate solution, and as a consequence, there is no damping in the transient waveforms.

To provide an indication of the correct response, Figure 4.27 also plots the transient current as computed from an integral equation solution for the same antenna. For early times, the approximate responses are reasonably accurate, but the quality of the solution degrades as time increases, due primarily to the lack of radiation loss in the solution.

Reference [4.54] discusses the possibility of "correcting" the transient responses in Eqs. (4.126) and (4.129b) by using a more accurate expression for the complex natural resonances of the wire antenna. The complex natural resonances of a straight thin wire have been computed by searching for the zeros of the system impedance matrix resulting from the integral equation for the wire current [4.52]. This results in several layers of poles in the s-plane, as shown in Figure 4.28. The poles closest to the $j\omega$-axis provide the dominant contributions to the response, in both the frequency and time domains, and they correspond to the poles of Eq. (4.125a). However, they have a negative real part, which provides radiation damping to the solution.

As discussed in [4.54], various approximations to the first layer of poles

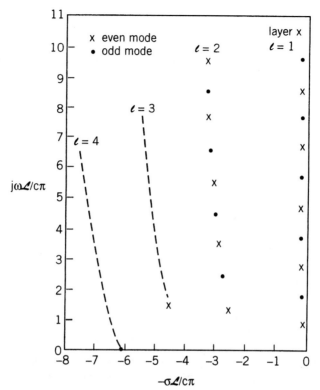

FIGURE 4.28 Locations of the natural frequencies of a thin wire of length $\mathcal{L}$ with $\Omega = 2\ln(\mathcal{L}/a) = 10.6)$. (From [4.52]. © 1973 Institute of Electrical and Electronics Engineers.)

can be made. One approximation, which is attributed to Oseen [4.55], provides the natural resonances as

$$s_\alpha = \sigma_\alpha + j\omega_\alpha \approx \frac{c}{\Omega_0 \mathcal{L}}[\text{Ci}(2\alpha\pi) - \ln(|\alpha|2\pi\Gamma)] + j\frac{c}{\mathcal{L}}\left[\alpha\pi - \frac{\text{Si}(2\alpha\pi)}{\Omega_0}\right]$$

(4.130)

where Ci and Si are the cosine and sine integrals [4.27], and $\Gamma = 1.781\ldots$. With these modified natural resonances, the exponentials corresponding to the inverse pole terms in Eq. (4.98) have complex arguments, and this results in exponentially damped transient responses for each pole pair. Thus the impulse- and step-function responses for the antenna become

$$i_{\text{impulse}}(z_0; t) = \frac{8\pi}{Z_0 \Omega_0} \frac{c}{\mathcal{L}} \sum_{\alpha=1}^{\infty} \sin\left(\frac{\alpha \pi}{\mathcal{L}} z_s\right) \sin\left(\frac{\alpha \pi}{\mathcal{L}} z_0\right) \cos(\omega_\alpha t) e^{\sigma_\alpha t}$$
(4.131a)

$$i_{\text{step}}(z_0; t) = \frac{8 V_0}{Z_0 \Omega_0} \sum_{\alpha=1}^{\infty} \sin\left(\frac{\alpha \pi}{\mathcal{L}} z_s\right) \sin\left(\frac{\alpha \pi}{\mathcal{L}} z_0\right) \cos(\omega_\alpha t) e^{\sigma_\alpha t}$$
(4.131b)

Figure 4.29 presents the step-function response for the same antenna as treated in Figure 4.27, but with modified antenna resonances. From this plot is seen that the late-time behavior of the solution is improved dramatically. For early times, however, the inaccuracy in the approximate solution is about the same—a bit less than a factor of 2 lower than the actual response.

4.5.7.2 Induced Current for the Scattering Problem.

The approximate solution for the thin wire can also be used to illustrate SEM concepts for the scattering problem. Evaluating Eq. (4.111) using the antenna SEM parameters in Eq. (4.125) yields the following expression for the class 2 coupling

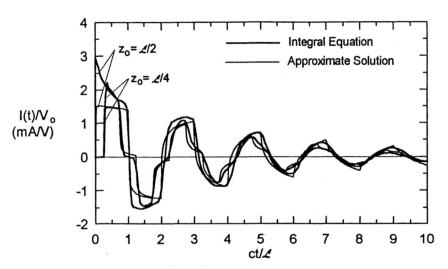

FIGURE 4.29 Current on a thin center-fed antenna $[\Omega = 2 \ln(\mathcal{L}/a) = 10.6]$ for a step-function excitation, using modified natural resonances.

coefficients for the scatterer:

$$\eta_\alpha(s) = jE^{\text{inc}} \sin \theta_i \frac{s_\alpha c}{s^2 \cos^2\theta_i - s_\alpha^2} [(-1)^\alpha e^{(s\mathcal{L}/c)\cos\theta_i} - 1] \qquad (4.132)$$

with $s_\alpha = j\alpha\pi c/\mathcal{L}$. At the pole, the residue of the SEM expansion $\eta_\alpha(s_\alpha)$ is the class 1 coupling coefficient, given by

$$\eta_\alpha(s_\alpha) = \frac{-jE^{\text{inc}}}{\sin \theta_i} \frac{c}{s_\alpha} [(-1)^\alpha e^{j\alpha\pi \cos\theta_i} - 1] \qquad (4.133)$$

With these last two equations, an explicit expression for the elusive entire function for the current on the scatterer in Eq. (4.115) can be developed.

REFERENCES

4.1. Harrington, R. F., *Field Computation by Moment Methods*, reprinted by the author, Syracuse University, Syracuse, NY, 1968.
4.2. Kunz, K. S., and R. J. Luebbers, *The Finite Difference Time Domain Method for Electromagnetics*, CRC Press, Boca Raton, FL, 1993.
4.3. Plonsey, R., and R. E. Collin, *Principles and Applications of Electromagnetic Fields*, McGraw-Hill, New York, 1961.
4.4. Stratton, J. A., *Electromagnetic Theory*, McGraw-Hill, New York, 1941.
4.5. Harrington, R. F., *Time Harmonic Electromagnetic Fields*, McGraw-Hill, New York, 1961.
4.6. Balanis, C., *Advanced Engineering Electromagnetics*, Wiley, New York, 1989.
4.7. Ramo, S., J. R. Whinnery, and T. Van Duser, *Fields and Waves in Communication Electronics*, 2nd Ed., Wiley, 1989.
4.8. Pocklington, H. C., "Electrical Oscillations in Wires," *Cambridge Philos. Soc. Proc.*, Vol. 9, 1897, pp. 324–332.
4.9. Tesche, F. M., "The Effect of the Thin-Wire Approximation and the Source Gap Model on the High Frequency Integral Equation Solution of Radiating Antennas," *IEEE Trans. Antennas Propag.*, Vol. AP-20, March 1972.
4.10. Jones, D. S., *The Theory of Electromagnetism*, Pergamon Press, London, 1964.
4.11. Hildebrand, F. B., *Advanced Calculus for Applications*, Prentice Hall, Englewood Cliffs, NJ, 1963.
4.12. Lee, S. W., W. R. Jones, and J. J. Campbell, "Convergence of Numerical Solutions of Iris-Type Discontinuity Problems," *Digest of the Joint URSI/IEEE PGAP Symposium*, IEEE 70C36-AP, Columbus, Ohio, September 1970.
4.13. Amitay, N., and V. Galindo, "On Energy Conservation and the Method of Moments in Waveguide Discontinuity and Scattering Problems," *URSI Symposium Digest*, Washington DC, 1969.
4.14. Burke, G. J., and A. J. Poggio, "Numerical Electromagnetic Code (NEC):

Method of Moments," *NOSC Technical Document 116*, Naval Ocean Systems Center, San Diego, CA, January 1980.

4.15. Mei, K. K., "On the Integral Equations of Thin Wire Antennas," *IEEE Trans. Antenna Propag.*, Vol. AP-13, No. 3, May 1965, pp. 374–378.

4.16. Schelkunoff, S. A., "Concerning Hallén's Integral Equation for Cylindrical Antennas," *Proc. IRE*, December 1945, pp. 872–878.

4.17. Bedrosian, G., "Stick-Model Characterization of the Natural Frequencies and Natural Modes of the Aircraft", *AFWL EMP Interaction Note, 326*, Air Force Weapons Laboratory, Kirtland AFB, NM, September 14, 1977.

4.18. Van Bladel, J., *Electromagnetic Fields*, McGraw-Hill, New York, 1964.

4.19. Jordan, E. C., and K. G. Balmain, *Electromagnetic Waves and Radiating System,"* Prentice Hall, Englewood Cliffs, NJ, 1968.

4.20. Tesche, F. M., "On the Behavior of Thin-Wire Antennas and Scatterers Arbitrarily Located Within a Parallel-Plate Region," *IEEE Trans. Antennas Propag.*, Vol. AP-20, July 1972.

4.21. Taylor, C. D., "Thin Wire Receiving Antenna in a Parallel Plate Waveguide," *IEEE Trans. Antennas Propag.*, Vol. AP-15, July 1967, pp. 572–576.

4.22. Collin, R. E., *Field Theory of Guided Waves*, McGraw-Hill, New York, 1960.

4.23. Tai, C. T., and P. Rozenfeld, "Different Representations of Dyadic Green's Functions for a Rectangular Cavity," *IEEE Trans. Microwave Theory Tech.*, Vol. MTT-24, No. 6, September 1976.

4.24. Rahmat-Samii, Y., "On the Question of Computation of the Dyadic Green's Function at the Source Region in Waveguides and Cavities," *IEEE Trans. Microwave Theory Tech.*, Vol. MTT-23, No. 6, September 1975.

4.25. Wu, D., and D. C. Chang, "An Investigation of a Ray-Mode Representation of the Green's Function in a Rectangular Cavity," *NBS Technical Note 1312*, National Bureau of Standards, Washington, DC, September 1987.

4.26. Du, L., and C. T. Tai, "Radiation Patterns for Symmetrically Located Sources on a Perfectly Conducting Sphere," *Report 1691-10*, Ohio State University Research Foundation, Columbus, OH, December 15, 1964.

4.27. Abramowitz, M., and I. A. Stegun, eds., *Handbook of Mathematical Functions*, National Bureau of Standards Publication, Washington, DC, June 1964.

4.28. Tesche, F. M., and A. R. Neureuther, "The Analysis of Monopole Antennas Located on a Spherical Vehicle, Parts 1 and 2," *IEEE Trans. Electromagn. Compat.*, Vol. EMC-18, No. 1, February 1976.

4.29. Baños, A., Jr., *Dipole Radiation in the Presence of a Conducting Half-Space*, Pergamon Press, Oxford, 1966.

4.30. Wait, J. R., *Electromagnetic Waves in Stratified Media*, Pergamon Press, Elmsford, 1962.

4.31. King, R. W. P., M. Owens, and T. T. Wu, *Lateral Electromagnetic Waves*, Springer-Verlag, New York, 1992.

4.32. Norton, K. A., "The Propagation of Radio Waves over the Surface of the Earth and in the Upper Atmosphere," *Proc. IRE*, Vol. 24, 1936 and Vol. 25, 1937.

4.33. Vance, E. F., *Coupling to Shielded Cables*, R. E. Krieger, Melbourne, 1987.

4.34. Mittra, R., "Integral Equation Methods for Transient Scattering," in *Topics in Applied Physics*, L. B. Felsen, ed., Vol. 10, Springer-Verlag, Berlin, 1976.

4.35. Jecko, B., "Etude d'interactions des ondes électromagnétiques sur des structures métalliques ou diélectriques en régime impulsionnel," Ph.D. thesis 8-79, Université de Limoges, 1979.

4.36. Dafif, O., and B. Jecko, "Diffraction d'impulsions électromagnétiques (IEM) par des structures filaires en présence du sol," *Actes 2ème Colloque National et Exposition sur la Compatibilité Electromagnétique*, Trégastel, France, June 1–3, 1983.

4.37. Miller, E. K., A. J. Poggio, G. L. Burke, and E. S. Selden, "Analysis of Wire Antennas in the Presence of a Conducting Half-Space," *Can. J. Phys.*, Vol. 50, 1972.

4.38. Van Blaricum, M., "A Numerical Technique for the Time-Dependent Solution of Thin-Wire Structures with Multiple Junctions," M.S. thesis, Electrical Engineering Department, University of Illinois, 1972.

4.39. Landt, J. A., E. K. Miller and M. Van Blaricum, "A Computer Program for the Time-Domain Electromagnetic Response of Thin-Wire Structures," *Report UCRL 51585*, Lawrence Livermore Laboratory, Livermore, CA, 1974.

4.40. Dafif, O., "Etude de la diffraction d'ondes électromagnétiques en régime transitoire par des structures filaires de formes quelconques en présence du sol," *Thèse de 3ème cycle 8-83*, Université de Limoges, February 14, 1983.

4.41. Bardet, C., O. Dafif, and B. Jecko, "Time-Domain Analysis of Large EMP Simulators," *IEEE Trans. Electromagn. Compat.*, Vol. EMC-29, No. 1, February 1987.

4.42. Dafif, O., A. Reineix, B. Jecko and J. J. Rodaro, "Behavior and Optimization of Transmission Line NEMP Simulators," *Proceedings of EUROEM '94*, Bordeaux, France, June 1–3, 1994.

4.43. Baraton, Ph., A. Zeddam, O. Dafif, and B. Jecko, "Détermination de la tension transitoire induite aux extrémités d'un câble coaxial par une IEMN. Analyse de la protection par un circuit non-linéaire", *Actes 5ème Colloque International en Langue Française sur la Compatibilité Electromagnétique*, Evian, September 12–14, 1989.

4.44. Raingeaud, A. J., and B. Jecko, "Analysis of the Coupling of an Electromagnetic Waveform with an Aircraft," *Proceedings of the 10th International Symposium and Technical Exhibition on EMC*, Zurich, March 9–11, 1993.

4.45. Recrosio, N., G. Fine, and M. Hélier, "Analysis of Radiation Characteristics of Distribution Line Carriers with the NEC Code," *IEEE Trans. Electromagn. Compat.*, Vol. EMC-35, No. 1, February 1993.

4.46. Baum, C. E., "The Singularity Expansion Method," Chapter 3 in *Transient Electromagnetic Fields*, L. B. Felsen, ed., Springer-Verlag, Berlin, 1976.

4.47. Pearson, L. W., ed., Special Issue on the Singularity Expansion Method, *Electromagnetics*, Vol. 1, No. 6, 1981.

4.48. Schelkunoff, S. I., and H. T. Friis, *Antennas: Theory and Practice*, Wiley, New York, 1952.

4.49. Cheng, D. C., *Analysis of Linear Systems*, Addison-Wesley, Reading, MA, 1959.

4.50. Marin, L., "Natural Mode Representation of Transient Scattered Fields," *IEEE Trans. Antennas Propag.*, Vol. AP-21, No. 6, November, 1973, pp. 809–818.

4.51. Goursat, E., *Functions of a Complex Variable*, Dover Publications, New York, 1959.

4.52. Tesche, F. M., "On the Analysis of Scattering and Antenna Problems Using the Singularity Expansion Technique," *IEEE Trans. Antenna Propag.*, Vol. AP-21, No. 1, January 1973.

4.53. Tesche, F. M., "The Far Field Response of a Step Excited Linear Antenna Using SEM," *IEEE Trans. Antennas Propag.* Vol. AP-23, No. 5, November 1975.

4.54. Marin, L., and T. K. Liu, "A Simple Way of Solving Transient Thin-Wire Problems," *Radio Sci.*, Vol. 11, No. 2, Feburary 1967, pp. 149–155.

4.55. Oseen, C. W., "Über die elektromagnetischen Schwingunden an dünnen Stàben, *Ark. Mat. Astron. Fys.*, Vol. 9, 1914, pp. 1–27.

PROBLEMS

4.1 Show that both the scalar and vector potentials are solutions to the Helmholtz equations (4.6) and (4.7).

4.2 Calculate and plot the far-zone E-field of a center-fed linear antenna whose length is one wavelength ($L = \lambda$) for the following cases:
   (a) The antenna is parasitically excited so that its current distribution is $I(z) = I_0 \sin kz$ for $-L/2 \leq z \leq L/2$.
   (b) The antenna is driven at the center ($z = 0$) so that the current distribution is $I(z) = I_0 \sin k|z|$.

4.3 The current on a particular half-wave linear antenna is determined to be given by $I(z) = I_0 \cos^2 kz$. Determine an expression for the far-zone E-field for this antenna, and evaluate the radiation resistance.

4.4 Outline a proof showing that for a horizontal or vertical current radiator located over perfectly conducting ground, the effect of the ground plane can be calculated by removing the ground and considering an *image* current source, as indicated in Figure P4.4. (*Hint:* It is

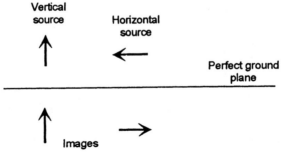

FIGURE P4.4 Vertical and horizontal current sources over a perfectly conducting ground plane.

sufficient to show that the total field due to the current element plus its image satisfies Maxwell's equations and obeys the condition that $E^{\tan} = 0$ at the location of the ground surface.)

4.5 Repeat Problem 4.4 for two current loops over the ground plane.

4.6* Consider a tall, thin building over a highly conducting ground plane, with an antenna mounted on top. Assuming that a vertically polarized incident EM field excites the building:

(a) Plot qualitatively the induced current as a function of length of the building, making sure to obey the boundary conditions at the ends of the building, for the first few natural resonances.

(b) In the same manner, plot the induced charge density along the building.

(c) Describe the EM field environment seen by the antenna on the roof.

4.7* Consider a missile that is tapered to a point at the nose. If the base of the missile is 1 m in diameter, and its length is 6 m, what is the lowest natural resonance frequency of the missile?

4.8* Discuss and illustrate the natural frequencies and the current and charge distributions on an in-flight missile, including the EM effects of its exhaust plume.

4.9* A large aircraft has a high-frequency (HF) wire antenna (2 to 30 MHz) installed for long-range communications. The wire radius is $a = 2$ mm and the antenna dimensions are given in Figure P4.9. The end of the antenna at point C is connected to the vertical stabilizer, and the input of the antenna is between terminals A and B.

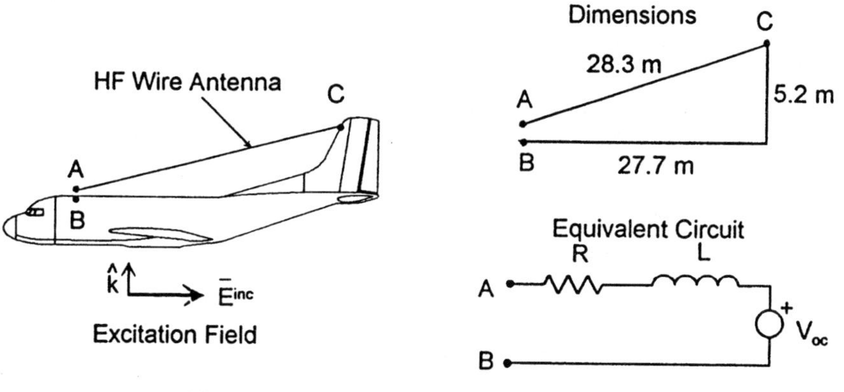

FIGURE P4.9 Aircraft with a HF antenna.

* Problems denoted by an asterisk were used in EMP short courses conducted by the Summa Foundation in 1983 through 1989. These contributions by C. Baum and the other lecturers are gratefully acknowledged.

(a) Show that for frequencies lower than 1.5 MHz, a small loop antenna approximation is valid.

(b) Compute the input impedance of the antenna, Z_{in}, at 10 kHz, 100 kHz, and 1 MHz.

(c) A cloud-to-cloud lightning discharge parallel to the aircraft produces an excitation electromagnetic field as in the figure. Outline procedure for finding the source term V_{oc} for the equivalent circuit shown in the figure, for frequencies lower than 1.5 MHz for the given incident wave.

Useful formulas: Radiation resistance = $320\pi^4(A/\lambda^2)^2$ ohms, and inductance of the loop = $(\mu_0/2\pi)P \ln[P/(2\pi a)]$ henrys, where A is the loop area, a the conductor radius, λ the wavelength, and P the loop perimeter.

4.10* For the aircraft in Figure P4.9 with the HF antenna removed, describe the natural current modes on the airframe. Discuss and illustrate the changes in the natural frequencies and the current modes if a very long trailing wire were to be attached to the aircraft. (*Hint:* See C. D. Taylor, "External Interaction of the Nuclear EMP with Aircraft and Missiles," *IEEE Trans. Electromagn. Compat.,* Vol. EMC-20, No. 1, February, 1978.)

4.11* A conical communication antenna is mounted on the roof of a building over a metallic baseplate having a radius of 0.6 m. The antenna has a total conical angle of 30° and a height of 0.3 m, as illustrated in Figure P4.11. A shielded cable 4 m long connects the antenna to a shielded facility situated inside the building. The cable has the sheath connected to the metallic plate under the antenna; it has a characteristic impedance of $Z_c = 50\,\Omega$ and is matched on the facility side. Estimate the current that can be induced on the internal conductor of the antenna by an electric field of 10 kV/m (peak value), 1 ms rise time and 100 ms fall time, generated by a lightning of average intensity (this field results from a $I_{peak} = 25$ kA lightning bolt striking the ground 100 m from the building). (*Hint:* See K. S. H.

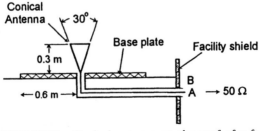

FIGURE P4.11 Conical antenna on the roof of a facilty.

Lee, ed., *EMP Interaction: Principles, Techniques and Reference Data*, Hemisphere, New York, 1989, Section 3.2.3.).

4.12* Discuss the transmission of energy mechanisms through the coaxial cavities shown in Figure P4.12.

4.13* Assuming that you have a sufficient supply of resistors and copper wires, suggest an effective way of using them to change the resonant behavior of the cavities defined in Problem 4.12. Discuss the effects on the natural frequencies.

4.14* In the configuration of Figure P4.12(*a*), suggest penetration treatments at the entry and exit of the cavity that will permit an audio frequency signal to pass through the cavity undistorted, but which will eliminate a fast-transient surge on the coaxial line.

4.15* You are asked to conduct some EM coupling measurements in a shielded room of dimension $6 \times 4 \times 2$ m.
  (a) Compute the first 10 cavity resonances and plot them on a frequency plot.
  (b) Discuss how the presence of the shield will affect the quality of your measurements.

4.16 Using the NEC code or another antenna analysis code of your choice, conduct an analysis of a thin, center-fed, linear antenna 1.5 m long with radius $a = 1$ mm. Assume that the antenna is fed with a voltage source $V_s = 1$ V and that the frequency is $f = 100$ MHz ($\lambda = 3$ m).
  (a) Compute the actual current distribution and compare it with the commonly used approximate current distribution of Eq. (4.20b).
  (b) Compute and plot the input admittance $Y_{in} = I_s/V_s$ for a range of frequencies from 10 to 500 MHz, being careful not to miss any of the peaks in the spectrum. Why doesn't the first antenna peak at approximately $f \approx 100$ MHz correspond exactly to the $\lambda/2$ frequency?
  (c) Locate an identical antenna parallel to the first such that the separation between the two antennas is $d = 10$ cm. At $f = 100$ MHz, compute the change in the first antenna's current distribution (i) for the case of the second antenna unexcited, and

FIGURE P4.12 Two cavities fed by a coaxial line.

(ii) for the case of the second antenna having the same excitation as the first.

(d) Compute and plot (on a log scale) the input admittance magnitude of the first antenna with the passive (unexcited) antenna nearby for antenna separations varying from 1 cm to 10 m for the frequency 100 MHz.

(ii) for the case of the second antenna having the same excitation as the first.

(d) Compute and plot (on a log scale) the input admittance magnitude of the first antenna with the passive (unexcited) antenna nearby for antenna separations varying from 1 cm to 10 m for the frequency 100 MHz.

CHAPTER 5
Radiation, Diffraction, and Scattering Models for Apertures

Although the wire antenna models discussed in Chapter 4 can be very useful for EMC analysis purposes, they are limited because they only represent one-dimensional structures. The more general class of aperture antennas is frequently encountered in practical problems, where the radiating structure has two principal dimensions that are comparable to the wavelength of the EM field. As a consequence, the EM fields on and near the apertures are also two-dimensional and are not defined as simply as for the wire antennas. In addition to the deliberate radiation problem, there is the issue of field leakage, or penetration, through holes or other imperfections in a shield boundary. The use of the same modeling techniques used for the aperture antenna problems can be helpful in predicting the reception and transmission of EM energy through such holes. In this chapter we discuss the subject of radiation and diffraction from apertures.

5.1 INTRODUCTION

There are various types of two-dimensional structures that are of interest for EMC studies. Figure 5.1a to c portray aperture-type radiation, or antenna, structures, while parts (d) and (e) of the figure illustrate field penetrations through apertures. In this chapter we describe several methods for treating aperture problems. As in the case of the thin-wire antennas, the fundamental requirement for conducting the analysis is knowledge of the surface current flowing on the electrical conductors in the problem. Unfortunately, the determining of these currents is difficult, if not impossible, for most realistic problems. As a result, approximate analysis methods have been developed.

Consider the practical problem of determining the radiation field at point P from the horn antenna in Figure 5.2a. The antenna is fed by an internal voltage source, and this results in electrical surface currents **J** flowing on the inside and outside of the conductors comprising the waveguide and horn. If these currents are known, direct integration of the form of Eq. (4.10) followed by the differential operation of Eq. (4.5) will yield the E-field at

184 RADIATION, DIFFRACTION, AND SCATTERING MODELS FOR APERTURES

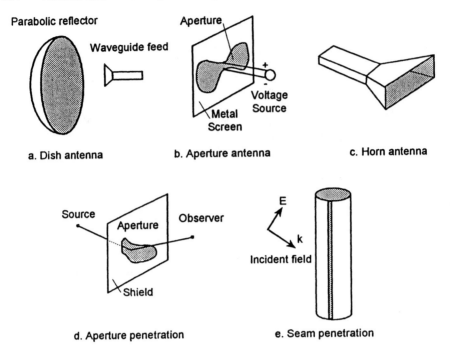

FIGURE 5.1 Examples of aperture antennas and penetrations.

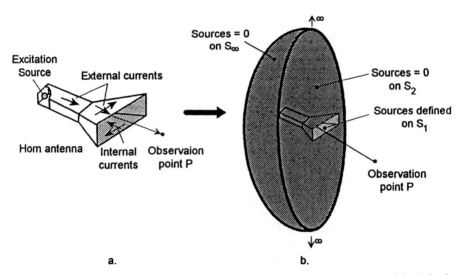

FIGURE 5.2 Example of the calculation of the field from a horn antenna: (*a*) original problem; (*b*) application of the field equivalence principle.

point P. There is an alternative to computing the detailed current distribution, however, and this involves the use of the field equivalence principle [5.1], which states that if the antenna is surrounded by an imaginary surface S, it is possible to postulate a set of *equivalent sources* on this surface which radiate and produce the same field at P as does the antenna. These equivalent sources are also unknown, but often by the careful choice of the shape of the surface S, it is possible to approximate the equivalent sources and compute the radiated field more easily.

For example, Figure 5.2b shows one choice of the closed surface S which is located just in front of the aperture of the antenna and extends to infinity where it is closed so that it encloses the antenna. By assuming that the tangential E and H fields over the front of the antenna (surface S_1) are equal to the unperturbed field distribution of the internal waveguide horn and that the fields are zero on S_2 and S_∞, the field at P can be approximated. In this chapter we discuss the field equivalence principle in more detail, starting with the simple scalar wave diffraction from an aperture, and then generalizing to the more interesting case of vector fields.

5.2 EM FIELD PENETRATION THROUGH APERTURES

5.2.1 Scalar Diffraction Theory

Although the treatment of radiation and scattering from aperture-type antennas using scalar theory is clearly approximate because the vector nature of the fields is neglected, it does provide some insight into the important aspects of the problem. Furthermore, it helps lay the foundation for an understanding of the more complex vector treatment to be considered in the next section. Early work in this field was performed by Kirchhoff, and it still forms the basis for many approximate, yet practical solutions for aperture radiation.

To illustrate the use of the scalar diffraction methods, consider representing an important component of the **E** or **B** field by a scalar field denoted by the wavefunction $\psi(\mathbf{r})$. As in the earlier chapters, a time dependence of $e^{j\omega t}$ is assumed so that ψ is a phasor quantity. Consider a closed, source-free volume V bounded by surface S as illustrated in Figure 5.3a. The scalar field ψ at an observation location P defined by the vector $\mathbf{r}$ is produced by an arbitrary set of electric and magnetic sources **J** and **M** outside the volume, and must satisfy the scalar Helmholtz wave equation

$$(\nabla^2 + k^2)\psi(\mathbf{r}) = 0 \tag{5.1}$$

where $k = \omega/c$ and ∇^2 is the differential Laplacian operator defined as $\nabla^2 = \partial^2/\partial x^2 + \partial^2/\partial y^2 + \partial^2/\partial z^2$.

Green's second identity (i.e., Green's theorem) [5.2] states that for two

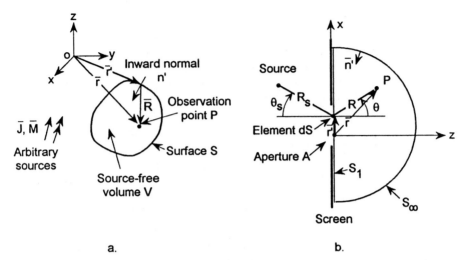

FIGURE 5.3 Diagram for the scalar formulation of aperture penetration: (*a*) arbitrary source-free region V surrounded by surface S; (*b*) application of the volume V to the problem of an aperture in an infinite screen.

scalar functions ψ and ϕ within volume V, the following equation holds:

$$\int_V (\phi \nabla^2 \psi - \psi \nabla^2 \phi)\, dV = -\oint_S \left(\phi \frac{\partial \psi}{\partial n} - \psi \frac{\partial \phi}{\partial n} \right) dS \qquad (5.2)$$

where $\hat{n}$ is the *inward* normal to the surface S. Introducing a Green's function $G(\mathbf{r}, \mathbf{r}')$, which is a solution to the wave equation with an impulse forcing function

$$(\nabla^2 + k^2) G(\mathbf{r}, \mathbf{r}') = -\delta(\mathbf{r} - \mathbf{r}') \qquad (5.3)$$

and letting G be the function ϕ in Eq. (5.2), the following integral expression results for ψ:

$$\psi(\mathbf{r}) = \int_S (\psi(\mathbf{r}')\hat{n}' \cdot \nabla' G(\mathbf{r}, \mathbf{r}') - G(\mathbf{r}, \mathbf{r}')\hat{n}' \cdot \nabla' \psi(\mathbf{r}'))\, dS' \qquad (5.4)$$

In this expression, the primed coordinates represent the source (or integration) coordinates. Consequently, the operator ∇' signifies a derivative operation on the source coordinates.

Many different solutions of Eq. (5.3) are possible, all of which are linear combinations of ingoing and outgoing waves of the form $e^{\pm jkR}/4\pi R$, where

$$R = |\mathbf{r} - \mathbf{r}'| = \sqrt{(x - x')^2 + (y - y')^2 + (z - z')^2}$$

For this present development, we require only outwardly propagating fields. Consequently, the Green's function has the form [5.3]

$$G(\mathbf{r},\mathbf{r}') = \frac{e^{-jkR}}{4\pi R} \tag{5.5}$$

Inserting this into Eq. (5.4) gives

$$\psi(\mathbf{r}) = -\int_S \frac{e^{-jkR}}{4\pi R} \hat{n}' \cdot \left[\nabla'\psi(\mathbf{r}') - jk\left(1 + \frac{1}{jkR}\right)\hat{R}\psi(\mathbf{r}')\right] dS' \tag{5.6}$$

where $\hat{R} = \mathbf{R}/|\mathbf{R}|$ is a unit vector from the integration point to the observer.

Consider applying this integral expression to the aperture penetration problem shown in Figure 5.3b, in which the sources $\mathbf{J}$ and $\mathbf{M}$ are on one side of a perfectly conducting, infinite sheet containing an aperture. To determine the transmitted field at P, we deform the surface S to conform to the infinite plate (surface S_1) with the closure of the surface being at infinity (surface S_∞). Using the Sommerfeld radiation conditions [5.4] on the function ψ

$$\lim_{R \to \infty} \psi \propto \frac{e^{-jkR}}{R} \quad \text{and} \quad \lim_{R \to \infty} \frac{1}{\psi}\frac{\partial \psi}{\partial r} \propto jk + \frac{1}{R} \tag{5.7}$$

the integral over S_∞ in Eq. (5.6) can be shown to vanish, and we are left with the Kirchhoff integral relationship for ψ. This involves integrating only over *equivalent sources* ψ and $\partial\psi/\partial n$ on the surface S_1. The main difficulty, of course, is that these functions on the surface are unknown and must be determined, either by an independent solution, or by a suitable approximation.

5.2.1.1 Kirchhoff Approximation.
The approach used by Kirchhoff for determining the values of ψ and $\partial\psi/\partial n$ on S_1 is straightforward. Both ψ and $\partial\psi/\partial n$ are chosen to be zero on the screen away from the aperture, and in the aperture they are set equal to ψ^{inc} and $\partial\psi^{inc}/\partial n$, which are the values provided by the incident field. Thus the field at P is given approximately as

$$\psi(\mathbf{r}) \approx -\frac{1}{4\pi}\int_{\text{aperture}} \frac{e^{-jkR}}{R} \hat{n}' \cdot \left[\nabla'\psi^{inc}(\mathbf{r}') - jk\left(1 + \frac{1}{jkR}\right)\hat{R}\psi^{inc}(\mathbf{r}')\right] dS' \tag{5.8}$$

That this is an approximation is clearly evident, as the correct values of the fields in the aperture actually should depend on the details of the aperture shape and geometry instead of simply being the incident quantities.

As noted in [5.3], the assumption that both ψ and $\partial\psi/\partial n = 0$ on a finite region poses serious mathematical problems, as it can be shown that this assumption leads to the conclusion that the only rigorous solution is $\psi \equiv 0$

everywhere. Of course, the real (i.e., the correct) solution will not have both ψ and $\partial\psi/\partial n = 0$ at the same point on the screen. Moreover, by assuming a value of ψ^{inc} and setting $\partial\psi^{inc}/\partial n = 0$ in the aperture and using Eq. (5.8) to evaluate the value of ψ on the screen does not yield the correct result. The results are inconsistent, due to the fact that the assumed distribution is not correct in the first place.

5.2.1.2 Dirichlet Solution.
The fundamental difficulty of specifying both ψ and $\partial\psi/\partial n$ on the surface S_1 can be eliminated by choosing an alternative Green's function for the development of the integral expression of Eq. (5.8). To do this, we define a Dirichlet Green's function $G_1(\mathbf{r}, \mathbf{r}')$, which is a solution to the inhomogeneous wave equation (5.3), but with the added constraint that $G_1(\mathbf{r}, \mathbf{r}') = 0$ for $\mathbf{r}'$ located on the surface S_1. This amounts to adding a solution to the homogeneous wave equation to Eq. (5.5). Because of the fact that S_1 is an infinite plane, this alternative Green's function can be constructed by image theory to give

$$G_1(\mathbf{r}, \mathbf{r}') = \frac{e^{-jkR}}{4\pi R} - \frac{e^{-jkR'}}{4\pi R'} \qquad (5.9)$$

where $R' = \sqrt{(x-x')^2 + (y-y')^2 + (z+z')^2}$. For the integration (source) point located on the screen, $z' = 0$ and Eq. (5.9) is seen to be identically zero.

Inserting Eq. (5.9) into Eq. (5.4) and again using the radiation conditions to eliminate the integrations on the surface S_∞ gives the modified integral relationship for ψ:

$$\psi(\mathbf{r}) = \int_{S_1} \psi(\mathbf{r}') \hat{n}' \cdot \nabla' G_1(\mathbf{r}, \mathbf{r}') \, dS \qquad (5.10)$$

Furthermore, by assuming that $\psi = 0$ on S_1 away from the aperture and that $\psi = \psi^{inc}$ over the aperture A, the integration of Eq. (5.10) is limited to only over the aperture area, resulting in the expression

$$\psi(\mathbf{r}) \approx \frac{jk}{2\pi} \int_{\text{aperture}} \left(1 + \frac{1}{jkR}\right) \hat{n}' \cdot \hat{R} \psi^{inc}(\mathbf{r}') \frac{e^{-jkR}}{R} \, dS' \qquad (5.11)$$

5.2.1.3 Neumann Solution.
The dual to the Dirichlet solution is obtained by using the Neumann Green's function $G_2(\mathbf{r}, \mathbf{r}')$, which satisfies the condition that $\partial G_2(\mathbf{r}, \mathbf{r}')/\partial n' = 0$ on S_1. This solution for the infinite screen also can be obtained by image theory, with the result

$$G_2(\mathbf{r}, \mathbf{r}') = \frac{e^{-jkR}}{4\pi R} + \frac{e^{-jkR'}}{4\pi R'} \qquad (5.12)$$

As a consequence, the integral expression for the function ψ in Eq. (5.4) simplifies to contain only the term involving ψ in the integrand:

$$\psi(\mathbf{r}) = -\int_{S_1} \frac{\partial \psi(\mathbf{r}')}{\partial n'} G_2(\mathbf{r}, \mathbf{r}') \, dS' \tag{5.13}$$

We now assume that $\partial \psi / \partial n' = 0$ on S_1 away from the aperture, and that $\partial \psi / \partial n' = \partial \psi^{inc} / \partial n'$ over the aperture A. In this manner the integration in Eq. (5.13) is reduced to a simple integration over the aperture area, as

$$\psi(\mathbf{r}) \approx -\frac{1}{2\pi} \int_{\text{aperture}} \frac{\partial \psi^{inc}(\mathbf{r}')}{\partial n'} \frac{e^{-jkR}}{R} \, dS' \tag{5.14}$$

5.2.1.4 Discussion of the Scalar Solutions.

As a consequence of the preceding mathematical development, we have three possible ways of calculating the scalar field penetrating the aperture: Eqs. (5.8), (5.11), and (5.14). By the uniqueness theorem, there can be only one solution to the physical problem, so we must conclude that these equations only provide approximations to the correct solution for the penetrating field. To gain an understanding of the differences between these three representations for the penetrating field, let us consider the case when the observation distance R in Figure 5.3b is electrically far from the aperture (i.e., $kR \gg 1$). Furthermore, we assume that the source distribution produces a spherical plane wave propagating towards the aperture of the form

$$\psi^{inc} = \psi_0 \frac{e^{-jkR_s}}{R_s} \tag{5.15}$$

Under these assumptions, we find that Eqs. (5.8), (5.11), and (5.14) all have the common form

$$\psi(\mathbf{r}) \approx \frac{jk\psi_0}{2\pi} \int_{\text{aperture}} \frac{e^{-jkR_s}}{R_s} \frac{e^{-jkR}}{R} \mathcal{O}(\theta, \theta_s) \, dS' \tag{5.16}$$

where $\mathcal{O}(\theta, \theta_s)$ is the *obliquity factor*, defined as

$$\mathcal{O}(\theta, \theta_s) = \begin{cases} \frac{1}{2}(\cos \theta + \cos \theta_s) & \text{for Kirchhoff approximation} \\ \cos \theta_s & \text{for } (\partial \psi / \partial n) \text{ defined on } S_1 \\ \cos \theta & \text{for } \psi \text{ defined on } S_1 \end{cases} \tag{5.17}$$

The angles θ and θ_s are illustrated in Figure 5.3b and arise from the normal derivatives in the various expressions for ψ.

Clearly, each of these solutions is different. However, if we limit our interest to fields that are almost normally incident ($\theta_s \approx 0$) and look at the fields that are transmitted with an angle $\theta \approx 0$, the obliquity factors are all about the same and equal to unity. Consequently, it really does not matter which approximation is used to compute the transmitted field: They all give practically the same result.

As an example of the calculation of the transmitted fields through the aperture, consider the coordinate system shown in Figure 5.4, which illustrates the source and observation points S and P and the aperture plane. The source location is defined by the angles χ_s and ξ_s (or equivalently, by θ_s and ϕ_s), together with the distance r_s. Similarly, the observation point is defined by the angles χ and ξ (or θ and ϕ) and the distance r.

For diffraction in the far field, the distances R and R_s in the denominator of Eq. (5.16) can be approximated by r and r_s and removed from the integral [5.5]. In the exponential functions, these distances are approximated by $R \approx r - x'\sin\chi - y'\sin\xi$ and $R_s \approx r_s - x'\sin\chi_s - y'\sin\xi_s$, where x' and y' are the local coordinates on the aperture screen. Thus the expression for the field ψ at point P becomes

$$\psi(P) \approx \frac{jk\psi_0}{2\pi} \mathbf{0}(\theta, \theta_s) \frac{e^{-jk(r_s+r)}}{r_s r} \int_{\text{aperture}} e^{jk[x'(\sin\chi+\sin\chi_s)+y'(\sin\xi+\sin\xi_s)]} dx' dy' \quad (5.18)$$

where the obliquity factor has been assumed to be a constant and is removed from the integral.

It is customary to define the normalized aperture diffraction function $\mathcal{G}(S,P)$ as

$$\mathcal{G}(S, P) \approx \frac{1}{A} \int_{\text{aperture}} e^{jk[x'(\sin\chi+\sin\chi_s)+y'(\sin\xi+\sin\xi_s)]} dx' dy' \quad (5.19)$$

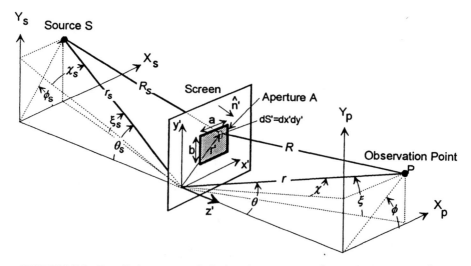

FIGURE 5.4 Detailed source and observation geometry for aperture penetration.

where A is the area of the aperture. Note that $\mathcal{G}$ is a function of both the source- and observation-point locations. Using this notation, the expression for the transmitted field becomes

$$\psi(P) \approx \frac{jkA\psi_0}{2\pi} \mathcal{O}(\theta, \theta_s) \frac{e^{-jk(r_s+r)}}{r_s r} \mathcal{G} \qquad (5.20)$$

Frequently, it is inconvenient to deal with the complex-valued field function ψ for describing the behavior of the aperture. An alternative is to define the irradiance at point P as $I(P) = |\psi(P)|^2$, in which case Eq. (5.20) yields

$$I(P) \approx \left[\frac{kA\psi_0}{2\pi R_{s0} R_0} \mathcal{O}(\theta, \theta_s) \right]^2 |\mathcal{G}|^2$$

$$\equiv I_0 |\mathcal{G}|^2 \qquad (5.21)$$

In this equation the term I_0 is defined to represent the bracketed term in Eq. (5.21).

5.2.1.4.1 Rectangular Aperture.
The expression for the aperture irradiance in Eq. (5.21) can be applied to the case of a rectangular aperture having a dimension a in the x-direction and b in the y-direction, as depicted in Figure 5.4. For this aperture the function $\mathcal{G}$ can easily be evaluated to be

$$\mathcal{G} = \frac{1}{ab} \int_{-a/2}^{a/2} e^{jkx'(\sin\chi + \sin\chi_s)} dx' \int_{-b/2}^{b/2} e^{jky'(\sin\xi + \sin\xi_s)} dy'$$

$$= \frac{\sin\pi\alpha}{\pi\alpha} \frac{\sin\pi\beta}{\pi\beta} \qquad (5.22a)$$

where the definitions for α and β are

$$\alpha = \frac{ka}{2\pi}(\sin\chi + \sin\chi_s) \quad \text{and} \quad \beta = \frac{kb}{2\pi}(\sin\xi + \sin\xi_s) \qquad (5.22b)$$

Thus the irradiance pattern for this rectangular aperture is given by

$$I(P) = I_0 \frac{\sin^2\pi\alpha}{(\pi\alpha)^2} \frac{\sin^2\pi\beta}{(\pi\beta)^2} \qquad (5.23)$$

5.2.1.4.2 Circular Aperture.
Another aperture of interest is a circular hole having a diameter d. For the *special case* of normal incidence ($\chi_s = \xi_s = 0$),

Eq. (5.19) becomes

$$\mathcal{G}(S, P) \approx \frac{1}{\pi d^2/4} \int_{\text{aperture}} e^{jk[\rho' \cos \phi'(\sin \chi) + \rho' \sin \phi'(\sin \xi_s)]} \rho' \, d\rho' \, d\phi'$$

$$= 2 \frac{J_1(\pi \alpha)}{\pi \alpha} \qquad (5.24)$$

where $\alpha = (kd/2\pi)\sin \chi$ and $J_1(\cdot)$ is a Bessel function of order 1. Note that because of circular symmetry, this pattern function does not depend on the observation angle ξ. Consequently, it depends only on one parameter, α. The more general case of an arbitrary incident field must be treated by a numerical integration.

Figure 5.5 illustrates these radiation patterns, expressed in decibels (dB) relative to I_0, as a function of the parameters α and β. Notice that for the circular aperture in part (b), the pattern depends only on the angle χ or the parameter α. The irradiance in this case is plotted as a surface to compare and contrast the pattern with that of the rectangular aperture. For an electrically small rectangular aperture, ka and kb are both $\ll 1$ and α and β are both small, resulting in an almost constant $\mathcal{G}$ function as the observation angles χ and ξ vary. As the aperture size increases, however, α and β can take on larger values and the peaks and nulls in the $\mathcal{G}$ function in the figure result in the well-known lobe pattern of the penetrating field.

5.2.2 General Vector Field Diffraction

As electromagnetic phenomena are inherently vector in nature, a correct description involves the use of vector fields. The more general treatment of vector field diffraction from an aperture is treated in this section.

5.2.2.1 Fundamentals. With reference to Figure 5.3a, it is well known (see [5.1]) that if there are no sources within volume V, the vector EM field within the volume arises from the tangential components of the E and H fields on the bounding surface S. In a manner analogous to that used to develop Eq. (5.4), these fields can be represented [5.6] by the equations†

$$\mathbf{E}(\mathbf{r}) = -\int_S (\hat{n}' \times \mathbf{E}) \cdot \nabla' \times \overline{\overline{\Gamma}}(\mathbf{r}'; \mathbf{r}) - j\omega\mu(\hat{n}' \times \mathbf{H}) \cdot \overline{\overline{\Gamma}}(\mathbf{r}'; \mathbf{r}) \, dS' \quad (5.25a)$$

$$\mathbf{H}(\mathbf{r}) = -\int_S (\hat{n}' \times \mathbf{H}) \cdot \nabla' \times \overline{\overline{\Gamma}}(\mathbf{r}'; \mathbf{r}) + j\omega\epsilon(\hat{n}' \times \mathbf{E}) \cdot \overline{\overline{\Gamma}}(\mathbf{r}'; \mathbf{r}) \, dS' \quad (5.25b)$$

† Note that for free space, the Green's dyad $\overline{\overline{\Gamma}}(\mathbf{r}';\mathbf{r})$ in Eqs. (5.25a) and (5.25b) are the same. However, when scattering objects are placed in the problem volume and appropriate boundary conditions are imposed on $\overline{\overline{\Gamma}}(\mathbf{r}';\mathbf{r})$, the dyads are different for each equation (see [5.6] for details).

5.2 EM FIELD PENETRATION THROUGH APERTURES   193

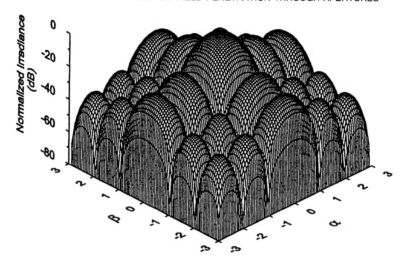

a. Rectangular aperture

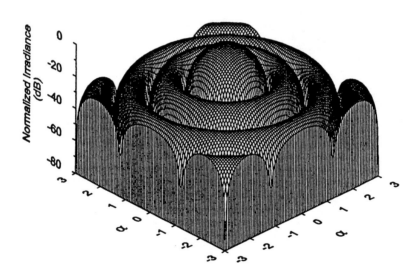

b. Circular aperture

FIGURE 5.5 Normalized irradiance patterns $I(P)/I_0$, for (a) the rectangular aperture shown as a function of the parameters α and β, and (b) the circular aperture shown as a function of α.

194 RADIATION, DIFFRACTION, AND SCATTERING MODELS FOR APERTURES

where again $\hat{n}'$ is the inward-directed unit normal vector to the surface. In the development that follows, we discuss only the behavior of the E-field diffraction, as the analysis of the H-field behavior is similar.

In Eq. (5.25) the term $\overline{\overline{\Gamma}}(\mathbf{r}';\mathbf{r})$ is the free-space dyadic Green's function, defined to be the solution to the *vector* wave equation‡

$$\nabla' \times \nabla' \times \overline{\overline{\Gamma}}(\mathbf{r}';\mathbf{r}) - k^2 \overline{\overline{\Gamma}}(\mathbf{r}';\mathbf{r}) = -\overline{\overline{I}}\delta(\mathbf{r}'-\mathbf{r}) \qquad (5.26)$$

subject to appropriate boundary conditions. This equation is the vector analog to the scalar equation (5.3). For free space, the necessary boundary conditions are the Sommerfeld radiation conditions discussed previously, and the solution to Eq. (5.25) is of the form

$$\overline{\overline{\Gamma}}(\mathbf{r}';\mathbf{r}) = -\frac{1}{4\pi}\left(\overline{\overline{I}} + \frac{1}{k^2}\nabla\nabla\right)\frac{e^{-jk|\mathbf{r}-\mathbf{r}'|}}{|\mathbf{r}-\mathbf{r}'|} \qquad (5.27)$$

where $k = \omega/c$, $\overline{\overline{I}}$ is the unit dyadic, and the operator $\nabla \mathbf{a}$ is interpreted as a dyad. Reference [5.7] discusses the development of this and other dyadic Green's functions in more detail.

An alternative form of Eq. (5.25), which does not use the dyadic notation and shows the e^{-jkR}/R terms explicitly, has been developed in [5.8]. This equivalent expression is

$$\mathbf{E}(\mathbf{r}) = \frac{1}{4\pi}\int_S (\hat{n}' \cdot \mathbf{E})\nabla'\frac{e^{-jkR}}{R} + (\hat{n}' \times \mathbf{E}) \times \nabla'\frac{e^{-jkR}}{R} - j\omega\mu(\hat{n}' \times \mathbf{H})\frac{e^{-jkR}}{R} dS' \qquad (5.28)$$

where $R = |\mathbf{r} - \mathbf{r}'|$. The term involving the normal component of $\mathbf{E}$ can be eliminated as shown in Eq. (8.102) of [5.8] and this results in another expression involving only the tangential fields:

$$\mathbf{E}(\mathbf{r}) = \frac{1}{4\pi}\left[\nabla \times \int_{S_1}(\hat{n}' \times \mathbf{E})\frac{e^{-jkR}}{R}dS' + \frac{1}{j\omega\epsilon}\nabla \times \nabla \times \int_{S_1}(\hat{n}' \times \mathbf{H})\frac{e^{-jkR}}{R}dS'\right] \qquad (5.29)$$

where the differential operators ∇ are now operating on the unprimed (i.e., observation) coordinates.

5.2.2.2 Application to the Aperture Penetration Problem. To use Eqs. (5.25), (5.28), or (5.29) to treat the case of field penetration through an

---

‡ Another definition of the Green's dyadic is frequently found in the literature, namely the solution to the equation $\nabla' \times \nabla' \times \overline{\overline{\Gamma}}(\mathbf{r}';\mathbf{r}) - k^2\overline{\overline{\Gamma}}(\mathbf{r}';\mathbf{r}) = \overline{\overline{I}}\delta(\mathbf{r}' - \mathbf{r})$. In our development here, we use the notation of Eq. (5.26) as used by Jones [5.6].

aperture, the surface S is deformed into a surface S_1 just over the conducting plane and aperture and another surface S_∞, as shown in Figure 5.3b for the scalar theory. Because of the radiation condition, the portion of the integration over S_∞ vanishes, and the only integral that remains is over the infinite surface S_1.

At this point, various assumptions can be made about the nature of the values of the equivalent sources ($\hat{n} \times \mathbf{E}$ and $\hat{n} \times \mathbf{H}$) over S_1. In an approach analogous to the Kirchhoff approximation for the scalar theory, the fields within the aperture can be approximated by the unperturbed, incident fields, and the tangential fields elsewhere on the screen set to zero. Using Eq. (5.25) for calculating the field, this approximation to the aperture fields leads to the expression

$$\mathbf{E}(\mathbf{r}) \approx -\int_{\text{aperture}} (\hat{n}' \times \mathbf{E}^{\text{inc}}) \cdot \nabla' \times \overline{\overline{\Gamma}}(\mathbf{r}'; \mathbf{r}) - j\omega\mu(\hat{n}' \times \mathbf{H}^{\text{inc}}) \cdot \overline{\overline{\Gamma}}(\mathbf{r}'; \mathbf{r}) \, dS' \tag{5.30}$$

As in the scalar case, it is possible to invoke another solution for the Green's function of Eq. (5.26) which satisfies both the radiation condition at infinity and the condition

$$\hat{n}' \times \overline{\overline{\Gamma}}_1(\mathbf{r}'; \mathbf{r}) = 0 \quad \text{for the source coordinate } \mathbf{r}' \text{ on } S_1 \tag{5.31}$$

This Green's dyad is denoted by $\overline{\overline{\Gamma}}_1$ to distinguish it from the free-space dyad $\overline{\overline{\Gamma}}$. With this modified Green's function, the integral involving the H-field in Eq. (5.25) can be shown to vanish. Assuming again that the tangential E-field over the plane S_1 is equal to the incident E-field in the aperture and zero elsewhere gives the expression

$$\mathbf{E}(\mathbf{r}) \approx -\int_{\text{aperture}} (\hat{n}' \times \mathbf{E}^{\text{inc}}) \cdot \nabla' \times \overline{\overline{\Gamma}}_1(\mathbf{r}'; \mathbf{r})_1 \, dS' \tag{5.32}$$

Notice that this is the vector analog to Eq. (5.11).

In a similar manner, a second modified Green's dyad $\overline{\overline{\Gamma}}_2$ can be developed which obeys the boundary condition

$$\hat{n}' \times \nabla' \times \overline{\overline{\Gamma}}_2(\mathbf{r}'; \mathbf{r}) = 0 \quad \text{for the source coordinate } \mathbf{r}' \text{ on } S_1 \tag{5.33}$$

and the resulting expression for the E-field at P becomes

$$\mathbf{E}(\mathbf{r}) \approx j\omega\mu \int_{\text{aperture}} (\hat{n}' \times \mathbf{H}^{\text{inc}}) \cdot \overline{\overline{\Gamma}}_2(\mathbf{r}'; \mathbf{r})_2 \, dS' \tag{5.34}$$

which is analogous to Eq. (5.14) for the scalar case.

The use of Eq. (5.30), (5.32), or (5.34) will give different answers for the field at P, due to the fact that the fields over S_1 (and the aperture) are not known, but are assumed. If the true fields were known, each of these equations would give consistent results. In the literature, use is frequently made of Eq. (5.30), with the assumption that the tangential E and H fields are related by an impedance Z. This approach is particularly useful if the source side is a waveguide feed in which the characteristic wave impedance is different from the value of 377 Ω for free space. More will be said about this in Section 5.2.3.

The use of the integral relationship in Eq. (5.28) is exact if the correct field distribution is known everywhere on the infinite surface S_1. As discussed by Silver [5.9] and by Collin and Zucker [5.10], when this particular formulation is used for calculating the radiated field from a finite aperture, errors can arise in the solution. This is due to the fact that the equivalent currents on the screen do not obey the continuity equation right at the edge of the aperture, where the fields are assumed to jump abruptly to zero. Because this formulation explicitly contains a charge-density term [related to the $(\hat{n}' \cdot \mathbf{E})$ term in the equation], we expect that this solution will be sensitive to rapid fluctuations of the charge at this point.

To correct this problem, it is possible to add a suitably chosen distribution of electric and magnetic charge on the perimeter of the aperture. This adds an additional term to Eq. (5.28) (and another to the corresponding equation for the magnetic field). After some manipulation the resulting equation for the aperture field becomes

$$\mathbf{E}(\mathbf{r}) = \frac{1}{4\pi} \int_{\text{aperture}} (\hat{n}' \times \mathbf{E}) \times \nabla \frac{e^{-jkR}}{R} dS'$$

$$+ \frac{1}{j\omega\epsilon} \int_{\text{aperture}} (\hat{n}' \times \mathbf{H}) \cdot (k^2 \bar{\bar{I}} + \nabla\nabla) \frac{e^{-jkR}}{R} dS' \quad (5.35)$$

Noting the definition of $\bar{\bar{\Gamma}}$ in Eq. (5.27), it can easily be shown that Eq. (5.35) is identical to eq. (5.25). Thus the dyadic expressions of Eq. (5.25) are the preferable formulas for calculating the radiated fields from the aperture.

5.2.3 Far-Field Vector Field Diffraction

In many cases we are interested in the *far-field* expressions for the E-field from the aperture. Extracting the $1/r$ radiation zone components of Eq. (5.25) as discussed in [5.9] results in the following expression for the E-field in terms of the tangential aperture fields:

$$\mathbf{E}(\mathbf{r}) = \frac{-jk}{4\pi r} e^{-jkr} \hat{r} \times \int_{\text{aperture}} [\hat{n}' \times \mathbf{E} - Z_0 \hat{r} \times (\hat{n}' \times \mathbf{H})] e^{jk\mathbf{r}' \cdot \hat{r}} dS' \quad (5.36)$$

In this expression, r is the distance from the coordinate origin located in the aperture to the field observation point, as illustrated in Figure 5.3*b*. The term $\hat{r}$ is the unit vector in this direction, $\mathbf{r}'$ is the vector from the origin to the elementary source location dS' on the aperture, and $\hat{n}'$ is the unit normal to the aperture pointing towards the field point. Z_0 is the free-space impedance of 377 Ω.

Using geometrical optics approximation, the incident H-field in the aperture can be related to the E-field through an appropriate impedance Z as

$$\mathbf{H}^{inc} = \frac{1}{Z}(\hat{k} \times \mathbf{E}^{inc}) \qquad (5.37)$$

where $\hat{k}$ is a unit vector in the direction of propagation of the assumed aperture excitation field. The impedance Z is equal to Z_0 for the case of free space on the source side of the aperture. However, if the aperture is fed by a waveguide, this impedance is the approximate impedance for the waveguide mode exciting the aperture. With this assumption for the aperture excitation field, the following far-field expression results:

$$\mathbf{E}(\mathbf{r}) = \frac{-jk}{4\pi r} e^{-jkr} \hat{r} \times \int_{\text{aperture}} \left[\hat{n}' \times \mathbf{E}^{inc} \right.$$
$$\left. - \frac{Z_0}{Z} \times [\hat{r} \cdot (\hat{k} \times \mathbf{E}^{inc})\hat{n}' - (\hat{k} \times \mathbf{E}^{inc})(\hat{n}' \cdot \hat{r})] \right] e^{jkr' \cdot \hat{r}} dS' \qquad (5.38)$$

This expression can be simplified further for the case of a normally incident field, for which $\hat{k} = \hat{n}'$. In this case, the vector component $\hat{r} \times \hat{n}' \times \mathbf{H} = (1/Z)\hat{r} \times \hat{n}' \times \hat{n} \times \mathbf{E} = -(1/Z)\hat{r} \times \mathbf{E}$ in Eq. (5.36), and the far field becomes

$$\mathbf{E}(\mathbf{r}) \approx \frac{-jk}{4\pi r} e^{-jkr} \hat{r} \times [\hat{n}' + \alpha \hat{r}] \times \mathbf{N} \qquad (5.39)$$

where $\alpha = Z_0/Z$ and the vector $\mathbf{N}$ is defined as

$$\mathbf{N} = \int_{\text{aperture}} \mathbf{E}^{inc} e^{jkr' \cdot \hat{r}} dS' = \int_{\text{aperture}} \mathbf{E}^{inc}(x', y') e^{jk(x' \sin \chi + y' \sin \xi)} dx' dy'$$
$$(5.40a)$$

with χ and ξ being the angles defined in the coordinate system in Figure 5.4. Alternatively, $\mathbf{N}$ can be expressed in terms of the polar angles θ and ϕ as

$$\mathbf{N} = \int_{\text{aperture}} \mathbf{E}^{inc}(x', y') e^{jk(x' \sin \theta \cos \phi + y' \sin \theta \sin \phi)} dx' dy' \qquad (5.40b)$$

The resulting radiated E-field in polar components can be obtained from Eq.

(5.39) as

$$E_\theta(\mathbf{r}) = \frac{jk}{4\pi r} e^{-jkr}(1 + \alpha \cos\theta)(N_x \cos\phi + N_y \sin\phi) \quad (5.41a)$$

$$E_\phi(\mathbf{r}) = \frac{-jk}{4\pi r} e^{-jkr}(\cos\theta + \alpha)(N_x \sin\phi - N_y \cos\phi) \quad (5.41b)$$

where the angles θ and ϕ are the standard polar angles relative to the x'–y' coordinate system centered on the aperture plane, as shown in Figure 5.4.

5.2.3.1 Example for a Rectangular Aperture.
As an example of the vector fields' diffraction, consider the same rectangular aperture of Figure 5.4, which was previously analyzed using the scalar theory. We assume that the incident excitation field is a plane wave with amplitude E_0 polarized in the $\hat{x}$ direction, and that there is free space on the source side of the aperture so the factor $\alpha = 1$. Equations (5.41) become

$$E_\theta(\mathbf{r}) = \psi(P) \cos\phi \quad (5.42a)$$

$$E_\phi(\mathbf{r}) = -\psi(P) \sin\phi \quad (5.42b)$$

where the wavefunction $\psi(P)$ is expressed as

$$\begin{aligned}\psi(P) &= \frac{jk}{4\pi r} e^{-jkr}(1 + \cos\theta) \int_{\text{aperture}} E_0 e^{jk(x' \sin\chi + y' \sin\xi)} \, dx' \, dy' \\ &= \frac{jkabE_0}{2\pi r} e^{-jkr} \frac{1 + \cos\theta}{2} \mathcal{G}\end{aligned} \quad (5.43)$$

and $\mathcal{G}$ is the aperture function defined in Eq. (5.22) with the angles χ_S and $\xi_S = 0$.

Notice that Eq. (5.43) is identical to the scalar wavefunction in Eq. (5.20) if it is assumed that the source S in Figure 5.4 is at infinity and the excitation field is a normally incident plane wave, such that χ_S and $\xi_S = 0$. In this case the relevant terms in Eq. (5.20) become

$$\psi_0 \frac{e^{-jkr_s}}{r_s} \to E_0, \qquad O(\theta, \theta_s) \to \tfrac{1}{2}(1 + \cos\theta)$$

which immediately shows the correspondence between (5.43) and (5.20). Thus, after all of the mathematical development to derive the vector form of the aperture fields, we see that the simple scalar theory provides the same result in the far field. This explains why the scalar theory is so widely used in EMC problems.

The vector formulation is not completely meaningless, however, because the general form of the solution in Eq. (5.35) can be used to study the behavior of the near-zone fields for the aperture. The inherent limitation of both the scalar and vector theory, of course, is that we actually do not know the aperture fields. This is analogous to the wire antenna problem, in which we can easily compute the radiated or scattered field *if* we know the current on the wire. As in the case of the thin wire, knowledge of the aperture fields can be obtained through the solution of a suitable integral equation in the aperture. This is discussed in the next section.

5.2.4 Aperture Integral Equation

A much more accurate solution for the EM field penetration through an aperture can be obtained through the use of an integral equation solution for the tangential E and H fields in the aperture. As we have seen, if either $\mathbf{n}' \times \mathbf{E}$ or $\mathbf{n}' \times \mathbf{H}$, or both, are known on the surface S_1, the field at point P can be evaluated by Eq. (5.25), using a suitably chosen Green's function. Such an integral equation cannot be solved analytically and a numerical solution must be employed, often using the method of moments as applied to a two-dimensional problem.

To develop an integral equation for the aperture fields, consider the planar perfectly conducting sheet with the aperture shown in Figure 5.6. Region 1 to the left of the sheet is assumed to have some distant sources that produce *incident* EM fields, denoted by $\mathbf{E}^{inc}$ and $\mathbf{H}^{inc}$. Frequently, these fields are assumed to be a plane wave, but there is no requirement that this be the case. If the aperture were removed (i.e., the hole is filled with conductor), *reflected* EM fields from the conducting sheet, denoted by $\mathbf{E}^{ref}$ and $\mathbf{H}^{ref}$, will be present. The total field in region 1 then will be the sum of these incident and reflected components. In region 2 the field is identically zero when the aperture is filled.

When the aperture is present, there is a tangential E-field and H-field component existing in the aperture, and additional components of E and H in regions 1 and 2 due to the aperture. In region 1 this additional component is the scattered field from the aperture, and in region 2 the fields are the transmitted fields. Note that these scattered fields are symmetric about the conducting sheet.

Equation (5.25a) can be used to represent the E-field produced by the unknown aperture E-field $\mathbf{E}^a$ by choosing a Green's dyad $\overline{\overline{\Gamma}}_1$ which satisfies the boundary condition in Eq. (5.31) on the conducting sheet. In this way the

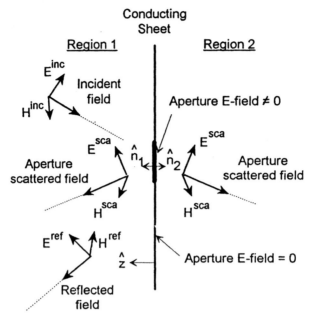

FIGURE 5.6 Geometry for the aperture integral equation.

total E-fields in both regions can be written as

$$\mathbf{E}_1^{tot}(\mathbf{r}_1) = \mathbf{E}^{inc}(\mathbf{r}_1) + \mathbf{E}^{ref}(\mathbf{r}_1) - \int_{aperture} (\hat{n}_1' \times \mathbf{E}^a) \cdot \nabla' \times \overline{\overline{\Gamma}}_1(\mathbf{r}'; \mathbf{r}_1) \, dS'$$
(5.44a)

$$\mathbf{E}_2^{tot}(\mathbf{r}_2) = -\int_{aperture} (\hat{n}_2' \times \mathbf{E}^a) \cdot \nabla_\times' \overline{\overline{\Gamma}}_1(\mathbf{r}'; \mathbf{r}_2) \, dS'$$
(5.44b)

where $\hat{n}_1$ and $\hat{n}_2$ are the unit normals on either side of the sheet pointing into regions 1 and 2, and $\mathbf{r}_1$ and $\mathbf{r}_2$ are vectors to the observation points in regions 1 and 2, respectively. Similarly, Eq. (5.25b) can be used to compute the H-fields in both regions due to the aperture E-field as

$$\mathbf{H}_1^{tot}(\mathbf{r}_1) = \mathbf{H}^{inc}(\mathbf{r}_1) + \mathbf{H}^{ref}(\mathbf{r}_1) - j\omega\epsilon \int_{aperture} (\hat{n}_1' \times \mathbf{E}^a) \cdot \overline{\overline{\Gamma}}_2(\mathbf{r}'; \mathbf{r}_1) \, dS'$$
(5.45a)

$$\mathbf{H}_2^{tot}(\mathbf{r}_2) = -j\omega\epsilon \int_{aperture} (\hat{n}_1' \times \mathbf{E}^a) \cdot \overline{\overline{\Gamma}}_2(\mathbf{r}'; \mathbf{r}_2) \, dS'$$
(5.45b)

where the Green's dyad $\overline{\overline{\Gamma}}_2$ now satisfies Eq. (5.33).

The boundary conditions on the E and H fields at the aperture are that the tangential components of the total fields are continuous in passing through the aperture. This condition on the E-field has already been used implicitly in Eqs. (5.44) and (5.45) in assuming that the equivalent source $\hat{n} \times \mathbf{E}^a$ is the same on each side of the aperture. For the magnetic field, the required continuity equation in the aperture is expressed as

$$(\hat{n}_1 \times \mathbf{H}_1^{\text{tot}})_{\mathbf{r}_1 \text{ on aperture}} = -(\hat{n}_2 \times \mathbf{H}_2^{\text{tot}})_{\mathbf{r}_2 \text{ on aperture}} \tag{5.46a}$$

Noting that $\hat{n}_1 = -\hat{n}_2$ and substituting Eqs. (5.45a) and (5.45b) into Eq. (5.46a) yields the following expression:

$$\hat{n}_1 \times (\mathbf{H}^{\text{inc}}(\mathbf{r}_1) + \mathbf{H}^{\text{ref}}(\mathbf{r}_1)) - j\omega\epsilon\, \hat{n}_1 \times \int_{\text{aperture}} (\hat{n}_1' \times \mathbf{E}^a) \cdot \overline{\overline{\Gamma}}_2(\mathbf{r}'; \mathbf{r}_1)\, dS'$$

$$= j\omega\epsilon\, \hat{n}_1 \times \int_{\text{aperture}} (\hat{n}_1' \times \mathbf{E}^a) \cdot \overline{\overline{\Gamma}}_2(\mathbf{r}'; \mathbf{r}_2)\, dS' \tag{5.46b}$$

Using the fact that $\mathbf{H}^{\text{ref}}(\mathbf{r}_1)$ has been defined so that $\hat{n}_1' \times (\mathbf{H}^{\text{inc}}(\mathbf{r}_1) + \mathbf{H}^{\text{ref}}(\mathbf{r}_1)) = 2\hat{n}_1' \times \mathbf{H}^{\text{inc}}(\mathbf{r}_1)$ for $\mathbf{r}_1$ everywhere on the conducting sheet containing the aperture, the following integral equation for the unknown tangential E-field results:

$$\hat{n}_1 \times \mathbf{H}^{\text{inc}}(\mathbf{r}_1) = j\omega\epsilon\, \hat{n}_1 \times \int_{\text{aperture}} (\hat{n}_1' \times \mathbf{E}^a) \cdot \overline{\overline{\Gamma}}_2(\mathbf{r}'; \mathbf{r}_1)\, dS'$$

(for $\mathbf{r}_1$ on aperture) (5.47)

The dyadic notation of Eq. (5.47) is useful for manipulation of the various equations but is not particularly illuminating for numerical calculations. Equation (5.27) can be expanded to provide an explicit expression for the free-space dyadic Green's function in Cartesian coordinates as

$$\overline{\overline{\Gamma}}(\mathbf{r}'; \mathbf{r}) = -\frac{1}{4\pi} \begin{pmatrix} 1 + \dfrac{1}{k^2}\dfrac{\partial^2}{\partial x^2} & \dfrac{1}{k^2}\dfrac{\partial^2}{\partial x\, \partial y} & \dfrac{1}{k^2}\dfrac{\partial^2}{\partial x\, \partial z} \\[6pt] \dfrac{1}{k^2}\dfrac{\partial^2}{\partial x\, \partial y} & 1 + \dfrac{1}{k^2}\dfrac{\partial^2}{\partial y^2} & \dfrac{1}{k^2}\dfrac{\partial^2}{\partial y\, \partial z} \\[6pt] \dfrac{1}{k^2}\dfrac{\partial^2}{\partial x\, \partial z} & \dfrac{1}{k^2}\dfrac{\partial^2}{\partial y\, \partial z} & 1 + \dfrac{1}{k^2}\dfrac{\partial^2}{\partial z^2} \end{pmatrix} \frac{e^{-jk|\mathbf{r}-\mathbf{r}'|}}{|\mathbf{r}-\mathbf{r}'|}$$

(5.48)

The corresponding magnetic field Green's dyad that satisfies Eq. (5.33) can be written by inspection using image theory. It is

$$\bar{\bar{\Gamma}}_2(\mathbf{r}';\mathbf{r}) =$$

$$-\frac{1}{4\pi}\begin{pmatrix} 1+\dfrac{1}{k^2}\dfrac{\partial^2}{\partial x^2} & \dfrac{1}{k^2}\dfrac{\partial^2}{\partial x\,\partial y} & \dfrac{1}{k^2}\dfrac{\partial^2}{\partial x\,\partial z} \\ \dfrac{1}{k^2}\dfrac{\partial^2}{\partial x\,\partial y} & 1+\dfrac{1}{k^2}\dfrac{\partial^2}{\partial y^2} & \dfrac{1}{k^2}\dfrac{\partial^2}{\partial y\,\partial z} \\ \dfrac{1}{k^2}\dfrac{\partial^2}{\partial x\,\partial z} & \dfrac{1}{k^2}\dfrac{\partial^2}{\partial y\,\partial z} & 1+\dfrac{1}{k^2}\dfrac{\partial^2}{\partial z^2} \end{pmatrix}\left(\dfrac{e^{-jk|\mathbf{r}-\mathbf{r}'|}}{|\mathbf{r}-\mathbf{r}'|} + \dfrac{e^{-jk|\mathbf{r}-\mathbf{r}''|}}{|\mathbf{r}-\mathbf{r}''|}\right)$$

(5.49)

where $\mathbf{r}''$ is a vector from the origin of the coordinate system to the image of the source point $\mathbf{r}'$ in the conducting screen. A similar equation can also be written for $\bar{\bar{\Gamma}}_1$, but with a minus sign in the last term in parentheses.

For the coordinate system shown in Figure 5.6, in which the z-direction is normal to the aperture, the pertinent components of the aperture E-field are in the x and y directions. Extracting these two components from Eq. (5.47) and noting that for both the source and observation points being located on the aperture, the term

$$\frac{e^{-jk|\mathbf{r}-\mathbf{r}'|}}{|\mathbf{r}-\mathbf{r}'|} + \frac{e^{-jk|\mathbf{r}-\mathbf{r}''|}}{|\mathbf{r}-\mathbf{r}''|} \Rightarrow 2\,\frac{e^{-jk\sqrt{(x-x')^2+(y-y')^2}}}{\sqrt{(x-x')^2+(y-y')^2}}$$

and the following set of coupled integrodifferential equations results:

$$\left(\frac{\partial^2}{\partial x^2}+k^2\right)F_x(x,y) + \frac{\partial^2}{\partial x\,\partial y}F_y(x,y) = -j\omega\mu H_x^{\text{inc}}(x,y) \qquad \text{(on aperture)}$$

(5.50a)

$$\left(\frac{\partial^2}{\partial y^2}+k^2\right)F_y(x,y) + \frac{\partial^2}{\partial y\,\partial x}F_x(x,y) = -j\omega\mu H_y^{\text{inc}}(x,y) \qquad \text{(on aperture)}$$

(5.50b)

Here the functions F are defined as

$$F_x(x,y) = \frac{1}{2\pi}\int_{\text{aperture}} (\mathbf{E}_{\tan}^a(x',y'))_x \frac{e^{-jk\sqrt{(x-x')^2+(y-y')^2}}}{\sqrt{(x-x')^2+(y-y')^2}}\,dx'\,dy'$$

(5.51a)

$$F_y(x,y) = \frac{1}{2\pi}\int_{\text{aperture}} (\mathbf{E}_{\tan}^a(x',y'))_y \frac{e^{-jk\sqrt{(x-x')^2+(y-y')^2}}}{\sqrt{(x-x')^2+(y-y')^2}}\,dx'\,dy'$$

(5.51b)

Equations (5.50) are the integral equations for the tangential components of the aperture E-field and are identical to those derived in [5.11]. They are the two-dimensional analog to the E-field integrodifferential equation (4.30) for the thin-wire antenna. Note that the same equations can be derived for the complementary problem of determining the induced surface current on a perfectly conducting scatterer having the same shape as the aperture by using Babinet's principle [5.10]. It should be noted in passing that there is another form of the aperture integral equation, as developed by Bouwkamp [5.12] and discussed further in [5.11]. This results in a different vector integral equation for the aperture field plus an unknown function defined along the perimeter of the aperture.

5.2.4.1 Example of Aperture Field Calculation.
The solution of Eq. (5.50) for the aperture E-field requires the use of the method of moments and, in many cases, can be quite demanding of computer resources. Assuming that 10 basis functions per wavelength are needed, a simple $3\lambda \times 3\lambda$ aperture will require the filling and inversion of a 300×300 full matrix. Clearly, for high-frequency aperture problems, in which the aperture may be hundreds of wavelengths across, this method is difficult to realize with present-day computers.

As an example of numerical results from an integral equation calculation, [5.13] presents the E_x and E_y field distribution in a $1\lambda \times 1\lambda$ square aperture in an infinite screen for a normally incident E-field polarized in the y-direction. These results are reproduced in Figure 5.7. Notice that E_y is the primary component in the aperture, since the incident field is also y-directed. However, unlike the assumptions for the Kirchhoff approximation, this field is not a constant given by E_y^{inc}. At the $x = 0$ and $x = 1\lambda$ edges of the aperture, the field goes to zero as required by the boundary condition that the tangential E-field along a perfect conductor vanish. Along the $y = 0$ and $y = 1\lambda$ edges, the field increases (to an infinite value in the ideal case), due to the edge condition for normal E-fields. Moreover, there is a cross-polarized component of the aperture field, as shown in Figure 5.7b. In the Kirchhoff approximation, this field is nonexistent.

It is useful to examine the need for a numerical solution of aperture penetration problems and to put the results of the Kirchhoff approximation into perspective. Figure 5.8, reproduced from [5.13], presents the square of the normalized E-field ratio $|E_y|^2/|E^{inc}|^2$ as a function of normalized distance d/λ for a square and circular aperture [5.14]. The three curves show basically the same behavior for $d/\lambda > 1.5$. The conclusion that is drawn from the result is that although the E-field distribution may differ considerably from that of the incident field, if the observer is sufficiently far from the aperture, the exact details of the aperture field are not very important in determining the far field.

5.2.5 Equivalent Area of an Aperture

The development in previous sections permits the calculation of the spatial dependence of the EM fields that penetrate through an aperture. In any EMC

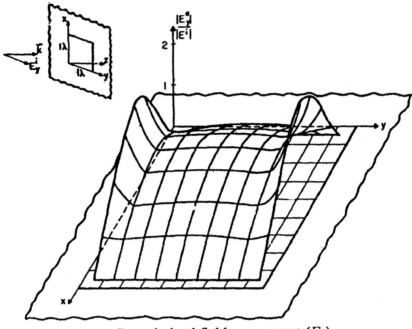

a. Co-polarized field component (E_y)

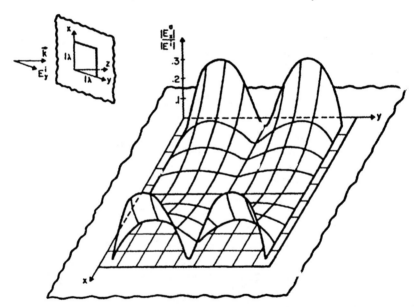

b. Cross-polarized field component (E_x)

FIGURE 5.7 Three-dimensional plots of the aperture field distribution in a $1\lambda \times 1\lambda$ square aperture. (From [5.13]. © 1978 Institute of Electrical and Electronics Engineers.)

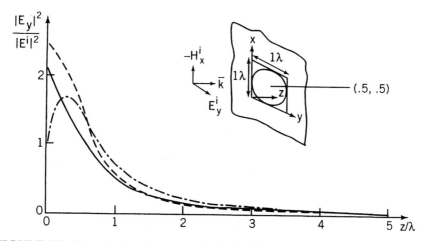

FIGURE 5.8 Intensity distribution of the E_y-field normal to a 1λ square and circular aperture. Integral equation solution for the square aperture (solid line), Kirchhoff approximation for the square aperture (long-short dashed line), measured results for circular aperture (dashed line). (From [5.13]. © 1978 Institute of Electrical and Electronics Engineers.)

problems, however, detailed knowledge of the E or H fields behind the aperture is not needed. Sometimes, a simple estimate of the fraction of EM power penetrating through the aperture will be sufficient.

To do this, we define an *equivalent area* of the aperture A_{eq} by means of the equation

$$\mathcal{P}_t = A_{eq}(\hat{n} \cdot \mathbf{P}^{inc}) \tag{5.52}$$

where $\mathcal{P}_t$ is the total power *transmitted* through the aperture and $\mathbf{P}^{inc}$ is the average *power density* in the incident field, which for an assumed plane wave is given by

$$\mathbf{P}^{inc} = \tfrac{1}{2}\,\mathrm{Re}(\mathbf{E}^{inc} \times (\mathbf{H}^{inc})^*) = \tfrac{1}{2} Z_0 |\mathbf{E}^{inc}|^2 \hat{k} \tag{5.53}$$

where the vector $\hat{k}$ is in the direction of propagation of the incident field and may not necessarily be normal to the aperture.

The calculation of the transmitted power can be accomplished by evaluating the integral

$$\mathcal{P}_t = \tfrac{1}{2}\,\mathrm{Re}\int_{\mathrm{surface}} \mathbf{E}^{sca} \times (\mathbf{H}^{sca})^* \cdot d\mathbf{S} \tag{5.54}$$

where the surface of integration is an open surface surrounding the aperture. With reference to Figure 5.3, this surface could be S_1, S_∞, or any other intermediate surface. If the plane of the screen is chosen, the benefit is that the region of integration is limited to only over the aperture. The difficulty with this, however, is that the aperture fields are not well known and the integrations may be in error. Choosing the surface S_∞ requires integrating over an infinite surface, but on this surface, E and H are far-field quantities and the field expressions are relatively simple [see Eq. (5.41)].

Reference [5.15] presents the equivalent areas for circular, elliptical, and rectangular apertures using the scalar aperture formulation discussed in Section 5.2.1, and shows that there is a good agreement with the more accurate treatments for apertures with diameters larger than about 0.8λ. For a circular aperture of radius a, the normalized equivalent area is given in closed form as

$$\frac{A_{eq}}{A_0} = 1 - \frac{2J_1(4\pi a/\lambda)}{4\pi a/\lambda} \tag{5.55}$$

where $J_1(\cdot)$ is the Bessel function of order 1, $\lambda = 2\pi c/\omega$, and $A_0 = \pi a^2$ is the geometrical area of the aperture. Figure 5.9 illustrates the behavior A_{eq}/A_0 for the circular aperture as a function of the aperture size in terms of wavelengths.

For the elliptical aperture having principal radii a and b, the normalized equivalent area is given by an integral,

$$\frac{A_{eq}}{A_0} = 1 - \frac{ab}{\pi^2 \lambda^2} \int_0^{\pi/2} \frac{J_1[4\pi\sqrt{(a/\lambda)^2 \cos^2\zeta + (b/\lambda)^2 \sin^2\zeta}]}{[(a/\lambda)^2 \cos^2\zeta + (b/\lambda)^2 \sin^2\zeta]^{3/2}} d\zeta \tag{5.56}$$

which must be evaluated numerically. Similarly, the equation for a rectangular aperture with side dimensions a and b (like that shown in Figure 5.4) is given

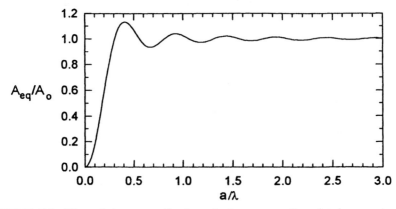

FIGURE 5.9 Plot of the normalized equivalent area of a circular aperture of radius a.

by the double integral

$$\frac{A_{eq}}{A_0} = \frac{ab}{\lambda^2} \int_0^{\pi/2} \int_0^{\pi/2} \frac{\sin([(a\pi/\lambda)\sin\theta\cos\phi]}{(a\pi/\lambda)\sin\theta\cos\phi} \frac{\sin[(b\pi/\lambda)\sin\theta\cos\phi]}{(b\pi/\lambda)\sin\theta\cos\phi} \sin\theta \, d\theta \, d\phi$$

(5.57)

Unfortunately, this last expression cannot be integrated analytically, nor can it be reduced to a single integral as in Eq. (5.56). A numerical integration is needed. Reference [5.15] presents several curves showing the differences in the equivalent areas of these apertures.

5.3 RADIATION FROM EXTENDED ANTENNAS

Although the integral field representation of Eq. (5.35) has been discussed with the application of aperture field penetration in mind, this methodology can also be applied to the problem of computing the radiation pattern from extended antennas, such as the horn-fed dish antenna shown in Figure 5.1a. Assuming that this antenna is a perfect conductor and neglecting the effects of the feed horn (other than illuminating the dish), Eq. (5.35) can be used to compute the E-field at a point P by integrating over an equivalent surface current density $\mathbf{J}_s(\mathbf{r}') = \hat{n}' \times \mathbf{H}$, where $\mathbf{H}$ is the magnetic field existing over the dish. Because the tangential E-field is zero on the dish surface, Eq. (5.35) becomes

$$\mathbf{E}(\mathbf{r}) = \int_{antenna} \left(-j\omega\mu \mathbf{J}_s(\mathbf{r}') \frac{e^{-jk|\mathbf{r}-\mathbf{r}'|}}{4\pi|\mathbf{r}-\mathbf{r}'|} + \frac{1}{j\omega\epsilon} \mathbf{J}_s(\mathbf{r}')\nabla'\nabla' \frac{e^{-jk|\mathbf{r}-\mathbf{r}'|}}{4\pi|\mathbf{r}-\mathbf{r}'|} \right) dS'$$

(5.58)

The usual geometrical optics approximation is used to estimate the antenna current by assuming that it is locally flat and invoking the flat-plate condition that $\mathbf{J}_s(\mathbf{r}') = \hat{n}' \times \mathbf{H} \approx 2\hat{n}' \times \mathbf{H}^{inc}$. Although this integration uses an assumed current distribution, it still can be time consuming for electrically large antennas.

One useful simplification for this equation is to extract the far-field components. When this is done, the E-field expression becomes [5.10]

$$\mathbf{E}(\mathbf{r}) = -j\omega\mu \frac{e^{-jkr}}{4\pi r} \int_{antenna} [\mathbf{J}_s(\mathbf{r}') - (\hat{r} \cdot \mathbf{J}_s(\mathbf{r}'))\hat{r}] e^{jk\hat{r}\cdot\mathbf{r}'} \, dS' \quad (5.59a)$$

The corresponding far-field expression for the H-field is

$$\mathbf{H}(\mathbf{r}) = jk \frac{e^{-jkr}}{4\pi r} \int_{antenna} (\mathbf{J}_s \times \hat{r}) e^{jk\hat{r}\cdot\mathbf{r}'} \, dS' \quad (5.59b)$$

Additional details on the analysis of this type of radiating antenna are provided in [5.16] and [5.17].

5.4 LOW-FREQUENCY APPROXIMATION

There are two practical difficulties in the application of the preceding models for EMC problems: The aperture field distribution is unknown and a time-consuming aperture integration must be performed to obtain the penetrating fields. However, in cases for which the frequency is sufficiently low, it is possible to represent the effects of the aperture by equivalent electric and magnetic dipoles which produce the same spatial distribution as the actual penetrating field. Moreover, these dipole moments can be simply related to the exciting E and H fields through quantities known as the aperture polarizabilities, which are functions of the aperture shape.

5.4.1 Dipole Moments

In Chapter 4 the electric dipole moment of a Hertzian current in a free-space element was defined as $\mathbf{p} = I\,d\mathbf{l}/j\omega$ and the dual magnetic dipole moment arising from a circulating loop of current was $\mathbf{m} = I\,d\mathbf{S}$. A more general volume distribution of current $\mathbf{J}$ can be expanded into a set of electric and magnetic multipole moments, each of which has a unique multilobe radiation pattern [5.3]. At low frequencies, all of the multipole terms exist in the current distribution, but the dipole terms dominate. These dipole moments in the general case are defined as

$$\mathbf{p} = \int_V \mathbf{r}\rho\,dV = \frac{1}{j\omega}\int_V \mathbf{J}\,dV \qquad (5.60)$$

$$\mathbf{m} = \frac{1}{2}\int_V (\mathbf{r}\times\mathbf{J})\,dV \qquad (5.61)$$

Equations (4.12) and (4.19) present the E and H fields radiated by electric and magnetic sources $I\,dl$ and $I\,dS$, which correspond to z-directed dipole moments. A more general expression for these fields at an observation point $\mathbf{r}$ due to a set of arbitrarily oriented dipole sources at location $\mathbf{r}'$ is provided in [5.18] as

$$\mathbf{E}(\mathbf{r}) = -\frac{1}{\epsilon}\nabla\times\left[\mathbf{p}(\mathbf{r}')\times\nabla'\frac{e^{-jkR}}{4\pi R}\right] + j\omega\mu\left[\mathbf{m}(\mathbf{r}')\times\nabla'\frac{e^{-jkR}}{4\pi R}\right] \qquad (5.62a)$$

$$\mathbf{H}(\mathbf{r}) = -\nabla\times\left[\mathbf{m}(\mathbf{r}')\times\nabla'\frac{e^{-jkR}}{4\pi R}\right] - j\omega\left[\mathbf{p}(\mathbf{r}')\times\nabla'\frac{e^{-jkR}}{4\pi R}\right] \qquad (5.62b)$$

where $R = |\mathbf{r} - \mathbf{r}'|$.

5.4.2 Aperture Polarizabilities

For the aperture problem, the electric current exists everywhere on the sheet and not over the aperture. As discussed in [5.18], this is the *dual* problem to that of a magnetic current $\mathbf{J}_m = (\hat{n} \times \mathbf{E})$ and magnetic charge $\rho_m = -(1/j\omega)\nabla \cdot \mathbf{J}_m$ located over the aperture, and as such, the aperture can be represented by *equivalent* aperture dipole moments $\mathbf{p}_a$ and $\mathbf{m}_a$, as shown in Figure 5.10. The electric dipole moment $\mathbf{p}_a$ is perpendicular to the aperture, and the magnetic dipole moment $\mathbf{m}_a$ lies in the aperture plane. Thus, for a coordinate system with the z-axis perpendicular to the aperture (as in Figure 5.3b), these dipoles are represented as $\mathbf{p}_a = p_{az}\hat{z}$ and $\mathbf{m}_a = m_{ax}\hat{x} + m_{ay}\hat{y}$.

The strengths of the aperture dipole moments are proportional to the E and H fields existing on the illuminated side of the aperture, with the aperture filled. These are the *short-circuited* fields, and because the plane of the aperture is assumed to be perfectly conducting, they consist only of a normal E-field, $E_{sc}\hat{z}$, and a tangential H-field, $\mathbf{H}_{sc} = H_{scx}\hat{x} + H_{scy}\hat{y}$. The relationships between the dipole and the short-circuit fields are given through the aperture polarizabilities, which are actually dyads because they relate one vector (the dipole moment) to another vector (the field). However, by virtue of the nature of the fields, only three independent components exist and these relationships are given conveniently in component form as

$$p_{az} = \alpha_e \epsilon_0 E_{sc}$$
$$m_{ax} = -\alpha_{mx} H_{scx} \quad (5.63a)$$
$$m_{ay} = -\alpha_{my} H_{scy}$$

Care must be used in interpreting the polarizabilities of Eq. (5.63a), because the equivalent aperture dipole moments $\mathbf{p}_a$ and $\mathbf{m}_a$ are located in the presence of the perfectly conducting screen with the aperture short-circuited. The screen will influence the fields produced by these sources. The usual approach for computing the radiated fields is to *image* the dipoles in the

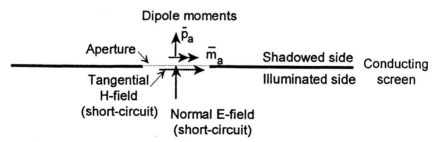

FIGURE 5.10 Equivalent dipole moments of the aperture on the top surface of a conducting plate.

screen. This provides *total* electric and magnetic dipole moments **p** and **m** as

$$\mathbf{p} = 2\mathbf{p}_a \quad \text{and} \quad \mathbf{m} = 2\mathbf{m}_a \tag{5.63b}$$

which are considered to be *radiating in free space*. As a consequence, Eq. (5.62) [or equivalently, Eqs. (4.12) and (4.19)] can be used to calculate the fields produced by these equivalent aperture sources.

The imaging of the dipole moments provides an additional factor of 2 in the dipole strengths of Eq. (5.63b), and as discussed in [5.18], this is the reason for some of the differences in the polarizability expressions in the literature. Following [5.18], we will define the aperture polarizabilties as in Eq. (5.63) and will require that the resulting dipole moments must be imaged in the screen to compute the radiated fields. Various expressions and curves are available for the polarizabilities of simple apertures in [5.18] and [5.19]. Table 5.1 lists the polarizabilities of several simple shapes that are useful approximations to realistic apertures found in EMC applications.

5.4.2.1 Corrections of the Polarizabilities for Aperture Loading.

The aperture polarizabilities of Eq. (5.63) were developed under the assumption that free space exists on either side of the aperture. In some cases there can be impedance loading over the aperture (like an imperfectly conducting screen covering the hole), or there can be materials of different dielectric constant on either side of the aperture. In these cases it is necessary to modify the

TABLE 5.1 Polarizabilities (to Be Imaged) for Selected Apertures[a]

Aperture Shape	α_e	α_{mx}	α_{my}
Circle (diameter d)	$\dfrac{1}{12}d^3$	$\dfrac{1}{6}d^3$	$\dfrac{1}{6}d^3$
Ellipse	$\dfrac{\pi}{24}\dfrac{w^2\ell}{E(e)}$	$\dfrac{\pi}{24}\dfrac{e^2\ell^3}{K(e)-E(e)}$	$\dfrac{\pi}{24}\dfrac{e^2\ell^3}{(\ell/w)^2 E(e)-K(e)}$
Narrow ellipse ($w \ll \ell$)	$\dfrac{\pi}{24}w^2\ell$	$\dfrac{\pi}{24}\dfrac{\ell^3}{\ln(4\ell/w)-1}$	$\dfrac{\pi}{24}w^2\ell$
Narrow slit ($w \ll \ell$)	$\dfrac{\pi}{16}w^2\ell$	$\dfrac{\pi}{24}\dfrac{\ell^3}{\ln(4\ell/w)-1}$	$\dfrac{\pi}{16}w^2\ell$

Source: [5.18].

[a] K and E are the complete elliptical integrals:

$$E(e) = \int_0^{\pi/2} (1 - e^2 \sin^2\xi)^{1/2}\, d\xi \quad K(e) = \int_0^{\pi/2} (1 - e^2 \sin^2\xi)^{-1/2}\, d\xi$$

Ellipse eccentricity $e = \sqrt{1 - (w/\ell)^2}$.

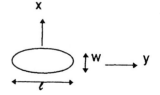

polarizabilities. Reference [5.18] discusses correction factors to account for these loading effects for both the magnetic and electric polarizabilities.

For the case of an impedance-loaded aperture, the presence of the conducting material will cause a reduction of the penetrating magnetic flux and consequently, will reduce the magnetic polarizability. Each component of the magnetic polarizability for the loaded aperture, denoted by α_m, can be expressed approximately in terms of the polarizabilities for the unloaded aperture, α_{m0}, by

$$\alpha_m = \left(1 + \frac{2}{3\pi} \frac{j\omega L_a}{Z_s}\right)^{-1} \alpha_{m0} \quad (5.64)$$

where L_a is an effective inductance of the aperture given by

$$L_a \approx \mu_0 \frac{\text{area}}{\text{perimeter}} = \mu_0 \frac{a}{2} \quad (5.65)$$

and Z_s is the effective sheet impedance loading of the aperture in units of ohms/square. For a reasonably good conductor covering the aperture the surface impedance is given as [5.20]

$$Z_s \approx (1+j)\sqrt{\frac{\omega\mu}{2\sigma}} \quad \Omega/\text{square} \quad (5.66)$$

where σ and μ are the conductivity and permeability of the cover material, respectively.

It is also possible to include the effects of a dielectric backings and coatings of an aperture. In Figure 5.10, consider the case when there is a material with a relative dielectric constant ϵ_{r1} in the region above the screen (i.e., in the observer, or shadowed, region), and a different material with dielectric constant ϵ_{r2} below the aperture (in the illuminated region). In this case, effective electric polarizability of the aperture α_e, referenced to the shadowed region, can be expressed in terms of the polarizability for the aperture with no dielectric present (denoted by α_{e0}) as

$$\alpha_e = \frac{2\epsilon_{r1}}{\epsilon_{r1} + \epsilon_{r2}} \alpha_{e0} \quad (5.67)$$

5.5 WIDEBAND AND TRANSIENT RESPONSES OF APERTURES

At times, there may be a requirement to calculate the penetration of transient EM fields through apertures for an EMP or lightning assessment. This requires the knowledge of the wideband response of the aperture, together with an inverse Fourier transform to obtain the transient response. Alternatively, a direct time-domain calculation can be used.

5.5.1 Wideband Responses

The wideband calculation of an aperture poses some difficulties, as the models that we have discussed are valid only for a limited frequency range. For example, if the aperture model uses any sort of a far-field approximation, the resulting calculation will be valid only down to a low frequency where $kR \approx 10$. Furthermore, the use of the Kirchhoff approximation in determining the equivalent aperture sources provides a fundamental limitation in the accuracy of the solution. As we will see, this approximation leads to a noncausal penetrating aperture field.

To illustrate the calculation of the transient response of an aperture, consider the rectangular aperture of Figure 5.4 with a normally incident transient plane-wave excitation ($\chi_s = \xi_s = 0$), which is polarized in the x-direction and is denoted by $\mathbf{E}^{inc} = E_x^{inc}\hat{x}$. For this example, we will be interested in calculating the penetrating field along the X_p axis in Figure 5.4, that is to say, the transmitted E_θ field component, as a function of the observation angle θ, with $\phi = 0$. Eqs. (5.42) and (5.43) provide this field component in the frequency domain as

$$E_\theta(\theta) = \frac{jk}{4\pi r} e^{-jkr}(1 + \cos\theta) \int_{\text{aperture}} E_x^{inc} e^{jk(x'\sin\theta)} \, dx' \, dy'$$

$$= \frac{jkab E_x^{inc}}{2\pi r} e^{-jkr} \frac{1 + \cos\theta}{2} \frac{\sin[(ka/2)\sin\theta]}{(ka/2)\sin\theta} \quad (5.68)$$

A suitable measure of the radiated field is the normalized function rE_θ / bE_x^{inc}, which represents the impulse response spectrum. Removing the exponential function e^{-jkr} in Eq. (5.68), which amounts to shifting $t = 0$ to the observer, yields the expression

$$\left.\frac{rE_\theta(\theta)}{bE_x^{inc}}\right|_{\text{impulse}} = \frac{jka}{4\pi}(1 + \cos\theta)\frac{\sin[(ka/2)\sin\theta]}{(ka/2)\sin\theta} \quad (5.69)$$

Figure 5.11 plots the magnitude of this spectrum as a surface function of the angle of observation θ and the parameter ka, which is the normalized frequency ($2\pi f/c$). Notice that for $\theta = 0$ the transmitted field is simply proportional to $j\omega$, which suggests that the transmitted transient field in the normal direction will appear as the *derivative* of the incident field. For other observation angles, notice that the $\sin x/x$ term causes ripples in the frequency response, indicating that the inverse Fourier transform of this response will be different from a simple doublet distribution (i.e., different from the derivative of a delta function).

It is known that for a causal, real-valued function $f(t)$, the resulting Fourier

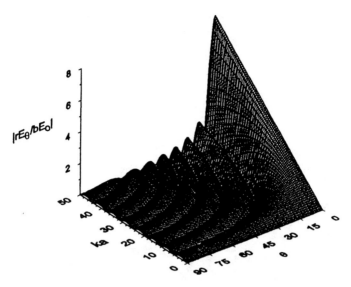

FIGURE 5.11 Surface plot of the magnitude of the normalized impulse response spectrum of the transmitted aperture field, as a function of the angle of observation θ and normalized frequency ka.

transform spectrum $F(\omega)$ must satisfy the Hilbert transform relations [5.21]

$$\mathrm{Re}[F(\omega)] = \frac{2}{\pi} \int_0^\infty \frac{\zeta \, \mathrm{Im}[F(\zeta)]}{\omega^2 - \zeta^2} \, d\zeta + K_0 \qquad (5.70a)$$

$$\mathrm{Im}[F(\omega)] = \frac{2\omega}{\pi} \int_0^\infty \frac{\mathrm{Re}[F(\zeta)]}{\omega^2 - \zeta^2} \, d\zeta \qquad (5.70b)$$

where Re and Im denote the real and imaginary parts of the spectrum F, and K_0 is a constant. For the case when $f(t) = \delta(t)$, the spectrum is $F(\omega) = 1$, so that $\mathrm{Re}[F(\omega)] = K_0 = 1$ and $\mathrm{Im}[F(\omega)] = 0$. Similarly, for a doublet function $f(t) = \delta'(t)$, the spectrum is $F(\omega) = j\omega$, with the consequence that $\mathrm{Im}[F(\omega)] = \omega$ and K_0 is equal to minus the integral in Eq. (5.70a), so that the resulting $\mathrm{Re}[F(\omega)] = 0$.

For other spectra not involving distributions, we see that the implication of Eqs. (5.70) is that for a causal function, both the real and imaginary parts of the spectrum must exist simultaneously. However, we note that the spectrum for the transmitted aperture field in Eq. (5.69) contains only an imaginary part. We must conclude that for observation angles other than zero, the spectrum is noncausal and the resulting transient response will be in error. This noncausal nature of the fields arises from the approximations made in developing a "simple" solution for the complex physical problem.

As an example of a transient field penetrating the aperture, we assume that

214 RADIATION, DIFFRACTION, AND SCATTERING MODELS FOR APERTURES

the incident field is a *step function* in time, so that the frequency-domain spectrum is

$$\mathbf{E}^{inc}(\omega) = E_0 \frac{1}{j\omega} \hat{x} \qquad (5.71)$$

With this excitation, the normalized transmitted field spectrum in Eq. (5.69) becomes

$$\left.\frac{rE_\theta(\theta)}{bE_0}\right|_{step} = \frac{a}{4\pi c}(1+\cos\theta)\frac{\sin[(\omega a/2c)\sin\theta]}{(\omega a/2c)\sin\theta} \qquad (5.72)$$

The inverse Fourier transform of Eq. (5.72) can be obtained analytically, to yield the transient step-function response

$$\frac{rE_\theta(\theta;t)}{bE_0} = \frac{1}{4\pi}\frac{1+\cos\theta}{\sin\theta}\left[U\left(t+\frac{a}{2c}\sin\theta\right) - U\left(t-\frac{a}{2c}\sin\theta\right)\right] \qquad (5.73)$$

where $U(\tau)$ is the unit step function defined as $U(\tau) = 0$ for $\tau < 0$ and $U(\tau) = 1$ for $\tau \geq 0$. Note that the time t in Eq. (5.73) is the retarded time, so that $t = 0$ is when the transmitted wave first arrives at the observation location. The noncausal nature of this transmitted waveform is evident. The first function U contributes to the response at time $t = -(a/2c)\sin\theta$, which is clearly before the wave is supposed to arrive at the observer.

To illustrate the temporal behavior of the transmitted field, Figure 5.12a presents a surface plot of the noncausal normalized field as a function of normalized time $t' = ct/a$. Notice that at $\theta = 0$, the transient response appears as an impulse function—the derivative of the incident step function—and the noncausal nature of the response is not readily apparent. As the angle of observation increases, however, it is clear that the initial response is occurring too soon in time.

There are several ways of enforcing causality in this solution. The best, of course, is to formulate the problem accurately in the first place by using a numerical integral equation solution for the aperture field and a numerical Fourier transform of the resulting penetrating field spectrum. Other alternative techniques are possible, however, with considerably less effort. One way suggested in [5.21] is to use the Hilbert transform relations in Eq. (5.70b) to compute a companion imaginary spectral component using Eq. (5.72) for the real part, and then compute the time-domain response numerically. An alternative is to compute the minimum phase spectral response, considering Eq. (5.72) to represent the magnitude of the desired spectrum. Each approach will provide a slightly different transient field response.

An alternative to these numerically intensive approaches simply is to shift the transient response in Figure 5.12a by the time $t = (a/2c)\sin\theta$, which will make the response causal. This is not a rigorous solution, but at this point in the analysis, one questions the advisability of extensive numerical "fixes" to a

5.5 WIDEBAND AND TRANSIENT RESPONSES OF APERTURES   215

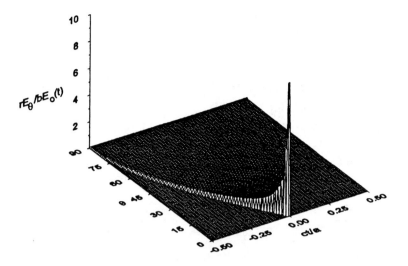

a. Non-causal

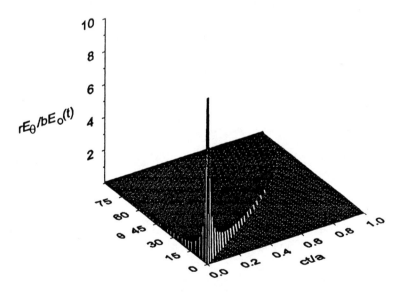

b. Causal

FIGURE 5.12 Surface plot of the normalized transient step-function response of the transmitted aperture field, shown as a function of the angle of observation θ and normalized time frequency ct/a.

solution that is so approximate in the first place. Figure 5.12*b* illustrates the surface plot of this modified, causal response.

5.5.2 Direct Time-Domain Calculations

An alternative to using the frequency-domain integral equation and its various approximations for treating transient aperture penetration problems is to use a direct time-domain solution of Maxwell's equations. This is referred to as the finite-difference, time-domain approach (FDTD), because the differential operators in Maxwell's equations are represented by finite-difference operators in the time domain.

Early work in this area was performed by Yee [5.22], who numerically implemented a "leapfrog" finite-difference algorithm for solving the transient fields in a bounded region. This approach involves dividing the volume of a problem space into many cells (on the order of thousands) surrounding a scattering structure, and for each cell, constructing a finite-difference relationship between the local E and H fields. This results in a large set of simultaneous equations that may be solved by stepping along in time. Inherent in such calculations is the need for imposing suitable boundary conditions at the outer boundary of the mesh. Several investigators, including Mur [5.23], have studied methods to simulate these conditions.

Details of this subject are beyond the scope of this book. The interested reader is referred to [5.24] for an in-depth discussion of this topic, together with a list of useful references and computer code listings.

REFERENCES

5.1. Harrington, R. F., *Time Harmonic Electromagnetic Fields*, McGraw-Hill, New York, 1961.
5.2. Born, M., and E. Wolf, *Principles of Optics*, Pergamon Press, Oxford, 1965.
5.3. Jackson, J. D., *Classical Electrodynamics*, Wiley, New York, 1975.
5.4. Sommerfeld, A, *Partial Differential Equations in Physics*, Academic Press, 1964.
5.5. Stone, J. M., *Radiation and Optics*, McGraw-Hill, New York, 1963.
5.6. Jones, D. S., *The Theory of Electromagnetism*, Pergamon Press, Oxford, 1964.
5.7. Tai, C.-T., *Dyadic Green's Functions in Electromagnetic Theory*, Intext Publishers, Scranton, PA, 1971.
5.8. Van Bladel, J., *Electromagnetic Fields*, McGraw-Hill, New York, 1964.
5.9. Silver, S., *Microwave Antenna Theory and Design*, Dover Publications, New York, 1965.
5.10. Collin, R. E., and F. J. Zucker, *Antenna Theory*, McGraw-Hill, New York, 1969.
5.11. Wilton, D. R., and O. C. Dunaway, "Electromagnetic Penetration Through Apertures of Arbitrary Shape: Formulation and Numerical Solution Procedure,"

Report AFWL-TR-74-194, Air Force Weapons Laboratory, Kirtland AFB, NM, August 1975.
5.12. Bouwkamp, C. J., "Diffraction Theory," *Rep. Prog. Phys.*, Vol. 17, 1954, 35–100.
5.13. Butler, C. M., et al., "Electromagnetic Penetration Through Apertures in Conducting Surfaces," *IEEE Trans. Antenna Propag.*, Vol. AP-26, No. 1, January 1978.
5.14. Andrews, C. L., "Diffraction Pattern in a Circular Aperture Measured in the Microwave Region," *J. Appl. Phys.*, Vol. 22, 1950, pp. 761–767.
5.15. Koch, G. F., and K. S. Kölbig, "The Transmission Coefficient of Elliptical and Rectangular Apertures for Electromagnetic Waves," *IEEE Trans. Antennas Propag.*, Vol. AP-16, No. 1, January 1968.
5.16. Balanis, C. A., *Antenna Theory: Analysis and Design*, Wiley, New York, 1982.
5.17. Elliot, R. S., *Antenna Theory and Design*, Prentice Hall, Englewood Cliffs, NJ, 1981.
5.18. Lee, K. S. H., ed., *EMP Interaction: Principles, Techniques and Reference Data*, AFWL-TR-80-402, Air Force Weapons Laboratory, Kirtland AFB, NM, 1981. Reprinted by Hemisphere, New York, 1989.
5.19. Collin, R. E., *Field Theory of Guided Waves*, McGraw-Hill, New York, 1960.
5.20. Ramo, S., J. R. Whinnery, and T. Van Duser, *Fields and Waves in Communication Electronics*, 2nd ed., Wiley, 1989.
5.21. Tesche, F. M., "On the Use of the Hilbert Transform for Processing Measured CW Data," *IEEE Trans. Electromagn. Compat.*, Vol. EMC-34, No. 3, August 1992.
5.22. Yee, K. S., "Numerical Solution of Initial Boundary Value Problems Involving Maxwell's Equations in Isotropic Media," *IEEE Trans. Antennas Propag.*, Vol. AP-14, No., 1966.
5.23. Mur, G., "Absorbing Boundary Conditions for Finite-Difference Approximation of the Time-Domain Electromagnetic Field Equations," *IEEE Trans. Electromagn. Compat.*, Vol. EMC-23, No. 9, November 1981, pp. 1073–1077.
5.24. Kunz, K. S., and R. J. Luebbers, *The Finite Difference Time Domain Method for Electromagnetics*, CRC Press, Boca Raton, FL, 1993.

PROBLEMS

5.1 Consider a circular aperture of radius a illuminated by a plane wave with normal incidence and with the excitation E-field E_0 polarized in the y_s direction (see Figure 5.4). From Eq. (5.20) and assuming that the obliquity factor $O = 1$, show that the E-field passing through the aperture at a distant point r in the $\phi = \pi/2$ plane is

$$E_\theta\big|_{\phi=\pi/2} \approx \frac{jk}{r} E_0 e^{-jkr} a^2 \frac{J_1(ka\sin\theta)}{ka\sin\theta}$$

5.2 As noted in Eq. (5.17), different formulations for the aperture penetration result in different obliquity factors O. For the case of normal incidence ($\theta_s = 0$) on the circular aperture in Problem 5.1, compute and plot the three different obliquity factors and discuss the effect on the results of the previous Problem 5.1.

5.3 Aperture antennas are often characterized by a *directive gain*, $G_d(\theta, \phi)$, which gives an indication of the spatial variation of the radiated power pattern. This gain is defined as the radiated power density at a particular location, divided by the average radiated power density from the antenna. Thus

$$G_d(\theta, \phi) = \frac{P(\theta, \phi)}{P_t/(4\pi r^2)} = \frac{1}{2} \frac{E_\theta H_\phi^* - E_\phi H_\theta^*}{P_t/(4\pi r^2)} = \frac{1}{2Z_0} \frac{4\pi r^2 (|E_\theta|^2 + |E_\phi|^2)}{P_t}$$

where P_t denotes the total power radiated by the aperture and Z_0 is the impedance of free space. For the circular aperture of Problem 5.1, show that:
   (a) The total radiated power from the aperture (assuming a uniform illumination) is given by the expression $P_t = (1/2Z_0)E_0^2 \pi a^2$.
   (b) The *directivity* of the antenna, G_0, which is defined as the *maximum value* of the directive gain $G_d(\theta, \phi)$, occurs in the direction $\theta = 0$ and is given by $G_0 = k^2 a^2$.

5.4 Compute and compare the directivities of a circular aperture and a square aperture, both having the same area A.

5.5 Consider a uniformly illuminated circular aperture of radius a_0 which is partially blocked by a smaller circular disk of radius a_1, as shown in Figure P5.5. Assuming that the disk does not affect the field distribution significantly elsewhere in the aperture, compute and plot the radiated field from this aperture. (*Hint:* Consider the use of the superposition principle.)

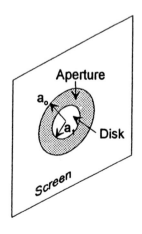

FIGURE P5.5. Partially blocked circular aperture.

5.6* Identical apertures are arranged in a row, forming an array of apertures. Discuss the relationship of polarizabilities (electric and magnetic) of single and arrayed apertures, considering the example of elliptical and of rectangular apertures.

5.7* Some apertures are unavoidable in any system (e.g., windows, doors, and ventilation openings). Discuss a few ways of aperture control, including aperture loading.

5.8* Find the electric and magnetic polarizabilities for a square aperture of dimensions 100 mm by 100 mm. Suppose that the aperture is now covered by a wire mesh dividing it into $n \times n$ smaller apertures. Develop the relationship for the polarizabilities of the hardened aperture by summing the smaller aperture effects.

5.9 A small circular aperture of radius a is located in a conducting screen. Determine:
   (a) The highest-frequency f_{max} for which the aperture polarizabilities may be used to describe the field penetration.
   (b) For a normally incident plane-wave E_0 on one side of the screen with $f < f_{max}$, compute and plot the behavior of the radiated (far-field) E and H components penetrating through the aperture.

5.10 Develop a computer program to numerically evaluate the normalized equivalent areas A_{eq}/A_0 for the ellipse and the rectangular apertures given in Eqs. (5.56) and (5.57). Plot these functions in a manner similar to Figure 5.9 for various ratios of the aperture parameters a and b.

---

* Problems denoted by an asterisk were used in EMP short courses conducted by the Summa Foundation in 1983 through 1989. These contributions by C. Baum and the other lecturers are gratefully acknowledged.

PART IV
TRANSMISSION LINE MODELS

CHAPTER 6

Transmission Line Theory

In Chapter 3, conducted interference was modeled using circuit theory. This approach is valid for low frequencies where the wavelength is much larger than the size of the system being considered. At high frequencies, however, wave-propagation effects begin to be important and the simple circuit models break down. In this case, conducted interference can often be treated by using transmission line models. This chapter will begin the discussion of transmission line theory and models for lines having lumped (discrete) excitation sources. We start with the introduction of the telegrapher's equations and then illustrate how these equations may be solved for the voltages and currents injected into electrical loads connected to the line. In this manner, these models can be applied to a variety of EMC problems. In Chapter 7 the important subject of transmission line excitation by a distributed EM field is discussed.

6.1 OVERVIEW OF TRANSMISSION LINE MODELS

Earlier it has been assumed that the dimensions of the circuits with respect to the wavelength of the disturbance are sufficiently small to neglect wave propagation effects. Such circuits are called *electrically small circuits*. For *electrically large circuits*, their size is large compared with the wavelength of the disturbance, and wave propagation effects must be taken into account. In this book we have adopted the convention of referring to a problem involving an electrically small circuit as a *low-frequency* problem and the case of an electrically large circuit as a *high-frequency* problem.

6.1.1 Lumped and Distributed Circuit Parameters

For the low-frequency cases described in Sections 3.2 and 3.3, the electrical parameters of the circuit (i.e., the resistance R, the self and mutual inductances L_i and M_{ij}, the partial capacitances C_{ij}, and the conductance G) can be regarded as lumped circuit elements. The fact that at high frequencies the propagation of transient phenomena in a circuit or along a conductor must be taken into account makes it necessary to model the conductor using

224 TRANSMISSION LINE THEORY

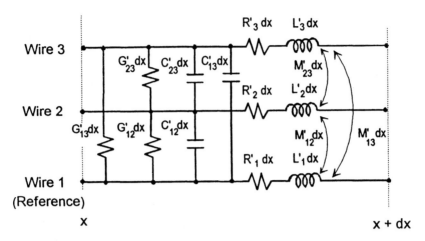

FIGURE 6.1 Distributed parameters of an infinitesimal section dx of a three-conductor transmission line with wire 1 serving as a reference conductor.

distributed parameters. For a three-conductor line, Figure 6.1 illustrates a section of line having an infinitesimal length dx. If the line is uniform along its length, the electrical parameters of the line can be considered as being uniformly distributed and are denoted by constant per-unit-length values R', L' (or M'), C', and G'. Note that these and other per-unit-length (i.e., distributed) line parameters in this book are denoted by a prime (') to distinguish them from the lumped circuit elements.

For a transmission line to be considered as uniform, two conditions are necessary:

1. The conductors comprising the line must be mutually parallel, and parallel to the reference conductor (ground plane) if one exists near the line.
2. Any dielectric material in the vicinity of the line must exhibit translational symmetry along the length of the line (e.g., two parallel dielectric jacketed lines with air separation between them).

6.1.2 Lumped and Distributed Excitations

A circuit or a system can be considered to be excited by one of several different mechanisms:

- By one (or more) lumped voltage sources located along the conductors
- By one (or more) lumped current sources located between two conductors
- By a distributed electromagnetic field in the form of a plane wave

- By general near fields from a nearby source

The lumped voltage or current excitation sources along the line can be:

- Inadvertently produced by unwanted EM field coupling inside the system, or
- Deliberately applied to the line, by a source injection at one point on the system

When we speak of a lumped excitation of a circuit in this book, we refer to a primary excitation injected onto or produced in the circuit itself at a point location. This discrete excitation will give rise to a response (either transient or time harmonic, depending on the nature of the source) within the entire circuit. This circuit can be coupled by an electric or magnetic field to other neighbor circuits, inducing a secondary response in the circuits.

A distributed excitation is given by an electromagnetic field that is illuminating the circuit or the system. The case most often considered is plane-wave excitation, in which the incident field appears as a transverse EM field propagating with a specific direction relative to the line. This excitation, which is always distributed along the conductor, will act on all the elements of a single circuit and on all the circuits if there are more than one.

In this chapter only lumped excitation is considered. Distributed excitation modeling is treated in Chapter 7.

6.1.2.1 Examples of System Excitation. *Lumped* voltage or current source excitation often can be useful in representing the effects of the following phenomena internal to a system:

- Surges produced by switching operations in electrical power systems, or electrical circuits
- Noise produced by the clock generator in computer systems
- Interference between high-frequency analog signals and digital circuits on PC boards
- An electrical short produced by a conductor fault in a system
- An impulsive static discharge within a PCB

The lumped excitation can also arise from sources external to the system, for instance, from:

- A direct lightning stroke on an aerial or buried line, or on the metallic structure of a building
- An electrostatic discharge on a circuit or on a ground conductor

Electromagnetic fields providing a *distributed* excitation can be produced by:

- An indirect nearby lightning stroke
- The radiation of the busbars of a GIS on data transmission lines
- The radiation of radio emitters, mobile radio communication or radar on the electronic circuits of household apparatus (such as radio, TV, videotape recorders, etc.), cars, and aircraft

6.1.3 Two-Conductor and Multiconductor Systems

In the previous discussion of lumped and distributed parameters shown in Figure 6.1, the special case of a three-conductor system has been illustrated. It is possible to consider a more general case involving many conductors, as illustrated in Figure 6.2. Part (*a*) of the figure illustrates an $(n + 1)$ conductor line which contains n signal wires and a reference return conductor. Alternatively, part (*b*) shows n wires located over a reference ground conductor, and part (*c*) shows n conductors surrounded by the reference conductor. Each of these multiconductor lines is referred to as an *n-wire multiconductor line*.

6.1.4 Transmission Line and Antenna Mode Responses

A single voltage source on a transmission line can excite different kinds of line responses. Consider the case of the two-wire line shown in Figure 6.3 with a single voltage source exciting the line at a location along the line. This

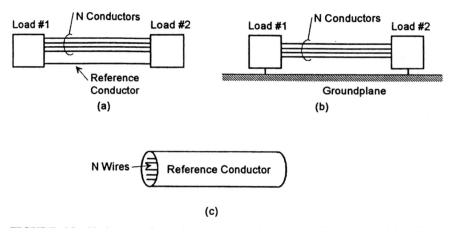

FIGURE 6.2 Various configurations of an *n*-wire multiconductor line; (*a*) isolated bundle of *n* wires plus a return (ground) conductor; (*b*) *n* wires over a ground plane; (*c*) *n* conductors enclosed by a cylindrical shield conductor.

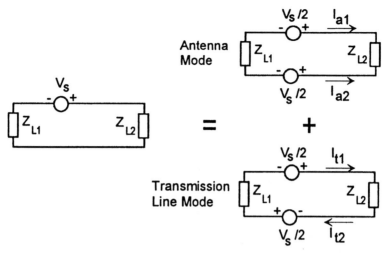

FIGURE 6.3 Excitation of the antenna and transmission line mode currents on a two-wire line.

excitation, and the subsequent line response, can be divided into common-mode and differential-mode components, which are called the antenna mode and transmission line mode, respectively.

The *antenna mode* consists of currents in each conductor flowing in the same direction on the line and vanishing at each end of the line. This behavior is similar to the antenna currents studied in Chapter 4, and as a consequence, this set of currents will radiate energy very well. As noted from the figure, this antenna mode current is excited by the symmetric component of the voltage source: that is, by a common-mode voltage source of $V_s/2$ located in each conductor.

The *transmission line mode* consists of equal and opposite currents flowing in each conductor. This mode is excited by the opposing voltage sources of strength $V_s/2$, as indicated in the figure. This transmission line mode also radiates energy away from the line, but because the currents flow in the opposite direction, the radiation is much less than that of the antenna mode.

If one is interested in the behavior of the current at a general point on the line, it is necessary to consider *both* modes in determining the response. However, in many practical EMC problems, only the responses at the loads of the line are desired. Since the antenna mode vanishes at these locations, the transmission line mode then becomes the dominate contributor to the response.

Under certain circumstances, the antenna mode can be zero due to symmetries in the transmission line geometry. Consider the case of Figure 6.4a, which shows a single-wire line over a perfectly conducting ground plane serving as the return conductor. Using image theory, the source voltage V_s

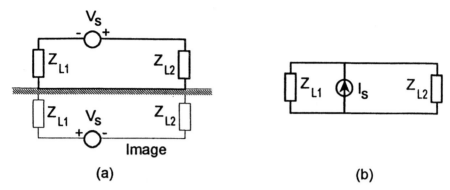

FIGURE 6.4 Line examples where no antenna mode is present: (a) single-conductor line over a ground plane; (b) two-wire line excited by a current source.

on the line has an image source of $-V_s$ on the image line. This collection of sources provides no contribution at all to the antenna mode, and consequently, only the transmission line mode will exist on the line.

This concept of antenna mode and transmission line mode decomposition of the induced current on a line can be extended to the case of a multiconductor transmission line. The analysis of the propagation of the transmission line mode voltages and currents along the multiconductor line requires the assumption that in any cross section of the line, the *total* current on the wires *and* on the reference conductor is zero. For a multiconductor line, several different transmission line modes satisfying this requirement may exist on the line, and as will be discussed shortly, each of these transmission line modes can contribute to the total transmission line response.

In this chapter we concentrate on determining the transmission line mode responses on various conductor configurations. It should be remembered that the solutions to be discussed here form only *part* of the total solution for a line. The antenna mode component can also exist, and in some cases, this may be much larger than the transmission line component.

6.1.5 Telegrapher's Equations for a Two-Conductor System

To facilitate understanding of the physical process involved in the transmission line theory, we start by considering the propagation phenomena on a *two-conductor* system, and then we generalize the result to a multiconductor line. Away from any sources on the line, the behavior of the transient voltage $v(x,t)$ and current $i(x,t)$ on a lossy two-conductor line excited by a lumped excitation source is described by the time-dependent telegrapher's equations

$$\frac{\partial v(x,t)}{\partial x} + R'i(x,t) + L'\frac{\partial i(x,t)}{\partial t} = 0 \tag{6.1}$$

$$\frac{\partial i(x,t)}{\partial x} + G'v(x,t) + C'\frac{\partial v(x,t)}{\partial t} = 0 \tag{6.2}$$

where x denotes the longitudinal direction of the line and R', L', G', and C' are the constant per unit length resistance, inductance, conductance, and capacitance, respectively, of the two-conductor line. The derivation of these equations is described in Chapter 7.

Figure 6.5 illustrates a differential section of the transmission line. For the moment we assume that the frequency is sufficiently high so that it is necessary to model the line by these distributed parameters. Later, we determine when lumped circuit parameters can be used.

The two-conductor system can consist of two isolated wires or a single conductor with a return path that can be the metallic ground plane of an electronic circuit, or the earth for the case of an overhead power line or a telecommunications line. For this line the excitation will be assumed to be due to a lumped voltage source v_0 or a lumped current source i_0 located on the conductor. The time harmonic (frequency-domain) response is examined in Section 6.2, and the transient (time-domain) response, in Section 6.3.

6.1.5.1 Evaluation of the Line Parameters. The distributed line parameters R', L', G', and C' can be determined either by measurement or calculation. For most insulated transmission lines, the distributed conductance term is $G' \approx 0$. The per-unit-length resistance R' can be calculated using the dimensions of the conductor and its electrical properties. For slowly varying signals (or equivalently, for low frequencies) this resistance is a constant and is calculated simply by dividing the bulk resistivity of the wire material by the area of the wire cross section. For faster transient signals (i.e., at higher frequencies) this resistance becomes a time-varying function due to the diffusion time of the current into the conductor. This is a result of

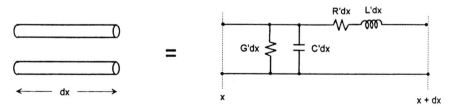

FIGURE 6.5 Distributed parameters of a two-conductor system.

the skin effect, in which the current tends to exist on or near the surfaces of conductors at high frequencies.

A general expression in the frequency domain for the internal impedance of a single conductor of radius a and conductivity σ_w is given in [6.1] and is discussed in more detail in Section 7.2.1.3. Generally, this impedance is a complex-valued, frequency-dependent function. However, at very low frequencies this impedance becomes mainly resistive and is given by the per-unit-length resistance of the wire:

$$R' = \frac{1}{\pi a^2 \sigma_w} \quad \Omega/m \tag{6.3}$$

For a two-conductor line as illustrated in Figure 6.5, the total line resistance will consist of two such terms, one for each conductor.

The inductance and capacitance values can be either measured or calculated using methods discussed in more detail in Sections 6.4 and 6.5. For an isolated two-wire line having the cross section shown in Figure 6.6a and immersed in a homogeneous dielectric region with dielectric constant ϵ, the per-unit-length capacitance is expressed as [6.1]

$$C' \approx \frac{2\pi\epsilon}{\ln(d^2/a_1 a_2)} \quad F/m \tag{6.4a}$$

If the region surrounding the line is free space, the parameter ϵ takes the value $\epsilon = \epsilon_0 \approx 1/(36\pi) \times 10^{-9}$ F/m. Here it is assumed that a_1 and $a_2 \ll d$.

For the case of a single wire of radius a at a height h over a perfect ground shown in Figure 6.6b, the line capacitance can be inferred from image theory to be

$$C' \approx \frac{2\pi\epsilon}{\ln(2h/a)} \quad F/m \quad (\text{for } a \ll h) \tag{6.4b}$$

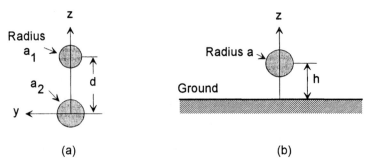

FIGURE 6.6 Transmission line cross section: (*a*) two-wire line; (*b*) single-wire line over a ground plane.

Similarly, the per-unit-length external inductance of the two-wire line is expressed as

$$L' = \frac{\mu}{2\pi} \ln \frac{d^2}{a_1 a_2} \quad \text{H/m} \tag{6.5a}$$

and the inductance for the single wire line is

$$L' = \frac{\mu}{2\pi} \ln \frac{2h}{a} \quad \text{H/m} \tag{6.5b}$$

where μ is the magnetic permeability of the material around the wires. For free space this parameter takes the value $\mu = \mu_0 = 4\pi \times 10^{-7}$ H/m.

For two lossless conductors in a homogeneous medium, the per-unit-length inductance and capacitance parameters are related by

$$L'C' = \mu\epsilon = \frac{1}{v^2} \tag{6.6}$$

where μ is the permeability of the surrounding medium, ϵ is the material permittivity, and v is the resulting velocity of wave propagation in the medium. Thus if either the capacitance or inductance is determined (e.g., by calculation) Eq. (6.6) permits the calculation of the other line parameter under the assumptions that there are no losses and that the propagation velocities on the transmission line and in the media surrounding the line are the same. This assumption can be valid for many practical cases involving isolated transmission line sections. However, some transmission lines might involve an inhomogeneous dielectric material, which means that the foregoing relationship between L' and C' is not valid. An example of this is a simple two-wire line having insulated jackets. In this case the wave velocity for the voltages and currents on the transmission line will be different from the wave velocity in the space away from the line. A more detailed discussion of this case is provided in Section 6.2.

6.2 FREQUENCY-DOMAIN RESPONSES

For an analysis conducted in the frequency domain, the temporal variation is assumed to be $e^{j\omega t}$, and all time derivatives in Eqs. (6.1) and (6.2) are expressed by the quantity $j\omega$. In this manner, the telegrapher's equations become

$$\frac{dV(x)}{dx} + Z'I(x) = 0 \tag{6.7a}$$

$$\frac{dI(x)}{dx} + Y'V(x) = 0 \tag{6.7b}$$

where the parameters Z' and Y' are per-unit-length impedance and admittance parameters of the line, defined as

$$Z' = R' + j\omega L' \tag{6.8a}$$

$$Y' = G' + j\omega C' \tag{6.8b}$$

Using the two-port notation from Chapter 3, the source-free telegrapher's equations (6.7) can be written in matrix form as

$$\frac{d}{dx}\begin{bmatrix} V(x) \\ I(x) \end{bmatrix} = \begin{bmatrix} 0 & -Z' \\ -Y' & 0 \end{bmatrix}\begin{bmatrix} V(x) \\ I(x) \end{bmatrix} \tag{6.9}$$

6.2.1 Solution of the Telegrapher's Equations for a Two-Conductor Line

If the line is uniform according to the conditions defined in Section 6.1.1, the line parameters Z' and Y' are not a function of position along the line. Equation (6.9) can be further manipulated into two wave equations, one for the voltage and another for the current on the line. These wave equations have solutions in the form of individual traveling waves propagating in the positive and negative directions on the line. With reference to Figure 6.7, the positive traveling voltage and current waves are [6.2]

$$V^+(x) = ae^{-\gamma x} \tag{6.10a}$$

$$I^+(x) = \frac{1}{Z_c}ae^{-\gamma x} \tag{6.10b}$$

and the negative traveling waves are

$$V^-(x) = be^{+\gamma x} \tag{6.11a}$$

$$I^-(x) = -\frac{1}{Z_c}be^{+\gamma x} \tag{6.11b}$$

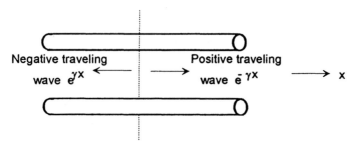

FIGURE 6.7 Positive and negative traveling waves.

where a and b are complex constants and the time-harmonic factor $e^{j\omega t}$ has been suppressed.

The term γ is the propagation constant along the transmission line, given as

$$\gamma = \sqrt{Z'Y'} \tag{6.12}$$

and Z_c is the ratio of the magnitudes of the positive traveling voltage and current waves, and is

$$Z_c = \frac{V^+(x)}{I^+(x)} = \sqrt{Z'/Y'} \tag{6.13}$$

The general solution for the line voltage or current is given by the sum of the two traveling waves

$$V(x) = V^+ + V^- = ae^{-\gamma x} + be^{\gamma x} \tag{6.14a}$$

$$I(x) = I^+ + I^- = Y_c(ae^{-\gamma x} - be^{\gamma x}) \tag{6.14b}$$

where Y_c is the characteristic admittance of the line given by $Y_c = 1/Z_c$.

6.2.1.1 Chain Parameter Representation of a Two-Wire Line.
Since the line of Figure 6.7 is infinite in length, the waves propagating to $\pm\infty$ continue with no reflections. This is also true if the line is finite in length but is terminated in the characteristic impedance Z_c of the line at each end. If the line is finite and not matched, there will be reflections at the ends of the line.

To determine expressions for the line voltage and current in this case, the finite line of length $\mathscr{L}$ can be represented by an equivalent two-port network as shown in Figure 6.8. For this network, it is possible to develop expressions for the chain ($ABCD$) parameters, and with these expressions, the responses of the line with arbitrary loading and excitation at the ports $x = 0$ and $x = \mathscr{L}$ can be determined as was done for the low-frequency circuits in Chapter 3.

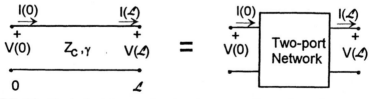

FIGURE 6.8 Equivalent two-port network representing a two-wire transmission line.

Using a matrix notation, Eqs. (6.14) can be written as [6.3]

$$\begin{bmatrix} V(x) \\ I(x) \end{bmatrix} = \begin{bmatrix} 1 & 1 \\ Y_c & -Y_c \end{bmatrix} \begin{bmatrix} e^{-\gamma x} & 0 \\ 0 & e^{+\gamma x} \end{bmatrix} \begin{bmatrix} a \\ b \end{bmatrix} \quad (6.15)$$

At the port at $x = 0$, this expression is

$$\begin{bmatrix} V(0) \\ I(0) \end{bmatrix} = \begin{bmatrix} 1 & 1 \\ Y_c & -Y_c \end{bmatrix} \begin{bmatrix} a \\ b \end{bmatrix} \quad (6.16)$$

and at the port at $x = \mathcal{L}$ it is

$$\begin{bmatrix} V(\mathcal{L}) \\ I(\mathcal{L}) \end{bmatrix} = \begin{bmatrix} 1 & 1 \\ Y_c & -Y_c \end{bmatrix} \begin{bmatrix} e^{-\gamma \mathcal{L}} & 0 \\ 0 & e^{+\gamma \mathcal{L}} \end{bmatrix} \begin{bmatrix} a \\ b \end{bmatrix} \quad (6.17)$$

Elimination of a and b between Eqs. (6.16) and (6.17) and simplifying the result yields a relationship between the output and input voltages and currents of the equivalent two-port network

$$\begin{bmatrix} V(\mathcal{L}) \\ I(\mathcal{L}) \end{bmatrix} = \begin{bmatrix} \cosh \gamma \mathcal{L} & -Z_c \sinh \gamma \mathcal{L} \\ -Y_c \sinh \gamma \mathcal{L} & \cosh \gamma \mathcal{L} \end{bmatrix} \begin{bmatrix} V(0) \\ I(0) \end{bmatrix} \quad (6.18)$$

This expression may be inverted as

$$\begin{bmatrix} V(0) \\ I(0) \end{bmatrix} = \begin{bmatrix} \cosh \gamma \mathcal{L} & Z_c \sinh \gamma \mathcal{L} \\ Y_c \sinh \gamma \mathcal{L} & \cosh \gamma \mathcal{L} \end{bmatrix} \begin{bmatrix} V(\mathcal{L}) \\ I(\mathcal{L}) \end{bmatrix} \quad (6.19)$$

which is seen to be the chain parameter relationship for the two-port network as defined in Eq. (3.15), with $A = D = \cosh \gamma \mathcal{L}$, $B = Z_c \sinh \gamma \mathcal{L}$, and $C = Y_c \sinh \gamma \mathcal{L}$.

6.2.1.2 Other Two-Port Representations for the Two-Wire Line.
Using the two-port conversion chart in Table 3.2, the chain parameters of Eq. (6.19) can be easily converted to the impedance or admittance parameters. These matrices are

$$[Z] = \frac{Z_c}{\sinh \gamma \mathcal{L}} \begin{bmatrix} \cosh \gamma \mathcal{L} & 1 \\ 1 & \cosh \gamma \mathcal{L} \end{bmatrix} \quad (6.20)$$

$$[Y] = [Z]^{-1} = \frac{Y_c}{\sinh \gamma \mathcal{L}} \begin{bmatrix} \cosh \gamma \mathcal{L} & -1 \\ -1 & \cosh \gamma \mathcal{L} \end{bmatrix} \quad (6.21)$$

Notice that when $\gamma \mathcal{L} = jn\pi$ (where n an integer), the term $\sinh \gamma \mathcal{L} = 0$ and the impedance or admittance formalisms for representing the line behavior becomes impossible. In this case the chain parameter representation may still be used.

6.2.1.3 Applications of Two-Port Representations.
The two-port chain parameter representation of the transmission line is useful when a voltage or current source is present at the line input and it is desired to calculate the voltage or current at the end of the line. Figure 6.9 illustrates a two-wire line excited by a Thévenin source at $x = 0$ and loaded by an impedance Z_L at $x = \mathscr{L}$. This model is frequently encountered in power system transmission lines, where switching transients occur at one end of a line.

For this transmission line, Eq. (6.19) defines the *ABCD* parameters, and Eq. (3.17) can be used to directly calculate the load current $I(\mathscr{L})$ as

$$I(\mathscr{L}) = \frac{V_s}{Z_L(\cosh \gamma \mathscr{L} + Z_s Y_c \sinh \gamma \mathscr{L}) + Z_c(\sinh \gamma \mathscr{L} + Z_s Y_c \cosh \gamma \mathscr{L})}$$

(6.22)

In addition to this single section of transmission line, multiple sections of lines can be cascaded together as in Eq. (3.18). Thus the parameters of Eq. (6.19) provide the building blocks for constructing more complicated transmission line structures.

6.2.1.4 The Π and T Equivalent Circuits of the Two-Wire Line.
An alternative representation of the two-wire transmission line is by the equivalent Π circuit shown in Figure 3.8. In this figure, the individual admittance elements of the circuit, Y_a, Y_b, and Y_c are given in terms of the admittance matrix parameters y_{11}, y_{12}, y_{21}, and y_{22} as

$$Y_a = y_{11} + y_{12} \qquad (6.23a)$$

$$Y_b = y_{22} + y_{12} \qquad (6.23b)$$

$$Y_c = -y_{12} \qquad (6.23c)$$

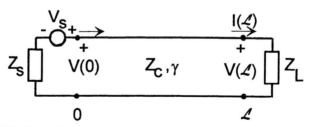

FIGURE 6.9 Two-wire transmission line excited at $x = 0$ and loaded at $x = \mathscr{L}$.

The y-matrix parameters for the line are given in Eq. (6.21) as

$$y_{11} = y_{22} = \frac{Y_c \cosh \gamma \mathcal{L}}{\sinh \gamma \mathcal{L}} \tag{6.24a}$$

$$y_{12} = y_{21} = \frac{-Y_c}{\sinh \gamma \mathcal{L}} \tag{6.24b}$$

These equations may be used directly in Eqs. (6.23) to obtain the lumped elements of the Π network as

$$Y_a = Y_b = \frac{1}{Z_c} \tanh \frac{\gamma \mathcal{L}}{2} \tag{6.25a}$$

$$Y_c = \frac{1}{Z_c \sinh \gamma \mathcal{L}} \tag{6.25b}$$

At low frequencies when $\gamma \mathcal{L}$ is small, the small argument approximations to the sinh and tanh terms in Eqs. (6.25) can be used to approximate the low-frequency Π parameters for the transmission line as

$$Y_a = Y_b \approx \frac{1}{2} Y' \mathcal{L} = \frac{1}{2}(G' + j\omega C')\mathcal{L} \tag{6.26a}$$

$$Y_c \approx \frac{1}{Z' \mathcal{L}} = \frac{1}{(R' + j\omega L')\mathcal{L}} \tag{6.26b}$$

For cases where the line length is such that $\mathcal{L} \leq \lambda/10$, the error in approximating the sinh and tanh terms by their arguments is *at the most* about 7%. In some engineering applications, this is an acceptable error and the results obtained by neglecting transmission line propagation may be acceptable.

If the condition $\mathcal{L} \leq \lambda/10$ is not fulfilled for a long conductor, the line can be subdivided into a number of smaller sections of length ℓ which do satisfy the requirement that $\ell \leq \lambda/10$, and each of these sections is represented by equivalent Π circuits [6.4]. This is shown in Figure 6.10. Note that each Π

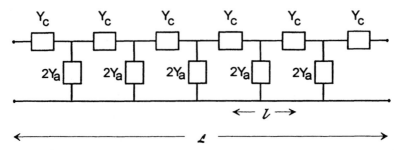

FIGURE 6.10 Two-conductor circuit divided into n equivalent Π circuits.

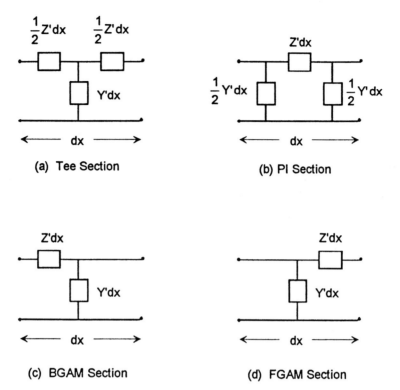

FIGURE 6.11 Lumped-model approximations of a two-conductor transmission line.

circuit has two shunt admittances Y_a and Y_b that are combined together into a single element.

In addition to the Π model for the line, other lumped-model approximations can be used. Figure 6.11 summarizes the possible circuit models for the differential line section, with the circuit elements expressed in terms of the per-unit-length impedance and admittance (Z' and Y') and the length of the line section, dx.

It has been shown [6.5] that the load of the circuit has little influence on predictions when using the lumped-model approximations (within an error limit of ±3 dB) and that the precision does not depend linearly on the addition (cascading) of more sections of the lumped model. This means that if one section yields accurate predictions up to a frequency f_1, two sections will not necessarily yield accurate predictions up to $2f_1$.

6.2.2 Lumped-Source Excitation of Transmission Lines

Up to now it has been assumed that an excitation source is present at one end of the line, and the equivalent two-port network representation permits the calculation of the voltage and current at the other end of the line. A simplified representation using the Π equivalent circuit has been introduced and its limits discussed. An EMI source, however, can be localized at an arbitrary position along the line, not just at one of the ends. To begin this discussion, consider an infinite line having a series lumped voltage source or a shunt lumped current source at $x = x_s$, as shown in Figure 6.12.

Consider first the effect of a voltage source in Figure 6.12a. Because the line is infinite in both directions, the line will have only outgoing voltage (and current) waves from the source. This means that for $x > x_s$ only positive propagating waves V^+ and I^+ exist on the line, and for $x < x_s$, only the negative propagating V^- and I^- exist. The complete solution is determined from the general equations (6.14) by calculating the constants a and b from the boundary conditions at $x = x_s$ (i.e., that the total line current is continuous through the source and the total line voltage is discontinuous by the amount V_0).

The resulting solutions for the line voltage and current at an observation location x are

$$V(x) = \frac{V_0}{2} e^{-\gamma(x-x_s)} \qquad \text{for } x > x_s \qquad (6.27a)$$

$$V(x) = \frac{-V_0}{2} e^{\gamma(x-x_s)} \qquad \text{for } x < x_s \qquad (6.27b)$$

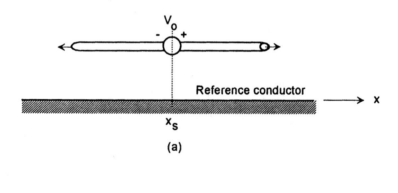

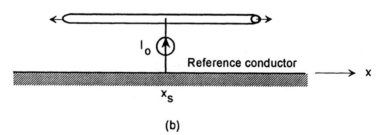

FIGURE 6.12 Single-source excitation of an infinite line.

and

$$I(x) = \frac{V_0}{2Z_c} e^{-\gamma(x-x_s)} \quad \text{for } x > x_s \qquad (6.28a)$$

$$I(x) = \frac{V_0}{2Z_c} e^{\gamma(x-x_s)} \quad \text{for } x < x_s \qquad (6.28b)$$

Similarly, the dual current source I_0 at x_s shown in Figure 6.12b has the following line responses:

$$V(x) = \frac{I_0 Z_c}{2} e^{-\gamma(x-x_s)} \quad \text{for } x > x_s \qquad (6.29a)$$

$$V(x) = \frac{I_0 Z_c}{2} e^{\gamma(x-x_s)} \quad \text{for } x > x_s \qquad (6.29b)$$

and

$$I(x) = \frac{I_0}{2} e^{-\gamma(x-x_s)} \quad \text{for } x > x_s \qquad (6.30a)$$

$$I(x) = \frac{-I_0}{2} e^{\gamma(x-x_s)} \quad \text{for } x < x_s \qquad (6.30b)$$

6.2.3 Terminated Lines: Voltage Reflection Coefficient

Realistic lines are not infinite in length. Consider the case of a finite line extending from $x = 0$ to $x = \mathcal{L}$ as shown in Figure 6.13. At the end of a line there is an impedance element Z_L, which is assumed to be a complex-valued quantity relating the total load voltage V_L to the load current I_L. A traveling wave propagating in the positive direction and incident on the load will create a reflected wave that propagates in the negative direction away from the load. Due to this reflection, at any point on the line there will be a combination of both positive and negative propagating waves. It is useful to define the ratio of the reflected *voltage* wave at any point along the line to

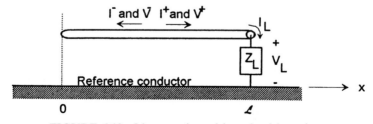

FIGURE 6.13 Line terminated by a load impedance.

the incident *voltage* wave as a general reflection coefficient as

$$\rho(x) = \frac{V^-}{V^+} = \frac{be^{\gamma x}}{ae^{-\gamma x}} \qquad (6.31)$$

The voltage and current at any point on the line can now be expressed in terms of the voltage reflection coefficient as

$$V(x) = ae^{-\gamma x} + be^{\gamma x} = ae^{-\gamma x} + \rho(x)ae^{-\gamma x} = ae^{-\gamma x}[1 + \rho(x)] \qquad (6.32a)$$

$$I(x) = Y_c(ae^{-\gamma x} - be^{\gamma x}) = Y_c(ae^{-\gamma x} - \rho(x)ae^{-\gamma x}) = Y_c ae^{-\gamma x}[1 - \rho(x)] \qquad (6.32b)$$

The ratio of line voltage to current defines an impedance quantity which is the input impedance of the line at the location x looking toward the load [6.1]. This is expressed as

$$Z_{in}(x) = \frac{V(x)}{I(x)} \qquad (6.33)$$

and, on using (6.32), this expression becomes

$$Z_{in}(x) = Z_c \frac{1 + \rho(x)}{1 - \rho(x)} \qquad (6.34)$$

At the load, the input impedance is just the load impedance

$$Z_{in}(\mathcal{L}) = Z_L = Z_c \frac{1 + \rho(\mathcal{L})}{1 - \rho(\mathcal{L})} \qquad (6.35)$$

and from this expression, the reflection coefficient at the load can be determined to be

$$\rho(\mathcal{L}) = \frac{Z_L - Z_c}{Z_L + Z_c} \qquad (6.36)$$

For the particular case of the matched line, $Z_L = Z_c$ and $\rho(\mathcal{L}) = 0$: The reflected wave from the load is zero.

6.2.4 General Solution for a Terminated Line

Consider a finite line terminated at both ends with load impedances and having both a voltage and current excitation source at $x = x_s$, as shown in Figure 6.14. In this case a voltage reflection coefficient is defined at each load

$$\rho_i = \frac{Z_{Li} - Z_c}{Z_{Li} + Z_c} \qquad \text{for } i = 1, 2 \qquad (6.37)$$

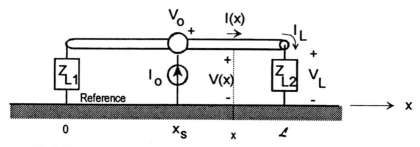

FIGURE 6.14 Line terminated at both ends with arbitrary lumped voltage and current sources.

The general solutions for the voltage and current at an arbitrary point of a line with a series voltage and a parallel current source at $x = x_s$ can be derived [6.6]. For $x > x_s$ the resulting expressions are

$$V(x) = \frac{e^{-\gamma x} + \rho_2 e^{\gamma(x-2\mathcal{L})}}{2(1 - \rho_1\rho_2 e^{-2\gamma\mathcal{L}})} [(e^{\gamma x_s} - \rho_1 e^{-\gamma x_s})V_0 + (e^{\gamma x_s} + \rho_1 e^{-\gamma x_s})Z_c I_0]$$

(6.38a)

$$I(x) = \frac{e^{-\gamma x} - \rho_2 e^{\gamma(x-2\mathcal{L})}}{2Z_c(1 - \rho_1\rho_2 e^{-2\gamma\mathcal{L}})} [(e^{\gamma x_s} - \rho_1 e^{-\gamma x_s})V_0 + (e^{\gamma x_s} + \rho_1 e^{-\gamma x_s})Z_c I_0]$$

(6.38b)

and for $x < x_s$,

$$V(x) = \frac{e^{\gamma(x-\mathcal{L})} + \rho_1 e^{-\gamma(x+\mathcal{L})}}{2(1 - \rho_1\rho_2 e^{-2\gamma\mathcal{L}})} [-(e^{\gamma(\mathcal{L}-x_s)} - \rho_2 e^{-\gamma(\mathcal{L}-x_s)})V_0$$
$$+ (e^{\gamma(\mathcal{L}-x_s)} + \rho_2 e^{-\gamma(\mathcal{L}-x_s)})Z_c I_0] \quad (6.39a)$$

$$I(x) = \frac{e^{\gamma(x-\mathcal{L})} - \rho_1 e^{-\gamma(x+\mathcal{L})}}{2Z_c(1 - \rho_1\rho_2 e^{-2\gamma\mathcal{L}})} [-(e^{\gamma(\mathcal{L}-x_s)} - \rho_2 e^{-\gamma(\mathcal{L}-x_s)})V_0$$
$$+ (e^{\gamma(\mathcal{L}-x_s)} + \rho_2 e^{-\gamma(\mathcal{L}-x_s)})Z_c I_0] \quad (6.39b)$$

6.2.4.1 Example of Line Response for a Voltage Source Excitation. Figure 6.15 shows the case of a voltage excited transmission line. A unit voltage source ($V_0 = 1$) is placed at location $x_s = 2.5$ m. The line length and other parameters are indicated on the figure. Using Eqs. (6.38b) and (6.39b), the spatial distribution of the line current can be evaluated. Figure 6.16a shows the resulting normalized current distribution magnitude $|I(x)/V_0|$ for different frequencies of the voltage source (i.e., for 20 MHz and 50 MHz). Equation (6.38b) can also be used to compute the load current in

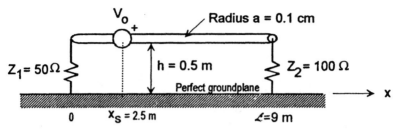

FIGURE 6.15 Single-conductor line over a perfect ground and excited by a lumped voltage source.

Z_{L2} by setting $x = \mathcal{L} = 9$ m. Figure 6.16b shows the normalized current magnitude in the load as a function of frequency. Periodic resonances are noted in the response, and these correspond to wave reflections from the loads.

6.2.5 Load Responses for a Finite Line

In many cases, only the transmission line load responses are needed and the general equations for the line responses in Eqs. (6.38) and (6.39) can be reduced in complexity. By setting $x = 0$ or $x = \mathcal{L}$ in these equations, the line voltage and line current responses at the two loads of the lines are given as [6.7] at $x = 0$,

$$V(0) = \frac{(1+\rho_1)e^{-\gamma\mathcal{L}}}{2(1-\rho_1\rho_2 e^{-2\gamma\mathcal{L}})}[-(e^{\gamma(\mathcal{L}-x_s)} - \rho_2 e^{-\gamma(\mathcal{L}-x_s)})V_0$$
$$+ (e^{\gamma(\mathcal{L}-x_s)} + \rho_2 e^{-\gamma(\mathcal{L}-x_s)})Z_c I_0] \quad (6.40a)$$

$$I(0) = \frac{(1-\rho_1)e^{-\gamma\mathcal{L}}}{2Z_c(1-\rho_1\rho_2 e^{-2\gamma\mathcal{L}})}[-(e^{\gamma(\mathcal{L}-x_s)} - \rho_2 e^{-\gamma(\mathcal{L}-x_s)})V_0$$
$$+ (e^{\gamma(\mathcal{L}-x_s)} + \rho_2 e^{-\gamma(\mathcal{L}-x_s)})Z_c I_0] \quad (6.40b)$$

and at $x = \mathcal{L}$,

$$V(\mathcal{L}) = \frac{(1+\rho_2)e^{-\gamma\mathcal{L}}}{2(1-\rho_1\rho_2 e^{-2\gamma\mathcal{L}})}[(e^{\gamma x_s} - \rho_1 e^{-\gamma x_s})V_0 + (e^{\gamma x_s} + \rho_1 e^{-\gamma x_s})Z_c I_0]$$

$$(6.41a)$$

$$I(\mathcal{L}) = \frac{(1-\rho_2)e^{-\gamma\mathcal{L}}}{2Z_c(1-\rho_1\rho_2 e^{-2\gamma\mathcal{L}})}[(e^{\gamma x_s} - \rho_1 e^{-\gamma x_s})V_0 + (e^{\gamma x_s} + \rho_1 e^{-\gamma x_s})Z_c I_0]$$

$$(6.41b)$$

6.2 FREQUENCY-DOMAIN RESPONSES

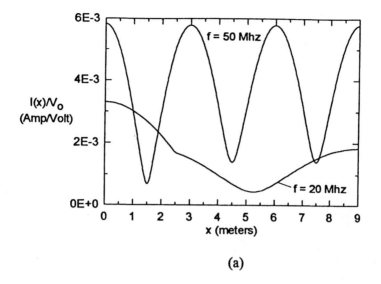

(a)

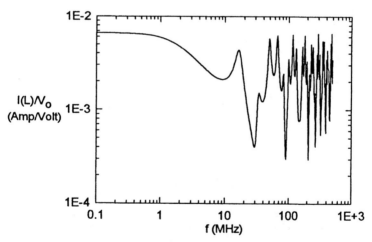

FIGURE 6.16 Line current magnitudes for a lumped voltage source excitation: (a) current distribution along line; (b) load current spectral responses at $x = 9$ m.

Note that these responses are the *line* voltages and currents. At $x = 0$, the *load* current is the negative of the line current, and at $x = \mathcal{L}$, the load current equals the line current.

6.2.5.1 BLT Equation. Expressions (6.40) and (6.41) can be written in a compact matrix form for the *load* voltages and currents. These are

$$\begin{pmatrix} V(0) \\ V(\mathcal{L}) \end{pmatrix} = \begin{pmatrix} 1+\rho_1 & 0 \\ 0 & 1+\rho_2 \end{pmatrix} \begin{pmatrix} -\rho_1 & e^{\gamma\mathcal{L}} \\ e^{\gamma\mathcal{L}} & -\rho_2 \end{pmatrix}^{-1} \begin{pmatrix} e^{\gamma x_s}(V_0 + Z_c I_0)/2 \\ -e^{\gamma(\mathcal{L}-x_s)}(V_0 - Z_c I_0)/2 \end{pmatrix}$$

(6.42a)

$$\begin{pmatrix} I(0) \\ I(\mathcal{L}) \end{pmatrix} = \frac{1}{Z_c} \begin{pmatrix} 1-\rho_1 & 0 \\ 0 & 1-\rho_2 \end{pmatrix} \begin{pmatrix} -\rho_1 & e^{\gamma\mathcal{L}} \\ e^{\gamma\mathcal{L}} & -\rho_2 \end{pmatrix}^{-1} \begin{pmatrix} e^{\gamma x_s}(V_0 + Z_c I_0)/2 \\ -e^{\gamma(\mathcal{L}-x_s)}(V_0 - Z_c I_0)/2 \end{pmatrix}$$

(6.42b)

This matrix form is known as the BLT (Baum–Liu–Tesche) equation [6.8].

In Eqs. (6.42) the effects of excitation, line resonances, and load responses have been factored into separate terms. By defining the following two-element vectors,

$$\bar{I} = \begin{pmatrix} I(0) \\ I(\mathcal{L}) \end{pmatrix} \quad \bar{V} = \begin{pmatrix} V(0) \\ V(\mathcal{L}) \end{pmatrix} \quad \bar{V}_s = \begin{pmatrix} e^{\gamma x_s}(V_0 + Z_c I_0)/2 \\ e^{\gamma(\mathcal{L}-x_s)}(V_0 - Z_c I_0)/2 \end{pmatrix} \quad (6.43)$$

and the following four-element matrices,

$$\bar{\bar{\Gamma}} = \begin{pmatrix} \rho_1 & 0 \\ 0 & \rho_2 \end{pmatrix} \quad \bar{\bar{D}} = \begin{pmatrix} -\rho_1 & e^{\gamma\mathcal{L}} \\ e^{\gamma\mathcal{L}} & -\rho_2 \end{pmatrix} \quad \bar{\bar{U}} = \begin{pmatrix} 1 & 0 \\ 0 & 1 \end{pmatrix} \quad (6.44)$$

the BLT equation for the load responses is given simply as

$$\bar{V} = [\bar{\bar{U}} + \bar{\bar{\Gamma}}]\bar{\bar{D}}^{-1}\bar{V}_s \quad (6.45a)$$

$$\bar{I} = \frac{1}{Z_c}[\bar{\bar{U}} - \bar{\bar{\Gamma}}]\bar{\bar{D}}^{-1}\bar{V}_s \quad (6.45b)$$

The BLT equation also permits a concise representation of the response of multiconductor lines to lumped (Section 6.2.7) or distributed (Chapter 7) source excitations.

6.2.5.2 Example of Frequency-Domain Voltage Response of a Line.
The BLT equations can be used to obtain either the load voltage or current responses. The open-circuit voltage response at $x = \mathcal{L}$ for a line excited by a voltage source at $x = 0$ is shown in Figure 6.17. The line geometry is shown in part (*a*) and the frequency response is plotted in part (*b*) of the figure. At low frequencies, the voltage response at the end of the line is simply the source voltage. As the frequency increases, the line length effects appear and

6.2 FREQUENCY-DOMAIN RESPONSES   245

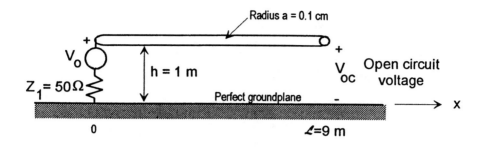

(a)

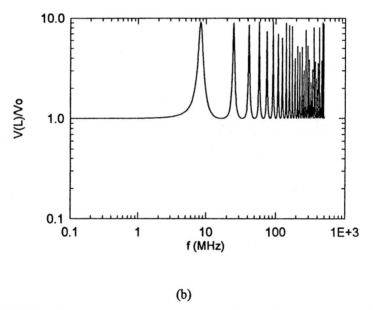

(b)

FIGURE 6.17 Open-circuit voltage response of a line in the frequency domain: (*a*) line geometry; (*b*) open-circuit voltage spectral magnitude, as computed with the BLT equation.

cause resonances. A $\lambda/2$ resonance can be seen around 16 MHz, followed by higher-order resonances.

6.2.5.3 Validation of the Transmission Line Models.
To validate the TL models described above, a comparison with results obtained using an integral equation solution (see Chapter 4) can be performed. Consider the line geometry shown in Figure 6.18*a*, with the load current being the desired observation quantity. This response is calculated in the frequency domain

246 TRANSMISSION LINE THEORY

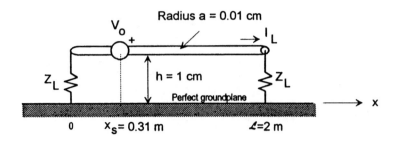

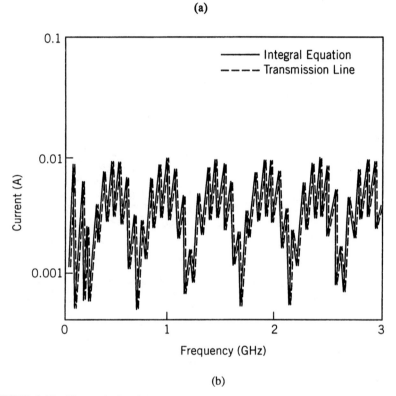

FIGURE 6.18 Transmission line model validation showing a comparison between the load current magnitude using the TL model and an integral equation (NEC) solution: (*a*) problem geometry; (*b*) comparisons for 50-Ω loads; (*c*) comparisons for 318-Ω (matched) loads.

using a computer program based on the BLT equation and with the NEC (Numerical Electromagnetic Code) [6.9].

The results of the comparison of the load currents for two different load values (50 Ω and a matched load 318 Ω) are shown in Figures 6.18*b* and *c*. It is noted that for both loads the results obtained using the TL model are in

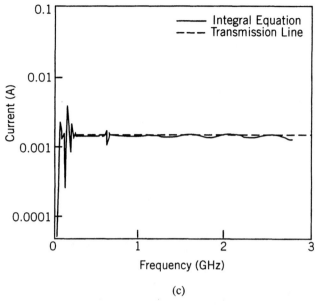

(c)

FIGURE 6.18 (*Continued*)

very good agreement with those given by the integral equation solution. The discrepancies at low frequencies between the two responses is due to numerical instabilities in the NEC code. If a double-precision calculation were performed using NEC, these differences would disappear.

6.2.6 Multiconductor Transmission Lines

For a system of n conductors (plus a reference conductor) which are mutually coupled, the telegrapher's equations (6.7) can be written in matrix form as

$$\frac{d[V(x)]}{dx} + [Z'][I(x)] = 0 \qquad (6.46a)$$

$$\frac{d[I(x)]}{dx} + [Y'][V(x)] = 0 \qquad (6.46b)$$

where $[V(x)]$ and $[I(x)]$ are the vectors of the n voltages and currents on the lines and $[Z']$ and $[Y']$ are the $n \times n$ matrices of the line impedances and admittances. The solution of Eqs. (6.46) is complicated by the fact that these two matrices are full matrices. This is due to the mutual inductive and capacitive coupling between the n lines of the system.

The two coupled first-order differential equations in Eqs. (6.46) can be used to derive uncoupled, second-order equations of the form

$$\frac{d^2[V(x)]}{dx^2} + [P][V(x)] = 0 \qquad (6.47a)$$

$$\frac{d^2[I(x)]}{dx^2} + [R][I(x)] = 0 \qquad (6.47b)$$

where the matrices $[P]$ and $[R]$ are full matrices defined as

$$[P] = [Z'][Y'] \qquad (6.48a)$$

$$[R] = [Y'][Z'] \qquad (6.48b)$$

6.2.6.1 Impedance and Admittance Matrices.
The n-conductor system plus the additional conductor serving as a ground return is useful for modeling a bundle of telecommunication wires in a nonshielded cable, an aerial power line, or a data transmission line linking two electronic circuits. In these cases the $n \times n$ per-unit-length impedance matrix is

$$[Z'] = \begin{bmatrix} R'_1 + R'_{g11} + j\omega L'_{11} & R'_{g12} + j\omega M'_{12} & \cdots & \cdot \\ R'_{g21} + j\omega M'_{21} & R'_2 + R'_{g22} + j\omega L'_{22} & \cdots & \cdot \\ \cdot & \cdot & \cdots & \cdot \\ \cdot & \cdot & \cdots & R'_n + R'_{gnn} + j\omega L'_{nn} \end{bmatrix} \qquad (6.49)$$

and

$$[Y'] = j\omega \begin{bmatrix} C'_{11} & -C'_{12} & \cdots \\ -C'_{21} & C'_{22} & \cdots \\ \cdot & \cdot & \cdots \\ \cdot & \cdot & C'_{nn} \end{bmatrix} \qquad (6.50)$$

where R'_i are the resistances of each of the line conductors, R'_{gij} is the resistance of the common ground, L'_{ii} and M'_{ij} are the self and mutual inductances of and between the conductors, and C'_{ij} are the capacitive coefficients for the conductors (see also Sections 6.4 and 6.5). In the expression for $[Y']$, the effects of any shunt conductance between the conductors has been neglected. It can be seen that the off-diagonal terms arise from mutual coupling effects and the common ground resistance. The differences between the diagonal terms are due to geometry differences between each of the conductors.

In many practical circumstances where there is *no loss* on the line and when there is *no variation* in the parameters μ and ϵ around the line, the matrices $[P]$ and $[R]$ of Eq. (6.48) become *diagonal* matrices given by

$$[P] = [Z'][Y'] = -\frac{\omega^2}{v^2}[U] \qquad (6.51a)$$

$$[R] = [Y'][Z'] = -\frac{\omega^2}{v^2}[U] \qquad (6.51b)$$

where v is the velocity of propagation along the line.

The diagonal form for $[P]$ and $[R]$ implies that each equation for the voltage or current in Eq. (6.47) is decoupled from the others and they may be solved independently, with solutions consisting of n forward and backward propagating modes on the line [6.7]. Each mode has the same propagation speed, and from Eq. (6.51) it is possible to see that the relationship between the $[L']$ and $[C']$ matrices of the line is

$$[C'] = \frac{1}{v^2}[L']^{-1} \qquad (6.52)$$

which is analogous to Eq. (6.6) for the single-conductor case. Given the fact that the $[L']$ matrix is easily computed since the self and mutual inductance terms do not depend on the presence of the other conductors, the $[C']$ matrix may be more easily computed from the equation than by evaluating the capacitive coefficients directly.

Some cases involving lossless multiconductor lines do involve conductors covered with dielectric jackets. As a consequence, the simple relationship between $[L']$ and $[C']$ in Eq. (6.52) is not obeyed and the $[P]$ and $[R]$ matrices are not diagonal. The implication of this is that the transmission line voltages (and currents) are coupled together and Eq. (6.47) cannot be solved in a simple, scalar manner.

In this case, there are generally n different voltage (and also current) modes which exist on the line, and each can have its own propagation velocity. Depending on the details of the line configuration, these modal velocities can be distinct or identical. As discussed in [6.10], the $[L']$ and $[C']$ matrices can be obtained from measurements of the various modal velocities and the characteristic admittance matrix $[Y_c]$ of the multiconductor line. Letting the matrix $[v]$ be a diagonal matrix containing each of the modal velocities along the diagonal, the $[L']$ and $[C']$ matrices can be determined as

$$[L'] = [v]^{-1}[Y_c]^{-1} \qquad (6.53a)$$

$$[C'] = [Y_c][v]^{-1} \qquad (6.53b)$$

The characteristic-admittance matrix $[Y_c]$ of the line can be obtained from time-domain reflectometer (TDR) measurements as discussed in [6.11], and once the $[L]$ and $[C]$ matrices are found, the multiconductor transmission line equations can be solved using the modal theory, which is described in the following sections. However, use of the relations in Eqs. (6.53) is usually not recommended, since the TDR method is difficult to implement for a multiconductor line having a large number of conductors. This is because it is based on a measurement of different time of arrival of reflected waves, due to different velocities on each conductor. If the different velocities are almost identical, it is difficult to separate out the effect of one mode from another [6.12].

To avoid the difficulty, cited above, several measurement methods permitting direct determination of the $[L']$ and $[C']$ values have been proposed and tested in [6.12]. An alternative approach is to calculate the values of these line parameters, and this is discussed in more detail later.

6.2.6.2 Natural Propagation Modes.
For the special case where the $[L']$ and $[C']$ matrices satisfy Eq. (6.52), the set of n uncoupled equations for the line voltages and currents can be written and solved by inspection [6.7]. Corresponding to each equation in the set of n equations, the response quantity (i.e., the conductor current or voltage relative to the reference conductor) can be thought of as a mode. An example of a current mode distribution in this case would be a current of I_0 in one conductor and 0 in all the other conductors. For this n-wire line (plus the reference conductor), there are n voltage modes and n current modes, which are related to each other by modal impedances. Thus if the n-vector representing the ith modal voltage distribution on the line is denoted by $[v_i]$ and the corresponding current mode is $[i_i]$, the relationship between the two is expressed by a scalar modal characteristic impedance z_i as

$$[v_i] = z_i [i_1] \qquad (6.54)$$

Note that in this discussion a lowercase quantity is used to distinguish a modal quantity from a physical (current of voltage) observable. The complete response of the line (i.e., the line currents or voltages on all the conductors) is determined as an appropriately weighted sum of the n individual modal responses.

For the more general case where $[P]$ and $[R]$ are not diagonal, the current or voltage mode structure is much more complicated. Physically, these modes correspond to the cross-sectional distribution of voltages or currents on the conductors that propagate at the same velocity on the line. For the n-wire line, each mode can have a distinct propagation velocity. If two or more modes have the same velocity, they are said to be *degenerate*. For degenerate modes, the resulting modal distributions are not unique, and any linearly independent set of modal distributions can be considered. As an example, for the lossless, single-velocity multiconductor line, it is possible to consider any distribution of current on the line to be a mode, as long as it is independent from the other modes.

6.2.6.3 Diagonalization of the [P] and [R] Matrices.

To obtain the modal voltages and currents and their propagation velocities in the most general case, the $[P]$ and $[R]$ matrices must be diagonalized. This can be done using similarity transformations involving a matrix $[S]$ that diagonalizes the matrix $[P]$ and a matrix $[T]$ that diagonalizes the matrix $[R]$. As discussed in [6.13], it is possible to show that the diagonalization of $[P]$ and $[R]$ gives the same matrix $[D]$. This fact can be also developed from the physical argument that voltage and current waves must have the same propagation characteristics on the line.

Applying the similarity transformations to $[P]$ and $[R]$ gives

$$[D] = [S][P][S]^{-1} = [T][R][T]^{-1} = \begin{bmatrix} \gamma_1^2 & 0 & 0 & \cdots & 0 \\ 0 & \gamma_2^2 & 0 & \cdots & 0 \\ 0 & 0 & \gamma_3^2 & \cdots & 0 \\ \cdot & \cdot & \cdot & \cdots & \cdot \\ 0 & 0 & 0 & \cdots & \gamma_n^2 \end{bmatrix} = \text{diag}(\gamma_n^2) \equiv [\gamma]^2$$

(6.55)

where the terms γ_i are the propagation constants for each mode. In the most general case, these constants are complex. However, for the case of a line with no loss, the γ_i are imaginary, equal to $\gamma_i = j\omega/v_i$, where v_i are the modal velocities. It can be shown that $[T]$ and $[S]$ are related by

$$[T]^t = [S]^{-1} \qquad (6.56)$$

6.2.6.4 Modal Voltages and Currents.

Operating on the matrix wave equations (6.47) by $[S]$ and $[T]$, respectively, gives two new equations in terms of the diagonal matrix $[D]$ (or $[\gamma]^2$) and *modal* voltages and currents as

$$\frac{d^2[v(x)]}{dx^2} + [\gamma]^2[v(x)] = 0 \qquad (6.57a)$$

$$\frac{d^2[i(x)]}{dx^2} + [\gamma]^2[i(x)] = 0 \qquad (6.57b)$$

Here $[v(x)]$ and $[i(x)]$ are the transformed modal vectors of the line voltages and currents defined as

$$[v(x)] = [S][V(x)] \tag{6.58}$$

$$[i(x)] = [T][I(x)] = \{[S]^{-1}\}'[I(x)] \tag{6.59}$$

6.2.6.5 Solution of the Modal Equations.
The fact that $[\gamma]^2$ is a diagonal matrix permits us to obtain simple solutions for Eqs. (6.57) [6.13]. These solutions are positive and negative propagating modes,

$$[v(x)] = [E^+(x)][a] + [E^-(x)][b] \tag{6.60a}$$

$$[i(x)] = [z_c]^{-1}\{[E^+(x)][a] - [E^-(x)][b]\} \tag{6.60b}$$

where

$$[E^\pm(x)] = \begin{bmatrix} e^{\mp\gamma_1 x} & 0 & \cdots & 0 \\ 0 & e^{\mp\gamma_2 x} & \cdots & 0 \\ \vdots & \vdots & \cdots & \vdots \\ 0 & 0 & \cdots & e^{\mp\gamma_n x} \end{bmatrix} \equiv e^{\mp[\gamma]x} \tag{6.61}$$

and $[z_c]$ is a modal characteristic impedance matrix. This matrix can be evaluated using Eqs. (6.58) and (6.59) and substituting Eqs. (6.60) into Eq. (6.46a) to yield

$$[z_c] = [\gamma]^{-1}[S][Z'][T]^{-1} \tag{6.62}$$

The n-vectors $[a]$ and $[b]$ contain the amplitudes of the forward and reverse propagating voltage modes. These constant vectors must be determined from the boundary conditions and line excitation functions, as discussed later. From the modal voltage and current values, the *physical* voltages and currents are obtained using the inverse of Eqs. (6.58) and (6.59) as

$$[V(x)] = [S]^{-1}[v(x)] = [S]^{-1}\{[E^+(x)][a] + [E^-(x)][b]\} \tag{6.63}$$

$$[I(x)] = [T]^{-1}[i(x)] = [T]^{-1}[z_c]^{-1}\{[E^+(x)][a] - [E^-(x)][b] \tag{6.64}$$

Notice that in Eq. (6.63) the first term, $[S]^{-1}[E^+(x)][a]$, corresponds to a set of positive traveling voltages on the line, and the second term, $[S]^{-1}[E^-(x)][b]$, corresponds to a set of negative traveling voltages. These are referred to as the *physical voltages* in that they are the measurable quantities on the line, unlike the modes, which are mathematically introduced quantities.

Similarly, Eq. (6.64) consists of a forward-traveling physical current wave $[T]^{-1}[z_c]^{-1}[E^+(x)][a]$ and a negative traveling current $-[T]^{-1}[z_c]^{-1}[E^-(x)][b]$. As in the case of a single-conductor line in Eq. (6.14), it is possible to relate the amplitudes of these current waves to the corresponding voltage waves through a characteristic admittance $[Y_c]$ (or impedance $[Z_c] = [Y_c]^{-1}$) of the line. In this way, Eq. (6.64) can be put into a slightly different form,

$$[I(x)] = [Y_c][S]^{-1}\{[E^+(x)][a] - [E^-(x)][b]\} \qquad (6.65)$$

where the characteristic impedance matrix $[Y_c]$ is given by

$$[Y_c] = [T]^{-1}[z_c]^{-1}[S] \qquad (6.66)$$

Noting the definition of the modal characteristic impedance matrix $[z_c]$ in Eq. (6.62), the characteristic admittance matrix of Eq. (6.66) becomes

$$[Y_c] = [Z']^{-1}[S]^{-1}[\gamma][S] \qquad (6.67)$$

and its inverse is the line characteristic impedance matrix $[Z_c]$:

$$[Z_c] = [Y_c]^{-1} = [S]^{-1}[\gamma]^{-1}[S][Z'] \qquad (6.68)$$

6.2.6.6 Calculation of the Propagation Matrix and Diagonalization Matrix Elements. In performing the multiconductor analysis, it is necessary to determine the diagonalization matrices $[S]$ and $[T]$ and the elements of the propagation matrix $[\gamma]$. The matrix $[\gamma]^2$ can be calculated by solving the eigenvalue equation

$$\det([P] - [\gamma]^2) = 0 \qquad (6.69)$$

and the diagonalization matrix satisfies the relation

$$[S][P][S]^{-1} = [\gamma]^2 \qquad (6.70)$$

To satisfy this matrix identity, each element of the matrix resulting from (6.70) must be equal to zero. This gives a system of homogeneous equations in which the values of S_{ii} and S_{ij} are not all independent [6.14]. One of these values must be specified to obtain the others. The elements of the $[T]$ matrix are obtained in a similar way.

From a computational standpoint, the determination of these quantities can be tedious. For the most general case, the matrix $[P]$ in Eq. (6.69) is a complex-valued function of frequency and the eigenvalues may also be complex. As a result, the complex eigenvalues of this equation must be determined numerically at each frequency, and this is normally done using a matrix eigenvalue routine. The eigenmodes corresponding to each of the n eigenvalues are also computed at the same time, and these modes are used to define the columns of the matrix $[S]$.

As an example of numerical results, consider the case of the ribbon cable

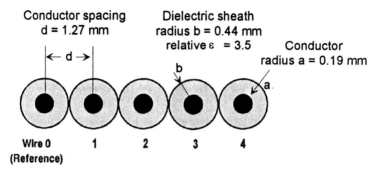

FIGURE 6.19 Cross section of a five-conductor multiwire line.

shown in Figure 6.19. Such cables are typically used to interconnect computer or electronics assemblies and carry switching signals. The cable under discussion here is assumed to consist of five individual conductors, each separated by a distance $d = 1.27$ mm. The individual conductors have a wire radius of $a = 0.19$ mm, and each are covered by a dielectric jacket of radius $b = 0.44$ mm. The relative dielectric constant of the jacket is $\epsilon = 3.5$, a value which is typical for polyvinyl chloride insulation. For this example, conductor and dielectric losses are neglected, so that the line is described only by the per-unit-length inductance and capacitance matrices. As a consequence, the modal propagation constants are imaginary.

For this line, the reference conductor is taken to be the wire at the left end of the bundle. The remaining four wires will then have a voltage relative to the reference conductor, and these are described by the previously defined telegrapher's equations. Under the assumptions and limitations of the transmission line theory, the total current in all five conductors must sum to zero.

The per-unit-length capacitance and inductance matrix of this line may be evaluated using the techniques described in Sections 6.4 and 6.5, or elsewhere [6.15]. The 4×4 inductance matrix of this line has the numerical values that are *unaffected* by the dielectric jackets; each element of this matrix may be evaluated independently from the others, and the matrix has the numerical value

$$L' = \begin{bmatrix} 0.74834 & 0.50711 & 0.45527 & 0.43295 \\ 0.50711 & 1.0132 & 0.71984 & 0.64569 \\ 0.45527 & 0.71984 & 1.1738 & 0.85842 \\ 0.43295 & 0.64569 & 0.85842 & 1.2914 \end{bmatrix} \mu H/m \quad (6.71)$$

Using the relationship that $[L'][C'] = \mu_0 \epsilon_0 = 1/c^2$, the capacitance matrix *for the case of no dielectric surrounding the wires* is

$$C_0' = \begin{bmatrix} 23.345 & -8.9057 & -2.1907 & -1.9178 \\ -8.9057 & 23.615 & -8.9057 & -2.9018 \\ -2.1907 & -8.9057 & 23.345 & -10.331 \\ -1.9178 & -2.9018 & -10.331 & 17.577 \end{bmatrix} \text{ pF/m} \quad (6.72)$$

Note that these matrices have the required symmetry about the diagonal elements.

For the wires with the dielectric sheaths, a numerical procedure is required to determine the capacitance matrix. For the wire bundle under discussion, [6.15] gives this matrix as

$$C' = \begin{bmatrix} 38.152 & -15.974 & -2.2829 & -2.0343 \\ -15.974 & 38.401 & -15.974 & -3.2263 \\ -2.2829 & -15.974 & 38.152 & -17.861 \\ -2.0343 & -3.2263 & -17.861 & 26.017 \end{bmatrix} \text{ pF/m} \quad (6.73)$$

For the case of the wire with the dielectric present, using Eqs. (6.51a) and (6.52), the eigenvalue equation (6.70a) becomes

$$-\omega^2 [S][L'][C'][S]^{-1} = [\gamma]^2 \quad (6.74a)$$

or

$$[S][L'][C'][S]^{-1} = \text{diag}[1/v_i^2] \quad (6.74b)$$

where v_i are the modal velocities (for $i = 1$ to 4) and $[S]$ contains the *voltage* eigenvectors, or voltage modes, with each column containing one of the modes. For the inductance and capacitance matrices of Eqs. (6.71) and (6.73), the modal velocities are given in Table 6.1.

For this particular multiconductor cable, the current modes (the columns of $[T]$) and the voltage modes (the columns of $[S]$) are shown in Figures 6.20 and 6.21, respectively.

TABLE 6.1 Computed Modal Velocities for the Five-Conductor Line

Mode	v_1 (m/s)	v_i/c
1	2.65×10^8	0.88
2	2.46×10^8	0.82
3	2.35×10^8	0.78
4	2.29×10^8	0.76

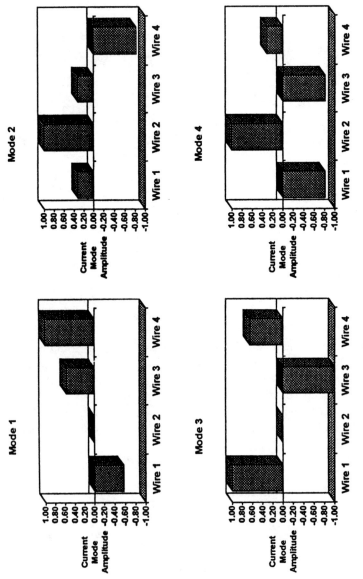

FIGURE 6.20 Current-mode distributions for the multiconductor ribbon cable.

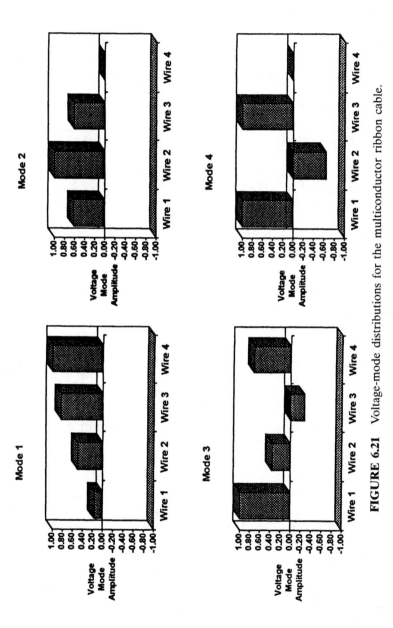

FIGURE 6.21 Voltage-mode distributions for the multiconductor ribbon cable.

6.2.6.7 Open-Circuit Voltage of a Semi-infinite Multiconductor Line Excited by a Voltage Source.
As an example of the use of the multiconductor analysis, consider the case of a semi-infinite multiconductor line with lumped voltage source excitations $[V_0]$ located at $x = x_s$, as shown in Figure 6.22. The line is open circuited at $x = 0$, where it is desired to determine the open-circuit voltage.

For this analysis, divide the line into two parts: section 1 to the left of the source $(0 > x > x_s)$ and section 2 to the left of the source $(x_s > x > \infty)$. Applying Eqs. (6.63), (6.65), and (6.61) in section 1 gives

$$[V(x)] = [S]^{-1}\{e^{-[\gamma]x}[a_1] + e^{[\gamma]x}[b_1]\} \tag{6.75a}$$

$$[I(x)] = [Y_c][S]^{-1}\{e^{-[\gamma]x}[a_1] - e^{[\gamma]x}[b_1]\} \tag{6.75b}$$

In section 2, as the line is semi-infinite in extent, there is no reflected wave in the negative direction of the x-axis, so the expression for the line voltage and current is

$$[V(x)] = [S]^{-1}e^{-[\gamma]x}[a_2] \tag{6.76a}$$

$$[I(x)] = [Y_c][S]^{-1}\{e^{-[\gamma]x}[a_2]\} \tag{6.76b}$$

where the indices 1 and 2 refer to the sections of the line.

From the boundary that $[I(x)] = 0$ at $x = 0$, we have the requirement that

$$[a_1] = [b_1] \tag{6.77}$$

which implies that in section 1 the voltage is

$$[V(x)] = [S]^{-1}[a_1]\{e^{-[\gamma]x} + e^{[\gamma]x}\} \tag{6.78}$$

At the source location ($x = x_s$) the line voltage must be discontinuous by the

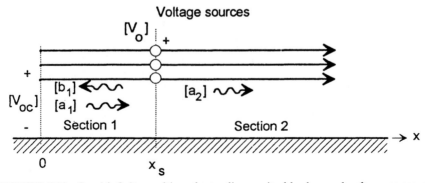

FIGURE 6.22 Semi-infinite multiconductor line excited by lumped voltage sources.

source voltage $[V_0]$:

$$V(x_{s+}) - V_0 = V(x_{s-}) \tag{6.79}$$

Using Eqs. (6.75a) and (6.76a), this becomes

$$[S]^{-1}[a_2]e^{-[\gamma]x_s} - [V_0] = [S]^{-1}\{[e^{[\gamma]x_s} + e^{-[\gamma]x_s}][a_1]\} \tag{6.80}$$

Moreover, at $x = x_s$ the current is continuous. This gives the relationship

$$[a_2]e^{-[\gamma]x_s} = [a_1]\{e^{-[\gamma]x_s} - e^{[\gamma]x_s}\} \tag{6.81}$$

The coefficients $[a_1]$, $[b_1]$, and $[a_2]$ can now be determined from conditions (6.77), (6.80), and (6.81) as

$$[a_1] = [b_1] = \tfrac{1}{2} e^{-[\gamma]x_s}[S][V_0] \tag{6.82a}$$

$$[a_2] = \tfrac{1}{2} \{e^{-[\gamma]x_s} - e^{[\gamma]x_s}\}[S][V_0] \tag{6.82b}$$

Using these expressions, Eq. (6.75a) evaluated at $x = 0$ gives the desired open-circuit voltage n-vector in section 1 as

$$[V_{oc}] = [S]^{-1}e^{-[\gamma]x_s}[S][V_0] \tag{6.83}$$

6.2.6.8 Simplified Modeling by the Equal-Velocity Assumption.

By comparing the propagation velocities of the different modes on a bundle of five insulated conductors over a metallic plane, Poudroux [6.16] has found that the differences in these velocities are not very significant in some cases. A first-order simplification to the multivelocity transmission line theory involves assuming that all modes have the same propagation velocity as the common-mode velocity, which can be easily measured.

In making this approximation, the following simplifications result:

$$[S] = [T] = [U] \tag{6.84}$$

$$[\gamma] = \gamma_1[U] \tag{6.85}$$

where γ_1 is the propagation constant of the common mode. These approximations simplify considerably the expressions for the line voltage and current:

$$[V(x)] = [a]e^{-\gamma_1 x} + [b]e^{\gamma_1 x} \tag{6.86a}$$

$$[I(x)] = \gamma_1 [Z']^{-1}\{[a]e^{-\gamma_1 x} - [b]e^{\gamma_1 x}\} \tag{6.86b}$$

Applying these expressions to the same line treated in the preceding section gives a significantly simpler expression for the open-circuit voltage:

$$[V_{oc}] = e^{-\gamma x_s}[V_0] \tag{6.87}$$

As an example of the behavior of the multivelocity modes on the line and the implications of using a single-mode approximation, consider again the ribbon cable of Figure 6.19 in the configuration of Figure 6.22. The line is semi-infinite in length, and the source is at $x_s = 1$ m from the open end. For this example, the voltage source on wire 1 is assumed to be a unit step function in time, having a frequency-domain representation of $V_1(\omega) = 1/j\omega$. The voltage sources on the other conductors are identically zero.

Figure 6.23 shows a comparison of the frequency-domain responses for V_{oc} on each of the conductors. Part (*a*) shows results for the multimode analysis, as given by Eq. (6.83), and part (*b*) shows the same results for the approximate analysis leading to Eq. (6.87). Notice that at low frequencies, the responses for the voltage on wire 1 are almost identical. However, at higher frequencies, they begin to differ as the multimode structure begins to become important. For the parasitic wires 2, 3, and 4, the approximate analysis provides a response that is identically zero. The effects of any mutual coupling between the wires are not evident.

Another interesting way of viewing these results is to consider the transient responses. Figure 6.24*a* illustrates the step-function response of the voltage on wire 1 for both analysis methods, and Figure 6.24*b* shows the transient behavior of V_{oc} on the other wires for the multimode case. In part (*a*), the approximate analysis provides a clean step-function voltage response for wire 1. Note that the oscillations at the discontinuities of this step function arise from Gibbs phenomenon and are not part of the true step-function response. In part (*a*) of this figure, the approximate analysis occurs significantly earlier in time, due to the assumption that the wave propagation speed on the line is c. For the multimode analysis, however, the wave speed is slower due to the presence of the dielectric around the conductors, and this implies that the response occurs at a later time.

The multimode analysis, however, exhibits a varying wavefront, as the different modes arrive at the end of the line and contribute to the response. Although this difference in the waveforms may not be very important in practical cases, the fact that the approximate analysis fails to predict any cross-coupling of the voltage source on line 1 to the parasitic wires 2, 3, and 4 can be important.

6.2.7 BLT Equation for Multiconductor Lines.
For a finite length of multiconductor line terminated in generalized matrix load impedances $[Z_{L1}]$ and $[Z_{L2}]$ at the ends at $x = 0$ and $x = \mathcal{L}$, and excited by lumped voltage and current sources $[V_0]$ and $[I_0]$ at $x = x_s$, the BLT equation, introduced in

6.2 FREQUENCY-DOMAIN RESPONSES   261

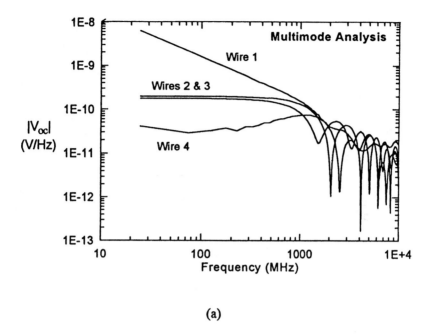

(a)

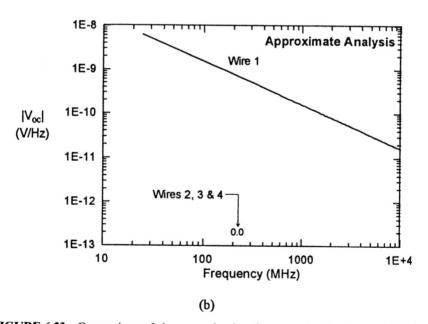

(b)

FIGURE 6.23 Comparison of the open-circuit voltage spectra for the semi-infinite ribbon cable: (a) results using multimode theory; (b) results using the single velocity approximation.

262 TRANSMISSION LINE THEORY

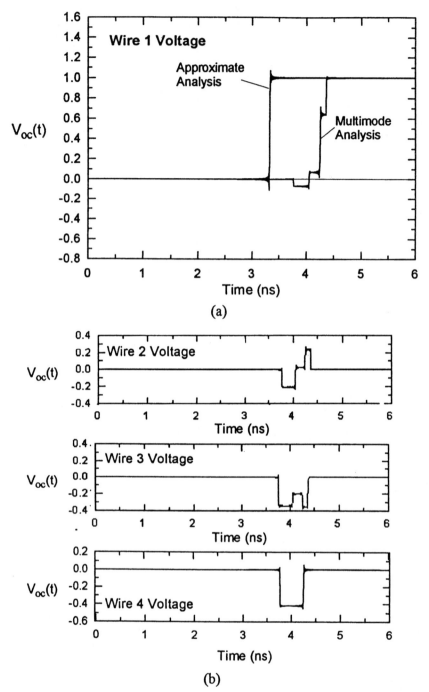

FIGURE 6.24 Comparison of the transient open-circuit voltages for the semi-infinite ribbon cable: (*a*) responses for wire 1; (*b*) multimode responses for wires 2, 3, and 4 (the approximate responses are identically zero).

6.2 FREQUENCY-DOMAIN RESPONSES

Section 6.2.5.1, can be generalized into a supermatrix expression:

$$\begin{pmatrix} [V(0)] \\ [V(L)] \end{pmatrix} = \begin{pmatrix} [U] + [\rho_1] & [0] \\ [0] & [U] + [\rho_2] \end{pmatrix} [D]^{-1} [V_s] \quad (6.88a)$$

$$\begin{pmatrix} [I(0)] \\ [I(L)] \end{pmatrix} = \begin{pmatrix} [Z_c]^{-1} & [0] \\ [0] & [Z_c]^{-1} \end{pmatrix} \begin{pmatrix} [U] - [\rho_1] & [0] \\ [0] & [U] - [\rho_2] \end{pmatrix} [D]^{-1} [V_s]$$

$$(6.88b)$$

where the supermatrices $[D]$ and excitation supervector $[V_s]$ are defined as

$$[D] = \begin{pmatrix} -[\rho_1] & [S]^{-1} e^{[\gamma] \mathcal{L}} [S] \\ [S]^{-1} e^{[\gamma] \mathcal{L}} [S] & -[\rho_2] \end{pmatrix} \quad (6.89)$$

and

$$[V_s] = \frac{1}{2} \begin{bmatrix} [S]^{-1} e^{[\gamma] x_s} [S] \{[V_0] + [Z_c][I_0]\} \\ -[S]^{-1} e^{[\gamma] (\mathcal{L} - x_s)} [S] \{[V_0] - [Z_c][I_0]\} \end{bmatrix} \quad (6.90)$$

As before, $[U]$ is the unit matrix, and the matrices $[\rho_i]$ are the multiconductor reflection coefficient matrices for load 1 or load 2, defined as

$$[\rho_i] = \{[Z_{Li}] - [Z_c]\} \{[Z_{Li}] + [Z_c]\}^{-1} \quad (6.91)$$

As an example of line responses computed using the BLT equation, consider the finite length ribbon cable geometry shown in Figure 6.25. The same line of Figure 6.22 is used, and again only wire 1 is excited by a step-function voltage. Each wire is connected to the reference conductor with a 50-Ω resistor at $x = 0$, and the open-circuit voltage at $x = \mathcal{L} = 1$ m is computed.

Figure 6.26a and b present the step-function spectral responses for the open-circuit voltage for both the multimode analysis (a) and for the approximate single-velocity analysis (b). For this line geometry, the V_{oc} responses for the approximate analysis are not identically zero, due to the

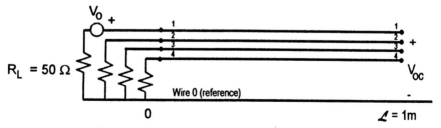

FIGURE 6.25 Finite ribbon cable with a source at one end.

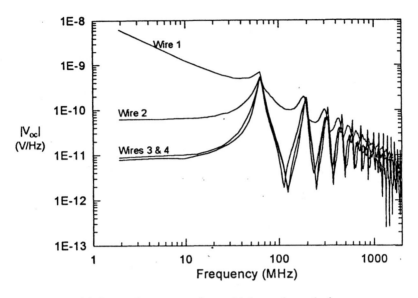

(a) Spectral responses for multiple mode analysis,

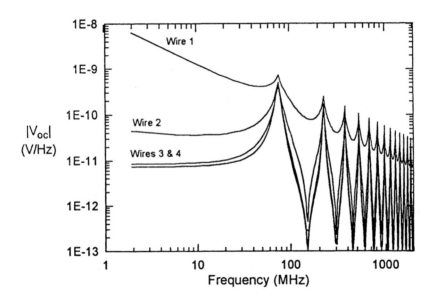

(b) Spectral responses for approximate single velocity analysis

FIGURE 6.26 Open-circuit voltage spectral responses for the ribbon line of Figure 6.25 for a step-function excitation.

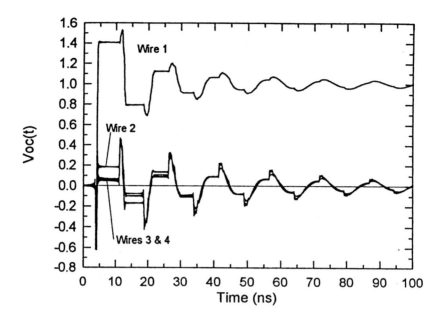

(c) Transient responses for multiple mode analysis

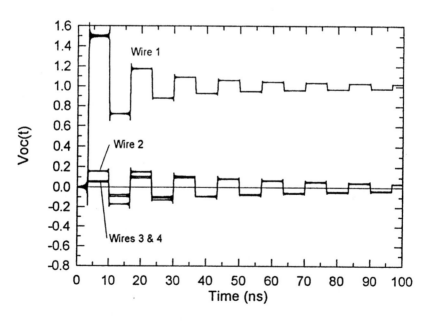

(d) Transient responses for approximate single velocity analysis

FIGURE 6.26 (*Continued*)

266 TRANSMISSION LINE THEORY

fact that at the $x = 0$ end of the line, there is a conversion of modes in the terminating impedance. That is, as the mode excited by the source on wire 1 reflects off the termination impedance, components of all the other modes are created, and these contribute to the other responses.

The corresponding step-function excited transient responses for V_{oc} are shown in Figure 6.26c and d for both single-velocity and multivelocity analysis. These responses exhibit periodic ringing of the response, due to the many reflections back and forth along the line. As there is loss in the 50-Ω termination resistors, the response eventually dies out, leaving a 1-V response on wire 1 and zero voltage on the other wires. The effects of using the approximate single-velocity analysis are noticeable, especially for the parasitic wire responses (2, 3, and 4), where a rather large voltage spike is missing in the approximate response.

6.2.7.1 Simplifications of the BLT Equation.
The BLT equations for the current and voltage in Eq. (6.88) can be simplified when the multivelocity nature of the modes is neglected. Assuming that the per-unit-length line parameters $[L']$ and $[C']$ satisfy Eq. (6.52) and are used to define the wave velocity on the line, the $[S]$, $[T]$, and $[\gamma]$ matrices are given by Eqs. (6.84) and (6.85). The scalar propagation constant $\gamma = j\omega/v$.

Under these conditions, the characteristic impedance matrix from Eq. (6.68) becomes

$$[Z_c] = v[L'] = \frac{1}{v}[C']^{-1} \tag{6.92}$$

and the reflection coefficients are

$$[\rho_i] = \{[Z_{Li}] - v[L']\}\{[Z_{Li}] + v[L']\}^{-1} \tag{6.93}$$

The matrix propagation term $[D]$ in Eq. (6.89) becomes

$$[D] = \begin{pmatrix} -[\rho_1] & e^{j\omega \mathcal{L}/v}[U] \\ e^{j\omega \mathcal{L}/v}[U] & -[\rho_2] \end{pmatrix} \tag{6.94}$$

and the source vector is

$$[V_s] = \frac{1}{2}\left[\begin{array}{c} e^{j\omega x_s/v}\{[V_0] + [Z_c][I_0]\} \\ -e^{j\omega(\mathcal{L}-x_s)/v}\{[V_0] - [Z_c][I_0]\} \end{array}\right] \tag{6.95}$$

6.2.8 Chain Parameters for a Multiconductor Line

As in the case of low-frequency circuits in Chapter 3 and the two-wire transmission line in Section 6.2.1.1, the two-port modeling concept can be generalized to describe a section of multiconductor line. For a system of n conductors (plus the reference conductor) of length $\mathcal{L}$, the expression relating the input to the output voltage and current of Eq. (3.23) becomes

$$\begin{bmatrix} [V(L)] \\ [I(L)] \end{bmatrix} = \begin{bmatrix} \cosh[\gamma]\mathscr{L} & -[Z_c]\sinh[\gamma]\mathscr{L} \\ -[Y_c]\sinh[\gamma]\mathscr{L} & \cosh[\gamma]\mathscr{L} \end{bmatrix} \begin{bmatrix} [V(0)] \\ [I(0)] \end{bmatrix} \quad (6.96)$$

Here $[V(\mathscr{L})]$, $[I(\mathscr{L})]$, $[V(0)]$, and $[I(0)]$, are the n vectors of the output and input line voltages and line currents, respectively, $[Z_c]$, $[Y_c]$, and $[\gamma]$ the $n \times n$ matrices of the characteristic impedances, admittances, and propagation constants of the lines. As in the case of the BLT equation, the expressions $\sinh([\gamma]\mathscr{L})$ and $\cosh([\gamma]\mathscr{L})$ must be evaluated in the same manner as we evaluated the expression $e^{-[\gamma]\mathscr{L}}$. This expression is analogous to the chain parameter representation for a two-port circuit and to Eq. (6.18) for the two-wire line.

6.2.9 Example of the Use of Multiconductor Line Models

The previously developed multiconductor line model can be applied to a wide variety of line problems. As an example, consider the crosstalk example portrayed in Figure 3.36 and discussed in Chapter 3. In this case, the BLT equations (6.88) can be used to evaluate any of the terminal responses in the lines. Assuming no losses on the line, the $[L']$ matrix is simply expressed as

$$[L'] = \begin{bmatrix} L_{11} & M \\ M & L_{22} \end{bmatrix} \quad (6.97)$$

where the elements L_{ii} are determined from Eq. (6.5b), and the mutual coupling parameter is determined from Eq. (3.49). The resulting inductance matrix is then used to compute the capacitance matrix using Eq. (6.52), with the assumption that the wave velocity on the line is $v = c = 3 \times 10^8$ m/s.

The characteristic impedance matrix then becomes

$$[Z_c] = c[L'] \quad (6.98)$$

and the reflection coefficients at each end of the line are

$$[\rho_1] = \left\{ \begin{bmatrix} Z_s & 0 \\ 0 & Z_a \end{bmatrix} - c \begin{bmatrix} L_{11} & M \\ M & L_{22} \end{bmatrix} \right\} \left\{ \begin{bmatrix} Z_s & 0 \\ 0 & Z_a \end{bmatrix} + c \begin{bmatrix} L_{11} & M \\ M & L_{22} \end{bmatrix} \right\}^{-1}$$

(6.99a)

$$[\rho_2] = \left\{ \begin{bmatrix} Z_L & 0 \\ 0 & Z_b \end{bmatrix} - c \begin{bmatrix} L_{11} & M \\ M & L_{22} \end{bmatrix} \right\} \left\{ \begin{bmatrix} Z_L & 0 \\ 0 & Z_b \end{bmatrix} + c \begin{bmatrix} L_{11} & M \\ M & L_{22} \end{bmatrix} \right\}^{-1}$$

(6.99b)

The propagation matrix is simply that given in Eq. (6.94), and the source vector may be expressed as

$$[V_s] = \frac{1}{2}\left[\begin{bmatrix} V_s \\ 0 \end{bmatrix} - \begin{bmatrix} e^{j\omega\mathcal{L}/c}V_s \\ 0 \end{bmatrix} \right] \qquad (6.100)$$

The BLT equation in (6.88) can be expanded algebraically to yield expressions as developed in [6.17] for the terminal voltages. Due to the complexity of these solutions, an algebraic manipulation computer program is useful for this task. The final solution, however, is sufficiently complex that it is difficult to gain insight into the behavior of the solution without resorting to a numerical evaluation of the equations. If a numerical solution is warranted, the BLT form of the solution is advantageous, due to the matrix form. It is very easy to program using the array features of scientific computer languages.

6.3 TIME-DOMAIN TRANSMISSION LINE RESPONSES

6.3.1 Time-Harmonic Excitation

In Section 6.2 we discussed the time-harmonic response of the transmission line, for which the time variation is represented as the function $e^{j\omega t}$. For this excitation, the set of coupled partial derivative equations (6.1) and (6.2) become coupled ordinary differential equations (6.7a) and (6.7b), and solutions for the voltage and current are of the form of traveling waves along the line, as given by Eqs. (6.10) and (6.11).

Although the analysis for this type of monochromatic excitation is relatively straightforward, the results are applicable only to a limited class of problems: those that do not involve a starting or stopping of the excitation. Physical excitations of electrical systems always involve a start and a stop of the excitation, and this implies that a transient analysis must be conducted at these times. Of course, if the excitation is a sinusoidal waveform that is turned on abruptly at an initial time $t = 0$, the time-harmonic response will usually be a very good approximation to the actual response for times after the initial switch-on transient has decayed. Another use of the pure sinusoidal wave response is in construction of a transient response by a Fourier series or integral. These subjects are discussed in this section.

6.3.2 Nonsinusoidal Traveling Waves

A general transient solution to the telegrapher's equations (6.1) and (6.2) cannot be obtained analytically. However, for the special (yet practical) case of no loss on the line, these equations become

$$\frac{\partial v(x,t)}{\partial x} + L'\frac{\partial i(x,t)}{\partial t} = 0 \qquad (6.101a)$$

$$\frac{\partial i(x,t)}{\partial x} + C'\frac{\partial v(x,t)}{\partial t} = 0 \qquad (6.101b)$$

General solutions to these equations consist of arbitrary functions f and g of the variables $x + ct$ and $x - ct$, where $c = 1/\sqrt{L'C'}$. The solutions to the voltage and current are

$$v(x,t) = f(x - ct) + g(x + ct) \qquad (6.102a)$$

$$i(x,t) = \frac{1}{Z_c}[f(x - ct) - g(x + ct)] \qquad (6.102b)$$

with $Z_c = \sqrt{L'/C'}$. The fact that Eqs. (6.102) are the required solutions can be verified by substitution. Notice that the function f is a forward ($+x$) propagating disturbance on the line, while g is a negative propagating disturbance. The determination of the two functions f and g must be done using the source excitations and boundary conditions specific to each particular problem.

6.3.3 Analytical Transformation from the Frequency Domain to the Time-Domain

Under certain circumstances it is possible to develop an analytical expression for the transient response of a transmission line. Consider the finite terminated line excited by either a voltage or a current source at an arbitrary location, x_s, as shown in Figure 6.14. As discussed earlier, explicit expressions of the form of Eqs. (6.40) and (6.41) can be developed for the voltage and current responses at each end of the line.

In the most general case, both the propagation constant γ and the reflection coefficients of the line ρ_i are complex functions of frequency. However, if we assume that there are no losses on the line and that the termination impedances are purely real, we have the following conditions:

$$\gamma = \frac{j\omega}{v} \qquad (6.103)$$

$$\rho_i = \text{constant} \quad (\text{for } i = 1, 2) \qquad (6.104)$$

Aside from the term $(1 - \rho_1\rho_2 e^{-2\gamma\ell})$ term in the denominator, all remaining terms of Eqs. (6.40) and (6.41) now have only a frequency dependence of the form $e^{-j\omega\tau}$, where τ is a constant. Since $|\rho_1\rho_2 e^{-2\gamma\mathcal{L}}| < 1$, the denominator term may be expanded as

$$\frac{1}{(1 - \rho_1\rho_2 e^{-j\omega 2\mathcal{L}/v})} = \sum_{n=0}^{\infty} (\rho_1\rho_2 e^{-j\omega 2\mathcal{L}/v})^n \tag{6.105}$$

which also has a frequency dependence containing only factors of the form $e^{-j\omega\tau}$.

Substituting Eqs. (6.103) to (6.105) into Eqs. (6.40a) and (6.41a), the time-harmonic expressions for the line voltages at $x = 0$ and $x = \mathcal{L}$ become

$$V(0) = (1 + \rho_1) \sum_{n=0}^{\infty} (\rho_1\rho_2)^n \left[\rho_2 e^{-j\omega[2(n+1)\mathcal{L} - x_s]/v} \frac{V_0(\omega) + Z_c I_0(\omega)}{2} \right.$$
$$\left. - e^{-j\omega(2n\mathcal{L} + x_s)/v} \frac{V_0(\omega) - Z_c I_0(\omega)}{2} \right]$$

(6.106a)

$$V(\mathcal{L}) = (1 + \rho_2) \sum_{n=0}^{\infty} (\rho_1\rho_2)^n \left[e^{-j\omega[2(n+1)\mathcal{L} - x_s]/v} \frac{V_0(\omega) + Z_c I_0(\omega)}{2} \right.$$
$$\left. - \rho_1 e^{-j\omega(2n\mathcal{L} + x_s)/v]} \frac{V_0(\omega) - Z_c I_0(\omega)}{2} \right]$$

(6.106b)

where the dependence on frequency of the voltage and current sources are shown explicitly. Factors of the form $e^{-j\omega\tau}$ in the frequency domain amount to a shift in time τ in the time domain. That is to say, if a function $f(t)$ has a Fourier transform $F(\omega)$, the shifted spectrum $e^{-j\omega\tau}F(\omega)$ has an inverse transform $f(t-\tau)$. This fact can be applied directly to Eqs. (6.106) to give the following analytical expressions for the transient voltage responses:

$$v(0, t) = (1 + \rho_1) \sum_{n=0}^{\infty} (\rho_1\rho_2)^n$$
$$\times \frac{1}{2} \left[\rho_2 v_0\left(t - \frac{2(n+1)\mathcal{L} - x_s}{v}\right) + Z_c i_0\left(t - \frac{2(n+1)\mathcal{L} - x_s}{v}\right) \right.$$
$$\left. - v_0\left(t - \frac{2n\mathcal{L} + x_s}{v}\right) - Z_c i_0\left(t - \frac{2n\mathcal{L} + x_s}{v}\right) \right]$$

(6.107a)

$$v(\mathcal{L}, t) = (1 + \rho_2) \sum_{n=0}^{\infty} (\rho_1 \rho_2)^n$$

$$\times \frac{1}{2} \left[\begin{array}{c} v_0\left(t - \dfrac{(2n+1)\mathcal{L} - x_s}{v}\right) + Z_c i_0\left(t - \dfrac{(2n+1)\mathcal{L} - x_s}{v}\right) \\ -\rho_1 v_0\left(t - \dfrac{2n+1)\mathcal{L} + x_s}{v}\right) - Z_c i_0\left(t - \dfrac{(2n+1)\mathcal{L} + x_s}{v}\right) \end{array} \right]$$

(6.107b)

These expressions consist of a series of non sinusoidal traveling waves, arriving at progressively later times. These correspond to multiple reflections on the transmission line. The $n = 0$ term is the direct contribution from the voltage source $v(t)$ and the current source, $i(t)$.

Similarly, the *line current* at $x = 0$ and $x = \mathcal{L}$ can be found from Eqs. (6.40b) and 6.41b), and these transient responses have the same form as Eqs. (6.107), but with the leading terms $(1 + \rho_1)$ and $(1 + \rho_2)$ replaced by $(1 - \rho_1)/Z_c$ and $(1 - \rho_2)/Z_c$, respectively. Recall that the load current at $x = 0$ is $-I(0)$, while the load current at $x = \mathcal{L}$ is simply the line current. As discussed by Paul [6.15], this formalism can be applied to multiconductor lines as well, as long as there is no dispersion in the propagation constants and the loads are not functions of frequency.

6.3.4 Numerical Transformation of the Solution from the Frequency Domain to the Time Domain

Many practical transmission line problems contain dispersion in the propagation constants or frequency-dependent loads, and the direct analytical solution in Section 6.3.3 cannot be used to compute the transient responses. A useful approach for obtaining transient responses in these cases is to perform calculations in the frequency domain and then to transform the solutions into the time domain using a numerical evaluation of the Fourier transform integral. Several examples of this approach have been provided in Section 6.2, in which a broadband, complex frequency spectrum is first calculated and the numerical transform taken using the fast Fourier transform (FFT) routine.

In doing this, the nonsinusoidal excitation (e.g., a step function) is first transformed in the frequency domain. Then, at each frequency point in the excitation function the frequency-domain computational approach presented in Section 6.2 is used to determine the response spectrum. The resulting response spectrum is converted into the time domain using an inverse FFT numerical approach, such as that described in [6.18]. The problems due to the choice of a correct window for the frequency domain can be avoided by the use of a piecewise transformation [6.19].

6.3.5 Numerical Solution of the Telegrapher's Equations in the Time Domain

An alternative technique for solving Eqs. (6.1) and (6.2) is to use a finite-difference method to approximate the partial derivatives [6.20]. This results in a set of algebraic equations in the variables x and t that may be solved by a variety of methods at each time step. Usually, the time variable is incremented or "stepped" by a value Δ.

Figure 6.27 illustrates the principle of this method. Each conductor is divided into alternating voltage and current nodes. Two nodes of the same type are separated by a spatial increment Δx, and the two extremities of the conductor are defined as voltage nodes. The voltage or the current at a point is calculated using the values determined in surrounding points at an earlier time.

With the notations of Figure 6.27, the finite-differences representation of the spatial derivative of a function $f(x, t)$ is written as

$$\frac{\partial f(x,t)}{\partial x} = \frac{f_{k+1}^{n+1} - f_k^{n+1}}{\Delta x} \tag{6.108}$$

and a temporal derivative is

$$\frac{\partial f(x,t)}{\partial t} = \frac{f_k^{n+1} - f_k^n}{\Delta t} \tag{6.109}$$

where k and n represent the position and time increments, respectively.

With these representations for the derivatives, the telegrapher's equations (6.1) and (6.2) can be written in discrete form, for points on the line

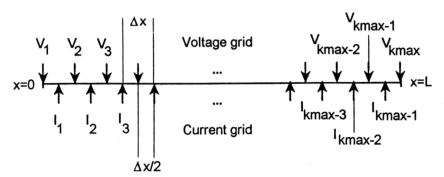

FIGURE 6.27 Voltage and current nodes for calculation using the point-centered finite-difference method.

excluding the ends, as follows:

$$\frac{v_{k+1}^{n+1} - v_k^{n+1}}{\Delta x} + R'_k \frac{i_k^{n+1} + i_k^n}{2} + L'_k \frac{i_k^{n+1} - i_k^n}{\Delta t} = 0 \quad \text{for } k = 1 \text{ to } k_{\max} - 1$$

(6.110a)

$$\frac{i_{k+1}^{n+1} - i_k^{n+1}}{\Delta x} + G'_k \frac{v_k^{n+1} + v_k^n}{2} + C'_k \frac{v_k^{n+1} - v_k^n}{\Delta t} = 0 \quad \text{for } k = 2 \text{ to } k_{\max} - 1$$

(6.110b)

where v_k^n represents the line voltages sampled at spatial locations $(k-1)\Delta x$, and at times $n \Delta t$:

$$v_k^n = v((k-1)\Delta x, n \Delta t)$$

(6.111a)

and i_k^n is the line current evaluated at locations $(k - \frac{1}{2})\Delta x$ and at times $(n + \frac{1}{2})\Delta t$

$$i_k^n = i((k - \tfrac{1}{2})\Delta x, (n + \tfrac{1}{2})\Delta t)$$

(6.111b)

Equations (6.110) are solved by iteration by extracting the terms i_k^{n+1} and v_k^{n+1}:

$$i_k^{n+1} = \left(\frac{L'_k}{\Delta t} + \frac{R'_k}{2}\right)^{-1} \left[\left(\frac{L'_k}{\Delta t} - \frac{R'_k}{2}\right)i_k^n - \frac{v_{k+1}^{n+1} - v_k^{n+1}}{\Delta x}\right] \quad \text{for } k = 1 \text{ to } k_{\max} - 1$$

(6.112a)

$$v_k^{n+1} = \left(\frac{C'_k}{\Delta t} + \frac{G'_k}{2}\right)^{-1} \left[\left(\frac{C'_k}{\Delta t} - \frac{G'_k}{2}\right)v_k^n - \frac{i_{k+1}^{n+1} - i_k^{n+1}}{\Delta x}\right] \quad \text{for } k = 2 \text{ to } k_{\max} - 1$$

(6.112b)

At each time step n the two finite-difference equations (6.112) are applied to all intermediate points of the line. Equation (6.112b) is not complete, however, since the voltages v_1^{n+1} and $v_{k\max}^{n+1}$ are not included. These are the voltages at the ends of the line, and they are expressed using the boundary, or termination, conditions at the loads. For resistive terminations of the line with a load R_0 at $x = 0$ and R_L at $x = \mathcal{L}$, the boundary conditions are expressed as

$$v_1^{n+1} = -R_0 i_1^{n+1}$$

(6.113a)

$$v_{k\max}^{n+1} = R_L i_{k\max - 1}^{n+1}$$

(6.113b)

The case of inductive and capacitive terminations is discussed in Section 6.3.6.

The solution for the line responses is achieved by sequentially stepping Eqs. (6.112) and (6.113) along in time by incrementing the index n. For the solution to remain stable, the time step must satisfy the Courant stability condition that

$$\Delta t < \frac{\Delta x}{v_p} \tag{6.114}$$

where v_p is the propagation velocity along the line. For lines where the per-unit-length parameters change as a function of distance, or when there are nonlinear or time-varying loads on the line, this convergence criterion becomes difficult to state exactly, and Eq. (6.114) serves only as a guide in determining the time step. As in most finite-difference schemes, a considerable amount of experimenting is required to obtain a stable and good solution.

6.3.6 Inductive and Capacitive Terminations in the Time Domain

Impedances with inductive and capacitive elements can be taken into account by introducing different boundary conditions at the line terminations. Consider, for example, the case of an RLC termination shown in Figure 6.28. For this circuit, Kirchhoff's equations yield the following differential equation relating the load voltage and current:

$$LC\frac{d^2v(t)}{dt^2} + RC\frac{dv(t)}{dt} + v(t) = L\frac{di(t)}{dt} + Ri(t) \tag{6.115}$$

In terms of finite differences, the first derivative with respect to the time is given by (6.109) and the second derivative may be expressed as

$$\frac{d^2f(t)}{dt^2} = \frac{f_k^{n+1} - 2f_k^n + f_k^{n-1}}{(\Delta t)^2} \tag{6.116}$$

In this manner, Eq. (6.115) can be written in terms of finite differences at the two line terminations and used as boundary condition for the numerical solution described in Section 6.3.6.

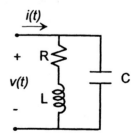

FIGURE 6.28 RLC circuit terminating the transmission line.

6.3.7 Bergeron's Graphical Solution in the Time Domain

As discussed in Section 6.3.2, the general form of the solution of the telegrapher's equations for a lossless two-conductor configuration consists of a combination of a forward-propagating wave $f(x - ct)$ and a backward-propagating wave $g(x + ct)$. A graphical solution that avoids explicit calculation of the functions f and g has been developed by Bergeron [6.21] (see also [6.22], pp. 35–39).

Depending of the nature of the problem, two similar approaches have been developed:

1. The method of the unique observer for homogeneous lines
2. The method of two simultaneous observers for nonhomogeneous lines

In development that follows, only the unique observer approach will be developed, as it has permitted the development of a numerical method for computing line responses.

6.3.7.1 Principle of Bergeron's Method.
Adding and subtracting Eqs. (6.102a) and (6.102b) provides the following expressions for the forward and backward wavefunctions

$$f(x - ct) = \tfrac{1}{2}(v(x,t) + Z_c i(x, t)) \qquad (6.117a)$$

$$g(x + ct) = \tfrac{1}{2}(v(x, t) - Z_c i(x, t)) \qquad (6.117b)$$

In these equations it is clear that f is constant if $x - ct =$ constant, and that g is constant if $x + ct =$ constant. The implication of these equations is that if a fictitious observer (called the direct Bergeron observer) moves along the line with a speed c in the forward direction, he or she will see that the voltage and current satisfy at each instant in time the relation

$$v(x, t) + Z_c i(x, t) = K_1 \qquad (6.118a)$$

Similarly, the backward Bergeron observer (who moves in the opposite direction with respect to the first and with the same speed c) will see the voltage and the current satisfying the relation

$$v(x, t) - Z_c i(x, t) = K_2 \qquad (6.118b)$$

In the v–i plane representing all possible values of voltage and current, the locus of points representing what the direct observer sees corresponds to the line $v + Z_c i = K_1$. For the backward observer, this is the line $v - Z_c i = K_2$. Figure 6.29 shows these lines, which have slopes of $-Z_c$ and $+Z_c$, respectively.

The values of the constants K_1 and K_2 can be determined using known voltage and current values in a certain point on the line and at a specified

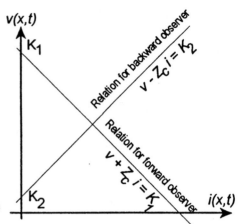

FIGURE 6.29 The v–i plane and the Bergeron lines.

time t. For instance, if the initial conditions at one of the two extremities of the line are zero at $t = 0$ (i.e., $v = 0$, $i = 0$), the respective constant K_1 or K_2 will be zero.

6.3.7.2 Numerical Application of Bergeron's Method.
The numerical application of Bergeron's method is achieved by using the fact that if $\tau = \ell/c$ is the time of travel of a wave on a line between its origin m and its end k (see Figure 6.30), the expression $(v + Z_c i)$ seen by an observer leaving node m at time $t - \tau$ will remain the same when the observer arrives at node k at time t. As discussed in [6.23], this results in the set of equations

$$v_m(t - \tau) + Z_c i_k(t - \tau) = v_k(t) + Z_c[-i_k(t)] \qquad (6.119a)$$

$$v_k(t - \tau) + Z_c i_k(t - \tau) = v_m(t) + Z_c[-i_m(t)] \qquad (6.119b)$$

for the voltages and currents at the two ends of the line segment. By defining the following equivalent current sources on the line,

$$I_k(t - \tau) = -\left[i_m(t - \tau) + \frac{v_m(t - \tau)}{Z_c}\right] \qquad (6.120a)$$

$$I_m(t - \tau) = -\left[i_k(t - \tau) + \frac{v_k(t - \tau)}{Z_c}\right] \qquad (6.120b)$$

one finds that Eq. (6.119a) becomes

$$i_k(t) = Y_c v_k(t) + I_k(t - \tau) \qquad (6.121)$$

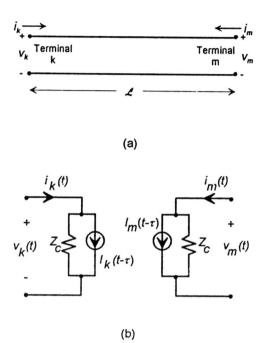

FIGURE 6.30 Equivalent diagram of a lossless line with distributed parameters: (a) lossless line; (b) equivalent impedance network.

where $Y_c = 1/z_c$. Similarly, Eq. (6.119b) becomes

$$i_m(t) = Y_c v_m(t) + I_m(t - \tau) \tag{6.122}$$

These relations permit the building of the equivalent circuit of the line shown in Figure 6.30, which fully describes the lossless line at its terminals. Topologically, the terminals are not connected: The conditions at the other end are seen only indirectly with a time delay τ through the equivalent current sources I.

This solution technique can be extended to a transmission line network as

$$[i(t)] = [Y][v(t)] + [I(t - \tau)] \tag{6.123}$$

where $[i(t)]$ is the column vector of the injected nodal currents, $[v(t)]$ represents the column vector of the node voltages at time t, $[Y]$ is the nodal admittance matrix, and $[I(t - \tau)]$ denotes the column vector of known equivalent current sources at a time $(t - \tau)$. The main advantage of this method compared to other approaches for solving the telegrapher's equations is that no reflection coefficients are needed.

To solve Eq. (6.123), the nodal admittance matrix must be determined. A network can consist of sections with distributed parameters but may also contain lumped parameters, as shown in Figure 6.31. If a lumped resistance

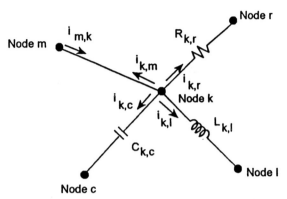

FIGURE 6.31 Network with lumped and distributed parameters.

is inserted in a transmission line segment denoted by *kr*, Ohm's law gives

$$i_{kr}(t) = \frac{v_k(t) - v_r(t)}{R_{kr}} \quad (6.124)$$

which corresponds to the equivalent diagram of Figure 6.32. If a lumped inductance is inserted in a transmission line segment denoted by *kl*, Ohm's law gives

$$v_k - v_l = L_{kl} \frac{di_{kl}}{dt} \quad (6.125)$$

and replacing the derivative by its finite-difference representation yields

$$i_{kl}(t) = i_{kl}(t - \Delta t) + \frac{1}{L_{kl}} \int_{t-\Delta t}^{t} (v_k - v_l) \, dt \quad (6.126)$$

The integral from Eq. (6.126) can be calculated numerically using the trapezoidal rule of integration:

$$i_{kl}(t) = i_{kl}(t - \Delta t) + \frac{\Delta t}{2L_{kl}} [v_k(t - \Delta t) - v_l(t - \Delta t) + v_k(t) - v_l(t)] \quad (6.127)$$

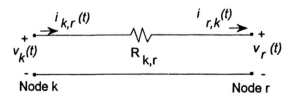

FIGURE 6.32 Equivalent diagram of a lumped resistance in the line.

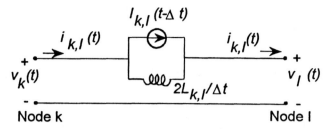

FIGURE 6.33 Equivalent diagram of a line section with a lumped inductance.

By defining

$$I_{kl}(t - \Delta t) \equiv i_{kl}(t - \Delta t) + \frac{\Delta t}{2L_{kl}}[v_k(t - \Delta t) - v_l(t - \Delta t)] \quad (6.128)$$

one finds that Eq. (6.127) becomes

$$i_{kl}(t) = \frac{\Delta t}{2L_{kl}}[v_k(t) - v_l(t)] + I_{kl}(t - \Delta t) \quad (6.129)$$

which corresponds to the equivalent diagram of Figure 6.33.

An equivalent approach for a section with a capacitance gives

$$i_{kc}(t) = \frac{2C_{kc}}{\Delta t}[v_k(t) - v_c(t)] + I_{kc}(t - \Delta t) \quad (6.130)$$

with

$$I_{kc}(t - \Delta t) \equiv -i_{kc}(t - \Delta t) - \frac{2C}{\Delta t}[v_k(t - \Delta t) - v_c(t - \Delta t)] \quad (6.131)$$

which corresponds to the equivalent diagram of Figure 6.34.

6.3.7.3 Solution of the Nodal Matrix Equation. The column voltage or current vectors of Eq. (6.123) contain known values of voltages and currents,

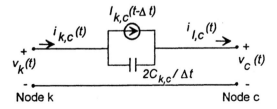

FIGURE 6.34 Equivalent diagram of a line section with a lumped capacitance.

representing lumped excitation sources. Subvector and submatrix notation can be used:

$$\begin{pmatrix} Y_{uu} & Y_{uk} \\ Y_{ku} & Y_{kk} \end{pmatrix} \begin{pmatrix} v_u \\ v_k \end{pmatrix} = \begin{pmatrix} i_u \\ i_k \end{pmatrix} - \begin{pmatrix} I_{k1} \\ I_{k2} \end{pmatrix} \quad (6.132)$$

where the index "k" denotes known quantities and the index "u," unknown values. The current sources $I_{kj}(t - \tau)$ are all known. For the first iteration step they are given by the boundary conditions. The indices pertaining to the nodal admittances do not indicate known and unknown quantities, but represent admittances connecting known and unknown values of the voltage and current. The system of equations (6.132) can be written as

$$[Y_{uu}] \cdot [v_u] + [Y_{uk}] \cdot [v_k] = [i_u] - [I_{k1}] \quad (6.133a)$$

$$[Y_{ku}] \cdot [v_u] + [Y_{kk}] \cdot [v_k] = [i_k] - [I_{k2}] \quad (6.133b)$$

From Eqs. (6.133), the unknown voltage and current values can be calculated as

$$[v_u] = [Y_{ku}]^{-1}([i_k] - [I_{k2}] - [Y_{kk}] \cdot [v_k]) \quad (6.134a)$$

$$[i_u] = [Y_{uu}] \cdot [Y_{ku}]^{-1}([i_k] - [I_{k2}] - [Y_{kk}] \cdot [v_k]) + [Y_{uk}] \cdot [v_k] + [I_{k1}] \quad (6.134b)$$

6.3.7.4 Example: Transient State of a Circuit After Closing Two Interrupters.
Consider the circuit of Figure 6.35 supplied by the voltage source V_0. The transient behavior of the branch currents after switches 1 and 2 are closed can be calculated using Bergeron's numerical method [6.24]. Using the approach introduced in Section 6.3.7.2, the electrical circuit can be replaced by the equivalent diagram of Figure 6.36. In each branch the

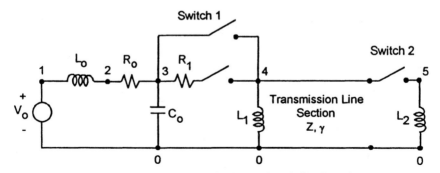

FIGURE 6.35 Electrical circuit with lumped and distributed parameters.

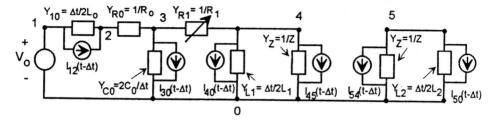

FIGURE 6.36 Equivalent diagram with equivalent current sources.

currents are

$$i_{12}(t) = Y_{L0}(V_0 - v_2) + I_{12}(t - \Delta t) \tag{6.135a}$$

$$i_{23}(t) = Y_{R0}(v_2 - v_3) \tag{6.135b}$$

$$i_{30}(t) = Y_{C0}v_3 + I_{30}(t - \Delta t) \tag{6.135c}$$

$$i_{34}(t) = Y_{R1}(v_3 - v_4) \tag{6.135d}$$

$$i_{40}(t) = Y_{L1}v_4 + I_{40}(t - \Delta t) \tag{6.135e}$$

$$i_{45}(t) = Y_Z v_4 + I_{45}(t - \Delta t) \tag{6.135f}$$

$$i_{54}(t) = Y_Z v_5 + I_{54}(t - \Delta t) \tag{6.135g}$$

$$i_{50}(t) = Y_{L2}v_5 + I_{50}(t - \Delta t) \tag{6.135h}$$

Writing the relations between the currents in nodes 2, 3, 4 and 5, the following matrix relation is obtained:

$$\begin{bmatrix} (Y_{L0} + Y_{R0}) & -Y_{R0} & 0 & 0 \\ Y_{R0} & -(Y_{R0} + Y_{R1} + Y_{C0}) & Y_{R1} & 0 \\ 0 & Y_{R1} & -(Y_{R1} + Y_{L1} + Y_Z) & 0 \\ 0 & 0 & 0 & -(Y_Z + Y_{L2}) \end{bmatrix} \begin{bmatrix} v_2 \\ v_3 \\ v_4 \\ v_5 \end{bmatrix}$$

(6.136)

$$= \begin{bmatrix} Y_{L0}V_0 \\ 0 \\ 0 \\ 0 \end{bmatrix} + \begin{bmatrix} I_{12}(t - \Delta t) \\ I_{30}(t - \Delta t) \\ I_{40}(t - \Delta t) + I_{45}(t - \Delta t) \\ I_{54}(t - \Delta t) + I_{50}(t - \Delta t) \end{bmatrix}$$

Using (6.136), the voltages v_2, v_3, v_4, and v_5 are calculated at time t. Then, with relations (6.135), it is possible to compute the currents. The iteration process is then continued with $t + \Delta t$ until $t = t_{max}$, when the steady state is reached.

6.3.8 Electromagnetic Transients Program

The numerical application of Bergeron's method has been developed into a computer code known as the *Electromagnetic Transients Program* (EMTP) [6.23]. This code, which is in use throughout the world, has been improved continuously from 1969 until the present. A few examples show various types of problems that have been solved with EMTP.

6.3.8.1 Transmission Line Response Using EMTP.
As an example of the use of EMTP, let us consider the pulse excitation voltage response of a transmission line connected to the ground at three points. As illustrated in Figure 6.37, the transmission line is assumed to have a wire diameter of 5 mm, length 10 m, and height 3 m, with the connections to ground as indicated. The terminations are nearly short-circuited ($R = 0.1\,\Omega$), and the middle point of the line is connected to the ground through a resistance of $0.1\,\Omega$ and an inductance of $1\,\mu H$. The line is excited at the left end by a voltage source, producing a half-sinusoidal voltage pulse of 20 ns duration and 1 V amplitude. Figure 6.38 shows the transient voltage at the $0.1\text{-}\Omega$ right-hand termination of the line, as calculated using the EMTP program.

6.3.8.2 Modeling a Test Installation with Two Parallel Lines.
One of the test installations used for estimating inductive and capacitive coupling between two lines is simply made of two parallel lines, one being the generator of disturbances and the second line being the victim. A surge generator injects pulse at one end of the first line, and the disturbance induced is measured with an oscilloscope at the other extremity of the second line. The two lines are inside a shielded cage where the measurement takes place, and the surge generator is outside the cage. This generator is connected to the entry feed-through of the cage by a coaxial cable of 1.5 m (PS to PS1 in Figure 6.39), and inside the cage another coaxial of 9.5 m (PS1 to LA) connects the first line. The second line is connected at its output to

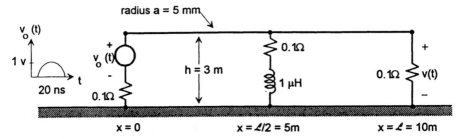

FIGURE 6.37 Transmission line, grounded at three points to a reference conductor and excited by a pulse voltage source.

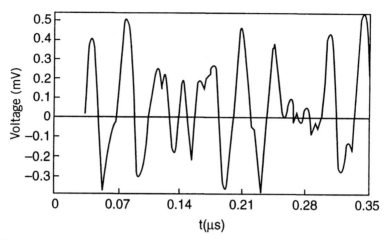

FIGURE 6.38 Transient voltage response of a line excited by a half-sinusoidal pulse, as calculated using the EMTP.

an oscilloscope through two coaxial cables (LB to PM and PM1 to PM), each of 1.5 m.

A waveform rise time $t_r = 5$ ns corresponds to a cutoff frequency of $f_c = 1/\pi t_r = 63$ MHz, with a minimum wavelength of $\lambda = 6$ m. To obtain cells with $l \leq \lambda/10$, the 4-m line must be divided into eight cells. A division into 16 cells has been chosen to obtain better precision. The model for calculations using EMTP is shown in Figure 6.39.

Figures 6.40 and 6.41 show a comparison between measurements and calculation using EMTP for the case when the victim line is terminated on a matched impedance at the left end and on a low impedance at the right end (1.7 Ω). The agreement is quite satisfactory.

6.4 DETERMINATION OF LINE INDUCTANCE PARAMETERS

The mutual and self inductances of different types of transmission lines can be determined either by measurement or by calculation. In what follows we present briefly one measurement method. The main part of this section, however, is dedicated to calculation methods.

If the return circuit (i.e., the neutral, or $(n + 1)^{th}$ conductor) of the line is bundled together with the other n conductors of the line, as shown in Figure 6.2a, calculation of the per-unit-length inductances for transmission line calculations is easier than the determination of inductances for low-frequency circuit coupling calculations, discussed in Chapter 3. This is related to the fact that:

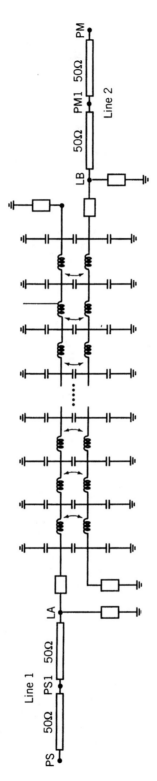

FIGURE 6.39 Model of the two parallel lines over a perfectly conducting ground.

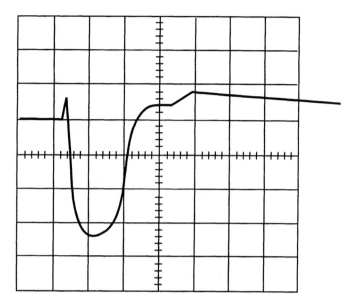

FIGURE 6.40 Oscillogram showing the voltage measured at the right end of the victim line. Left termination matched, right termination closed on a low impedance (1.7 Ω). Scale: horizontal, 20 ns/division; vertical, 1 V/division.

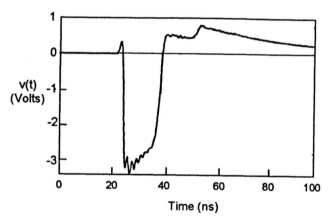

FIGURE 6.41 Voltage calculated at the right end (node PM) of the victim line using EMTP. Left termination matched, right determination closed on a low impedance (1.7 Ω).

- Transmission line conductors are parallel lines and are cylindrical in cross section.
- Transmission line problems generally concern long lines, in which length is usually larger than the dimensions of the circuits discussed in Chapter 3; consequently, end effects can be neglected.

The axially symmetric nature of the conductors makes solution of the integrals in the general formulas (3.43) or (3.44) easier to evaluate than for circular loops or segments displaced from each other (see Figure 3.29). However, difficulties arise in determining the inductance parameters if the return conductor is a lossy ground plane (i.e., soil with a finite conductivity), as shown in Figure 6.2b, or if the lines are not parallel.

After a brief introduction to a measurement method for determining the per-unit-length inductance parameters of lines, we describe inductance computational methods that apply to parallel lines with a return conductor having infinite conductivity. Then the case of lines over a soil with finite conductivity is discussed. The ground impedance of a real soil with finite conductivity is discussed further in Chapter 8.

6.4.1 Inductance Measurement

The measurement of the self and mutual inductance matrix of a multiconductor line of length $\mathcal{L}$ over a perfectly conducting ground can be performed by short-circuiting all the conductors at one end of the line to the reference conductor, as shown in Figure 6.42. At the other end, conductor i is excited by a time-harmonic voltage source, with all the other conductors open circuited. The open-circuit voltage induced by inductive coupling between conductors i and j is measured using a voltmeter.

The inductive coupling of conductor j to the other conductors in a system of n conductors, plus a return conductor, is described by the following equation:

$$\frac{dV_j}{dx} = Z'_{j1}I_1 + Z'_{j2}I_2 + \cdots + Z'_{jn}I_n \qquad (6.137)$$

If the wavelength of the harmonic source is large compared to the line length $\mathcal{L}$, Eq. (6.137) can be written as

$$V_j(0) - V_j(\mathcal{L}) = (Z'_{j1}I_1 + Z'_{j2}I_2 + \cdots + Z'_{jn}I_n)\mathcal{L} \qquad (6.138)$$

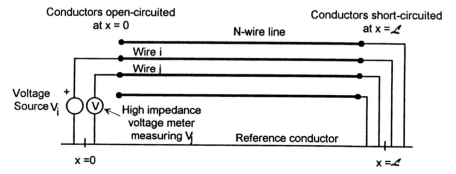

FIGURE 6.42 Configuration for inductance measurement.

Since all the lines are short circuited at termination $x = \mathcal{L}$, all $V_j(\mathcal{L}) = 0$, and because only conductor i is excited, all the currents $I_k = 0$ for $k \neq i$. Then as $Z'_{ij} = j\omega L'_{ij}$, the per-unit-length inductance parameter L'_{ij} is expressed as

$$L'_{ij} = \frac{1}{\mathcal{L}} \frac{1}{j\omega} \frac{V_j(0)}{I_i} \quad \text{H/m} \tag{6.139}$$

By changing the locations of the source and the current sensor to each conductor, all the elements of the inductance matrix can be determined. It should be noted that the $[L']$ matrix is symmetric about the diagonal. Thus, because $L'_{ij} = L'_{ji}$, only $n(n+1)/2$ unique matrix elements must be measured for the n-wire line.

6.4.2 Analytical Inductance Evaluation

By using relations (3.43) and (3.44), the mutual inductance between two circuits can be calculated analytically if the integrals in these two equations can be evaluated in closed form. In this section we present a similar approach that can be applied to conductors with dimensions of the same order of magnitude as the distances between them, using the concept of the geometrical mean distance (GMD) and the geometrical mean radius (GMR), introduced by Maxwell [6.25].

Let us consider a system of two circuits with four conductors of arbitrary shape, which have cross sections S_1, S_2, S_3, and S_4, as shown in Figure 6.43. The dimensions of the line cross sections are assumed to be of the same order of magnitude as the distances r_{ij} between them.

The current I_a flows through circuit a formed by conductors 1 and 2 and the current I_b through circuit b formed by conductors 3 and 4. These

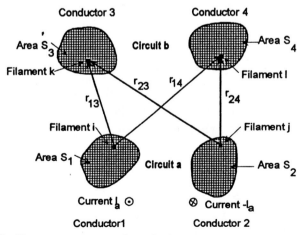

FIGURE 6.43 Transverse cross-section of a four-conductor system for two circuits.

currents are assumed to vary slowly enough with the time so that the skin effect and other proximity effects are negligible.

In an elementary current filament of cross section dS within the conductor, a current $dI = \mathbf{J} \cdot d\mathbf{S}$ flows. $\mathbf{J}$ is the current density in the conductor, which is assumed to be constant over the cross section S and is given by $|\mathbf{J}| = I/S$. Each filament of the circuit a couples its magnetic field to each element of the circuit b. Thus, for a filament area of dS_1 of conductor 1, a current $I_a\, dS_1/S_1$ flows, and in an element dS_2 of conductor 2, the current is $-I_a\, dS_2/S_2$.

The partial flux in a length Δx linking the filaments k and l of circuit b due to the current $(I_a\, dS_1/S_1)$ in element i of conductor 1 is equal to

$$d\Phi_{kl,i} = \frac{\mu_0 \Delta x}{2\pi} \frac{I_a}{S_1} dS_1 \int_{r_{13}}^{r_{14}} \frac{dr}{r} = \frac{\mu_0 \Delta x}{2\pi} \frac{I_a}{S_1} dS_1 (\ln r_{14} - \ln r_{13}) \quad (6.140)$$

while the contribution of the current element j of conductor 2 to the elementary flux between k and l is

$$d\Phi_{kl,j} = -\frac{\mu_0 \Delta x}{2\pi} \frac{I_a}{S_2} dS_2 \int_{r_{23}}^{r_{24}} \frac{dr}{r} = -\frac{\mu_0 \Delta x}{2\pi} \frac{I_a}{S_2} dS_2 (\ln r_{24} - \ln r_{23})$$

(6.141)

Each of these expressions for the partial fluxes consists of two terms, one containing the distances r_{13} and r_{23} between the circuit a and the cross section S_3, and the other having the distances r_{14} and r_{24} between the circuit a and the cross section S_4. To calculate the total flux between the elementary filaments k and l due to the complete circuit a, it is necessary to integrate over cross sections S_1 and S_2 in circuit a containing the current. This gives

$$(\Phi_{kl})_a = \frac{\mu_0 \Delta x}{2\pi} I_a \left[\int_{S_1} \frac{\ln r_{14}}{S_1} dS_1 - \int_{S_1} \frac{\ln r_{13}}{S_1} dS_1 - \int_{S_2} \frac{\ln r_{24}}{S_2} dS_2 + \int_{S_2} \frac{\ln r_{23}}{S_2} dS_2 \right]$$

(6.142)

The expression (6.142) permit calculation of the flux produced by filaments k and l in circuit b due to the total current flowing in circuit a. However, this flux is different, depending on the position of k and l inside cross sections S_3 and S_4. The total coupled flux $\Phi_{b,a}$ in circuit b arising from all the current in circuit a is the algebraic mean (or average) and is given by the integral of Eq. (6.141) over cross-sectional areas S_3 and S_4 as

$$\overline{\Phi}_{a,b} = \frac{\mu_0 \Delta x}{2\pi} I_a \left[\int_{S_4}\int_{S_1} \frac{\ln r_{14}}{S_4 S_1} dS_1\, dS_4 - \int_{S_3}\int_{S_1} \frac{\ln r_{13}}{S_3 S_1} dS_1\, dS_3 \right.$$
$$\left. - \int_{S_4}\int_{S_2} \frac{\ln r_{24}}{S_4 S_2} dS_2\, dS_4 + \int_{S_3}\int_{S_2} \frac{\ln r_{23}}{S_3 S_2} dS_2\, dS_3 \right] \quad (6.143)$$

The per-unit-length mutual inductance between circuits a and b due to the current I_a is then given as

$$M'_{a,b} = \frac{\overline{\Phi}_{a,b}}{I_a \Delta x}$$

$$= \frac{\mu_0}{2\pi} \left[\int_{S_4}\int_{S_1} \frac{\ln r_{14}}{S_4 S_1} dS_1\, dS_4 - \int_{S_3}\int_{S_1} \frac{\ln r_{13}}{S_3 S_1} dS_1\, dS_3 \right.$$

$$\left. - \int_{S_4}\int_{S_2} \frac{\ln r_{24}}{S_4 S_2} dS_2\, dS_4 + \int_{S_3}\int_{S_2} \frac{\ln r_{23}}{S_3 S_2} dS_2\, dS_3 \right] \quad (6.144)$$

From this expression it is possible to see that $M'_{ba} = M'_{ab}$.

6.4.2.1 Geometrical Mean Distance Between Two Circuits.
Starting from the analytical expressions of the flux linking two coupled circuits given in Eq. (6.144), it is possible to express the mutual inductances of the circuits in terms of the geometrical mean distance of the circuits. The geometrical mean distance (GMD) of cross section S_i to cross section S_j, is denoted by g_{ij} and is defined as

$$\ln g_{ij} = \frac{1}{S_i S_j} \int_{S_i}\int_{S_j} \ln r_{ij}\, dS_i\, dS_j \quad (6.145)$$

where r_{ij} is the distance between an element dS_i of the surface S_i to an element dS_j of the surface S_j.

6.4.2.2 Mutual Inductance per Unit Length.
Using the geometrical mean distance as defined by Eq. (6.145), the analytical expression of the mutual inductance between two circuits a and b in Figure 6.44 becomes

$$M'_{ab} = \frac{\mu_0}{2\pi} \ln \frac{g_{14} g_{23}}{g_{24} g_{13}} \quad \text{H/m} \quad (6.146)$$

Here g_{14}, g_{13}, g_{23}, and g_{24} are the geometrical mean distances between the conductors 1 and 4, 1 and 3, 2 and 3, and 2 and 4, respectively. Notice that if d_a and $d_b \ll D$, the GMDs have the following values:

$$g_{14} \approx r_{14} = D + \frac{d_a + d_b}{2}$$

$$g_{23} \approx r_{23} = D - \frac{d_a + d_b}{2}$$

$$g_{24} \approx r_{24} = D - \frac{d_a - d_b}{2} \quad (6.147a)$$

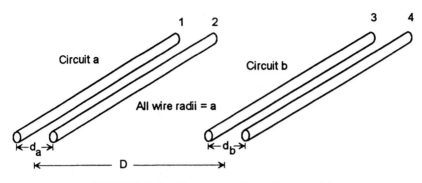

FIGURE 6.44 Two coupled circuits, a and b.

$$g_{13} \approx r_{13} = D + \frac{d_a - d_b}{2}$$

Furthermore, if the circuits are identical, $d_a = d_b = d$, and

$$g_{14} = D + d$$
$$g_{23} = D - d \qquad (6.147\text{b})$$
$$g_{24} = g_{13} = D$$

which results in the following expression for the mutual inductance of the two lines:

$$M'_{ab} = \frac{\mu_0}{\pi} \ln \frac{\sqrt{D^2 - d^2}}{D} \qquad \text{H/m} \qquad (6.148)$$

If the two circuits have a common ground return conductor n as in Figure 6.45, the mutual inductance between circuits i–n and j–n is

$$M'_{ij} = \frac{\mu_0}{2\pi} \ln \frac{g_{in} g_{jn}}{g_{ij} g_{nn}} \qquad \text{H/m} \qquad (6.149)$$

where g_{in}, g_{jn}, and g_{ij} are the GMD between conductors i–n, j–n, and i–j, respectively, and g_{nn} is the GMR of the neutral conductor n, which is defined as the geometrical mean distance of a surface S_n to itself:

$$\ln g_{nn} = \frac{1}{S_n^2} \int_{S_n} \int_{S_n} \ln r_{nn} \, dS_n \, dS_n \qquad (6.150)$$

and where r_{nn} takes all possible values between the separate elements dS_n of the surface S_n.

6.4 DETERMINATION OF LINE INDUCTANCE PARAMETERS

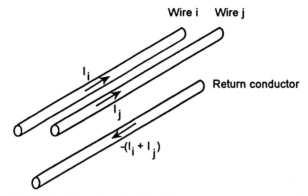

FIGURE 6.45 Coupled circuits with a common return conductor.

6.4.2.3 Self Inductance per Unit Length. The self inductance is defined as the flux linking a circuit due to the current flowing in the circuit divided by the same current. Therefore, the expression of the self inductance can be found by superposing the circuits a and b of Figure 6.43 and applying Eq. (6.150). In this case, $r_{14} = r_{23}$, $S_1 = S_3$, $S_2 = S_4$; that is, $g_{14} = g_{23} = g_{34} = g_{12}$, $r_{13} = r_{33} = r_{11}$ and the GMR of S_3 is g_{33}, which is calculated using Eq. (6.146). Furthermore, since $r_{24} = r_{44}$, the GMR of S_4 is g_{44} and is also calculated using Eq. (6.150). In evaluating these self-GMR terms, a mathematical problem can arise due to the fact that $\ln 0 = -\infty$. Consequently, the principal value of these singular integrals must be taken in Eq. (6.150).

The self inductance per unit length of circuit b in Figure 6.46 is then given by

$$M'_{bb} = \frac{\mu_0}{2\pi} \ln \frac{g_{34}^2}{g_{33} g_{44}} \quad H/m \qquad (6.151)$$

or more generally, the per-unit-length self inductance of a loop comprised of

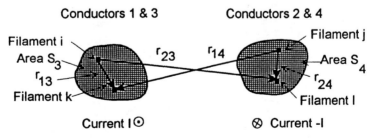

FIGURE 6.46 Transverse cross section through a single current loop resulting from the superposition of circuits a and b.

conductors i and a neutral (return) conductor n is

$$M'_{ii} = \frac{\mu_0}{2\pi} \ln \frac{g_{in}^2}{g_{ii}g_{nm}} \qquad \text{H/m} \qquad (6.152)$$

The GMR and GMD of the most usual conductors and circuit configurations are summarized in [6.26, 6.27] and also in Tables C.5 and C.6 in Appendix C.

6.4.2.4 Mutual and Self-Inductances of Lines with the Earth as a Return Conductor.
Two main difficulties arise in the introduction of the earth as a return conductor into analytical expressions:

- The shape and dimensions of the ground layer in which the return current is distributed cannot be well defined
- The ground itself is not uniform and other metallic conductors are often present in the soil (water and gas pipes, cables, galleries). This makes the current distribution not uniform in the ground.

6.4.2.4.1 Mutual and Self-Inductances of Lines over Perfectly Conducting Ground.
From the general expressions (6.149) and (6.152), we shall find the expressions of the mutual and self inductances of lines over a perfectly conducting ground. In this case, the image method can be used as shown in Figure 6.47. Replacing the GMD and GMR in Eq. (6.147) by the values given in Tables C.5 and C.6 in Appendix C for the configuration corresponding to the coupled image circuits ii^* and kk^* from Figure 6.47, the

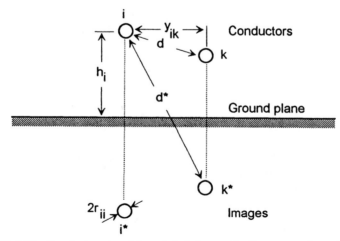

FIGURE 6.47 Conductors i and k and their images in the assumption of a perfectly conducting ground.

mutual inductance is

$$M'_{ik} = \frac{1}{2}\frac{\mu_0}{2\pi}\ln\frac{g_{ik*}g_{ki*}}{g_{ik}g_{i*k*}} \quad \text{H/m} \quad (6.153)$$

where $g_{ik*} = g_{ki*} = d*$ and $g_{ki} = g_{k*i*} = d$, respectively. Note that the leading factor of $\frac{1}{2}$ arises from the fact that we wish to consider only the component of mutual inductance arising from the flux *above* the ground plane. Equation (6.153) can be rewritten as

$$M'_{ik} = \frac{\mu_0}{4\pi}\ln\left(\frac{g_{ik*}}{g_{ik}}\right)^2 = \frac{\mu_0}{2\pi}\ln\frac{g_{ik*}}{g_{ik}} = \frac{\mu_0}{2\pi}\ln\frac{d*}{d} \quad \text{H/m} \quad (6.154)$$

For the self inductance, applying Eq. (6.153) for $i = k$ gives

$$M'_{ii} = \frac{\mu_0}{4\pi}\ln\frac{g_{ii*}^2}{g_{ii}g_{i*i*}} \quad \text{H/m} \quad (6.155a)$$

and since $g_{ii*} = 2h_i$ and $g_{ii} = g_{i*i*} = r_{ii}$, we have

$$M'_{ii} = \frac{\mu_0}{2\pi}\ln\frac{2h_i}{r_{ii}} \quad \text{H/m} \quad (6.155b)$$

The mutual and self inductances of lines over a finitely conducting ground is discussed in Chapter 8 as part of the concept of ground impedance.

6.5 DETERMINATION OF LINE CAPACITANCE PARAMETERS

As in the case of the per-unit-length inductance parameters, the per-unit-length capacitance of different types of transmission lines can be determined either by measurement or by calculation. A measurement method similar to that used for inductances will be presented. The focus, however, will be on computational methods.

Calculation of the per unit capacitances for transmission line calculations is usually easier than the determination of capacitances for low-frequency circuit coupling calculations (Chapter 3), due to two factors:

1. The problems concern generally long lines, in which the length is many times larger than the dimensions of the circuits discussed in Chapter 3, and consequently, the end effects can be neglected.
2. In some configurations (such at those for power transmission and distribution lines) the environment is uniform (air), the conductors have a cylindrical configuration, and simple expressions permit the calculation of the capacitance.

Difficulties arise in calculation of the line capacitance, however, when many conductors are present in the line, or when there is a region of multiple dielectric constants surrounding the wires, such as in a multiconductor line with dielectric jackets.

6.5.1 Measurement of the Capacitance Parameters

Measurement of the capacitances to ground and between the various conductors of a multiconductor line of length $\mathcal{L}$ can be performed by short-circuiting to the ground all the conductors except conductor j, which is supplied by a time-harmonic voltage source V_j. The other terminations at $x = \mathcal{L}$ are open circuited, as shown in Figure 6.48 [6.28]. The current $I_i(0)$ is measured using a current probe.

The capacitive coupling of conductor i to the other conductors in a system of n conductors plus the reference is described by the equation

$$\frac{dI_i}{dx} = Y'_{i1}V_1 + Y'_{i2}V_2 + \cdots + Y'_{in}V_n \tag{6.156}$$

If the wavelength of the harmonic source is large compared to the line length $\mathcal{L}$, Eq. (6.156) can be written as

$$I_i(0) - I_i(\mathcal{L}) \approx (Y'_{i1}V_1 + Y'_{i2}V_2 + \cdots + Y'_{ii}V_i + \cdots + Y'_{in}V_n)\mathcal{L} \tag{6.157}$$

As all the lines are open circuited at termination $x = \mathcal{L}$, all $I_i(\mathcal{L}) = 0$, and as only conductor j is supplied, all the voltages $V_k = 0$, for $k \neq j$. Then since $Y'_{ij} = j\omega C'_{ij}$, the capacitance is given by

$$C'_{ij} = \frac{1}{\mathcal{L}} \frac{1}{j\omega} \frac{I_i(0)}{V_j} \quad \text{F/m} \tag{6.158}$$

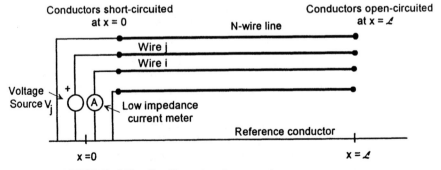

FIGURE 6.48 Configuration for capacitance measurement.

By changing the source and current sensor locations to all the conductors, the capacitances to ground and between the conductors can be determined. Although this method seems straightforward, there are some practical problems. At low frequencies, the term $j\omega$ can be small, and the resulting current $I_i(0)$ is also small. The ratio of these two small quantities may then contain a significant amount of noise, making an accurate determination of C'_{ij} difficult. If the frequency were increased to the point where the signal/noise ratio is high, the line length may then become comparable to the wavelength, making this method invalid.

6.5.2 Analytical Capacitance Evaluation

Let us consider a system of two perfect conductors in a homogeneous medium as shown in Figure 6.49. The potentials of the two conductors are V_A and V_B, respectively. From the definition of the per-unit-length capacitance

$$C' = \frac{q'}{V} = \epsilon_0 \epsilon_r \frac{\int_{S_A} \hat{n} \cdot \mathbf{E}\, dA}{\int_A^B \mathbf{E} \cdot d\hat{l}} \qquad \text{F/m} \qquad (6.159)$$

where q' is the per-unit-length charge density on the line and V is the line voltage, its value can be calculated if the electric field around the conductors is known. This means that to determine the capacitance, the electrostatic problem that permits electric field calculation must be solved.

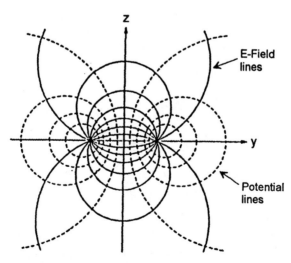

FIGURE 6.49 Two conductors in a homogeneous medium showing the E-field lines (solid) and the potential lines (dashed).

For a limited number of conductors (one or two conductors and a ground plane) the equivalent capacitances have been calculated using analytical methods [6.29]. The capacitance expressions for the most important simple configurations are listed in Appendix B.

6.5.2.1 Partial Capacitances.
For a system of many conductors, calculation of the capacitance is more involved. Consider a system of three charged conductors over a perfectly conducting plane, as illustrated in Figure 6.50. For this system the per-unit-length charges on the conductors are denoted by q_1', q_2', and q_3', respectively; the potential differences (i.e., the voltages) between the conductors are v_{12}, v_{23} and v_{31}, and the voltages between the conductors and the reference conductor are labeled as v_1, v_2, and v_3. Using the partial capacitances between the conductors and between the conductors and the ground as defined in the figure, the relations between the conductor charges and the voltages are

$$q_1' = \mathcal{C}_{12}' \cdot v_{12} + \mathcal{C}_{13}' \cdot v_{13} + \mathcal{C}_{11}' \cdot v_1 = \mathcal{C}_{12}'(v_1 - v_2) + \mathcal{C}_{13}'(v_1 - v_3) + \mathcal{C}_{11}' \cdot v_1$$
$$= (\mathcal{C}_{11}' + \mathcal{C}_{12}' + \mathcal{C}_{13}') \cdot v_1 - \mathcal{C}_{12}' \cdot v_2 - \mathcal{C}_{13}' \cdot v_3 \qquad (6.160a)$$

In a similar way one finds that

$$q_2' = -\mathcal{C}_{21}' \cdot v_1 + (\mathcal{C}_{21}' + \mathcal{C}_{22}' + \mathcal{C}_{23}') \cdot v_2 - \mathcal{C}_{23}' \cdot v_3 \qquad (6.160b)$$

$$q_3' = -\mathcal{C}_{31}' \cdot v_1 - \mathcal{C}_{32}' \cdot v_2 + (\mathcal{C}_{31}' + \mathcal{C}_{32}' + \mathcal{C}_{33}') \cdot v_3 \qquad (6.160c)$$

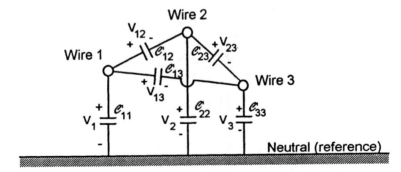

FIGURE 6.50 System of three charged conductors over a perfectly conducting plane.

Thus for a system of n conductors (plus the reference conductor), relations (6.160) can be generalized in a matrix form as

$$\begin{bmatrix} q'_1 \\ q'_2 \\ \vdots \\ q'_n \end{bmatrix} = \begin{bmatrix} \sum_{j=1}^{n} c'_{1j} & -c'_{12} & \cdots & -c'_{1n} \\ -c'_{21} & \sum_{j=1}^{n} c'_{2j} & \cdots & -c'_{2n} \\ \vdots & \vdots & \cdots & \vdots \\ -c'_{n1} & -c'_{n2} & \cdots & \sum_{j=1}^{n} c'_{nm} \end{bmatrix} \begin{bmatrix} V_1 \\ V_2 \\ \vdots \\ V_n \end{bmatrix} \quad (6.161)$$

6.5.2.2 Static Capacitances.
Relation (6.160) can also be written as

$$\begin{bmatrix} q'_1 \\ q'_2 \\ \vdots \\ q'_n \end{bmatrix} = \begin{bmatrix} C'_{11} & C'_{12} & \cdots & C'_{1n} \\ C'_{21} & C'_{22} & \cdots & C'_{2n} \\ \vdots & \vdots & \cdots & \vdots \\ C'_{n1} & C'_{n2} & \cdots & C'_{nn} \end{bmatrix} \begin{bmatrix} V_1 \\ V_2 \\ \vdots \\ V_n \end{bmatrix} \quad (6.162a)$$

and upon representing the per-unit-length charge and wire-neutral voltages as the vectors $[q']$ and $[V]$, respectively, this equation can be represented as

$$[q'] = [C'][V] \quad (6.162b)$$

By comparison with (6.160), we immediately see that

$$C'_{ii} = \sum_{j=1}^{n} c'_{ij} \quad \text{F/m} \quad (6.163a)$$

$$C'_{ij} = -c'_{ij} \quad \text{F/m} \quad (6.163b)$$

Note that all of the C'_{ii} elements are positive and all of the C'_{ij} elements for $i \neq j$ are negative. The $[C']$ matrix is called the *static* or *Maxwellian capacitance matrix*, or, in the power network literature, the *nodal capacitance matrix*. These capacitances have no physical meaning but can be used to calculate the physical partial capacitances between the conductors.

6.5.2.2.1 Sign of the Partial Capacitances.
Despite the negative sign on the off-diagonal elements in Eq. (6.161), the physical capacitances between the individual conductors in Figure 6.50 are positive, including the capacitance from each conductor to ground [6.30]. The resulting negative sign arises in changing the voltage reference from a wire-to-wire basis to a wire-to-ground (or reference conductor) basis.

6.5.3 Calculation of the Static Capacitances

To calculate the terms in the $[C']$ matrix the charges on the conductors and the resulting potential of the conductors must be known. The voltages are usually known quantities and can be measured; the charges are difficult to measure directly. Consequently, analytical or numerical methods for calculation are frequently used.

There is a problem with evaluating the matrix $[C']$ directly. Unlike the inductance matrix discussed in Chapter 5, where M'_{ij} was a function only of the ith and jth conductor geometry, the terms C'_{ij} depend on the geometry of the entire collection of conductors. By defining a potential coefficient matrix $[K']$ as

$$[V] = [K'][q'] \qquad (6.164)$$

the static capacitance matrix can be calculated by inversion from the matrix of the potential coefficients

$$[C'] = [K']^{-1} \qquad (6.165)$$

The advantage of using the potential coefficients is that they may be calculated analytically for a number of particular configurations. An example of such a configuration is a system of bare conductors over the ground.

6.5.3.1 Conductors in a Homogeneous Medium over the Ground.
Consider n bare cylindrical conductors in the air over a perfectly conducting ground, as shown in Figure 6.51. The advantage of this configuration is that by assuming that the ground is a perfect conductor, application of the image method permits us to obtain an analytical expression of the conductor potential as a function of the charges. Using the image method,

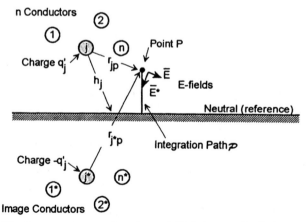

FIGURE 6.51 System of n conductors and their images in the ground plane.

the ground is replaced by the image of the n conductors. Let j and $j*$ be a pair of parallel conductors of infinite length far from any other conductor and $+q'_j$ and $-q'_j$ the charges per unit length of each conductor. At a point P situated at distances r_{jP} and r_{j*P}, respectively, from the two conductors, the total electric field from these two conductors is the vector sum of the fields due to the two charges

$$\mathbf{E}_t = \mathbf{E} + \mathbf{E}* = \frac{1}{2\pi\epsilon_0} \frac{q'_j}{r_{jP}} \hat{r}_{jP} - \frac{1}{2\pi\epsilon_0} \frac{q'_j}{r_{j*P}} \hat{r}_{j*P} \quad \text{V/m} \quad (6.166)$$

The potential of point P with respect to a ground plane due to this pair of charged conductors is obtained from the line integral $\int \mathbf{E}_t \cdot d\hat{l}$ between point P and the ground plane along the arbitrary integration path $\mathcal{P}$ shown in the figure. This potential is given by

$$V_P = \int_{r_{jP}}^{h_j} \frac{q'_j}{2\pi\epsilon_0 r} \, dr + \int_{h_j}^{r_{j*P}} \frac{q'_j}{2\pi\epsilon_0 r*} \, dr* \quad (6.167)$$

where h_j defines the ending location of the contour $\mathcal{P}$ on the ground plane. After integration this potential is

$$V_P = \frac{q'_j}{2\pi\epsilon_0} \ln \frac{h_j}{r_{jP}} - \frac{q'_j}{2\pi\epsilon_0} \ln \frac{r_{j*P}}{h_j} = \frac{q'_j}{2\pi\epsilon_0} \ln \frac{r_{j*P}}{r_{jP}} \quad \text{V} \quad (6.168)$$

For a configuration of n arbitrary conductors, the potential of the kth conductor with respect to the ground [taken as the $(n+1)$st conductor] due to the charges on *all* of the other conductors ($j = 1$ to n) is expressed as

$$V_{kn} = \frac{1}{2\pi\epsilon_0} \sum_{j=1}^{n} q'_j \ln \frac{r_{j*k}}{r_{jk}} \quad (6.169)$$

where r_{jk} and r_{j*k} are the distances between the geometrical axes of conductor k and the axes of conductors j and its image $j*$, respectively. Concerning the term $j = k$, $r_{k*k} = 2h_k$ represents the distance between conductor k and its image, where h_k is the height of this conductor with respect to the ground and r_{kk} the radius of conductor k.

Comparing the form of Eq. (6.169) with one row of the matrix equation (6.164), it is clear that the potential coefficient linking conductors k and j is

$$K'_{kj} = \frac{1}{2\pi\epsilon_0} \ln \frac{r_{j*k}}{r_{jk}} \quad \text{m/F} \quad (6.170)$$

This approach can be used if the conductors have a cylindrical configuration and are located in a homogeneous medium, so that the charge density around the wires is approximately constant. This is usually the case for aerial power or distribution lines.

6.5.3.2 Transmission Lines in Nonhomogeneous Media.
Frequently, the conductors comprising transmission lines are not bare conductors. This is usually the case in electrical circuits or assemblies, where the conductors are insulated wires with air interspaces. Moreover, the traces on PCBs have a rectangular cross section and are half embedded in a dielectric material, as shown in Figure 6.52.

For such configurations the potential coefficients do not have simple expressions, and a numerical method must be used to calculate the electric field configuration around the conductors and the resulting capacitance and inductance matrices. Appendix C presents approximations for these parameters for a single stripline and for a microstrip conductor, which have been based on a curve-fit solution of numerically computed data. For the multiconductor line shown in Figure 6.52, numerical methods must be used, and representing these results by simple equations becomes difficult. Paul [6.15] discusses determination of per-unit-length parameters for these striplines in more detail and provides computer software for their evaluation. For EMC problems involving transmission line models of PCB boards, evaluation of the line parameters in this manner is advisable.

6.5.3.3 Use of the Finite Element Method to Calculate Partial Capacitances.
One numerical method used for electric field calculations is the finite-element method. In this approach the electromagnetic problem is solved by dividing the region surrounding the conductors (i.e., in the y–z plane) into a set of arbitrary shapes known as finite elements. The problem is to find the electric potential for an inhomogeneous region, where permittivity $\epsilon(y, z)$ is a function of position. The partial differential equation describing the two-dimensional potential distribution

$$\frac{\partial^2 \Phi(y, z)}{\partial y^2} + \frac{\partial^2 \Phi(y, z)}{\partial z^2} = 0 \tag{6.171}$$

is associated to Dirichlet and Neumann boundary conditions. A technique involving first- and second-order finite elements is used [6.31].

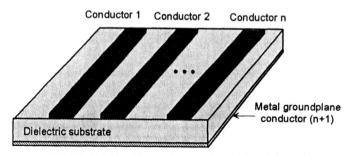

FIGURE 6.52 Traces on a printed circuit board.

Once the potential distribution is obtained, two methods can be used for calculation of the physical capacitances. From Eq. (6.162a) it can be seen that

$$C'_{ij} = \frac{q'_i}{V_j}\bigg|_{(V_1, \ldots, V_{(j-1)}, V_{(j+1)}, \ldots, V_n = 0)} \tag{6.172}$$

Since

$$q'_i = \int_{S_i} \mathbf{D} \cdot \hat{n} \, dS_i = \int_{S_i} \epsilon \mathbf{E} \cdot \hat{n} \, dS_i = -\int_{S_i} \epsilon \nabla \Phi \cdot \hat{n} \, dS_i \tag{6.173}$$

it is necessary to find the solution for the potential distribution $\Phi(y, z)$ $(n-1)$ times by using the finite-element method. Each time, a static potential is applied to only one of the conductors, while the rest are grounded. After the elements C'_{ij} are determined, the mutual physical capacitances are determined immediately using (6.163).

For the determination of the self-capacitance term C'_{ii}, an approach based on energy considerations can be used. The self-capacitance matrix of each conductor (or trace) with respect to ground can be obtained from the energy definition as

$$C'_{ii} = \frac{2W'_e}{V_{in}^2} \quad \text{F/m} \tag{6.174}$$

where W'_e is the energy per unit length given by [6.32]

$$W'_e = \frac{1}{2} \int_R \epsilon |\nabla \Phi|^2 \, dy \, dz \tag{6.175}$$

Using this approach, [6.32] has demonstrated that the rate of convergence of the solutions is much more rapid than by using the charge definition.

6.5.3.4 Integral Equation Evaluation of the per-Unit-Length Capacitance Matrix.
An alternative to the use of the finite-element method for evaluating the capacitance matrix is to use an integral equation solution. This technique may be applied to simple cases of isolated conductors in a homogeneous dielectric region [6.33], isolated conductors covered with a dielectric jacket in a homogeneous medium [6.34], or conductors in an inhomogenous dielectric region, such as strip lines in a PCB [6.15].

As a first example of the use of the integral equation solution, consider the simple case of an isolated multiconductor line in a homogeneous dielectric region, as shown in Figure 6.53. The line consists of $N+1$ conductors, each having its center at a vector position $\mathbf{r}_i$, a conductor radius a_i, and a per-unit-length charge q'_i. Wire 0 is taken to be the reference conductor for the bundle. As noted previously in this chapter, for a

302 TRANSMISSION LINE THEORY

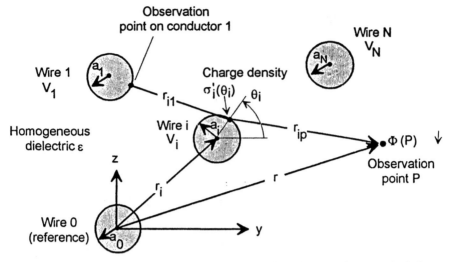

FIGURE 6.53 Cross section of a multiconductor line for capacitance calculation.

transmission line model this bundle of conductors is described by an $N \times N$ per-unit-length capacitance matrix $[C']$, as used in Eq. (6.162). Details of how this matrix may be determined numerically are described in the following section.

Because of the proximity to other wires, the charge q_i' on each wire will not be distributed uniformly around the circumference of the conductor. For the ith wire, the *circumferential* charge density is denoted by $\sigma_i'(\theta_i)$, where θ_i is the angular coordinate shown in the figure. The total per-unit-length charge on each conductor is given by the integral of the charge density as

$$q_i' = \int_0^{2\pi} \sigma_i'(\theta_i) a_i \, d\theta_i \qquad (6.176)$$

For the transmission line model, we require the relationship between the total line charges and the conductor voltages. Usually, we are not interested in the details of how this charge is distributed over the conductor. However, if the charge distribution is known, the total charge can be determined from Eq. (6.176), and relationships between the total charge on the line and the wire voltages can be determined.

The conductor charge distribution can be evaluated by formulating a set of coupled integral equations for the conductor charge densities and then solving them numerically. Using the expression for the electrostatic potential of a filament of charge as a Green's function [6.15], the electrostatic potential ϕ_i at point P due to the charge distribution on the ith wire can be written as

$$\phi_i(\mathbf{r}) = -\frac{1}{2\pi\epsilon} \int_0^{2\pi} \sigma_i'(\theta_i) \ln(r_{ip}) a_i \, d\theta_i \qquad (6.177)$$

where r_{ip} is the distance from the integration point on the ith wire to the point P, as shown in Figure 6.53. The total potential due to all the conductors (including the reference) is the sum of terms like Eq. (6.177):

$$\Phi(\mathbf{r}) = \sum_{i=0}^{N} \phi_i = -\frac{1}{2\pi\epsilon} \sum_{i=0}^{N} \int_0^{2\pi} \sigma_i' \ln(r_{ip}) a_i \, d\theta_i \qquad (6.178)$$

A set of $N+1$ integral equations for the distributions $\sigma_i'(\theta_i)$ can be formed by letting observation point P approach each conductor, on which the potential is assumed to be specified as a constant Φ_j. This yields the set of integral equations

$$\Phi_j = -\frac{1}{2\pi\epsilon} \sum_{i=0}^{N} \int_0^{2\pi} \sigma_i'(\theta_i) \ln(r_{ij}) a_i \, d\theta_i \qquad j = 0, 1, \ldots, N \qquad (6.179)$$

where the term r_{ij} denotes the distance from the integration point on the ith wire to the observation point on the jth wire shown in the figure. Since the observation point P can be located anywhere on the conductor circumference, Eq. (6.179) is valid for *any* value of r_{ij} as long as it terminates somewhere on the surfaces of wires i and j.

The set of integral equations for $\sigma_i'(\theta_i)$ can be solved using the method of moments [6.35], which was discussed in Chapter 4 in relation to the linear antenna. This procedure involves representing the unknown charges on each conductor in terms of an expansion of m basis functions with unknown coefficients, and then taking moments of the integral equations using k testing functions. This results in a system of linear equations of dimensions $[m \times (N+1)]$ by $[k \times (N+1)]$. Usually, the number of testing functions equals the number of expansion functions, so that the resulting matrix equation may be inverted using standard matrix inversion techniques.

Different choices of expansion and testing functions may be used, as discussed in [6.35]. The simplest is to represent the charge density by m pulse functions around the wire circumference and to use point matching, in which the equation is enforced at an equal number of points on the wire. An alternative method is to expand the charge in a Fourier series in the angle θ. In either solution, the specific charge distribution $\sigma_i'(\theta_i)$ on each of the i conductors results for a given set of specified potentials Φ_i on the conductors.

As discussed in [6.33], the logarithmic kernel in Eq. (6.179) becomes singular when the source and observation point are located on the same wire. This singularity is integrable, however, and poses no problem mathematically. From the numerical standpoint, however, care must be used in extracting the singular contribution to the integral in a way that does not affect the accuracy of the solution. Further details of this solution are provided in [6.33].

Once the charge distributions are known, the *total* per-unit-length charge on each of the wires can be evaluated using Eq. (6.176). By choosing different combinations of potentials Φ_i on the wires, the following generalized per-unit-length capacitive coefficient matrix for the $N+1$ set of conductors may be developed to relate this charge on each conductor to the conductor potentials:

$$\begin{bmatrix} q'_0 \\ q'_1 \\ q'_2 \\ \vdots \\ q'_N \end{bmatrix} = \begin{bmatrix} K'_{00} & K'_{01} & K'_{02} & \cdots & K'_{0N} \\ K'_{10} & K'_{11} & K'_{12} & \cdots & K'_{1N} \\ K'_{20} & K'_{21} & K'_{22} & \cdots & K'_{2N} \\ \vdots & \vdots & \vdots & & \vdots \\ K'_{N0} & K'_{N1} & K'_{N2} & \cdots & K'_{NN} \end{bmatrix} \begin{bmatrix} \Phi_0 \\ \Phi_1 \\ \Phi_2 \\ \vdots \\ \Phi_N \end{bmatrix} \quad (6.180)$$

Typically, this matrix is determined by setting all wire potentials to zero with the exception of the ith conductor, which has its potential set to unity. The integral equations (6.179) are then solved and the total line charges computed. This ordered set of charges forms the ith column in the $[K']$ matrix. This calculation is performed $N+1$ times, once for each conductor i to yield the entire $[K']$ matrix.

To determine the $N \times N$ capacitance matrix from the $(N+1) \times (N+1)[K']$ matrix, we first determine the *generalized* per-unit-length elastance matrix $[S']$ for the set of $N+1$ conductors. This is given by the inverse of Eq. (6.180) as

$$\begin{bmatrix} \Phi_0 \\ \Phi_1 \\ \Phi_2 \\ \vdots \\ \Phi_N \end{bmatrix} = [K']^{-1} \begin{bmatrix} q'_0 \\ q'_1 \\ q'_2 \\ \vdots \\ q'_N \end{bmatrix} \equiv \begin{bmatrix} S'_{00} & S'_{01} & S'_{02} & \cdots & S'_{0N} \\ S'_{10} & S'_{11} & S'_{12} & \cdots & S'_{1N} \\ S'_{20} & S'_{21} & S'_{22} & \cdots & S'_{2N} \\ \vdots & \vdots & \vdots & & \vdots \\ S'_{N0} & S'_{N1} & S'_{N2} & \cdots & S'_{NN} \end{bmatrix} \begin{bmatrix} q'_0 \\ q'_1 \\ q'_2 \\ \vdots \\ q'_N \end{bmatrix} \quad (6.181)$$

For the transmission line approximation, the total charge on this system of conductors is assumed to be zero (i.e., $\sum_{i=0}^{N} q'_i = 0$), and the voltages of wires 1 through N are all referred to reference conductor, as $V_i = \Phi_i - \Phi_0$. Subtracting the first row of Eq. (6.181) from all the other rows and using the fact that $q'_0 = -\sum_{i=1}^{N} q'_i$ yields the following $N \times N$ transmission line *elastance* matrix:

$$\begin{bmatrix} V_1 \\ V_2 \\ \vdots \\ V_N \end{bmatrix} = \begin{bmatrix} \boldsymbol{S}'_{11} & \boldsymbol{S}'_{12} & \cdots & \boldsymbol{S}'_{N1} \\ \boldsymbol{S}'_{21} & \boldsymbol{S}'_{22} & \cdots & \boldsymbol{S}'_{N2} \\ \vdots & \vdots & & \vdots \\ \boldsymbol{S}'_{N1} & \boldsymbol{S}'_{N2} & \cdots & \boldsymbol{S}'_{NN} \end{bmatrix} \begin{bmatrix} q'_1 \\ q'_2 \\ \vdots \\ q'_N \end{bmatrix} \quad (6.182a)$$

or, equivalently, in compact notation,

$$[V] = [\boldsymbol{S}'][q'] \quad (6.182b)$$

In the matrix $[\boldsymbol{S}']$, the individual elements are defined in terms of the $[S']$

matrix of Eq. (6.181) as

$$S'_{ij} = S'_{ij} + S'_{00} - S'_{0j} - S'_{i0} \quad \text{for } i, j = 1, \ldots, N \quad (6.183)$$

The inversion of the $[S']$ matrix provides the per-unit-length capacitance matrix $[C']$, so that

$$[q'] = [S']^{-1}[V] \equiv [C'][V] \quad (6.184)$$

As an example of the use of the integral equation formalism above, [6.34] has calculated the per-unit-length capacitance for a two-wire transmission line in free space ($\epsilon = \epsilon_0 \approx 8.854 \times 10^{-12}$ F/m) with equal radii of $a = 1$ cm and a close separation ($d = 2.1$ cm). Figure 6.54 presents the capacitance as calculated for several different types of basis functions for the integral equation, together with the exact solution for the capacitance which is given

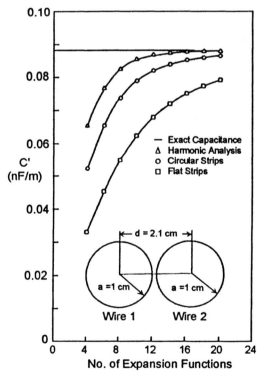

FIGURE 6.54 Convergence of the capacitance of two bare wires in free space with ($d/a = 2.1$) for different expansion functions. (From [6.34]. © 1975 Institute of Electrical and Electronics Engineers.)

in closed form as

$$C' = \frac{\pi\epsilon}{\text{arccosh}(d/2a)} \quad \text{F/m} \tag{6.185}$$

Clearly, care must be used in developing such numerical solutions to ensure that a sufficient number of basis functions is used and that the solution has converged adequately.

As discussed in [6.15], the integral equation solution can be extended to treat the case of dielectric cladding (insulation) covering the conductors. In this case, two integral equations for each conductor must be formulated and solved numerically.

To illustrate this solution, consider a typical wire i of the multiconductor bundle in Figure 6.53 to be surrounded by a dielectric coating of radius b_i and dielectric constant ϵ, as shown in Figure 6.55. In this case, two types of charges can exist: the freely flowing charge on the conducting wire, denoted by q'_i, and bound dielectric charges, q'_{ib} and q'_{ia}, which exist at the dielectric interface surfaces at the radii b_i and a_i, respectively. The latter charges arise from polarization of the dielectric material. Since these charges are distributed over the wire and insulation perimeters, the corresponding surface charge densities given by σ'_i, σ'_{ib}, and σ'_{ia} are needed to describe the potential.

As illustrated in Figure 6.55, the potential may be calculated by replacing the dielectric by the bound charges, σ'_{ib} and σ'_{ia} and calculating the potential as if all the charges were located in a homogeneous region with permittivity

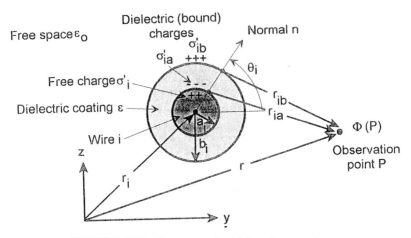

FIGURE 6.55 Geometry of a dielectric-coated wire.

ϵ_0. Thus, the potential ϕ_i at a point P due to the single wire is expressed as

$$\phi_i(\mathbf{r}) = -\frac{1}{2\pi\epsilon_0}\left[\int_0^{2\pi}\sigma'_i(\theta_i)\ln(r_{ia})a_i\,d\theta_i + \int_0^{2\pi}\sigma'_{ia}(\theta_i)\ln(r_{ia})a_i\,d\theta_i\right.$$
$$\left. + \int_0^{2\pi}\sigma'_{ib}(\theta_i)\ln(r_{ib})b_i\,d\theta_i\right] \quad (6.186)$$

where r_{ia} and r_{ib} are the distances from the particular integration points on the circular contours to the observation point, as indicated in the figure. Combining the bound and free charges at the wire surface into a single effective charge density as $\sigma'_{i(\text{eff})} = \sigma'_i + \sigma'_{ia}$ yields the following expression for the potential due to the single conductor:

$$\phi_i(\mathbf{r}) = -\frac{1}{2\pi\epsilon_0}\left[\int_0^{2\pi}\sigma'_{i(\text{eff})}(\theta_i)\ln(r_{ia})a_i\,d\theta_i + \int_0^{2\pi}\sigma'_{ib}(\theta_i)\ln(r_{ib})b_i\,d\theta_i\right]$$
$$(6.187)$$

This expression can be used as in Eq. (6.178) to find the potential due to all the conductors in the line by summing over each wire's contribution. Letting the observation point approach each of the conductors, a set of $N+1$ integral equations result:

$$\Phi_j = -\frac{1}{2\pi\epsilon_0}\sum_{i=0}^{N}\left[\int_0^{2\pi}\sigma'_{i(\text{eff})}(\theta_i)\ln(r_{ia})a_i\,d\theta_i\right.$$
$$\left. + \int_0^{2\pi}\sigma'_{ib}(\theta_i)\ln(r_{ib})b_i\,d\theta\right] \quad j=0,1,\ldots,N$$

J (6.188)

Because there are now two unknown charge distributions for each conductor, $\sigma'_{i(\text{eff})}$ and σ'_{ib}, it is necessary to find an additional set of $N+1$ conditions for the solution to be unique. This is accomplished by requiring that the electric flux (the normal component of **D**) be continuous in crossing the air/dielectric boundary at radius $\rho = b_i$. This condition is expressed for each wire i, as

$$\epsilon\left.\frac{\partial\Phi_i}{\partial n}\right|_{\rho=a_{i-}} = \epsilon_0\left.\frac{\partial\Phi_i}{\partial n}\right|_{\rho=a_{i+}} \quad (6.189)$$

where ρ denotes the radial distance from the center of the wire and n is in the direction normal to the surface.

Equations (6.188) and (6.189) provide the necessary information for uniquely determining the charge densities using the moment method. Once the charge densities $\sigma'_{i(\text{eff})}$ and σ'_{ib} are determined, the corresponding per-unit-length charges on the conducting wires $q'_{i(\text{eff})}$ and q'_{ib} can be determined

308 TRANSMISSION LINE THEORY

by Eq. (6.176). Assuming that the dielectric coating is electrically neutral, we require that the total bound charge on the dielectric surfaces cancel (i.e., $q'_{ib} = -q'_{ia}$). This permits the free charge on the conductor to be written as $q'_i = q'_{i(\text{eff})} - q'_{ia}$, or in terms of the quantities calculated from the integral equation, we have

$$q'_i = q'_{i(\text{eff})} + q'_{ib} \qquad (6.190)$$

For the transmission line model, the required line capacitance is the relationship between the per-unit-length free charge and the wire potentials. Thus, for the dielectric coated wires, Eq. (6.190) is used with Eqs. (6.180) to (6.184) to compute the necessary capacitance matrix. An example of capacitance calculation using this approach is shown in Figure 6.56.

6.5.3.5 Capacitance Calculation Using Inductance Values.
Use of finite elements and moment methods to determine line capacitance parameters permits us to obtain accurate results and provides a good way to determine the line values when it is necessary to take into account the various modal propagation constants. These approaches, however, need rather sophisticated numerical procedures. An alternative approach for multiconductor lines which combines modal measurements with calculations is also rigorous, except for measurement errors.

For an n-wire multiconductor line (plus a return conductor) in a nonhomogeneous environment, the modal velocities are usually different. Denoting each of these velocities as v_n, a diagonal modal velocity matrix can be defined: $[v_n]$. Using the diagonalization matrix $[S]$ defined under Section 6.2, a generalized velocity matrix $[v]$ can be defined as [6.10]

$$[v] = [S][v_n][S]^{-1} \qquad (6.191)$$

The capacitance matrix is now given by

$$[C'] = [Y_c] \cdot [v]^{-1} \qquad (6.192)$$

where $[Y_c]$ is the characteristic admittance matrix. The characteristic admittance values $[Y_c]$ and velocities v_n for each propagation mode can be determined using time-domain reflectometry.

This method can be used if the conductors from the bundle are rather unequally spaced from each other, giving strong inhomogeneities, and if they are sufficiently long (not less than 10 to 20 m), which permits a reasonable precision with the time-domain reflectometry.

When the nonhomogeneities are less strong (i.e., for equally spaced conductors all with the same diameter) and the length of the bundle is relatively short (a few meters), a simplified approach can be achieved by using the inductance values. In this environment, the differences between velocities are not very large (10 to 15% for a cable bundle of 2.66 m a few

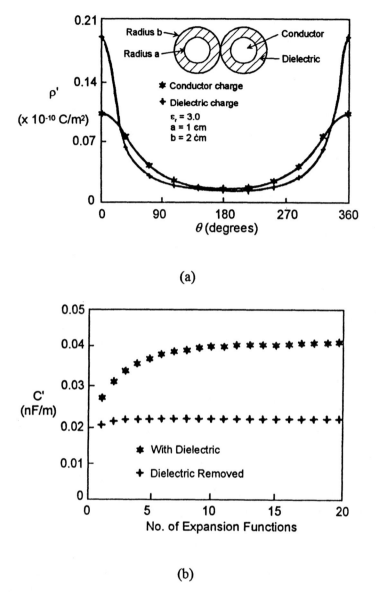

FIGURE 6.56 Convergence of the solution for the capacitance of two coated wires ($\epsilon_r = 3$) as a function of the number of expansion functions on each boundary: (*a*) geometry and charge distribution; (*b*) convergence of the capacitance. (From [6.34]. © 1975 Institute of Electrical and Electronics Engineers.)

centimeters above a ground plane), and a constant velocity equal to the common-mode velocity v_c can be assumed [6.36]. This common-mode velocity can be measured by reflectometry. The risk of errors due to very

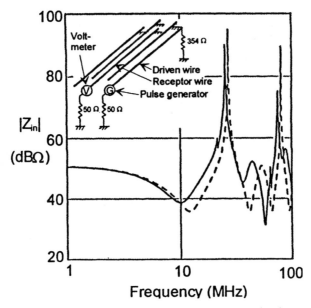

FIGURE 6.57 Variation of the input impedance magnitude $|Z_{in}|$ of the driven wire at the pulse generator input as a function of frequency as determined using modal theory (solid line) or a constant velocity (dashed line). (From [6.36]. Reprinted by permission.)

small differences between the modal velocities is avoided. The inductance matrix can then be calculated using analytical methods (see Section 6.4). By analogy with a homogeneous environment, the capacitance matrix is given by Eq. (6.52).

The effects of the error due to this simplified approach can be estimated by comparing the magnitude of the input impedance Z_{in} of one of the conductors calculated by both the simplified and modal methods. The comparison is shown in Figure 6.57 for a multiconductor cable consisting of five insulated conductors 2.66 m in length at a height of 100 mm from a ground metallic plane [6.36]. In this figure the impedance magnitude is presented in dB relative to $1\,\Omega$. Some differences can be seen at high frequencies, where resonances for the simplified approach are found at slightly different values than for the modal approach.

REFERENCES

6.1. Ramo, S., J. R. Whinnery, and T. Van Duser, *Fields and Waves in Communication Electronics*, 2nd ed., John Wiley and Sons, 1989.

6.2. Johnk, C. T. A., *Engineering Electromagnetic Fields and Waves*, John Wiley and Sons, New York, 1976.

6.3. Gardiol, F., *Electromagnétisme*, Traité d'Electricité, Vol. 3, Giorgi, St. Saphorin, Switzerland, 1977.
6.4. Colvin, D. H., "Computationally Efficient Method of Calculations Involving Lumped-Parameter Transmission-Line Models," *IEEE Trans. Electromagn. Compat.*, Vol. EMC-27, No. 1, February 1986.
6.5. Everett, W. W., "Lumped Model Approximations of Transmission Lines: Effect of Load Impedances on Accuracy," *Proceedings of the IEEE 1983 International Symposium on EMC*, Arlington, VA, August 23-25, 1983.
6.6. F. M. Tesche, "An Overview of Transmission Line Analysis," *Proceedings of EMC EXPO '87 International Conference on EMC*, San Diego, CA, May 19-21, 1987.
6.7. Liu, T. K., "Electromagnetic Coupling Between Multiconductor Transmission Lines in a Homogeneous Medium," *Report, AFWL-TR-76-333*, Air Force Weapons Laboratory, Kirtland AFB, NM, May 1977.
6.8. Tesche, F. M., and T. K. Liu, "Application of Multiconductor Transmission Line Network Analysis to Internal Interaction Problems," *Electromagnetics*, Vol. 6, No, 1, 1986, pp. 1-20.
6.9. Burke, G. J., A. J. Poggio, J. C. Logan, and J. W. Rockway, "Numerical Electromagnetics Code: A Program for Antenna System Analysis," *Proceedings 3rd International Symposium and Technological Exhibition on EMC*, Rotterdam, May 1-3, 1979.
6.10. Agrawal, A. K., H. M. Fowles, and L. D. Scott, "Experimental Characterization of Multiconductor Transmission Lines in Inhomogeneous Media Using Time-Domain Techniques,"*IEEE Trans. Electromagn. Compat.*, Vol. EMC-21, No. 1, February, 1979, pp. 28-32.
6.11. Carey, V. L., T. R. Scott, and W .T. Weeks, "Characterization of Multiple Parallel Transmission Lines Using Time-Domain Reflectometry", *IEEE Trans. Instrum. Meas.*, Vol. IM-18, September 1969, pp. 166-171.
6.12. Poudroux, C., "Étude de l'incidence des paramètres primaires des lignes couplées sur la précision de prediction de l'amplitude des parasites induits sur des torons multifilaires," Thèse 973, Université des Sciences et Technologies de Lille, September 30, 1992.
6.13. Eclangon, Ph., "Equations de Propagation sur Ligne Polyphasée-Applications," *Bull. Dir. Etudes Rech. EDF*, Sér. B, No. 1, 1968, pp. 27-44.
6.14. Wedepohl, L. M., "Application of Matrix Methods to the Solution of Traveling-Wave Phenomena in Polyphase Systems," *Proc. IEEE*, Vol. 110, 1963, pp. 2200-2212.
6.15. Paul, C. R., *Analysis of Multiconductor Transmission Lines*, Wiley, New York, 1994.
6.16. Poudroux, C., M. Rifi, B. Demoulin, and P. Degauque, "Evaluation de l'amplitude des parasites induits sur les câbles assemblés en toron," *6ème Colloque International et Exposition sur la CEM*, Ecole Centrale de Lyon, June 2-4, 1992.
6.17. Paul, C. R., "Computation of Crosstalk in a Multiconductor Transmission Line," *IEEE Trans. Electromagn. Compat.*, Vol. EMC-23, No. 4, November, 1981.
6.18. Press, W. H., et al, *Numerical Recipes*, Cambridge Press, Cambridge, 1986.
6.19. Arreghini, F., M. Ianoz, C. A. Nucci, and F. Rachidi, "Une comparaison entre

les méthodes de calcul temporel et fréquentiel appliqués au problème de couplage IEM: ligne aérienne," *Actes du 6ème Colloque International sur la CEM, Ecole Centrale de Lyon*, June 2–4, 1992, pp. 329–333.

6.20. Agrawal, A. K., H. J. Price, and S. Gurbaxani, "Transient Response of a Multiconductor Transmission Line Excited by a Nonuniform Electromagnetic Field", *IEEE Trans. Electromagn. Compat.*, Vol. EMC-22, May 1980, pp. 119–129.

6.21. Bergeron, L., "Propagation d'ondes le long de lignes électriques," *Bull. Soc. Fr. Electr.*, October 1937, pp. 579–1004.

6.22. Magnusson, P. C., G. C. Alexander, and V. K. Tripathi, *Transmission Lines and Wave Propagation*, CRC Press, Boca Raton, FL, 1992.

6.23. Dommel, H. W., "Digital Computer Solution of Electromagnetic Transients in Single- and Multiphase Networks", *IEEE Trans. Power Appar. Syst.*, Vol. PAS-88, No. 4, April 1969, pp. 388–399.

6.24. Sabonnadière, J.-C., and Ph. Auriol, "Surtensions de manoeuvre dans les réseaux HT et THT," *Rev. Gen. Electr.*, Vol. 82, No. 11, November 1973, pp. 718–727.

6.25. Maxwell, J. C., *A Treatise on Electricity and Magnetism*, 3rd ed., Oxford University Press, London, 1959.

6.26. Rosa, E. B., and F. W. Grover, "Formulas and Tables for the Calculation of Mutual and Self-inductance," *Scientific Paper 169*, National Bureau of Standards, Washington, DC, 1908.

6.27. Dwight, H. B., "Geometric Mean Distances for Rectangular Conductors," *Electr. Eng.*, Vol. 65, 1946, pp. 536–538.

6.28. Agrawal, A. K., K. M. Lee, L. D. Scott, and H. M. Fowles, "Experimental Characterization of Multiconductor Transmission Lines in the Frequency Domain," *IEEE Trans. Electromagn. Compat.*, Vol. EMC-21, No. 1, February 1979, pp. 20-27.

6.29. Walker, C. S., *Capacitance, Inductance and Crosstalk Analysis*, Artech House, Boston, 1990.

6.30. Boite, R., and J. Neirynck, *Théorie des Réseaux de Kirchhoff*, Traité d'Electricité, vol. 4, Giorgi, St. Saphorin, Switzerland, 1978.

6.31. Silvester, P., "Higher-Order Polynomial Triangular Finite Elements for Potential Problems," *Int. J. Eng. Sci.*, Vol. 7, 1969, pp. 849–861.

6.32. Khan, R. L., and G. Costache, "Finite Element Method Applied to Modeling Crosstalk Problems on Printed Circuit Boards," *IEEE Trans. Electromagn. Compat.*, Vol. 31, No. 1, February 1989, pp. 5–16.

6.33. Giri, D. V., F. M. Tesche, and S. K. Chang, "The Transverse Distribution of Surface Charge Densities on Multiconductor Transmission Lines," *IEEE Trans. Electromagn. Compat.*, Vol. EMC-21, No. 3, August 1979.

6.34. Clements, J. C., C. R. Paul, and A. T. Adams, "Computation of the Capacitance Matrix for Systems of Dielectric-Coated Cylindrical Conductors," *IEEE Trans. on Electromagn. Compat.*, Vol. 17, No. 4, November 1975, pp. 238–248.

6.35. Harrington, R. F., *Field Computation by Moment Methods*, Macmillan, New York, 1968.

6.36. Poudroux, C., M. Rifi, B. Demoulin, and P. Degauque, "Influence of the

Different Propagation Modes on the Response of a Multiconductor Transmission Line to a Disturbing Wave, *Proceedings 10th International Symposium and Technical Exhibition on EMC*, Zurich, March 9–11, 1993, paper 47H6.

PROBLEMS

6.1 A two-wire lossless transmission line has the per-unit-length parameters $L' = 1.75 \times 10^{-7}$ H/m and $C' = 6.60 \times 10^{-12}$ F/m. Evaluate the propagation velocity and the characteristic impedance of the line.

6.2 Develop expressions for the time-averaged power $P(x)$ on a transmission line in terms of the forward and backward traveling voltage and current waves $V^+(x)$, $I^+(x)$, $V^-(x)$, and $I^-(x)$.

6.3 For a particular section of coaxial cable, the following electrical parameters have been measured at a frequency of $f = 250$ kHz: characteristic impedance $Z_c = 93.0 e^{-j1.7°}$ Ω, line propagation constant $\beta = 6.28 \times 10^{-3}$ rad/m, line attenuation constant $\alpha = 4.60 \times 10^{-4}$ Np/m. (Note that $\gamma = \alpha + j\beta$). For this line, compute the per-unit-length parameters L', C', R', and G'.

6.4 Express the complex propagation constant $\gamma = \alpha + j\beta$ for a general transmission line in terms of the per-unit-length parameters L', C', R', and G'. Assuming that the line loss is low ($R' \ll \omega L'$ and $G' \ll \omega C'$), show that the propagation constant can be approximated as

$$\gamma \approx \frac{1}{2}\left(\frac{R'}{Z_0} + G'Z_0\right) + j\omega\sqrt{L'C'}$$

6.5 In Problem 6.4, notice that the attenuation constant α can be a function of frequency. Discuss the impact that this frequency dependence has on a pulse as it propagates down the line. What happens in the low-loss case?

6.6 Consider the special case of a lossy, yet distortionless line for which $R'/L' = G'/C'$. Compute the general complex propagation constant and show that it leads to a distortionless propagation, even though the losses can be large. This is known as the *Heaviside line*.

6.7 Verify by algebraic manipulation that the BLT equations of Eqs. (6.42a) and (6.42b) yield the terminal voltage and current expressions of Eqs. (6.40) and (6.41). (*Hint*: Consider using a symbolic algebra program such as Mathcad, Mathematica, or Maple.)

6.8 Consider two infinitely long parallel circuits 1 and 2, as shown in Figure P6.8. Circuit 1, consisting of branches A and B, is constructed

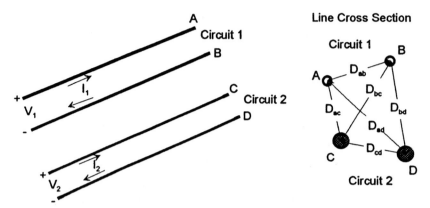

FIGURE P6.8 Two coupled circuits.

from a hollow conductor of diameter d_1, thickness $t \ll d_1$ and conductivity σ_1 S/m. Circuit 2, having branches C and D, is a solid conductor of diameter d_2 and conductivity σ_2 S/m. Assuming that the distances D_{jk} between the axes of different conductors are known:

(a) Evaluate the voltages V_1 and V_2 as a function of the currents I_1, and I_2, and the per-unit-length line parameters R_1, R_2, L_{11}, L_{22}, $M_{12} = M_{21}$, and deduce the analytical expressions of R_1, R_2, L_{11}, L_{22}, and M_{12}.

(b) What happens if $d_1 \to 0$; if $d_2 \to 0$; if d_1 and $d_2 \to 0$?

(c) How should the conductors be arranged to obtain $M_{12} = M_{21} < 0$, $M_{12} = M_{21} = 0.4$?

(d) What is the voltage between the points x_1 and x_2 situated on conductor A?

6.9 A buried computer center is to be built near a 220-kV power line. The line conductors are at a height of 20 m over the ground and have a diameter of 38 mm (see Figure P6.9). A short circuit on phase 1 occurs at a distance of 10 km from the power plant, and a large primary current flows in this phase. To evaluate the risk of failure within the computer center, it is necessary to calculate the induced *common-mode* voltage on a buried cable which runs parallel to the power line at a horizontal distance of $\ell = 30$ m from the short-circuited phase. Assuming that the computer cable is installed at a depth of $d = 10$ m below the ground surface, compute the induced open-circuit voltage on a 10-m section of this computer cable. Assume that one end of the cable is short-circuited to the ground reference conductor and that the other end is open-circuited. The ground may be assumed to be a conductor with a diameter equal to the skin depth at the current frequency (50 Hz, as the steady-state regime will be considered).

6.10 Compute the 5×5 capacitive coefficient matrix for the three-wire

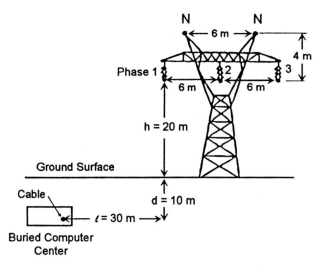

FIGURE P6.9 Power line tower and nearby buried computer center.

power transmission line, together with the two-wire neutral (shield) wires denoted by N, as shown in Figure 6.2. Assume that the earth is a perfect conductor.

6.11 Compute the per-unit-length inductance matrix for the line in Problem P6.9.

6.12 Under a vertically oriented three-phase 125-kV power line, a telecommunication line has been installed, as indicated in Figure P6.12.

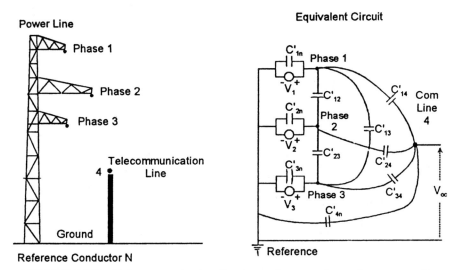

FIGURE P6.12 Telecommunication line located near a three-phase power line.

Maintenance work has been scheduled on the telecommunication line and it is desired to assess possible safety hazards. The load impedance of this line is large, so that the main coupling mode is through the mutual capacitance between the power line and the communication line. The equivalent diagram of the four parallel lines is shown in the figure.

(a) For this line configuration, compute the open-circuit voltage of the telecommunication line.

(b) Compute the short-circuit current through the body of a person inadvertently touching the communication line. (Assume a line of length ℓ and neglect the resistance of the body to ground.)

(c) Compute the minimum length of the line for which a current of 10 mA, dangerous for life, flows through the human body.

(d) Compute the minimum length of the line for which a current of 1 mA, which can give a shock, flows through the human body.

Hint: From the equivalent diagram in the figure, it can be seen that the per-unit-length capacitances C'_{12}, C'_{23}, C'_{31}, C'_{1n}, C'_{2n}, and C'_{3n}, are either short-circuited or in parallel with a voltage source (i.e., they can be neglected). Using the line geometry indicated in the figure, determine the other partial capacitances needed to solve the problem. Then, using the superposition principles and considering successively only phase 1, then phase 2, and then phase 3 energized with the others short-circuited, it is possible to compute the open-circuit voltage analytically or graphically. Note that the voltages of the three phases form a three-phase system with angles of 120° between them.)

6.13 To perform maintenance work in substation II, interrupters 1 and 2 have been opened and the busbars of substation II grounded, as shown in Figure P6.13. During this work, a very low current lightning strike (2.5 kA) strikes one conductor of the 200-km line connecting substations I and II at a distance of 50 km from substation II. To

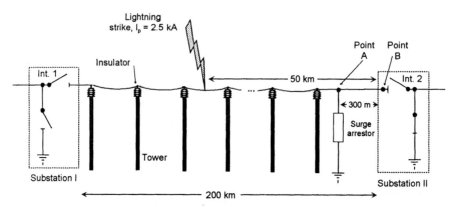

FIGURE P6.13 A 200-km power line struck by lightning.

protect the power equipment from such a surge, a lightning arrester has been installed 300 m from the entry of substation II. For this line, the assumed breakdown voltages of its different elements are:
- 1100 kV for the line-to-tower insulation
- 900 kV between the terminals of the open-circuited interrupters in the substation
- 600 kV for the lightning arrester.

The lightning current of 2.5 kA peak amplitude injected onto the line will be divided by 2 in the two directions of the line, and if a characteristic line impedance of 400 Ω is assumed, a traveling voltage wave having a peak value of 500 kV will propagate in both directions away from the strike point with the speed of $c \approx 3.0 \times 10^8$ m/s and arrive at the entries of the substations.

For simplicity, assume a triangular shape of the lightning-induced voltage waveform with a rise time of 1 μs and a fall time of 10 μs. Using a graphical method for drawing the forward and backward waves along the line at different moments in time, represent the voltages at points A and B as a function of time, first *without* taking into account the various breakdown levels and then taking them into account. Discuss the consequences of installing the arrester at a distance of 300 m from the substation entry, and propose a solution to avoid the breakdown of the open insulator in substation II.

6.14 A three-phase power line has the dimensions shown in Figure P6.9. The conductors have a radius of 3.8 cm. Line 1 carries a transient current due to a direct lightning stroke on the line of $I_0 e^{-t/\tau}$ with $I_0 = 1000$ A and $\tau = 100$ μs.
(a) What will be the current induced in lines 2 and 3?
(b) Discuss the effect of the height of the lines above the ground.
(c) Discuss qualitatively the effect of the soil conductivity. (This subject is developed more completely in Chapter 8.)

6.15 The NULINE code in Appendix F is a useful tool for performing a continuous-wave or transient analysis on a single-wire transmission line located over a ground plane. Using this program, compute the responses to the following problems and plot the quantities indicated.
(a) Consider the lossless transmission line shown in Figure P6.15a. Compute (independently from the computer code) the per-unit-length line inductance and capacitance, along with the propagation constant and the line impedance.
(b) Assuming that the line in part (a) is excited by a transient voltage pulse 10 ns wide, turning on at at $t = 0$, and with an amplitude of 1 V, compute and plot the open-circuit voltage at the far end of the line (at $x = 9$ m). Discuss the origin of each pulse obtained in the transient response. (*Hint:* The necessary input data set for this calculation is located in the data file EX1.CMD. See Appendix F

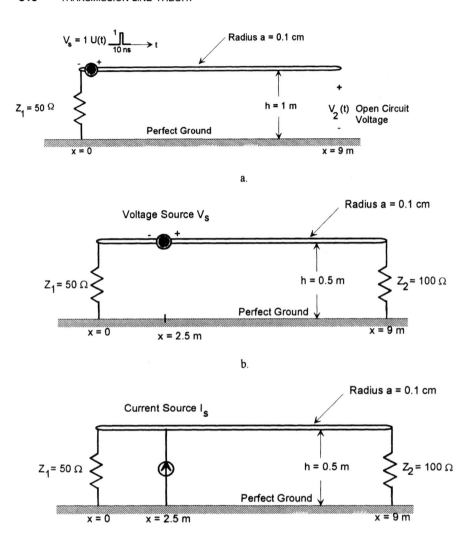

FIGURE P6.15 Transmission line configurations for lumped source excitations.

for details on how to execute NULINE using this data file. The transient response is computed through the use of a fast Fourier transform (FFT), in which a maximum time window T_{max} of 1 μs is used, together with 512 data points.)

(c) Change the number of time sample points to the maximum permitted, 2048, and recalculate the response for the line in part (b). Observe the better resolution of the pulses but with an over-

and undershoot of the pulses arising from Gibbs phenomenon. (*Hint:* The needed data set appears in the file EX1A.CMD.)

(d) Time shift the excitation pulse so that it turns on at $t = 0.4 \ \mu s$. Note that the resulting response waveform $V_2(t)$ has not died away at the end of the FFT time window, and that it "feeds through" into the early-time portion of the response. (This is because a discretely sampled spectrum provides a periodic transient response.) (*Hint:* Data file EX1B.CMD contains the necessary data.)

(e) Compute the frequency-domain response for the line voltage of Figure P6.15a, from a minimum frequency of 0.1 MHz to a maximum frequency of 200 MHz, with the voltage source $V_s(\omega) = 1$ for all frequencies. Plot the real and imaginary parts and the magnitude of the voltage response on both a log-log scale and a linear-linear scale. Note that there is a resonance in the line at a frequency of $f \approx 8.3$ MHz. What is the physical reason for this resonance? *Hint:* The file EX3.CMD contains the data needed for this calculation. What happens if $Z_1 = 0$ or $Z_1 \to \infty$? Try doing the calculation to check your answer.)

(f) Using the line configuration of Figure P6.5a, compute the transient line current and per-unit-length charge at a location $x = 3$ m, due to the 1-V pulsed source. Carefully examine the waveforms and explain where the various pulses come from. Why are these responses more "complicated" than those at the end of the line? (*Hint:* Use data file EX5.CMD.)

(g) For the transmission line in Figure P6.15b, which is excited by a time-harmonic voltage source at $x = 2.5$ m, determine the current and charge distributions $I(x)$ and $\rho(x)$ from $x = 0$ to $x = 9$ m. Assume that the frequency of the source is $f = 20$ MHz. Discuss why there is a discontinuity in the linear charge density in passing through the voltage source, and why the current is continuous. *Hint:* Data file EX9.CMD contains the data needed. Note that this calculation involves the "sweeping" of one of the parameters of the problem, namely, the observation point on the line. This results in output data files with a suffix *.FRS instead of *.FRQ. In addition, the format of the output data is slightly different from that of non-swept responses.)

(h) Change the operating frequency to 50 MHz in part (g) and recalculate the line current and charges. Discuss what is happening here. What will happen if the load resistance $Z_2 = 0$ or $Z_2 \to \infty$? Try it!

(i) Consider the same transmission line as before, but now with a time-harmonic current source of 1 A at $x = 2.5$ m, as shown in

Figure P6.15c. For a frequency of $f = 20$ MHz, calculate the line current and charge distributions and explain why these are different from the responses for the voltage source. (*Hint:* Use file EX10.CMD.)

CHAPTER 7
Field Coupling Using Transmission Line Theory

In addition to the lumped voltage or current sources discussed in Chapter 6, transmission lines can be excited by electromagnetic fields. Such fields can be produced by distant EMI sources and their effect is to induce currents and voltages on the line and in the load impedances at the ends. Generally, the behavior of the induced responses can be evaluated using EM scattering theory. However, in many cases of practical interest, simple transmission line models are sufficient. In this chapter we discuss these models and illustrate their use for some typical EMC problems.

In dealing with this subject, keep in mind that we are discussing transmission systems having *two* types of wave propagation: the free-space propagation of an incident EM field that excites the line, and the propagation of the induced voltage and current disturbances along the line. In most cases in Chapters 7 to 9, the incident field propagation will be assumed to be lossless, and consequently, the wave propagation will be described by the function e^{-jkx}, where $k = \omega/c$, with ω being the angular frequency and c the speed of light in the region. The wave propagation of the line disturbances, however, may contain the effects of loss due to a nonzero resistance or to other losses in the transmission line system. This gives rise to a complex-valued transmission line propagation constant $\gamma(\omega)$ with a wave propagation described by the function $e^{-\gamma(\omega)x}$.

7.1 INTRODUCTION

There are three different approaches for describing the coupling of an external EM field to a line using transmission line theory:

1. The line may be viewed as being excited by the incident magnetic flux linking the two conductors and the incident electric flux terminating on the two conductors that give rise to distributed voltage and current sources on the line. This method of analysis follows the developments of Taylor [7.1] and is referred to as the *Taylor approach*.

2. The problem can be considered to be an EM scattering process in which the tangential incident *E*-field along the conductors can be viewed as distributed voltage sources exciting the transmission line. This analysis has been developed by Agrawal [7.2] and is called the *Agrawal method*.
3. The line also can be considered to be excited only by the incident *B*-field components, which give rise only to distributed current sources on the line. This method has been developed by Rashidi [7.3].

Each of these coupling formulations gives the same response for the transmission line if they are used properly [7.4]. There are subtle differences in these techniques, however, and this has contributed to confusion about field excitation of lines. In this chapter we examine the first two of these formulations and illustrate typical transmission line responses to an incident plane-EM-wave excitation.

The most important aspect of this subject is to realize that transmission line theory does *not* provide a complete solution for the excitation of a line by an incident EM field. It gives only an approximate solution [7.5]. To illustrate this point, consider the hypothetical EM scattering problem shown in Figure 7.1. An *incident* EM field E^{inc} is assumed to illuminate a perfectly conducting body. A *scattered* electric field E^{sca} is created in such a way that the *sum* of the incident and scattered tangential *E*-fields is zero on the surface of the scatterer: $E^{inc}_{tan} + E^{sca}_{tan} = 0$. This scattered field is sustained by an induced current density $\mathbf{J}_s$ (and a companion surface charge density ρ_s) that exist on the surface of the scatterer.

The behavior of the induced surface current depends on the nature of the exciting field as well as on the shape and composition of the scattering body. Consider next the case of the conducting body being deformed into an isolated two-wire line as shown in Figure 7.2. The induced currents in each wire of the transmission line and in the two loads at the ends of the line must

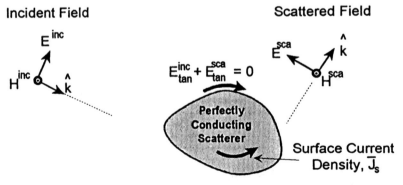

FIGURE 7.1 General EM scattering problem. (From [7.5]. Reprinted by permission from Academic Press, Inc.)

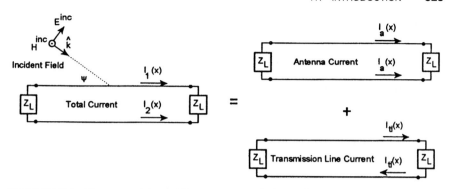

FIGURE 7.2 Decomposition of line current and charge into antenna and transmission line components. (From [7.5]. Reprinted by permission from Academic Press, Inc.)

also create a scattered E-field so that the total tangential E-field over all conductors is again zero.

The induced current in the line has two components: an "antenna mode" current I_a and a "transmission line mode" current I_{tl} [7.6,7.7]. The total currents I_1 and I_2 on conductors 1 and 2 can be written as a combination of these modal currents as

$$I_1 = I_a + I_{tl} \qquad I_2 = I_a - I_{tl}$$

Similarly, the per-unit-length charge density on the conductors q' is written as a combination of antenna mode and transmission line mode components as

$$q'_1 = q'_a + q'_{tl} \qquad q'_2 = q'_a - q'_{tl}$$

The transmission line component of the current produces a transverse electromagnetic (TEM) field, which means that there is no longitudinal component of the E- or H-field along the line. A question that frequently arises is: How is the boundary condition $E^{inc}_{tan} + E^{sca}_{tan} = 0$ satisfied along the conductors if the transmission line mode has no longitudinal component? The answer is that the transmission line current I_{tl} forms only *part* of the total induced current on the line. The antenna mode current also exists on the line, and it is the *combination* of these two components that provides the proper E-field boundary conditions on the line and at the ends of the line.

Generally, the use of transmission line models requires that the line length be significantly larger than the separation of the conductors. If this is not the case, the "line" appears more like a loop antenna and the transmission line models described in this chapter are not appropriate. By using these models we are able to compute only the transmission line component of the current. If we desire a knowledge of the *load* responses of the line, these models are usually adequate, because the antenna mode

324 FIELD COUPLING USING TRANSMISSION LINE THEORY

current response is small near the ends of the line [7.6]. However, if we require a calculation of the current *distribution* along the line, the transmission line component of the current is not accurate and a full field analysis may be required.

As an example of the comparison of the antenna and transmission line response components, consider the isolated two-conductor line shown in Figure 7.2 with a length $\mathcal{L} = 30$ m, the wire separation $d = 20$ cm, and equal wire radii of $a = 0.15$ cm. The wires are assumed to be perfectly conducting and the load impedances at the $x = 0$ and $x = \mathcal{L}$ ends are taken to be $Z_{L1} = Z_{L2} = 293\,\Omega$, which is approximately equal to one-half of the characteristic impedance of the line. This line is excited by an incident plane wave that propagates in the plane of the line and strikes the line with an angle of incidence $\psi = 60°$. For a frequency $f = 20$ MHz (wavelength $\lambda = c/f = 15$ m) the ratio $d/\lambda = 0.0133$ is sufficiently small for the transmission line model to be valid (see Section 7.2.1 for details). Figure 7.3 illustrates the magnitudes of the actual wire currents induced in each line, I_1 and I_2, together with the transmission line current component. The curves labeled I_1 and I_2 have been calculated by an integral equation solution using the method of moments, as implemented in the Numerical Electromagnetics Code (NEC) [7.8]. It is seen that the total current in each conductor is much larger than the transmission line component. Note that the currents under discussion here are complex-valued quantities, but the plot in Figure 7.3 shows only the current magnitudes.

Clearly, for this particular case the transmission line solution is not

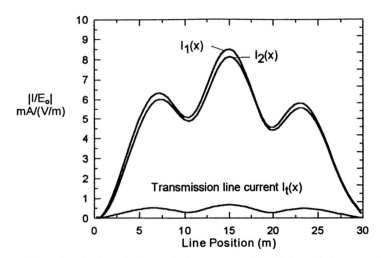

FIGURE 7.3 Magnitudes of the total wire currents I_1 and I_2, and the transmission line current component I_{t1} for the line of Figure 7.2, shown as a function of position along the line. (From [7.5]. Reprinted by permission from Academic Press, Inc.)

adequate for determining the current distribution along the line. Under certain circumstances, however, the transmission line solution can provide accurate results. Consider the case of a single conductor over a perfect ground plane, as shown in Figure 6.4a but with excitation arising from an incident EM plane wave. In this case the antenna mode response is nearly zero due to the structural symmetry of the line, its image in the ground plane, and the excitation field (i.e., the incident field plus a ground-reflected component).

Figure 7.4 illustrates the magnitude of the induced current for a line with length $\mathcal{L} = 30$ m located at a height $h = 10$ cm over a perfect ground. Note that $h = d/2$, where d is the distance between the wire and its image in the ground. As in the preceding example, the termination impedances are taken to be equal to $Z_c/2$, the angle of incidence $\psi = 60°$, and $f = 20$ MHz. The current response labeled "NEC" is again the result of using the Numerical Electromagnetics Code to calculate the actual current distribution on the line. For the case of the conductor near the perfectly conducting ground, we see that the transmission line model provides good accuracy in computing the field-induced currents.

In the remainder of this chapter, we neglect the antenna mode current response for a transmission line and concentrate on the transmission line mode. This provides accurate estimates for the load voltage and current responses, since the antenna mode responses are small at these locations. It must be remembered, however, that if accurate estimates of the current or charge distributions *along* the line are required, the antenna mode contribu-

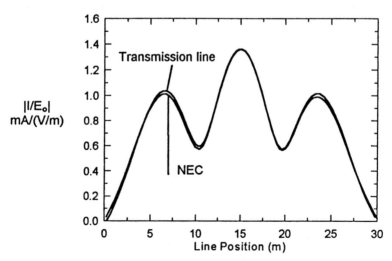

FIGURE 7.4 Magnitudes of the current distributions along a 30-m transmission line over a perfect ground using the transmission line theory and an integral equation solution (NEC). (From [7.5]. Reprinted by permission from Academic Press, Inc.)

tion usually must be considered together with the transmission line contribution, except for those special cases where the antenna mode is negligible due to symmetry.

7.2 THE TWO-WIRE TRANSMISSION LINE

In Chapter 6, a set of partial differential equations, referred to as the telegrapher's equations, was introduced without derivation for the purpose of describing the behavior of the line current and voltage on the line. In this section, the derivation of a similar set of equations will be presented, with the excitation of the line being due to an arbitrary distribution of voltage and/or current sources along the line. It will then be shown how these sources may be related to an incident EM field.

7.2.1 Derivation of the Telegrapher's Equations with an External Excitation

Figure 7.5 illustrates the cross section of a two-wire transmission line illuminated by an incident EM field. A vertically polarized field is considered here, but the development also is valid for horizontally polarized fields. The incident field induces a current and charge on the wires, and the charge results in a potential difference (i.e., a voltage) between the two wires. Initially in this development, the following assumptions are made:

- The two conductors are identical, perfectly conducting wires, each of radius a.
- The medium between the conductors is lossless dielectric.

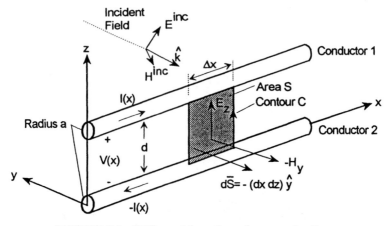

FIGURE 7.5 Differential section of a two-wire line.

- The separation of the wires d is large compared with a ($d \gg a$) and is small compared with the wavelength of the incident field ($d \ll \lambda$).
- Only the transmission line mode currents are considered (i.e., the current I on the top conductor is equal to $-I$ on the lower conductor).

Later in this chapter, the first two conditions are relaxed to permit lines with conductor losses and lossy dielectric material between the conductors (see Sections 7.2.1.3 and 7.2.1.4).

7.2.1.1 First Telegrapher's Equation.
Following the development of Taylor [7.1], the starting point for the development of the first of the two telegrapher's equations for the field-excited transmission line is Maxwell's equation,

$$\nabla \times \mathbf{E} = -j\omega\mu_0 \mathbf{H} \tag{7.1}$$

for the total electric and magnetic flux density fields. In this development it is assumed that the material surrounding the wires in nonmagnetic, with permeability μ equal to that of free space with $\mu = \mu_0 = 4\pi \times 10^{-7}$ H/m.

The total E and H fields may be decomposed into three different components:

1. The incident fields (E^{inc} and H^{inc}) which exist in the absence of the transmission line
2. The scattered fields (E^{sca} and H^{sca}) from the line which are created by the currents and charges flowing on the conductors
3. A quasi static field produced by the separation of charge and current around the circumference of the wire

The latter component of the fields is described in more detail in [7.9] and is important only when the wire radius becomes comparable to the conductor separation. In the present development this quasistatic term is negligible since we assume that $a \ll d$. Consequently, only the (E^{inc}, H^{inc}) and (E^{sca}, H^{sca}) components are used in the analysis.

To develop equations for the induced line current and voltage in terms of the primary excitation fields E^{inc} and H^{inc}, Stokes' theorem [7.9] is used. This theorem states that for a vector field $\mathbf{F}$, the following integral relationship holds:

$$\int_C \mathbf{F} \cdot d\mathbf{l} = \iint_S \nabla \times \mathbf{F} \cdot d\mathbf{S}$$

where C denotes a closed contour enclosing an area S, as shown in Figure 7.5. Letting $\mathbf{F}$ represent the E-field and applying this expression to Eq. (7.1) yields

$$\oint_C \mathbf{E} \cdot d\mathbf{l} = -j\omega\mu_0 \iint_S \mathbf{H} \cdot d\mathbf{s} \tag{7.2}$$

Notice that the contour in Figure 7.5 is defined in the counterclockwise direction, and the element of area dS is positive in the $-y$ direction. Equation (7.2) can be evaluated over the differential area dS as

$$\int_0^d [E_z(x + \Delta x, z) - E_z(x, z)]\, dz - \int_x^{x+\Delta x} [E_x(x, d) - E_x(x, 0)]\, dx$$

$$= -j\omega\mu_0 \int_0^d \int_x^{x+\Delta x} -H_y\, dx\, dz \tag{7.3}$$

The field quantities in Eq. (7.3) are the total fields. Since $d \ll \lambda$, the total line-to-line voltage at a position x can be defined in the quasistatic sense as

$$V(x) = -\int_0^d E_z(x, z)\, dz \tag{7.4}$$

The negative sign in this expression arises from the convention that if the voltage of the line at $z = d$ is taken to be positive with respect to the ground, the corresponding z component of the E-field is in the $-z$ direction.

On the perfectly conducting wires, the total tangential E-fields, $E_x(x, d)$ and $E_x(x, 0)$, are zero. Dividing Eq. (7.3) by Δx and taking the limit as Δx approaches zero gives the following differential equation for the line voltage:

$$\frac{dV(x)}{dx} = -j\omega\mu_0 \int_0^d H_y(x, z)\, dz$$

$$= -j\omega\mu_0 \int_0^d H_y^{\text{inc}}(x, z)\, dz - j\omega\mu_0 \int_0^d H_y^{\text{sca}}(x, z)\, dz \tag{7.5}$$

where the total H-field has been decomposed into the incident and scattered components.

The last integral in Eq. (7.5) represents magnetic flux between the two conductors produced by the currents flowing $I(x)$ in each of the conductors. Assuming that the current is uniformly distributed around the circumference of the wire (since $d \gg a$) and that the line separation is electrically small ($d \ll \lambda$), the magnetic flux density produced by this current can be calculated using the Biot–Savart law and the integral evaluated analytically. This results in a linear relationship between the magnetic flux linking the conductors and the line current. The proportionality constant between the flux Φ and the current I is the per-unit-length inductance of the transmission line L' and is defined through the expression [7.10]

$$\Phi(x) = \mu_0 \int_0^d H_y^{sca}(x, z) \, dz = L'I(x) \tag{7.6a}$$

where L' is evaluated from Eq. (6.150) to be

$$L' \approx \frac{\mu_0}{\pi} \ln \frac{d}{a} \quad \text{H/m} \tag{7.6b}$$

Inserting this inductance term into Eq. (7.5), the first telegrapher's equation for the case of a distributed field excitation of the line becomes

$$\frac{dV(x)}{dx} + j\omega L'I(x) = V'_{S1}(x) \tag{7.7}$$

where the distributed voltage source is

$$V'_{S1}(x) = -j\omega\mu_0 \int_0^d H_y^{inc}(x, z) \, dz \tag{7.8}$$

This term arises from the incident magnetic flux linking the two conductors. Note that if the source term is zero, Eq. (7.7) is the same as the frequency-domain telegrapher's equation in Eq. (6.7a) describing the v–i relationships on an unexcited wire.

7.2.1.2 Second Telegrapher's Equation. Assuming that the material surrounding the lines is a lossless dielectric with permittivity $\epsilon = \epsilon_r \epsilon_0$ with $\epsilon_0 = 8.854 \times 10^{-12}$ F/m, the second telegrapher's equation can be derived from the Maxwell equation

$$\nabla \times \mathbf{H} = j\omega\epsilon \mathbf{E} + \mathbf{J} \tag{7.9}$$

Noting that for a closed surface S, Stokes theorem applied to a vector function $\mathbf{F}$ becomes [7.11]

$$\oint_S \nabla_s \times \mathbf{F} \cdot d\mathbf{S} = 0$$

Letting $\mathbf{F}$ represent the H-field and using Eq. (7.9) with the closed surface S surrounding one of the conductors as shown in Figure 7.6, the following relationship is obtained:

$$I(x + \Delta x) - I(x) + j\omega\epsilon \iint_{S_1} E_r r \, d\phi \, dx = 0 \tag{7.10}$$

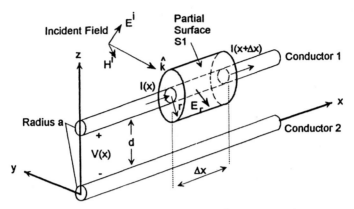

FIGURE 7.6 Closed surface surrounding one conductor.

where E_r is the total radial E-field in the vicinity of the wire and the partial surface S_1 is the cylindrical surface indicated in Figure 7.6.

The total E-field again can be separated into the incident and scattered components. Upon dividing by Δx and taking the limits as $r \to a$ and $\Delta x \to 0$, Eq. (7.10) becomes

$$\frac{dI(x)}{dx} + j\omega\epsilon \int_0^{2\pi} E_r^{sca}(x) a\, d\phi + j\omega\epsilon \int_0^{2\pi} E_r^{inc}(x) a\, d\phi = 0 \qquad (7.11)$$

Since $a \ll d$, the E-field in the vicinity of the wire can be assumed to be independent of the angle ϕ around the wire. Consequently, the first integral in Eq. (7.11) becomes

$$j\omega\epsilon \int_0^{2\pi} E_r^{sca}(x) a\, d\phi = j\omega 2\pi a \epsilon E_r^{sca}(x) = j\omega q'(x) \qquad (7.12)$$

where $q'(x)$ is the linear charge density along the conductor.

The second integral in Eq. (7.11) involving the incident field is identically zero because there is no free change in the vicinity of the wire with the wire removed. Consequently, the second telegrapher's equation becomes

$$\frac{dI(x)}{dx} + j\omega q'(x) = 0 \qquad (7.13)$$

This is simply the one-dimensional equation of continuity relating current and charge along the wire. It is customary to express this equation in terms of a voltage by introducing a per-unit-length capacitance C' of the line, which relates the line charge to the *scattered* component of the line voltage as

$$q'(x) = C' V^{sca}(x) \qquad (7.14)$$

It is important to note that the voltage in Eq. (7.14) is not the *total* line-to-line voltage but only the component due to the charge on the line. For the two-wire line under discussion here, the per-unit-length capacitance is obtained from a two-dimensional static calculation as discussed in Chapter 6 and has the value

$$C' = \frac{\pi \epsilon}{\ln(d/a)} \quad (a \ll d) \quad \text{F/m}$$

Remembering that the *total* line-to-line voltage is given as

$$V(x) = V^{inc}(x) + V^{sca}(x) = -\int_0^d E_z^{inc}(x, z)\, dz + V^{sca}(x)$$

Eqs. (7.13) and (7.14) can be combined to give the second telegrapher's equation,

$$\frac{dI(x)}{dx} + j\omega C' V(x) = I'_{S1}(x) \tag{7.15}$$

which contains a per-unit-length current source $I'_{S1}(x)$ given by

$$I'_{S1}(x) = -j\omega C' \int_0^d E_z^{inc}(x, z)\, dz \tag{7.16}$$

This second telegrapher's equation in Eq. (7.15) is equivalent to Eq. (6.7b) when there is no excitation present.

The source terms in Eqs. (7.8) and (7.15) can be thought of as arising from a set of distributed voltage and current generators located along the length of the line, as depicted in Figure 7.7. It is this set of equations,

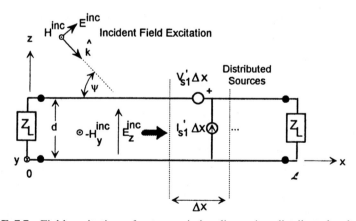

FIGURE 7.7 Field excitation of a transmission line using distributed voltage and current sources (i.e., the Taylor approach). (From [7.5]. Reprinted by permission from Academic Press, Inc.)

together with a specification of the voltage and current boundary conditions at the ends of the line, that forms the Taylor approach for solving for the line voltage and current.

7.2.1.3 Modification of the Telegrapher's Equations for a Finitely Conducting Wire.

If the transmission line wire is not perfectly conducting, the total tangential electric field on the conductor is not zero. In this case the telegrapher equation (7.7) can be modified to take into account the presence of a finite electrical conductivity in the wire. This is done by introducing a surface impedance approximation which relates the *total* local tangential electric field on the surface of the wire to the current flowing on the wire at the same point [7.12]. This relationship is expressed as

$$E_x(x) = -\frac{Z_{cw}}{2\pi a} \frac{I_0(\gamma_w a)}{I_1(\gamma_w a)} I(x) \equiv Z'_w(x) I(x) \qquad (7.17)$$

where Z'_w is the per-unit-length impedance of the wire. The terms $I_0(\cdot)$ and $I_1(\cdot)$ are modified Bessel functions, Z_{cw} is the wave impedance in the conducting material comprising the wire given by

$$Z_{cw} = \sqrt{\frac{j\omega\mu_w}{\sigma_w + j\omega\epsilon_w}}$$

and the term γ_w is the propagation constant in the wire material,

$$\gamma_w = \sqrt{j\omega\mu_w(\sigma_w + j\omega\epsilon_w)}$$

The electrical conductivity and dielectric constant of the wire are denoted by σ_w and ϵ_w. In this discussion, the wire is assumed to be magnetic with a permittivity μ_w.

Various approximations to this wire impedance are possible [7.13]. At low frequencies where $|\gamma_w a| \ll 1$, this impedance is

$$Z'_w \text{ (low frequency)} \approx \frac{1}{\pi a^2 \sigma_w} + j\omega \frac{\mu_w}{8\pi}$$

and at high frequencies, where $|\gamma_w a| \gg 1$, this impedance becomes

$$Z'_w \text{ (high frequency)} \approx \frac{1+j}{2\pi a} \sqrt{\frac{\omega\mu_w}{2\sigma_w}}$$

With the introduction of this wire impedance in the expressions for $E_x(x,d)$ and $E_x(x,0)$ in Eq. (7.3) and carrying out the same steps as done previously, the following telegrapher's equation for the lossy line results:

$$\frac{dV(x)}{dx} + Z'I(x) = V'_{S1}(x) \qquad (7.18)$$

where the per-unit-length impedance term Z' contains the original external inductance term $j\omega L'$ as well as the Z'_w contribution from each of the two wires of the transmission line. This total per-unit-length impedance is given by

$$Z' = j\omega L' + 2Z'_w \qquad (7.19)$$

For this addition of conductor loss, the distributed voltage source $V'_{S1}(x)$ is unmodified and is given again by Eq. (7.8).

7.2.1.4 Modification for a Lossy Medium Surrounding the Line.
If the material surrounding the transmission line has a finite conductivity σ_m and dielectric constant ϵ_m, there can be a conduction current flowing between the two lines. As discussed in [7.13], this can be modeled as a shunt per-unit-length conductance G', which has a value related to the per-unit-length capacitance by

$$G' = \frac{\sigma_m}{\epsilon_m} C'$$

Hence the resulting second telegrapher's equation becomes

$$\frac{dI(x)}{dx} + Y'V(x) = I'_{S1}(x) \qquad (7.20)$$

where the per-unit-length admittance of the line is the parallel combination of the G' and C' elements

$$Y' = j\omega C' + G' = \left(j\omega + \frac{\sigma_m}{\epsilon_m} \right) C' \qquad (7.21)$$

For this case, the distributed current source I'_{S1} is again given by Eq. (7.16).

7.2.2 Alternative Forms of the Telegrapher's Equations

7.2.2.1 Total Voltage Formulation.
For the case of a field-excited transmission line, the telegrapher's equations (7.18) and (7.20) for the line current $I(x)$ and the total (or observable) line-to-line voltage $V(x)$ are similar to the homogeneous telegrapher's equations in Eq. (6.9). However, they contain two source terms related to the incident field. These equations can be written conveniently in matrix form as

$$\frac{d}{dx}\begin{bmatrix} V(x) \\ I(x) \end{bmatrix} + \begin{bmatrix} 0 & Z' \\ Y' & 0 \end{bmatrix}\begin{bmatrix} V(x) \\ I(x) \end{bmatrix} = \begin{bmatrix} V'_{S1} \\ I'_{S1} \end{bmatrix} \qquad (7.22)$$

with the source terms being defined in Eqs. (7.8) and (7.16).

The solution to differential equations of this form are known to consist of a homogeneous solution (i.e., a solution with zero forcing function) plus a forced solution [7.14]. To obtain a unique solution for any particular problem, it is necessary to add appropriate boundary conditions relating V and I at the ends of the line. If the line is infinite in length, it is sufficient to require that the line voltage and current appear at outward-propagating traveling waves. For a finite line of length $\mathscr{L}$, terminated in load impedances Z_1 and Z_2 at each end, as depicted in Figure 7.7, the following relationships must be included with Eq. (7.22) for a unique solution:

$$V(0) = -Z_1 I(0) \quad \text{and} \quad V(\mathscr{L}) = Z_2 I(\mathscr{L}) \tag{7.23}$$

Note that the negative sign in Eq. (7.23) arises from the definition of positive current flow along the line.

7.2.2.2 Scattered Voltage Formulation.
An alternative form of the telegrapher's equations (7.22) is possible by considering the line current $I(x)$ and the *scattered* line voltage $V^{\text{sca}}(x)$ as the unknown quantities. This form of the solution is described by Agrawal [7.2]. Separating the incident and scattered field components in Eq. (7.3) results in the following expression:

$$\int_0^d [E_z^{\text{inc}}(x + \Delta x, z) - E_z^{\text{inc}}(x, z)]\, dz + \int_0^d [E_z^{\text{sca}}(x + \Delta x, z) - E_z^{\text{sca}}(x, z)]\, dz$$

$$- \int_x^{x+\Delta x} [E_x(x, d) - E_x(x, 0)]\, dx \tag{7.24}$$

$$= -j\omega\mu_0 \int_0^d \int_x^{x+\Delta x} -H_y^{\text{inc}}\, dx\, dz - j\omega\mu_0 \int_0^d \int_x^{x+\Delta x} -H_y^{\text{sca}}\, dx\, dz$$

Note that the E_x components in Eq. (7.24) are still the total field quantities, which are related to the line current through Eq. (7.17). Using the definitions of the scattered voltage through Eq. (7.4), the line inductance in Eq. (7.6a), and taking the limit as $\Delta x \to 0$, Eq. (7.24) becomes

$$\frac{dV^{\text{sca}}(x)}{dx} + j\omega L' I(x)$$

$$= -j\omega\mu_0 \int_0^d H_y^{\text{inc}}(x, z)\, dz + \lim_{\Delta x \to 0} \int_0^d [E_z^{\text{inc}}(x + \Delta x, z) - E_z^{\text{inc}}(x, z)]\, dz \tag{7.25}$$

By applying the Stokes theorem to the *incident* field components of Eq.

(7.25), we find that

$$-j\omega\mu_0 \int_0^d H_y^{\text{inc}}(x, z) \, dz + \lim_{\Delta x \to 0} [E_z^{\text{inc}}(x + \Delta x, z) - E_z^{\text{inc}}(x, z)] \, dz$$
$$= E_x^{\text{inc}}(x, d) - E_x^{\text{inc}}(x, 0)$$

and using this last expression in Eq. (7.25), the first telegrapher's equation becomes

$$\frac{dV^{\text{sca}}(x)}{dx} + j\omega L'I(x) = V'_{S2}(x) \tag{7.26a}$$

or, for a lossy line,

$$\frac{dV^{\text{sca}}(x)}{dx} + Z'I(x) = V'_{S2}(x) \tag{7.26b}$$

where the per-unit-length line impedance Z' is given by Eq. (7.19). The distributed source term $V'_{S2}(x)$ is different from that in Eq. (7.8) and is given by the difference of the tangential E-fields on the upper and lower wires:

$$V'_{S2}(x) = E_x^{\text{inc}}(x, d) - E_x^{\text{inc}}(x, 0) \tag{7.27}$$

The second telegrapher's equation is derived from Eqs. (7.13) and (7.14), which yields the following:

$$\frac{dI(x)}{dx} + j\omega C' V^{\text{sca}}(x) = 0 \tag{7.28a}$$

or if a line with lossy dielectric is considered, the equation becomes

$$\frac{dI(x)}{dx} + Y'V^{\text{sca}}(x) = 0 \tag{7.28b}$$

where Y' is defined by Eq. (7.21).

Equations (7.26b) and (7.28b) can again be put into matrix form as

$$\frac{d}{dx}\begin{bmatrix} V^{\text{sca}}(x) \\ I(x) \end{bmatrix} + \begin{bmatrix} 0 & Z' \\ Y' & 0 \end{bmatrix}\begin{bmatrix} V^{\text{sca}}(x) \\ I(x) \end{bmatrix} = \begin{bmatrix} V'_{S2} \\ 0 \end{bmatrix} \tag{7.29}$$

where now the only element of the source vector is given by Eq. (7.27), and it is equal to the *difference* of the tangential E-fields along the two conductors of the transmission line.

As in the previous form of the telegrapher's equations, it is necessary to impose the appropriate boundary conditions at $x = 0$ and $x = \mathcal{L}$ for a unique solution. In terms of the scattered voltages, Eq. (7.23) becomes

$$V^{sca}(0) = -Z_1 I(0) + \int_0^d E_z^{inc}(0, z)\, dz \qquad (7.30a)$$

$$V^{sca}(\mathcal{L}) = Z_2 I(\mathcal{L}) + \int_0^d E_z^{inc}(\mathcal{L}, z)\, dz \qquad (7.30b)$$

The integral terms in the end conditions of Eq. (7.30) can be thought of as arising from two additional voltage excitations sources at each end of the transmission line. Thus in the Agrawal formulation there are distributed voltage sources along the line given by Eq. (7.27), together with the two lumped sources at the ends given by

$$V_1 = -\int_0^d E_z^{inc}(0, z)\, dz \qquad (7.31a)$$

$$V_2 = -\int_0^d E_z^{inc}(\mathcal{L}, z)\, dz \qquad (7.31b)$$

as depicted in Figure 7.8. By the uniqueness theorem, the results of either formulation must give the same response for the line. Care must be used, however, in remembering that one formulation provides the *total* voltage and the other provides the *scattered* voltage.

7.2.2.3 Numerical Example of the Two Formulations.
As a comparison of the induced line responses as calculated by the Taylor and Agrawal formulations, [7.15] has examined the problem of a cloud–ground lightning discharge near a transmission line. In this reference, the load voltage

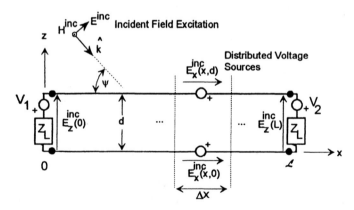

FIGURE 7.8 Field excitation of a transmission line using only voltage sources (i.e., the Agrawal approach). (From [7.5]. Reprinted by permission from Academic Press, Inc.)

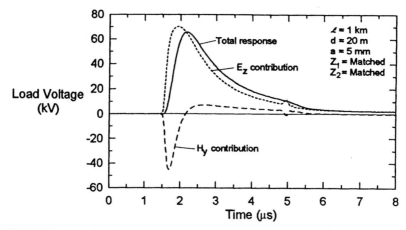

FIGURE 7.9 Load voltage on a 1-km line excited by a nearby lightning strike as computed by the Taylor formulation. (Data courtesy of F. Rashidi.)

responses are calculated. The line had a total length of $\mathcal{L} = 1000$ m, an effective conductor separation of 20 m, and conductor radii of 5 mm. The line was excited by a cylindrical incident EM field produced by a z-directed traveling wave lightning current source located 50 m away from the midpoint of the line.

Figure 7.9 illustrates the load voltage response for this line as computed with the Taylor formulation. The total response has a component due to the magnetic field H_y excitation and another due to the vertical electric field E_z. Both of these partial responses and the complete response are shown in the figure. For the Agrawal formulation, Figure 7.10 shows the voltage response, again illustrating the partial contributions to the total response. In this case,

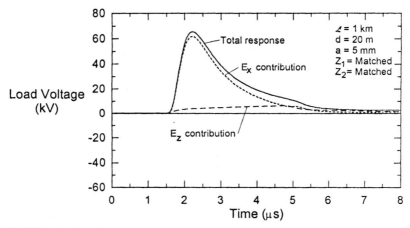

FIGURE 7.10 Load voltage on a 1-km line excited by a nearby lightning strike as computed by the Agrawal formulation. (Data courtesy of F. Rashidi.)

the tangential E-field components (E_x) along the line, and the components of E_z tangential to the loads on the line form the two sources. In comparing these responses, it is evident that both formulations yield the same total voltage response, yet the contributions from the partial sources are completely different. Thus it is evident that it does not matter which formulation is used in an analysis, as long as both of the excitation contributions are included.

7.2.3 Solution for the Line Current and Voltage

The telegrapher's equations (7.22) or (7.29), together with their boundary conditions, can be solved in a number of different ways. One way is to use the expressions for the line voltage and current for a point voltage and current source as Green's functions, and integrate over the distribution of sources. For the case of the scattered voltage formulation, only voltage sources exist on the line. The solution for the line current and voltage at an arbitrary location x on the line due to a voltage source at location x_s was given in Eqs. (6.38) and (6.39). Considering a voltage source of unit amplitude, these expressions can be put into the form of a Green's function as [7.5]

$$G_I(x;x_s) = \frac{e^{-\gamma \mathcal{L}}}{2Z_c(1 - \rho_1\rho_2 e^{-2\gamma \mathcal{L}})} [e^{-\gamma(x_> - \mathcal{L})} - \rho_2 e^{\gamma(x_> - \mathcal{L})}](e^{\gamma x_<} - \rho_1 e^{-\gamma x_<})$$

(7.32a)

$$G_V(x;x_s) = \frac{\delta e^{-\gamma \mathcal{L}}}{2(1 - \rho_1\rho_2 e^{-2\gamma \mathcal{L}})} [e^{-\gamma(x_> - \mathcal{L})} + \delta\rho_2 e^{\gamma(x_> - \mathcal{L})}](e^{\gamma x_<} - \delta\rho_1 e^{-\gamma x_<})$$

(7.32b)

where the standard notation $x_<$ represents the *smaller* of x or x_s, and $x_>$ represents the *larger* of x or x_s. The expression for G_I represents the Green's function for the line current, and G_V is for the line voltage.

The term δ is a sigmoidal function defined as $\delta = 2U(x - x_s) - 1$, where $U(x - x_s)$ is the unit Heaviside step function. Thus $\delta = 1$ for $x > x_s$ and $= -1$ for $x < x_s$. In these expressions, $\gamma = \sqrt{Z'Y'}$ is the complex propagation constant along the transmission line, and $Z_c = \sqrt{Z'/Y'}$ is the line's characteristic impedance. As in Eq. (6.42), the terms ρ_1 and ρ_2 in Eq. (7.32) are the voltage reflection coefficients at the loads of the transmission line given by Eq. (6.37).

Using the voltage source terms of Eqs. (7.27) and (7.31) for an arbitrary incident field exciting the line, the transmission line component of the line current and *scattered* voltage can be written as the following integrals of the

Green's functions:

$$I(x) = \int_0^{\mathcal{L}} G_I(x; x_s) V'_s \, dx_s - G_I(x; 0) V_1 + G_I(x; \mathcal{L}) V_2 \qquad (7.33)$$

$$V^{\text{sca}}(x) = \int_0^{\mathcal{L}} G_V(x; x_s) V'_s \, dx_s - G_V(x; 0) V_1 + G_V(x; \mathcal{L}) V_2 \qquad (7.34)$$

The total voltage at the location x is then determined from the scattered voltage by adding the contribution from the incident field as

$$V(x) = V^{\text{sca}}(x) + V^{\text{inc}}(x) = V^{\text{sca}}(x) - \int_0^d \mathbf{E}^{\text{inc}} \cdot d\hat{z}$$

7.2.4 Solution for the Load Currents and Voltages: The BLT Equation

As we have seen earlier, the transmission line response can be considerably different from the actual current on a two-wire line except at the load ends, where the antenna mode component becomes small. Consequently, we are usually interested only in the *terminal* responses of the line when we use the transmission line model. The BLT equations (6.42) provide a compact expression for the load current and voltage due to a lumped voltage and current source.

Considering the total voltage (Taylor) formulation, the load current and voltage also can be expressed as integrals over the distributed sources on the line. Integrating over the x_s coordinate in Eqs. (6.42) with the distributed field sources [7.15] provides the following BLT equation for the load currents and voltages:

$$\begin{bmatrix} I(0) \\ I(\mathcal{L}) \end{bmatrix} = \frac{1}{Z_c} \begin{bmatrix} 1-\rho_1 & 0 \\ 0 & 1-\rho_2 \end{bmatrix} \begin{bmatrix} -\rho_1 & e^{\gamma \mathcal{L}} \\ e^{\gamma \mathcal{L}} & -\rho_2 \end{bmatrix}^{-1} \begin{bmatrix} S_1 \\ S_2 \end{bmatrix} \qquad (7.35a)$$

$$\begin{bmatrix} V(0) \\ V(\mathcal{L}) \end{bmatrix} = \begin{bmatrix} 1+\rho_1 & 0 \\ 0 & 1+\rho_2 \end{bmatrix} \begin{bmatrix} -\rho_1 & e^{\gamma \mathcal{L}} \\ e^{\gamma \mathcal{L}} & -\rho_2 \end{bmatrix}^{-1} \begin{bmatrix} S_1 \\ S_2 \end{bmatrix} \qquad (7.35b)$$

The source vector is given by [6.8] as

$$\begin{pmatrix} S_1 \\ S_2 \end{pmatrix} = \begin{pmatrix} \frac{1}{2} \int_0^{\mathcal{L}} e^{\gamma x_s} [V'_{s1}(x_s) + Z_c I'_{s1}(x_s)] \, dx_s \\ -\frac{1}{2} \int_0^{\mathcal{L}} e^{\gamma(\mathcal{L}-x_s)} [V'_{s1}(x_s) - Z_c I'_{s1}(x_s)] \, dx_s \end{pmatrix} \qquad (7.36)$$

where the distributed voltage and current sources V'_{s1} and I'_{s1} are given by Eqs. (7.8) and (7.16), respectively.

Alternatively, if the scattered voltage (Agrawal) formulation is used, the load currents and the *total* voltages at the loads are given by Eqs. (7.35) with different source terms

$$\begin{pmatrix} S_1 \\ S_2 \end{pmatrix} = \begin{pmatrix} \frac{1}{2}\int_0^{\mathcal{L}} e^{\gamma x_s} V'_{s2}(x_s)\,dx_s - \frac{V_1}{2} + \frac{V_2}{2}e^{\gamma \mathcal{L}} \\ -\frac{1}{2}\int_0^{\mathcal{L}} e^{\gamma(\mathcal{L}-x_s)} V'_{s2}(x_s)\,dx_s + \frac{V_1}{2}e^{\gamma \mathcal{L}} - \frac{V_2}{2} \end{pmatrix} \quad (7.37)$$

Notice that in the BLT equations for the voltage, the contribution from the incident field has been included, so that Eq. (7.35b) always provides the total voltage response, regardless of the formulation used. The distributed voltage source $V'_{s2}(x)$ is now given by Eq. (7.27), and the two lumped voltage sources V_1 and V_2 occurring at the ends of the line are related to the integral terms in Eq. (7.31).

Either one of the BLT source representations in Eqs. (7.36) or (7.37) can be used for computing the terminal responses. As we have seen, both give the same answer if the various components of the sources are included properly. However, a common mistake in using the tangential E-field source model [i.e., the Agrawal formulation of Eq. (7.37)] is to neglect the two lumped voltage sources at the ends of the line.

7.2.5 Load Responses for Plane-Wave Excitation

The BLT equations in (7.35) can be used for studying the behavior of the terminal responses for many different types of excitation fields. For example, if a radiating antenna is located near the transmission line, the quasistatic, induction, and far-field components produced by the antenna will all excite the line. These field components must be integrated along the line according to the source terms in either Eqs. (7.36) or (7.37), and the terminal responses will be determined.

In the discussion that follows, we use the scattered field (Agrawal) formulation and evaluate Eq. (7.37) for the excitation terms, because this requires only a calculation of the E-field components. Unfortunately, for an arbitrary excitation field, the required integrations cannot be performed analytically and a relatively time-consuming numerical integration process must be used. However, for the special case of the excitation being a plane wave, the integrations can be performed and closed-form expressions for the load responses are possible.

Consider the excitation field to be an incident plane wave, as shown in Figure 7.11. This field is described by angles of incidence ψ and ϕ shown in the figure, as well as a polarization angle α, which defines the E-field vector direction relative to the vertical plane of incidence. For the vertically polarized incident field, the E-field vector lies in the plane of incidence with the angle $\alpha = 0°$. The horizontally polarized incident field has the E-field perpendicular to the plane of incidence and $\alpha = 90°$ for this case. We denote the magnitude of the incident E-field by E_0.

For this special class of excitation field, the distributed voltage source

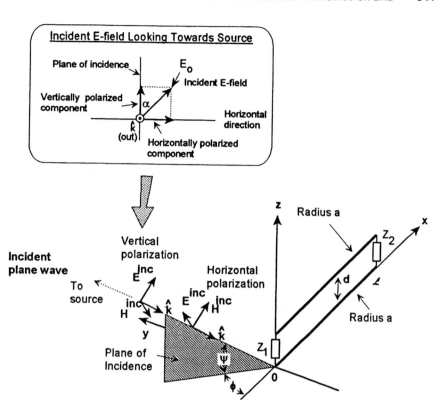

FIGURE 7.11 Isolated two-wire line excited by an incident plane wave.

$V'_{S2}(x)$ of Eq. (7.36) is expressed as

$$V'_{S2}(x) = E_x^{\text{inc}}(x, d) - E_x^{\text{inc}}(x, 0)$$
$$= E_0(\cos \alpha \sin \psi \cos \phi + \sin \alpha \sin \phi)(e^{jkd \sin \psi} - 1)e^{-jkx \cos \psi \cos \phi}$$
$$\approx E_0(\cos \alpha \sin \psi \cos \phi + \sin \alpha \sin \phi)jkd \sin \psi e^{-jkx \cos \psi \cos \phi} \quad (7.38)$$

where $k = \omega/c$ and the approximation $kd \ll 1$ has been used. As a consequence of the exponential dependence on the distance x, the two integrals indicated in Eq. (7.37) can be evaluated analytically. Furthermore, the two lumped voltage sources V_1 and V_2 in Eq. (7.37) are given by

$$V_1 = -\int_0^d \mathbf{E}^{\text{inc}}(0, 0, z) \cdot d\hat{z} \approx -E_z^{\text{inc}}(0, 0, 0)d = -E_0 d \cos \psi \cos \alpha$$

$$V_2 = -\int_0^d \mathbf{E}^{\text{inc}}(\mathcal{L}, 0, z) \cdot d\hat{z} \approx -E_z^{\text{inc}}(\mathcal{L}, 0, 0)d = -E_0 d \cos \psi \cos \alpha e^{-jk\mathcal{L} \cos \psi \cos \phi}$$

$$= V_1 e^{-jk\mathcal{L} \cos \psi \cos \phi} \quad (7.39b)$$

With these expressions, the resulting source vector in Eq. (7.39) is given in terms of the incident plane-wave parameters as

$$\begin{bmatrix} S_1 \\ S_2 \end{bmatrix} = \begin{bmatrix} -\frac{1}{2}\left(\frac{E_0(\cos\alpha\sin\psi\cos\phi + \sin\alpha\sin\phi)jkd\sin\psi}{\gamma - jk\cos\psi\cos\phi} - E_0 d\cos\psi\cos\alpha\right)(1 - e^{(\gamma - jk\cos\psi\cos\phi)\mathcal{L}}) \\ -\frac{1}{2}e^{\gamma\mathcal{L}}\left(\frac{E_0(\cos\alpha\sin\psi\cos\phi + \sin\alpha\sin\phi)jkd\sin\psi}{\gamma + jk\cos\psi\cos\phi} + E_0 d\cos\psi\cos\alpha\right)(1 - e^{-(\gamma + jk\cos\psi\cos\phi)\mathcal{L}}) \end{bmatrix}$$

(7.40)

Approximate analytical expressions for the load responses of a two-wire transmission line illuminated by a plane wave of EMP type described in Section 7.2.6.2 in the frequency and time domains have been developed in [7.16].

7.2.6 Examples of Line Responses

7.2.6.1 Frequency-Domain Responses.
As an example of the frequency-domain response calculated by the BLT equation (7.34a) for the two-wire transmission line, consider the line in Figure 7.11 with the following parameters: overall line length $\mathcal{L} = 30$ m, conductor separation $d = 20$ cm, and wire radii $a_1 = a_2 = 0.15$ cm. For this line geometry, the characteristic impedance is $Z_c \approx 586\,\Omega$. The termination impedances at $x = 0$ and $\mathcal{L}$ are set to the values $Z_1 = Z_2 = Z_c/2 \approx 293\,\Omega$. The excitation of this line is provided by a vertically polarized plane-wave field ($\alpha = 0°$), with angles of incidence $\psi = 60°$, $\phi = 0°$. Figure 7.12 shows the magnitude of the complex-valued

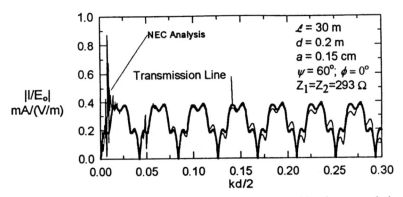

FIGURE 7.12 Current spectrum in load at $x = \mathcal{L}$ computed by the transmission line and integral equation models for a two-wire line. (From [7.5]. Reprinted by permission from Academic Press, Inc.)

frequency spectrum of the current flowing in load 2 at $x = \mathcal{L}$ for the normalized frequency parameter $(kd/2 = \pi fd/c)$ varying from 0 to 0.3. This current is normalized by the E-field magnitude E_0 so it corresponds to the response for a unit incident E-field. The curve marked "transmission line" is from a solution of the BLT equation (7.35a). Notice that this spectrum has periodic resonances due to the constructive interference of the traveling waves on the line at certain frequencies.

Also shown in this figure is a result computed using an integral equation solution (NEC). The latter solution contains both the transmission line modal component as well as higher-order, radiating modes which are excited on the line. Consequently, there are some differences between the two solutions, a fact that must be noted when one uses transmission line models. At very low frequencies, the NEC solution contains a significant amount of noise. This is a well-known problem with this code for loop structures at low frequencies and is not actually part of the solution. At higher frequencies, the difference between the two curves can be attributed to the inadequacies of the transmission line model.

7.2.6.2 Transient Response.
Responses in the time domain can be more useful than the spectral responses in understanding the external field coupling mechanism to a line. Although the BLT equation provides results in the frequency domain, Fourier transform techniques can be used to transform a wideband spectrum into the time domain. As an example, consider the incident E-field to be a double exponential waveform of unit amplitude and vertical polarization ($\alpha = 0°$), described by the expression $E^{inc}(t) = 1.05 \times (e^{-4 \times 10^6 t} - e^{-4.76 \times 10^8 t})$ V/m.

This waveform, when scaled in amplitude to 50 kV/m, is often used to represent a nuclear electromagnetic pulse (NEMP) [7.17,7.18]. It is shown in Figure 7.13 (right-hand scale). Taking the Fourier transform of the excitation

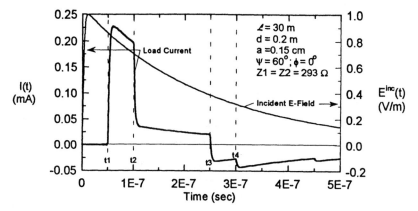

FIGURE 7.13 Plot of the early-time transient excitation waveform (right scale) and the induced transient current (left scale) at $x = \mathcal{L}$ for the two-wire line.

waveform using a standard FFT method and multiplying the resulting spectrum by the response spectrum in Figure 7.12 gives the current spectrum for this double exponential excitation. An inverse FFT can be performed to obtain the transient response shown in Figure 7.13 (left scale).

This current waveform exhibits a number of waveform discontinuities that result from contributions to the current arriving from different parts of the transmission line structure. Figure 7.14 shows several different signal paths that these response contributions take in arriving at the observation location at $x = \mathcal{L}$.

As shown in Figure 7.14, the incident field is assumed to strike the line at $t = 0$ at the $x = 0$ end of the line. At this instant, current is induced on the line at $x = 0$. However, at the $x = \mathcal{L}$ end, there is no response until time $t_1 = (\mathcal{L}/c) \cos \psi = 0.5 \times 10^{-7}$ s, when the wavefront arrives at the observer. Notice that the incident field sweeps across the line with an apparent velocity $v = c/\cos \psi$, which is faster than the speed of light. Of course, any useful information arriving at the observer is constrained to arrive with a velocity of c or less. The next waveform discontinuity arrives at a time $t_2 = \mathcal{L}/c = 1 \times 10^{-7}$ s. This is the time that the observer at $x = \mathcal{L}$ is first aware that the line is finite. Prior to this time, the response is that of a semi-infinite line.

After times t_1 and t_2, there are multiple reflections along the transmission line with round-trip time intervals of $2\mathcal{L}/c = 2 \times 10^{-7}$ s. Thus the response at time t_3 occurs at $t_1 + 2 \times 10^{-7}$ s or $t_3 = 2.5 \times 10^{-7}$ s, and this corresponds to one round-trip from the $x = \mathcal{L}$ end. The response at time t_4 corresponds to $1\frac{1}{2}$ round trips from the $x = 0$ end, and is $t_4 = 3 \times 10^{-7}$ s. Each of these waveform times is clearly seen in Figure 7.13, and this provides a good way of checking calculations to ensure that they are being done correctly.

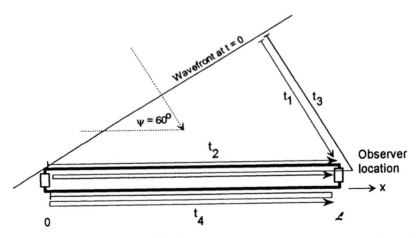

FIGURE 7.14 Early-time contributions to the current response at $x = \mathcal{L}$ due to the incident field excitation.

7.3 SINGLE LINE OVER A PERFECTLY CONDUCTING GROUND PLANE

Transient results for such transmission line problems are generally much easier to interpret than are frequency-domain responses.

7.3 SINGLE LINE OVER A PERFECTLY CONDUCTING GROUND PLANE

Frequently, a transmission line structure will consist of a single conductor near a conducting ground plane that serves as the return conductor. As discussed earlier and illustrated in Figure 7.3, the current excited by an incident field on this line consists primarily of the transmission line mode, with the antenna mode being small due to the symmetry of the excitation. In this section we discuss the derivation of the telegrapher's equations for this single conductor.

Figure 7.15 illustrates the geometry of an infinite line over a conducting ground plane. For the moment we assume that the ground is perfectly conducting. In Chapter 8, the effects of having a lossy ground as a return conductor are discussed. In Figure 7.15 the conductor is assumed to have a radius a and is at a height h over the ground. The line is illuminated by an incident E-field and the response of the line is the line current $I(x)$ and the line-to-ground plane voltage $V(x)$.

7.3.1 Derivation of the Telegrapher's Equations

The behavior of the induced current and voltage on the line can be analyzed using the same set of telegrapher's equations as for the two-wire line in Section 7.2.1, but with two differences. The *excitation E* and *H* fields (i.e.,

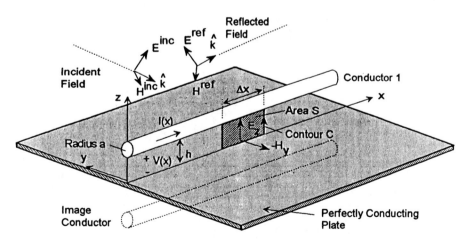

FIGURE 7.15 Geometry of an infinite conductor over a perfectly conducting ground plane.

346 FIELD COUPLING USING TRANSMISSION LINE THEORY

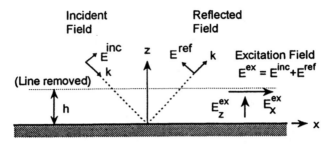

FIGURE 7.16 Excitation E-field for the single line over a perfect conductor. (From [7.5]. Reprinted by permission from Academic Press, Inc.)

the fields existing near the line but with the line removed) now consist of the incident plus ground-reflected field components. This is illustrated in Figure 7.16. On the perfectly conducting ground, the boundary condition for the E-field requires that both $E_x = 0$ and $E_y = 0$. Consequently, only E_z exists. At a height h over the ground, the E_x component of the field is small, but nonzero. Similarly, the boundary condition for the H-field is $H_z = 0$ on the ground surface, with the H_x and H_y components having nonzero values. In addition to these differences in the excitation fields, the per-unit-length parameters of the line are different from those of the isolated two-conductor line.

7.3.1.1 Total Voltage Formulation. The derivation of the telegrapher's equations for this case parallels that of the two-wire case described in Section 7.2.1. The Maxwell curl equation (7.1) can be integrated over the shaded area shown in Figure 7.15. Instead of dividing the H-field into incident and scattered components as done for Eq. (7.5), the field is divided into the *excitation* component $H^{ex} = (H^{inc} + H^{ref})$ and the line-scattered component H^{sca}. Using the same steps in the derivation, this results in the first of the telegrapher's equations

$$\frac{dV(x)}{dx} + j\omega L' I(x) = V'_{S1}(x) \qquad (7.41)$$

where the per-unit-length external inductance L' is now one-half of the value in Eq. (7.6):

$$L' = \frac{\mu_0}{2\pi} \ln \frac{2h}{a} \quad \text{H/m} \qquad (7.42)$$

The distributed voltage source for Eq. (7.40) is given by

$$V'_{S1}(x) = -j\omega\mu_0 \int_0^h H_y^{ex}(x, z)\, dz$$
$$= -j\omega\mu_0 \int_0^h [H_y^{inc}(x, z) + H_y^{ref}(x, z)]\, dz \qquad (7.43)$$

which involves both the incident and the ground-reflected H-fields. The details of this derivation are left as an exercise for the reader.

In an analogous manner, the second telegrapher's equation is derived by defining the *excitation* E-field as $E^{\text{ex}} = E^{\text{inc}} + E^{\text{ref}}$ and then applying Eq. (7.9) to a closed surface surrounding the wire, as shown in Figure 7.7. This again results in the continuity equation (7.13). With a per-unit-length capacitance of the conductor relative to the conducting ground given by

$$C' = \frac{2\pi\epsilon}{\ln(2h/a)} \quad \text{F/m} \tag{7.44}$$

the continuity equation becomes the second telegrapher's equation

$$\frac{dI(x)}{dx} + j\omega C' V(x) = I'_{S1}(x) \tag{7.45}$$

where the distributed current source $I'_{S1}(x)$ is now given in terms of the incident and ground-reflected E-fields as

$$\begin{aligned} I'_{S1}(x) &= -j\omega C' \int_0^h E_z^{\text{ex}}(x, z)\, dz \\ &= -j\omega C' \int_0^h [E_z^{\text{inc}}(x, z) + E_z^{\text{ref}}(x, z)]\, dz \end{aligned} \tag{7.46}$$

Thus, for the case of a single wire over a planar return conductor, the telegrapher's equations (7.41) and (7.45) using the Taylor formulation have the same form as for the two-wire case. They can be put into the matrix of Eq. (7.22), but with different Z' and Y' parameters, and different excitation sources given by Eqs. (7.43) and (7.46). Solution of these telegrapher's equations again requires a knowledge of the termination conditions at the ends of the line.

For this formulation, the view of the excitation provided by Figure 7.7 for the two-conductor line is again valid, with the wire on the bottom representing the ground conductor and the wire on the top representing the physical conductor. Distributed voltage and current sources both exist along the line, and this formulation yields the line current and the total line-to-ground voltage.

7.3.1.2 Scattered Voltage Formulation.
The development of the scattered voltage (Agrawal) version of the telegrapher equations for the single wire over the ground plane also parallels that discussed for the two-wire case in Section 7.2.2. As it is easy to confuse the roles that the various field components play in this formulation, the derivation is presented below.

Applying Stokes' theorem to the Maxwell curl-E equation (7.1) over contour C shown in Figure 7.15, an integral expression similar to Eq. (7.3) is obtained:

$$\int_0^h [E_z(x+\Delta x, z) - E_z(x,z)]\,dz - \int_x^{x+\Delta x} [E_x(x,d) - E_x(x,0)]\,dx$$

$$= -j\omega\mu_0 \int_0^h \int_x^{x+\Delta x} -H_y\,dx\,dz \qquad (7.47)$$

In Eq. (7.47), E and H are the *total* field quantities. These may be decomposed into two parts: the *excitation fields* (E^{ex}, H^{ex}), which exist above the ground when the line is removed, and the *scattered fields* (E^{sca}, H^{sca}), which are produced by the induced currents and charges on the line when it is present. This field decomposition is similar to what was done for Eq. (7.25) for the two-wire case, except that the excitation field was *only* the incident field. Applying this field decomposition to the transverse E and H fields only, Eq. (7.47) becomes

$$\int_0^h [E_z^{ex}(x+\Delta x, z) - E_z^{ex}(x,z)]\,dz + \int_0^h [E_z^{sca}(x+\Delta x, z) - E_z^{sca}(x,z)]\,dz$$

$$- \int_x^{x+\Delta x} [E_x(x,d) - E_x(x,0)]\,dx$$

$$= -j\omega\mu_0 \int_0^h \int_x^{x+\Delta x} -H_y^{ex}\,dx\,dz - j\omega\mu_0 \int_0^h \int_x^{x+\Delta x} -H_y^{sca}\,dx\,dz \qquad (7.48)$$

Note that the third integral in Eq. (7.48), involving $E_x(x,d)$ and $E_x(x,0)$, is still written in terms of the *total* E-field components. On the ground, the total tangential E-field is identically zero, as the ground is assumed to be a perfect conductor. Furthermore, on the surface of the wire, the total E_x-field is also zero for an assumed perfectly conducting wire. Thus $E_x(x,0) = E_x(x,d) = 0$ and the third integral vanishes. Defining the scattered voltage of the line by the integral

$$V^{sca}(x) = -\int_0^h E_z^{sca}(x,z)\,dz$$

and taking the limit of Eq. (7.48) as $\Delta x \to 0$ gives

$$\frac{dV^{sca}(x)}{dx} + j\omega L'I(x) = -j\omega\mu_0 \int_0^h H_y^{ex}(x,z)\,dz$$

$$+ \lim_{\Delta x \to 0} \int_0^h [E_z^{ex}(x+\Delta x, z) - E_z^{ex}(x,z)]\,dz \qquad (7.49)$$

Because the excitation fields E^{ex} and H^{ex} satisfy Maxwell's equations independently, an application of Stokes' theorem to these fields over the contour shown in Figure 7.15 yields the expression

$$\int_0^h [E_z^{ex}(x + \Delta x, z) - E_z^{ex}(x, z)] \, dz - \int_x^{x+\Delta x} [E_x^{ex}(x, d) - E_x^{ex}(x, 0)] \, dx$$

$$= -j\omega\mu_0 \int_0^h \int_x^{x+\Delta x} -H_y^{ex} \, dx \, dz \qquad (7.50)$$

Note that $E_x^{ex}(x, 0) = 0$ on the ground plane because the excitation field must satisfy the same boundary condition on the ground plane as the total field does. Taking the limit of Eq. (7.50) as $\Delta x \to 0$ gives

$$-j\omega\mu_0 \int_0^h H_y^{ex}(x, z) \, dz + \lim_{\Delta x \to 0} \int_0^h [E_z^{ex}(x + \Delta x, z) - E_z^{ex}(x, z)] \, dz = E_x^{ex}(x, d)$$

This expression can be substituted into the right-hand source term of Eq. (7.49) to give the first telegrapher's equation for the Agrawal formulation:

$$\frac{dV^{sca}(x)}{dx} + j\omega L' I(x) = V'_{S2}(x) \qquad (7.51a)$$

The excitation source $V'_{S2}(x)$ is now given by the tangential excitation (incident plus ground-reflected) E-fields along just the single conductor comprising the line:

$$V'_{S2}(x) = E_x^{ex}(x) = E_x^{inc}(x) + E_x^{ref}(x) \qquad (7.51b)$$

The second telegrapher's equation, involving only the scattered voltage on the line, again is derived directly from Eqs. (7.13) and (7.14). It is given by Eq. (7.28a) with the line capacitance C' being given by Eq. (7.44). This equation, together with Eq. (7.51a), can be put into the same matrix form as Eq. (7.29):

$$\frac{d}{dx}\begin{bmatrix} V^{sca}(x) \\ I(x) \end{bmatrix} + \begin{bmatrix} 0 & j\omega L' \\ j\omega C' & 0 \end{bmatrix}\begin{bmatrix} V^{sca}(x) \\ I(x) \end{bmatrix} = \begin{bmatrix} V'_{S2} \\ 0 \end{bmatrix} \qquad (7.52)$$

As in the case of the Agrawal formulation for the two-wire line, the termination conditions relating the total line voltage to the line current at each end of the line must be used in the solution of the telegrapher's equations (7.52). For this formulation, the voltage sources defined in Eqs. (7.31) and used in the end conditions of Eqs. (7.30) will involve the z-directed excitation E-field at the end of the line instead of the incident field. The end sources are

$$V_1 = -\int_0^h E_z^{ex}(0, z)\, dz = -\int_0^h [E_z^{inc}(0, z) + E_z^{ref}(0, z)]\, dz \qquad (7.53a)$$

$$V_2 = -\int_0^h E_z^{ex}(\mathcal{L}, z)\, dz = -\int_0^h [E_z^{inc}(\mathcal{L}, z) + E_z^{ref}(\mathcal{L}, z)]\, dz \qquad (7.53b)$$

Figure 7.17 illustrates the resulting voltage sources for the single-wire line over the ground. It is similar to that of Figure 7.9 in that there are two lumped voltage sources at the ends of the line and distributed voltage sources along the length of the line. It is different from the previous model, however, in that there is *no* voltage source in the ground conductor. This is because the tangential component of the excitation field ($E^{inc} + E^{ref}$) is identically zero on the ground, as discussed previously.

7.3.1.3 Comments on the Line Excitation from an EM Scattering Viewpoint.
The view of the excitation of the single line in Figure 7.17 is consistent with the treatment of a scattering problem for a perfectly conducting object located over a ground plane, as shown in Figure 7.18. In this scattering problem, the incident field induces current (and charge) on both the body and on the ground plane. We are interested primarily in the current on the body.

A rigorous solution for the currents on the body J_{body} and on the ground J_{ground}, can be developed by enforcing the fact that the total tangential E-fields on all conductors must be zero. As shown in the figure, this total field consists of an incident component, a component reflected from the ground plane, and a component scattered from the body itself. The solution

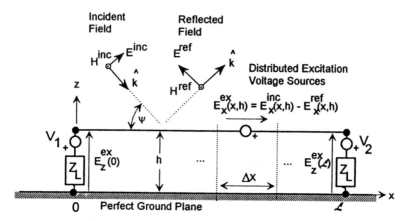

FIGURE 7.17 Equivalent voltage source excitation for the single conductor over a perfect ground.

7.3 SINGLE LINE OVER A PERFECTLY CONDUCTING GROUND PLANE

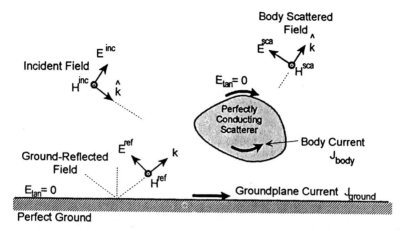

FIGURE 7.18 Scattering problem involving a perfectly conducting body near a perfectly conducting ground.

to the problem is provided by an integral equation having two parts:

$$-E_{\tan}^{\text{inc}}(\mathbf{r}) = \hat{n} \times \left[\iint_{\text{body}} \mathbf{J}_{\text{body}} \cdot \overline{\overline{\Gamma}}_0 \, ds + \iint_{\text{ground}} \mathbf{J}_{\text{ground}} \cdot \overline{\overline{\Gamma}}_0 \, ds \right] \quad (7.54)$$

The first integral is evaluated over the surface of the scattering body, and the second is evaluated over the infinite ground surface. In the integrals, there is a kernel (or Green's function) Γ_0 which is related to the expression of the E-field components produced by a small current source radiating in free space, and the excitation function $E_{\tan}^{\text{inc}}(\mathbf{r})$ is the tangential component of the incident E-field, with the field point $\mathbf{r}$ being located on both the body and the ground plane.

Although this equation can be solved by numerical methods, the infinite extent of the ground plane poses some difficulties. Consequently, it is helpful to introduce a different Green's function Γ_1, which is derived in such a way that the integral over the ground in Eq. (7.54) vanishes and the integral equation becomes

$$-[E_{\tan}^{\text{inc}}(\mathbf{r}) + E_{\tan}^{\text{ref}}(\mathbf{r})] = \hat{n} \times \iint_{\text{body}} \mathbf{J}_{\text{body}} \cdot \overline{\overline{\Gamma}}_1 \, ds \quad (7.55)$$

The mathematical details of this process are discussed in Section (4.2.4) and the interested reader is referred elsewhere for more information [7.9, 7.19, 7.20, 7.21, or 7.22].

Although this alternative Green's function simplifies the integrations in Eq. (7.55), some additional complications arise:

352 FIELD COUPLING USING TRANSMISSION LINE THEORY

- A more complicated Green's function Γ_1, which is related to the E-fields produced by a small current element *radiating in the presence of the ground plane*, must be evaluated.
- The excitation function is now the *sum* of the incident and the ground-reflected E-fields tangential to the scattering body.
- The current flowing on the ground plane is not directly calculated.

We see from this scattering problem that there is a direct analogy to the case of the single transmission line over a ground problem:

- The presence of the groundplane in the transmission line problem has been accounted for in determining the modified per-unit-length line parameters (i.e., the line capacitance and line inductance).
- The excitation field in the transmission line problem also consists of the incident plus ground-reflected E-fields.
- Instead of computing the current distribution in the ground plane and in the wire, we obtain only the currents in the wire.

Although this correspondence between the transmission line problem and the scattering problem may seem trivial in the present case, it is useful in explaining excitation mechanisms in the more difficult case when the ground plane becomes lossy, as treated in Chapter 8.

7.3.1.4 Modifications of the Telegrapher's Equations.
As in the case of the two-wire line, the telegrapher's equations may be modified slightly to account for the effects of an imperfectly conducting conductor or the presence of loss in the medium surrounding the line. In this instance, the line is described by per-unit-length impedance and element given by

$$Z' = j\omega L' + Z'_w \qquad (7.56)$$

where Z'_w is the internal impedance of the single conductor defined by Eq. (7.17) and L' is given by Eq. (7.42). Note that since there is only a single conductor over a perfectly conducting ground plane, only one Z'_w term is present in Eq. (7.56) instead of two Z'_w terms as for the two-wire line in Eq. (7.19). Similarly, for the line immersed in a lossy material with parameters σ_m and ϵ_m, its per-unit-length admittance is

$$Y' = j\omega C' + G' = \left(j\omega + \frac{\sigma_m}{\epsilon_m}\right) C' \qquad (7.57)$$

where C' is the appropriate per-unit-length capacitance given by Eq. (7.44). With these changes, the terms $j\omega L'$ and $j\omega C'$ in Eq. (7.52) should be replaced by Z' and Y', respectively.

7.3.2 Solution to the Telegrapher's Equations for Load Responses

The telegrapher's equations for the case of a single wire over the ground are similar to those of the two-wire line in Eqs. (7.7) and (7.15), with the following exceptions:

- The per-unit-length inductance of the line is one-half that of the two-wire case.
- The per-unit-length capacitance is twice that of the two-wire case.
- The excitation field is now the incident plus reflected field from the ground rather than just the incident field.

As a consequence of the first two points, we find that the propagation constant on the single-conductor line $\gamma = \sqrt{Z'Y'}$ is the same as for the two-wire line, and the characteristic impedance $Z_c = \sqrt{Z'/Y'}$ is one-half that of the two-conductor line. These observations are consistent with the view of replacing the ground plane by the image of the wire, as suggested in Figure 7.15. Thus the solution of the terminal voltages and currents in this case is again given by the BLT equations (7.35).

7.3.2.1 BLT Sources.
For a general excitation EM field, the source vectors for the total voltage and the scattered voltage formulations of the BLT equation are given by Eqs. (7.36) and (7.37), respectively. The voltage or current sources involve the incident and ground-reflected fields, as given in Eqs. (7.43) and (7.46) or Eqs. (7.51b) and (7.53). The integrals in these expressions must be evaluated either analytically or numerically, depending on the details of the field.

7.3.3 Load Responses for Plane-Wave Excitation

Let us now consider the case of an incident plane wave exciting the line, as shown in Figure 7.19. As in the case of the two-wire line, the incident field arrives with angles of incidence ψ and ϕ, and the polarization of the incident E-field is described by angle α detailed in Figure 7.11. For ease in treating the reflected field in the ground plane, the incident field again is divided into a vertically polarized component and a horizontally polarized component, as illustrated in the figure.

For this incident plane wave and the perfectly conducting ground plane, the distributed source term $V'_{s2}(x)$ in Eq. (7.51b), given as [7.11]

$$V'_{s2}(x) = E_x^{ex}(x)$$
$$= E_0(\cos \alpha \sin \psi \cos \phi + \sin \alpha \sin \phi)$$
$$\times (e^{jkh \sin \psi} - e^{-jkh \sin \psi})e^{-jkx \cos \psi \cos \phi} \qquad (7.58)$$

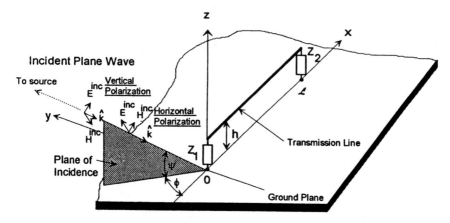

FIGURE 7.19 Single-wire line over a perfectly conducting ground excited by an incident plane wave.

Because of the requirement that the line height $h \ll \lambda$, a frequently made approximation is to consider the small argument expansion for the exponentials in Eq. (7.58), which gives

$$V'_{s2}(x) \approx jk \sin \psi \, 2hE_0(\cos \alpha \sin \psi \cos \phi + \sin \alpha \sin \phi)e^{-jkx \cos \psi \cos \phi}$$
(7.59)

The total vertical excitation E-field at a location x and height z over the perfect ground involves only the vertically polarized field component and is expressed as

$$E_z^{ex}(x, z) = E_0 \cos \alpha \cos \psi (e^{jkz \sin \psi} + e^{-jkz \sin \psi})e^{-jkx \cos \psi \cos \phi} \quad (7.60)$$

Using Eq. (7.53), the lumped voltage source at the $x = 0$ end of the line is evaluated as

$$V_1 = -\int_0^h E_z^{ex}(0, z) \, dz = -E_0 \cos \alpha \cos \psi \int_0^h (e^{jkz \sin \psi} + e^{-jkz \sin \psi}) \, dz$$

$$= -E_0 \cos \alpha \cos \psi \, \frac{e^{jkh \sin \psi} - e^{-jkh \sin \psi}}{jk \sin \psi}$$
(7.61)

At low frequencies this expression becomes

$$V_1 \approx -2E_0 h \cos \alpha \cos \psi$$
(7.62)

As noted from the form of the vertical field in Eq. (7.60), the lumped voltage source V_2 at the $x = \mathcal{L}$ end is simply V_1 multiplied by the phase-shift

7.3 SINGLE LINE OVER A PERFECTLY CONDUCTING GROUND PLANE

factor accounting for the propagation of the excitation field along the line:

$$V_2 = V_1 e^{-jk\mathcal{L}\cos\psi\cos\phi} \qquad (7.63)$$

For the present case of a perfectly conducting ground using the scattering formulation, the source vector for the BLT equation is given by Eq. (7.37). With the definitions of $V'_{s2}(x)$, V_1, and V_2 from Eqs. (7.59), (7.62), and (7.63), the source integrals can be evaluated analytically as was done for the two-wire case, and the following result is obtained:

$$\begin{bmatrix} S_1 \\ S_2 \end{bmatrix} =$$

$$\begin{bmatrix} -\left(\dfrac{E_0(\cos\alpha\sin\psi\cos\phi + \sin\alpha\sin\phi)jkh\sin\psi}{\gamma - jk\cos\psi\cos\phi} - E_0 h\cos\psi\cos\alpha\right)(1 - e^{(\gamma - jk\cos\psi\cos\phi)\mathcal{L}}) \\ -e^{\gamma\mathcal{L}}\left(\dfrac{E_0(\cos\alpha\sin\psi\cos\phi + \sin\alpha\sin\phi)jkh\sin\psi}{\gamma + jk\cos\psi\cos\phi} + E_0 h\cos\psi\cos\alpha\right)(1 - e^{-(\gamma + jk\cos\psi\cos\phi)\mathcal{L}}) \end{bmatrix}$$

$$(7.64)$$

The excitation sources in Eqs. (7.58) and (7.60) and the resulting BLT equation source in Eq. (7.64) are derived for the case of a perfect ground. As discussed in Chapter 8, similar expressions can be developed to take into account the effects of a ground plane having a finite conductivity.

7.3.3.1 Comparison with Two-Wire Results.
Note that the BLT source vector for the single-wire case in Eq. (7.64) is the same as the expression for the two-wire case in Eq. (7.40) if the wire height $h = d/2$. Thus, after all the work to formulate and solve the problem of a single wire over a perfect ground, we see that the BLT excitation sources are the same as for the two-wire problem. The BLT equations, repeated below for convenience,

$$\begin{bmatrix} I(0) \\ I(\mathcal{L}) \end{bmatrix} = \dfrac{1}{Z_c}\begin{bmatrix} 1 - \rho_1 & 0 \\ 0 & 1 - \rho_2 \end{bmatrix}\begin{bmatrix} -\rho_1 & e^{\gamma\mathcal{L}} \\ e^{\gamma\mathcal{L}} & -\rho_2 \end{bmatrix}^{-1}\begin{bmatrix} S_1 \\ S_2 \end{bmatrix}$$

$$\begin{bmatrix} V(0) \\ V(\mathcal{L}) \end{bmatrix} = \begin{bmatrix} 1 + \rho_1 & 0 \\ 0 & 1 + \rho_2 \end{bmatrix}\begin{bmatrix} -\rho_1 & e^{\gamma\mathcal{L}} \\ e^{\gamma\mathcal{L}} & -\rho_2 \end{bmatrix}^{-1}\begin{bmatrix} S_1 \\ S_2 \end{bmatrix}$$

are also of the same form as for the two-conductor line. The characteristic impedance of the single wire over the ground, however, is one-half that of the two-conductor case. If we set the termination impedances in the single-conductor-with-ground case to one-half those in the two-conductor problem, the value of the reflection coefficients ρ are the same in both problems. The line propagation constant γ remains unchanged with respect to the two-wire case, because L' is two times smaller than the corresponding inductance for the line over the ground, and the capacitance term C' is a factor of 2 larger than the corresponding single-line capacitance. Consequently, the only difference in the solutions is a factor of 2 in the current magnitude.

As a numerical verification of this fact, the single-conductor line shown in Figure 7.19 has been considered with the parameters $\mathcal{L} = 30$ m, $h = 0.1$ m, and $a = 0.15$ cm. The terminating loads are chosen to be $Z_1 = Z_2 = Z_c/2 \approx 147\,\Omega$, and the incident plane-wave field has the same characteristics as in the example in Section 7.2.5.1: a vertically polarized plane wave with angles of incidence $\psi = 60°$, $\phi = 0°$, $\alpha = 0°$. Figure 7.20 plots the load current spectra at $x = \mathcal{L}$, computed using the BLT equation, and also using the integral equation analysis (NEC). Again, at low frequencies the NEC solution agrees well with the TL solution (neglecting the noise in the NEC solution), and at high frequencies the solutions begin to diverge due to the radiation loss. Note that this current is a factor of 2 larger than for the corresponding two-wire case shown in Figure 7.12.

Summarizing these observations, we see that the line-to-line voltage is the *same* as the line-ground voltage, and the induced current in the single-wire case is double that of the two-wire case. These correspondences are valid only for the perfectly conducting ground plane. In Chapter 8, the ground will be permitted to have a finite conductivity, and this will cause significant differences between the two solutions.

7.3.4 Validation of the Coupling

As the field-to-transmission line coupling models are designed primarily for the calculation of the effects of transient fields, two types of impulsive field sources have been used for the validation of the coupling models: (1) EMP simulators and (2) artificial (triggered) or natural lightning. The first method

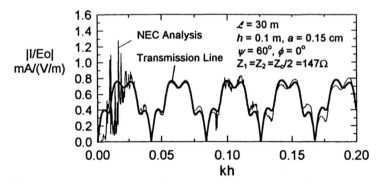

FIGURE 7.20 Current spectra in the load at $x = \mathcal{L}$ for a single line over a perfectly conducting ground, $Z_1 = Z_2 = Z_c/2 = 147\,\Omega$, $\psi = 60°$, $\phi = 0°$. (From [7.5]. Reprinted by permission from Academic Press, Inc.)

is easier to use because in an EMP simulator, the behavior of the EM fields can be controlled more precisely than in the case of lightning electromagnetic fields. An example of model validation using a parallel-plate EMP simulator is shown in Figure 7.21. Here the configuration of the experiment is a very simple one: The incident electromagnetic field is vertically polarized ($\alpha = 0$) with $\phi = \psi = 0$ (see Figure 7.11 for the angle definitions).

Figure 7.21a shows the arrangement of the line in the EMP simulator, the line, and the termination (load) parameters ($\mathcal{L} = 5$ m, $h = 1$ m, $Z_1 = Z_2 = Z_c = 470\,\Omega$, i.e. matched), together with the measured vertical electric field component. Figure 7.21b shows the comparison between the current induced at one end of the line measured and calculated using Agrawal model. The shape of the two curves is very similar, but a difference of about 14% can be seen in the peak values.

The validation using lightning-induced currents is much more sensitive to uncontrolled environment conditions, such as the presence of nearby elevated structures (trees, metallic towers, protection shelters). However, a few published results [7.24,7.25] show a reasonable good agreement between measurements and calculations using the Agrawal model.

7.3.5 Load Response for Non-Plane-Wave Excitation

A useful model for EMC purposes is one that can predict the effect of a nearby lightning strike on an overhead cable. As discussed in [7.26], natural lightning is a very complex and little understood phenomenon. However, certain aspects of the lightning interaction with a line can be modeled. Although the model may not be comprehensive or precise, it can be used in performing useful sensitivity studies to determine the most important parameters in the model. This in turn can provide guidance for future measurements and model refinements. Such studies have been performed by several authors [7.27–7.31].

Consider as in [7.31] the case of a nearby cloud-to-ground lightning strike, as illustrated in Figure 7.22. The lightning is assumed to strike at a distance y_c from the midpoint of the line. The lightning-return current is contained in a vertical channel between the ground and a distant point $\mathcal{H}$ above the ground. This case is interesting, in that the excitation field produced by the lightning is not a plane wave but a type of cylindrical wave propagating radially outward from the lightning channel. Consequently, the convenient expressions for the BLT equation sources for plane-wave excitation given by Eq. (7.64) cannot be used here. Assuming that the Agrawal formulation for the line coupling is to be used, it will be necessary to evaluate the required integrals in the source terms numerically.

In [7.31] it has been determined that if the lightning strike is close to the line, with the distance y_c being on the order of a few hundred meters, the earth can be modeled as a perfect conductor for computing the excitation

358 FIELD COUPLING USING TRANSMISSION LINE THEORY

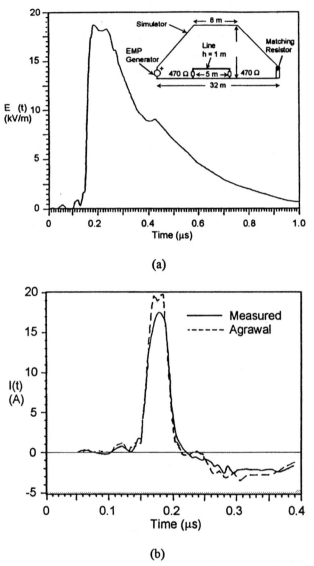

FIGURE 7.21 Experimental validation of the field-to-transmission line coupling models: (a) schematic layout of the experiment using an EMP simulator and the measured electric field inside the simulator volume; (b) solid line, current measured at the left termination of the line; dashed line, current predicted by the Agrawal coupling model. (Measurements courtesy of Electricite de France.)

fields. The lightning channel return current is modeled as an upward traveling wave plus its image in the ground plane, and it is assumed to be attenuated exponentially with distance. As a consequence, the transient

7.3 SINGLE LINE OVER A PERFECTLY CONDUCTING GROUND PLANE

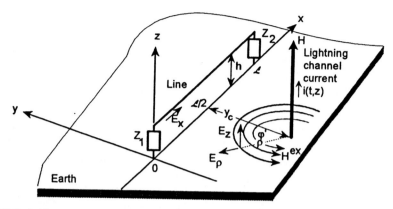

FIGURE 7.22 Transmission line excited by a nearby cloud-to-ground lightning discharge.

current distribution along the line is given by

$$i(t, z) = i(\tau)e^{-\alpha z} \quad (7.65)$$

where τ is the retarded time given by $\tau = t - z/v$. In this expression, α is an attenuation constant along the channel and is quoted in [7.31] to be on the order of 0.5 to 1 km^{-1}, and the effective channel velocity v can vary from 0.6 to 2.0×10^8 m/s. These values are estimated from measurements made of lightning discharges [7.32].

The temporal behavior of the current at the base of the channel $i(t)$ is assumed to be modeled by a sum of two functions, each of the form

$$i(t) = \frac{I_0}{\eta} \frac{(t/\tau_1)^n}{1 + (t/\tau_1)^n} e^{-t/\tau_2} \quad (7.66a)$$

where η is an amplitude normalization factor defined as

$$\eta = e^{-(\tau_1/\tau_2)(n\tau_2/\tau_1)^{1/n}} \quad (7.66b)$$

so that the peak value of the transient waveform is I_0. The parameters required for the two functions to resemble a measured lightning-return current waveform are given in [7.31] and are summarized in Table 7.1. The

TABLE 7.1 Parameters for Representing the Current Waveform at the Base of a Lightning Channel

Waveform	I_0 (kA)	t_1 (μs)	t_2 (μs)	n
1	10.7	0.25	2.5	2
2	7.5	2.1	230	2

resulting transient current waveform and its Fourier spectrum are shown in Figure 7.23.

In the frequency domain, the traveling-wave current distribution of Eq. (7.65) is a simple exponential traveling wave given by

$$I(z, \omega) = I(\omega)e^{-\gamma_c z} \qquad (7.67)$$

where the channel propagation constant $\gamma_c = a + j\omega/v$ and $I(\omega)$ is the Fourier transform of the sum of the two time functions in Eq. (7.61).

For Agrawal coupling formulation, vertical and horizontal E-field com-

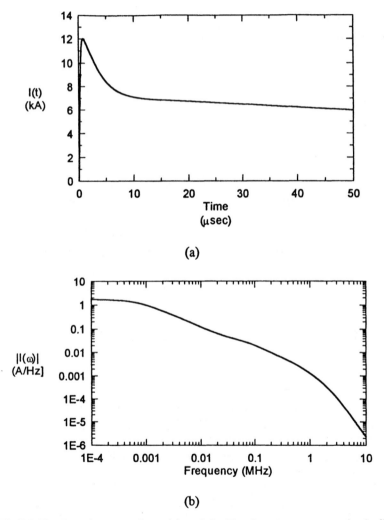

FIGURE 7.23 Transient waveform (*a*) and the Fourier spectrum magnitude (*b*) for representing the lightning return-stroke current at the base of the channel.

7.3 SINGLE LINE OVER A PERFECTLY CONDUCTING GROUND PLANE

ponents are needed. As discussed in Chapter 4, these fields can be calculated by integrating over the current distribution in the channel with the observation point at the transmission line. With the assumption of a perfectly conducting earth, this integration can be performed in the time or frequency domain. For the more general problem of the lightning channel over a lossy earth, the calculations are performed in the frequency domain as the earth effects are easier to express as functions of frequency.

In the frequency domain, the vector components of the differential E-field produced by a Hertzian dipole of strength $I\,dz'$ at the origin of a spherical coordinate system have been given in Chapter 4. Shifting the dipole location to a point z' in the coordinate system as shown in Figure 7.24 and expressing the E-fields at a point P at the coordinates (ρ, φ, z) in terms of the E_z (vertical) and E_ρ (horizontal) components, the following expressions can be developed:

$$dE_z = \frac{Z_0}{4\pi} I(z', \omega)\, dz' \left[\frac{2(z-z')^2 - \rho^2}{r^4} \left(1 + \frac{1}{jkr}\right) - jkr \frac{\rho^2}{r^4} \right] \quad (7.68a)$$

$$dE_\rho = \frac{Z_0}{4\pi} I(z', \omega)\, dz' \left[\frac{3\rho(z-z')}{r^4} \left(\frac{1}{jkr} + 1 + \frac{jkr}{3}\right) \right] \quad (7.68b)$$

where Z_0 is the free-space impedance of 377 Ω, $r = \sqrt{\rho^2 + (z-z')^2}$, and k has been defined under Eq. (7.38).

The complete excitation E-field is obtained by integrating Eqs. (7.68) with respect to the z' coordinate in the lightning channel from 0 to $\mathcal{H}$ and over the image of the channel in the ground, from 0 to $-\mathcal{H}$. As with the parameters of the lightning base current waveform in Table 7.1, the height of channel, $\mathcal{H}$, is also a parameter which must be set by comparing calculated results with measurements. Values of $\mathcal{H}$ can range upward to 7 km, but as long as $\mathcal{H} \gg 1/a$ [where a is the attenuation factor in Eq. (7.65)], this parameter is not significant in determining the radiated waveforms.

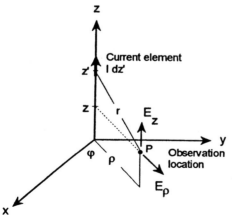

FIGURE 7.24 Elementary point dipole source at $(0, 0, z')$ and field point P at (ρ, ϕ, z).

Using Eq. (7.65), the excitation fields can be expressed as

$$E_z^{ex}(\rho, z) = \frac{Z_0}{4\pi} I(\omega) \int_{-\ell}^{\ell} e^{-|\gamma_c z'|} \left[\frac{2(z-z')^2 - \rho^2}{r^4} \left(1 + \frac{1}{jkr}\right) - jkr \frac{\rho^2}{r^4} \right] dz' \tag{7.69a}$$

$$E_\rho^{ex}(\rho, z) = \frac{Z_0}{4\pi} I(\omega) \int_{-\ell}^{\ell} e^{-|\gamma_c z'|} \left[\frac{3\rho(z-z')}{r^4} \left(\frac{1}{jkr} + 1 + \frac{jkr}{3}\right) \right] dz' \tag{7.69b}$$

The x-component of the horizontal excitation field E_ρ^{ex} at the height $z = h$ provides the excitation field along the line. Consequently, the distributed voltage source along the line is $V'_{s2}(x) = E_x^{ex}(x, h)$. As noted from the geometry in Figure 7.22, this can be expressed as

$$\begin{aligned} V'_{s2}(x) &= E_x^{ex}(x, h) = -E_\rho^{ex}(\rho, h) \sin \phi' \\ &= -E_\rho^{ex}(\rho, h) \frac{\ell/2 - x}{\rho} \end{aligned} \tag{7.70}$$

where $\rho = \sqrt{(\ell/2 - x)^2 + y_0^2}$. Furthermore, the vertical excitation sources V_1 and V_2 are given in terms of the vertical E-field by Eq. (7.31) as

$$V_1 = -\int_0^h E_z^{ex}(0, z) \, dz \quad \text{and} \quad V_2 = -\int_0^h E_z^{ex}(\ell, z) \, dz \tag{7.71}$$

The voltage sources in Eqs. (7.70) and (7.71) can be used in Eq. (7.37) to determine the BLT source terms for this problem. The integrals over the line in Eq. (7.37) cannot be evaluated analytically, due to the complicated dependence of the integrand on the parameter x through the term ρ. Consequently, a numerical evaluation is necessary. Because this is a double integration, a significant increase in computer time over that needed for a plane-wave excitation will be required.

An alternative to evaluating the excitation field in the frequency domain is to evaluate the fields directly in the time domain. Noting that the terms in Eq. (7.68) involving $j\omega I(\omega)$ correspond to $di(t)/dt$ in time, and $I(\omega)/j\omega$ corresponds to $\int_{-\infty}^{t} i(\tau) \, d\tau$, the transient differential fields produced by a current element $i(z', t) \, dz'$ are

$$dE_z = \frac{Z_0}{4\pi} e^{-az'} dz' \left\{ \frac{2(z-z')^2 - \rho^2}{r^4} \left[i(t^*) + \frac{c}{r} \int_0^t i(\tau^*) \, d\tau \right] - \frac{\rho^2}{cr^3} \frac{d}{dt} [i(t^*)] \right\} \tag{7.72a}$$

$$dE_\rho = \frac{Z_0}{4\pi} e^{-az'} dz' \left[\frac{3\rho(z-z')}{r^4} \left\{ i(t^*) + \frac{c}{r} \int_0^t i(\tau^*) \, d\tau + \frac{r^2}{3c} \frac{d}{dt} [i(t^*)] \right\} \right] \tag{7.72b}$$

where $t^* = (t - z'/v - r/c)$ and $\tau^* = (\tau - z'/v - r/c)$ where the fact that $i(t) = 0$ for $t < 0$ has been used to change the lower limit of the integrals.

7.3 SINGLE LINE OVER A PERFECTLY CONDUCTING GROUND PLANE

These differential fields must be integrated over the lightning channel and its image, as in Eq. (7.69). However, because the channel current is a traveling wave, the integrations over part of the channel will be identically zero if the current wave has not propagated to the integration point. Specifically, at a particular calculation time t_1, the upper and lower limits of the integrals $\pm \mathscr{H}$ in Eq. (7.69) can be replaced by $\pm z_1$, where z_1 is the solution to

$$\frac{z_1}{v} - \frac{\sqrt{(z-z_1)^2 + \rho^2}}{c} = t_1$$

Thus, by evaluating the excitation fields in the time domain, the integration over the lightning channel can be reduced significantly, especially for early time. Of course, these fields must then be converted to the frequency domain and the BLT equation sources evaluated by the integration over the line.

As an example of computed lightning-induced load voltages for this lightning channel model, [7.31] presents the load voltages at either end of a line having parameters $\mathscr{L} = 1$ km, $h = 10$ m, and $a = 5$ mm. The lightning channel was located a distance $y_c = 50$ m from the midpoint of the line and was assumed to have a total channel height of $\mathscr{H} = 7.5$ km. The return-stroke velocity was assumed to be $v = 1.3 \times 10^8$ m/s and the current decay constant in Eq. (7.65) was $\alpha = 0.588$ km^{-1}. The time-domain behavior of the lightning current at the channel base was similar to that of Figure 7.23a, but had three different values of the peak value of the channel current: 30, 12, and 4.6 kA. For each of these cases, the maximum derivative of the channel current was constrained to be equal to 40 kA/μs. Figure 7.25 illustrates the time history

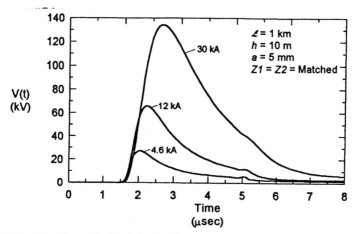

FIGURE 7.25 Plots of the lightning-induced voltages at the ends of the line for different peak values of the lightning return stroke current, with the maximum current derivative equal to 40 kA/μs. (Data courtesy of F. Rachidi).

of the load voltages (which are identical for both ends of the line due to symmetry) for these three different channel currents.

7.4 TREATMENT OF HIGHLY RESONANT STRUCTURES

7.4.1 Single-Wire Line

In examining the BLT equation (7.35), we see that the zeros of the determinant of the inverse matrix correspond to the natural resonances of the transmission line. These resonances occur at complex frequencies $s_n = \sigma_n + j\omega_n$, where

$$\det \begin{bmatrix} -\rho_1 & e^{\gamma(s_n)\mathcal{L}} \\ e^{\gamma(s_n)\mathcal{L}} & -\rho_2 \end{bmatrix} = \rho_1\rho_2 - e^{2\gamma(s_n)\mathcal{L}} = 0 \qquad n = 1, 2, \ldots$$

For most realistic cases where there is loss along the transmission line, or when the terminating impedances have loss, these resonant frequencies are complex. This does not pose a problem in evaluating the determinant for frequencies along the $j\omega$-axis of the complex frequency plane. However, if the line is *lossless* so that the propagation constant $\gamma(s) = j\omega/c$, and if the loads at both ends of the line are either open or short, so that ρ_1 and $\rho_2 = \pm 1$, the complex resonant frequencies lie on the $j\omega$-axis at locations given by

$$s_n = 0 + j\frac{n\pi c}{2\mathcal{L}} \quad (\text{for } \rho_1\rho_2 = 1) \qquad n = 0, 1, 2, \ldots$$

or

$$s_n = 0 + j\frac{n\pi c}{\mathcal{L}} \quad (\text{for } \rho_1\rho_2 = -1) \qquad n = 1, 2, \ldots$$

In evaluating the wideband spectral response of the BLT equation in this case, it is possible that a selected frequency might coincide with one of these natural resonances, and a singularity in the response will occur. This singularity is a consequence of the approximations made in the transmission line model (i.e., in neglecting the effects of radiation in determining the behavior of the induced current), and approximations made in modeling the structure, since all real structures have loss.

7.4.1.1 Numerical Example.
As an example of a highly resonant line response, consider the case of a single-wire line over a perfect ground as shown in Figure 7.19, with the following parameters: $\mathcal{L} = 30$ m, $h = 10$ cm, $a = 0.15$ cm, and loads $Z_1 = \infty$ and $Z_2 = 0\,\Omega$. This line, together with its image in the ground plane, is similar to the 30-m line examined in Section 7.2.6.1. Figure 7.26 plots the normalized current spectral magnitude in the

7.4 TREATMENT OF HIGHLY RESONANT STRUCTURES

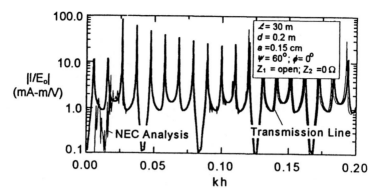

FIGURE 7.26 Current spectra in the load at $x = \mathcal{L}$ for a perfectly conducting ground, $Z_1 = \infty$, $Z_2 = 0\,\Omega$. (From [7.5]. Reprinted by permission from Academic Press, Inc.)

load at $x = \mathcal{L}$, again for the same incident with $\psi = 60°$, $\phi = 0°$, $\alpha = 0°$. In this plot we note the sharp periodic spikes corresponding to the transmission line resonances. Formally, these are infinite in height, but the numerically calculated frequency points were chosen carefully to avoid these singularities.

Also plotted on this figure are results calculated using the integral equation solution in the NEC code. At low frequencies, the NEC solution again exhibits a significant amount of "noise." At higher frequencies, there is a very slight deviation between the NEC and the transmission line solutions, due principally to the fact that the NEC solution accounts for radiation effects. This causes the true resonance frequencies of the line to become slightly lower and the widths of the resonances to become larger. However, we do see that the transmission line solution provides a very good alternative to the much more time consuming integral equation solution, as long as the singularities in the response are avoided.

In the time domain, highly resonant lines can pose problems, however, especially if the FFT is used to obtain the transient response from the frequency spectrum. As an example, using the double exponential waveform described in Section 7.2.6.1 for incident field, applying it to the spectrum in Figure 7.26, and taking its inverse FFT to obtain the transient response of the load current, we obtain the waveform shown in Figure 7.27. This response has an extremely large amplitude (800 A for a 1 V/m pulse is unrealistic!) and its early-time waveform is not what is expected from the ray-tracing arguments discussed earlier. The problem is that this response is contaminated by the late-time portions of the waveform "folding back" into the early time, a basic difficulty with the FFT.

To eliminate this problem, there are several possible approaches:

- Use a Fourier integral transform (FIT) instead of the FFT.

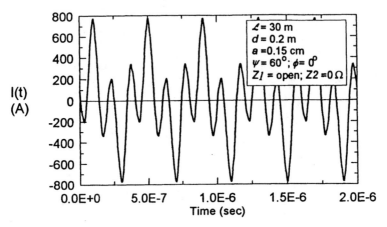

FIGURE 7.27 Incorrect transient current at $x = \mathcal{L}$ for the highly resonant transmission line as calculated using the FFT.

- Add a small amount of loss to the problem.
- Increase the maximum time window of the FFT to the point where the response has died away.
- Modify the BLT equation.

7.4.1.2 Expansion of the BLT Resonance Matrix.
This last approach is particularly interesting, as it permits us to select only a finite number of oscillations in the transient response, but *it is done in the frequency domain*. This is accomplished by first writing the inverted matrix in the BLT equation explicitly as

$$\begin{bmatrix} -\rho_1 & e^{\gamma\mathcal{L}} \\ e^{\gamma\mathcal{L}} & -\rho_2 \end{bmatrix}^{-1} = \frac{e^{-2\gamma\mathcal{L}}}{1 - \rho_1\rho_2 e^{-2\pi\mathcal{L}}} \begin{bmatrix} \rho_2 & e^{\gamma\mathcal{L}} \\ e^{\gamma\mathcal{L}} & \rho_1 \end{bmatrix} = 0 \qquad (7.73)$$

Letting $\zeta = \rho_1\rho_2 e^{-2\gamma(s)\mathcal{L}}$ and noting that $|\zeta| < 1$, the term $1/(1-\zeta)$ in Eq. (7.73) can be written in a power series as $1/(1-\zeta) = 1 + \zeta + \zeta^2 + \zeta^3 + \cdots$. Substituting this into the BLT equation for the current (7.35a) gives

$$\begin{bmatrix} I(0) \\ I(\mathcal{L}) \end{bmatrix} = \frac{1}{Z_c}[1 + (\rho_1\rho_2 e^{-2\gamma\mathcal{L}}) + (\rho_1\rho_2 e^{-2\gamma\mathcal{L}})^2 + \cdots]$$

$$\times e^{-2\gamma\mathcal{L}} \begin{bmatrix} 1-\rho_1 & 0 \\ 0 & 1-\rho_2 \end{bmatrix} \begin{bmatrix} \rho_2 & e^{\gamma\mathcal{L}} \\ e^{\gamma\mathcal{L}} & \rho_1 \end{bmatrix} \begin{bmatrix} S_1 \\ S_2 \end{bmatrix} \qquad (7.74)$$

with a similar equation resulting for the load voltages.

Thus the inverted matrix in the BLT equation can be approximated by a

7.4 TREATMENT OF HIGHLY RESONANT STRUCTURES

series of terms involving $e^{-2n\gamma\mathcal{L}}$ with $n = 1, 2, \ldots$. Each term corresponds to a reflection on the line, occuring at times that are multiples of $t = 2\mathcal{L}/c$. By truncating the series after the second or third term, the inverse transform of the spectrum provided by the BLT equation will give, in the time domain, only the contributions from these early-time reflections. Afterward, the response goes to zero, as the later-time contributions are not included in the spectrum. As an example of the transient response calculated using this method, Figure 7.28 shows the load current for the same problem treated in Figure 7.27, but with only three terms in the power series expansion. It is clear that the early-time transient response is now reasonable and correct up to a time of about 0.8 μs. In using this resonance modification method it is important to remember that the spectral response will be incorrect. If the spectrum is desired, the unmodified BLT equation must be calculated.

7.4.2 Extension to Multiconductor Lines

This expansion of the resonance terms of the BLT equation can be generalized for the treatment of resonances in multiconductor lines or networks. In this case we do not have a simple scalar resonance term as in Eq. (7.73), but a supermatrix to invert, which becomes singular at the natural resonances of the multiconductor line or network. As noted in the BLT equation for the load (terminal) voltage responses of a multiconductor line [see Eq. (6.88)], we must invert the supermatrix

$$\begin{bmatrix} -[\rho_1] & [S]^{-1}[e^{\gamma\mathcal{L}}][S] \\ [S]^{-1}[e^{\gamma\mathcal{L}}][S] & -[\rho_2] \end{bmatrix}^{-1} = \left[\begin{pmatrix} [0] & [S]^{-1}e^{[\gamma]\mathcal{L}}[S] \\ [S]^{-1}e^{[\gamma]\mathcal{L}}[S] & [0] \end{pmatrix} - \begin{pmatrix} [\rho_1] & [0] \\ [0] & -[\rho_2] \end{pmatrix} \right]^{-1}$$

$$\equiv [[A] - [B]]^{-1}$$

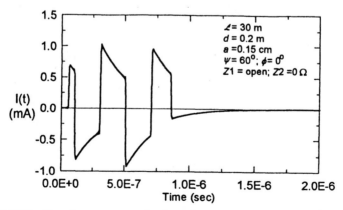

FIGURE 7.28 Transient response for load 2 current using the BLT equation with the modified resonance treatment.

Using the definitions of the matrices $[A]$ and $[B]$ above, the inverted matrix can be now written as

$$[[A]-[B]]^{-1} = [[A][[U]-[A]^{-1}[B]]]^{-1}$$
$$= [[U]-[\Gamma]]^{-1}[A]^{-1} \qquad (7.75)$$
$$\approx [[U]+[\Gamma]+[\Gamma]^2+[\Gamma]^3+\cdots][A]^{-1}$$

where $[\Gamma] = [A]^{-1}[B]$ and $[U]$ is the unit matrix. As in the scalar case, each term in the expansion in Eq. (7.75) also corresponds to a reflection contribution to the response, and each term contributes to a response occurring later in time. As a result, a relatively short calculation time window can be used for highly resonant responses on multiconductor lines or networks.

7.5 RADIATION FROM TRANSMISSION LINES

Although the transmission line equations are derived in such a way that radiation effects are neglected, there can be radiation from a transmission line. As discussed in Chapter 4, the radiation field from a specific current distribution can be evaluated by performing an integral over the current. The fact that radiation is neglected in the transmission line theory means only that the current distribution on the line is approximate. Such an approximate current distribution can and does radiate EM energy.

If a transmission line did not radiate energy, then by the principle of electromagnetics reciprocity, it would not receive energy from a plane wave exciting it from a distant source. Earlier in this chapter we developed expressions for the induced responses on a line due to an incident plane wave. From this solution we can conclude that such transmission lines do indeed radiate.

In this section we investigate how the plane-wave field excitation response of the line can be used to compute the radiated fields from the same line using the reciprocity theorem. In doing this we must again remember that there are both transmission line and antenna modes existing on a line, and that only the transmission line mode is calculated using our models. Consequently, if the transmission line geometry and excitation is such that a large antenna mode is excited by a local source on the line, the radiated field will be in error, since the antenna mode contribution to the field is not included. This is generally not the case for the single wire line over a ground plane, where the antenna mode is small. Consequently, we concentrate on this specific line geometry in computing the radiated fields.

7.5.1 Reciprocity Theorem

The starting point for calculating the radiating behavior of a transmission line excited by a local voltage source is the reciprocity theorem [7.10,7.33].

7.5 RADIATION FROM TRANSMISSION LINES

This theorem states that if there are *two* sets of electric and magnetic current sources $(\mathbf{J}_1, \mathbf{M}_1)$ and $(\mathbf{J}_2, \mathbf{M}_2)$ producing EM fields $(\mathbf{E}_1, \mathbf{H}_1)$ and $(\mathbf{E}_2, \mathbf{H}_2)$, then in an unbounded, linear, isotropic region, these quantities are related by the volume integral

$$\iiint_{volume} (\mathbf{E}_1 \cdot \mathbf{J}_2 - \mathbf{E}_2 \cdot \mathbf{J}_1 - \mathbf{H}_1 \cdot \mathbf{M}_2 + \mathbf{H}_2 \cdot \mathbf{M}_1) \, dv = 0$$

For the case where the electric and magnetic sources are localized in space (i.e., for individual current and voltage sources), [7.10] shows that these field relationships become reciprocal voltage–current relationships for a general N-port circuit. For the two-port network shown in Figure 7.29a, the reciprocity expression simplifies to

$$V_{a1}I_{a2} + V_{b1}I_{b2} = V_{a2}I_{a1} + V_{b2}I_{b1} \tag{7.76}$$

where a and b denote the two ports of the network, and 1 and 2 refer to the different source configurations.

Various versions of Eq. (7.76) can be useful in special cases of the network excitation. Consider the source configurations with $I_{b1} = 0$ in case 1 and $V_{a2} = 0$ in case 2, as shown in Figure 7.29b. In this case, the reciprocity relation in Eq. (7.76) becomes

$$\frac{V_{b1}}{V_{a1}} = -\frac{I_{a2}}{I_{b2}} \tag{7.77}$$

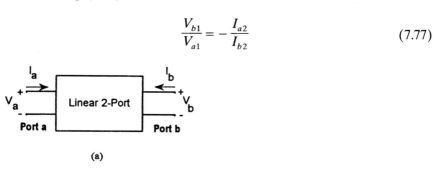

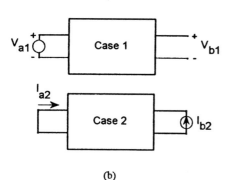

FIGURE 7.29 Reciprocal two-port network (a) and applications of the reciprocity theorem for $I_{b1} = 0$ and $V_{a2} = 0$ (b).

370 FIELD COUPLING USING TRANSMISSION LINE THEORY

This equation forms the basis for computing the radiation from the transmission line.

7.5.2 Radiating Transmission Line

The geometry for the radiating transmission line is shown in Figure 7.30a, where a single conductor line is located over a ground plane. A discrete voltage source V_s at x_s induces currents on the line and in the loads. Equations (6.38) and (6.39) can be used to compute these currents, and once the current distribution is known, the *radiated* E-field from this line can be found by using the approach discussed in Chapter 4. This involves directly integrating over the computed transmission line current distribution, taking into account the effects of the nearby ground. Doing this in a rigorous manner for the lossy ground is numerically difficult due to the ground-plane

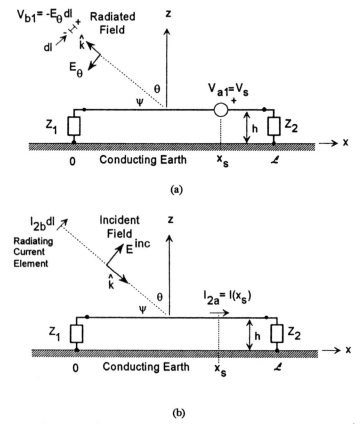

FIGURE 7.30 Transmission line geometry for reciprocity calculation: (*a*) radiation problem; (*b*) the receiving (scattering) problem. (From [7.5]. Reprinted by permission from Academic Press, Inc.)

7.5 RADIATION FROM TRANSMISSION LINES

effects and the requirement of evaluating the Sommerfeld integrals for the currents near the ground. The reciprocity concept provides an alternative solution to this problem.

To use reciprocity, consider case 1 of Figure 7.29b to represent the radiating line configuration shown in Figure 7.30a. We will let the excitation voltage source be $V_{a1} = V_s$ and will consider V_{b1} to be the resulting voltage induced across an infinitesimal distance dl located in the far field at angles $\psi = \pi/2 - \theta$ and ϕ, and oriented in the θ direction. This voltage is produced by the radiated E-field from the line and is given by $V_{b1} = -E_\theta \, dl$, where E_θ is the radiated E-field in the usual (r,θ,ϕ) polar coordinate system.

For case 2 we consider the scattering problem shown in Figure 7.30b. An infinitesimal current element, denoted by $I_{b2} \, dl$, is located at the same distant field point as in case 1 (at a distance r from the transmission line) and produces an incident E-field E^{inc}. This current source is assumed to be oriented in the $\hat{\theta}$ direction, and consequently, the radiated field from the current element given by Eq. (4.14a) with $\theta = 90°$ as

$$E^{\text{inc}} = jkZ_o I_{b2} \, dl \, \frac{e^{-jkr}}{4\pi r} \approx j \frac{60\pi}{\lambda} I_{b2} \, dl \, \frac{e^{-jkr}}{r} \tag{7.78}$$

where λ is the wavelength. This field is a plane wave with vertical polarization and an angle of incidence ψ, which is the same as the angle of radiation in case 1.

The incident field given in Eq. (7.78) for case 2 induces current on the line, and this may be evaluated at an arbitrary location using the results of Section 7.2.3. Equation (7.33) provides an integral expression for the current in terms of the Green's function of Eq. (7.32a), and the required voltage sources $V'_{s2}(x)$, V_1, and V_2. For the line over the ground, Eqs. (7.58), (7.61), and (7.63) provide these sources. As in the case of the BLT equations for the load responses, the integrals in Eq. (7.33) can be evaluated analytically for plane-wave excitation, and this results in the following expression for the induced current at a location x on the line:

$$I(x) = \frac{V'_{s2}(x)}{2Z_o(1 - \rho_1 \rho_2 e^{-2\gamma \mathcal{L}})}$$

$$\times \left[(1 - \rho_2 e^{-2\gamma(\mathcal{L}-x)}) \left(\frac{1 - e^{-\gamma_- x}}{\gamma_-} + \rho_1 e^{-2\gamma x} \frac{1 - e^{\gamma_+ x}}{\gamma_+} \right) \right.$$

$$\left. + (1 - \rho_1 e^{-2\gamma x}) \left(\frac{1 - e^{-\gamma_+ (\mathcal{L}-x)}}{\gamma_+} + \rho_2 e^{-2\gamma(L-x)} \frac{1 - e^{\gamma_- (\mathcal{L}-x)}}{\gamma_-} \right) \right]$$

$$+ \frac{V_1 e^{-jkx \cos \psi \cos \phi}}{2Z_o(1 - \rho_1 \rho_2 e^{-2\gamma \mathcal{L}})} \left[\begin{array}{c} (1-\rho_1)(e^{-\gamma_- x} - \rho_2 e^{-2\gamma \mathcal{L}} e^{\gamma_+ x}) \\ -e^{-\gamma_+ \mathcal{L}}(1-\rho_2)(e^{\gamma_+ x} - \rho_1 e^{-\gamma_- x}) \end{array} \right] \tag{7.79}$$

In this expression, the propagation terms $\gamma_\pm$ are defined as $\gamma_+ = \gamma + jk \cos \psi \cos \phi$ and $\gamma_- = \gamma - jk \cos \psi \cos \phi$.

Using Eq. (7.79), the field-induced current at the location x_s due to the incident E-field can be determined. This current is denoted by $I(x_s)$ and is the required current I_{a2} in Eq. (7.77). Inserting the forgoing voltages and currents into Eq. (7.77), dropping the phase term e^{-jkr} and canceling the terms dl results in the following expression for the radiated field at angles (θ, ϕ) from the transmission line:

$$rE_\theta(\theta, \phi) = j30\left(\frac{\omega}{c}\right)\frac{I(x_s; \theta, \phi)}{E_0}V_s \qquad (7.80)$$

The term $I(x_s; \theta, \phi)/E_0$ in Eq. (7.80) is simply the induced current in the transmission line at the voltage source location, due to a unit excitation field, and may be calculated using transmission line theory. Note that the factor $j\omega$ in Eq. (7.80) amounts to a time derivative. Consequently, the waveform of the radiated E-field will look like the derivative of the induced line current in the scattering problem. The corresponding H-field radiated from this line is orthogonal to the E-field, and is given as $|H| = |E|/377$ (A/m), due to the far-field relationship between E and H.

7.5.3 Example of the Radiation from a Transmission Line

As an example of the calculation of radiation from a transmission line, consider the geometry illustrated in Figure 7.31, with $\mathcal{L} = 30$ m, $h = 0.1$ m, and $a = 0.15$ cm. In this example the ground plane is assumed to be imperfectly conducting, with an electrical conductivity of $\sigma_g = 0.01$ S/m and relative dielectric constant $\epsilon_r = 10$. Exactly how these earth parameters enter into the transmission line calculation is examined in detail in Chapter 8.

The line is driven by a lumped voltage source at the $x = \mathcal{L}$ end of the line, and it is desired to compute the radiated E_θ field at the angles $\psi = 60°$, and $\phi = 0°$. In this $\phi = 0°$ plane, symmetry dictates that E_ϕ must be zero. However, for other angles ϕ, both E_θ and E_ϕ will exist. Since the line is open

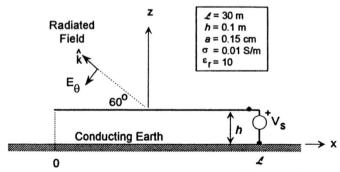

FIGURE 7.31 Geometry of line for calculation of radiated field example. (From [7.5]. Reprinted by permission from Academic Press, Inc.)

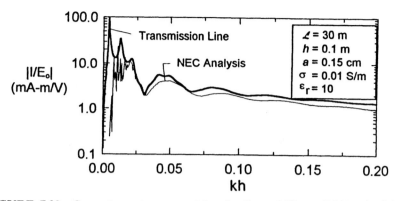

FIGURE 7.32 Current spectrum $x = \mathscr{L}$ for the line of Figure 7.31 excited by an incident field with $\psi = 60°$, $\phi = 0°$, $\alpha = 0°$ (From [7.5]. Reprinted by permission from Academic Press, Inc.)

circuited at $x = 0$, we have $Z_1 = \infty$, and for the shorted termination at $x = \mathscr{L}$, the load is $Z_2 = 0\,\Omega$. This structure is often called a traveling-wave or Beverage antenna.

To apply the reciprocity principle for computing the radiated field, we first analyze the scattering problem to determine the short-circuit current I_{2a} flowing at the source location due to an incident field at $\psi = 60°$, $\phi = 0°$, $\alpha = 0°$. Because the voltage source location is at the end of the line, the BLT equation (7.35a) can be used to evaluate the current instead of the more complicated equation for the arbitrary source location in Eq. (7.79). The resulting normalized current spectrum I_{2a}/E_0 is illustrated in Figure 7.32. As in the previous examples, a comparison of the transmission line results with the integral equation solution from NEC is shown. The NEC results agree quite well, except at low frequencies, where again the integral equation solution exhibits numerical instabilities.

This spectrum can be used directly in Eq. (7.80) to compute the spectrum of the radiated E-field. The resulting spectrum of the E-field (normalized so that rE_θ/V_s is the plotted quantity) is shown in Figure 7.33. Also shown in the figure is a direct calculation of the radiated field using the NEC code.

7.6 TRANSMISSION NETWORKS

The preceding development has focused on single transmission line sections. The propagation and field coupling concepts for the single line can be extended to include the case of transmission line networks. Figure 7.34 illustrates a possible network of single conductor lines over the earth, in which some lines are terminated at their ends in a single impedance element and others are connected to one or more additional lines. For a specified excitation of this network, either by an incident field or by discrete sources,

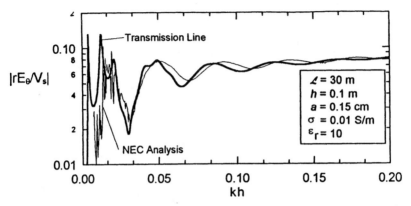

FIGURE 7.33 Radiated E_θ field at $\psi = 60°$, $\phi = 0°$ for the line of Figure 7.31. (From [7.5]. Reprinted by permission from Academic Press, Inc.)

it is desired to compute the voltage and current responses at the nodes of the network.

In formulating the solution to this problem, the mutual interaction between the currents and charges on each of the transmission line sections in the network is neglected. The only coupling between the line segments is through the voltage and current relationships at the junctions (i.e., Kirchhoff's voltage and current laws.) This is entirely consistent with the transmission line approximation on a single section of line, which neglects the mutual coupling between two current elements at different locations on the line.

There are two different approaches for treating such network problems. For a simple treelike network shown in Figure 7.34, it is possible to identify the observation location within the network and then to collapse the

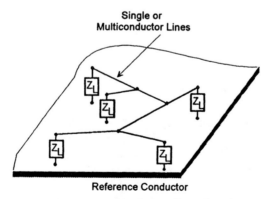

FIGURE 7.34 Interconnection of transmission lines forming a network. (From [7.5]. Reprinted by permission from Academic Press, Inc.)

network around it by using a series of Thévenin or Norton transformations on the individual line segments comprising the network. A more general approach, which is useful for analyzing networks containing loops, is to develop a generalized BLT equation for the entire network and then solve for the voltages and/or currents at each of the network nodes simultaneously. These approaches are discussed below.

7.6.1 Network Analysis by Thévenin Transformations

For simple treelike networks, distant branches of the network can be represented by their Thévenin equivalent circuits and then combined to find the response as a specified observation location. Consider the simple two-branch network illustrated in Figure 7.35a, in which the current through the impedance Z_3 is the desired response. As shown part (b) of the figure, branch 1 in the network can be replaced at cut A–A' by the equivalent Thévenin circuit containing an open-circuit voltage V_{oc1} and an input impedance Z_{in1}. Similarly, branch 2 can be represented by another equivalent circuit at cut B–B' with elements V_{oc2} and Z_{in2}. The combination of the two equivalent circuits provides the equivalent circuit shown in part (c), which permits calculation of the response at the element Z_3. This is most easily by transforming the two Thévenin equivalent circuits to Norton circuits, combining the current sources, and computing the response by inspection as

$$I_3 = (V_{oc1} Y_{in1} + V_{oc2} Y_{in2}) \frac{Y_3}{Y_{in1} + Y_{in2} + Y_3}$$
$$V_3 = \frac{I_3}{Y_3}$$
(7.81)

where $Y_3 = 1/Z_3$, $Y_{in1} = 1/Z_{in1}$, and $Y_{in2} = 1/Z_{in2}$.

Some networks require multiple Thévenin transformations to collapse the branches to the observation location. Consider the same network as in Figure 7.35, but with the observation location now at the element Z_2. In this case the series of transformations shown in Figure 7.36 are required. First branch 1 is collapsed to form an equivalent circuit connected to the end of branch 2. Then branch 2 is collapsed to load Z_2 and the response is determined. In doing this, branch 2 is excited by two mechanisms: the incident field excitation *plus* the excitation provided by the connection to branch 1.

For performing these network transformations, expressions for the three basic quantities shown in Figure 7.37 are needed. These are:

- The input impedance of a line section of length $\mathcal{L}$,

376 FIELD COUPLING USING TRANSMISSION LINE THEORY

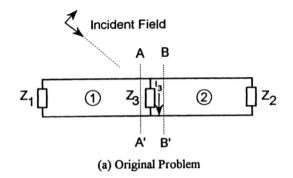

(a) Original Problem

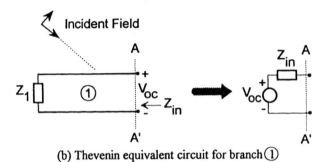

(b) Thevenin equivalent circuit for branch ①

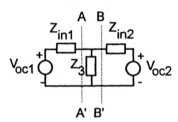

(c) Equivalent circuit for the load Z_3

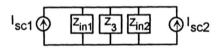

(d) Parallel form of equivalent circuit

FIGURE 7.35 Collapse of the network around element Z_3 using equivalent circuits.

- The open-circuit voltage at the end of a line of length $\mathcal{L}$ excited by a lumped voltage source at $x = 0$.
- The open-circuit voltage at the end of a line of length $\mathcal{L}$ excited by an incident field.

Each of these responses may be calculated as special cases using the BLT

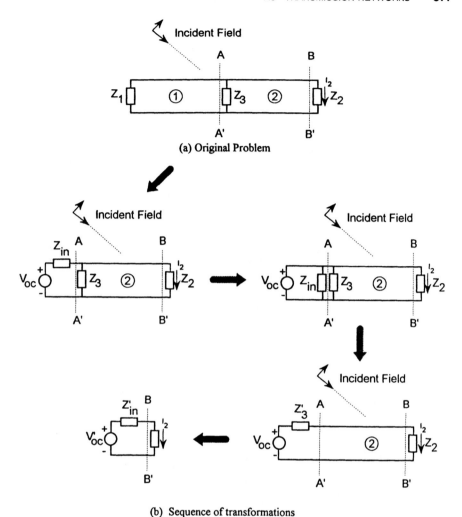

(b) Sequence of transformations

FIGURE 7.36 Collapse of the network around element Z_2 using equivalent circuits.

equation (7.35). For the determination of Z_{in} in Figure 7.37a, we can consider the case of a lumped current source I_s exciting the line at $x = \mathcal{L}$. For an open-circuit condition at $x = \mathcal{L}$, the reflection coefficient $\rho_2 = 1$ and the BLT equation can be used to evaluate the load voltage $V(\mathcal{L})$. The ratio of this open-circuit voltage to the excitation current is the definition of the input impedance and has the value

$$Z_{in} = \frac{V(\mathcal{L})}{I_0} = Z_c \frac{1 + \rho_1 e^{-2\gamma\mathcal{L}}}{1 - \rho_1 e^{-2\gamma\mathcal{L}}} \qquad (7.82)$$

Similarly, the open-circuit voltage of Figure 7.37b can be determined by

378 FIELD COUPLING USING TRANSMISSION LINE THEORY

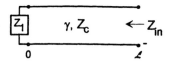

(a) Input impedance

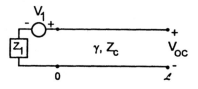

(b) Voltage source at x = 0

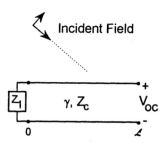

(c) Distributed field excitation

FIGURE 7.37 Required transmission line responses for the network response calculation.

calculating $V(\mathcal{L})$ due to a single lumped voltage source at $x = 0$, again with the condition that $\rho_2 = 1$. For this case the BLT equation yields

$$V_{oc} = V_0 \frac{e^{-\gamma \mathcal{L}}(1 - \rho_1)}{1 - \rho_1 e^{-2\gamma \mathcal{L}}} \quad (7.83)$$

The case of a distributed field excitation of the line is slightly more complicated, in that it is necessary to consider the integrated effects of the sources along the length of the line, plus the effects of the fields at each end of the line. The BLT equation for the load voltages for this case is given in Eq. (7.35b) with the source terms defined in Eq. (7.37). Setting $\rho_2 = 1$ and extracting the voltage response at $x = \mathcal{L}$ yields the expression

$$V_{oc} = \frac{2}{1 - \rho_1 e^{-2\gamma \mathcal{L}}} (e^{-\gamma \mathcal{L}} S_1 + \rho_1 e^{-2\gamma \mathcal{L}} S_2) \quad (7.84)$$

where the sources S_1 and S_2 are given in Eq. (7.36). For a line segment at a height h over a perfect ground plane, these source terms are given by Eq. (7.64).

7.6.1.1 Example of a Network Response Using Thévenin Transformations.
As an example of a network response computed by Thévenin transformations of the network, consider Figure 7.38, which illustrates the topology of a simple nine-node, eight-branch network. These lines have heights ranging from 5 to 15 m over the earth, which has a conductivity of $\sigma = 0.001$ S/m and $\epsilon_r = 10$. (The details of how the finite earth conductivity are included in this analysis are deferred until Chapter 8.)

The excitation field is a vertically polarized double exponential waveform expressed as $E^{inc}(t) = 52.5 \times (e^{-4 \times 10^6 t} - e^{-4.76 \times 10^8 t})$ (kV/m). This field strikes the network with angles of incidence $\psi = 45°$ and $\phi = 45°$.

For convenience, the load impedance at each of the nodes containing only one TL section (i.e., nodes 1, 5, 6, 7, 8, and 9) are taken to be matched to the line characteristic impedance. The observation location is assumed to be on line 3 at node 4. Figure 7.39a shows the calculated current spectrum at this load point. The corresponding transient response, shown in Figure 7.39b, exhibits an initial early-time response that is due to the interaction of the incident field with the network in the vicinity of the observation point. At later times, contributions arrive from other parts of the network, giving rise to a more complicated waveform. The localized nature of the early-time response suggests that for cases where only the peak value or rate of rise of the waveform is desired, it may not be necessary to model the entire network. Only the local geometry of the line conductors must be modeled out to some specified distance. Of course, if the spectral response is desired, the entire network must be modeled, as the line response at a specific frequency depends on contributions from the entire network.

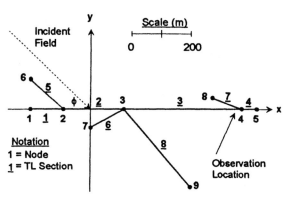

FIGURE 7.38 Transmission line network for example, as seen from above. (From [7.5]. Reprinted by permission from Academic Press, Inc.)

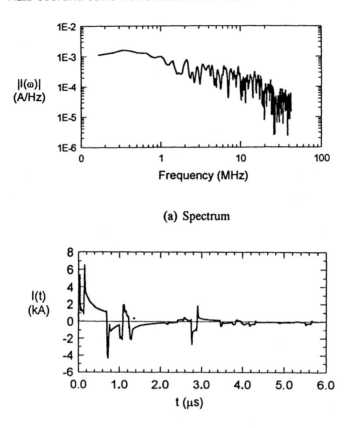

(a) Spectrum

(b) Transient response

FIGURE 7.39 Field-induced current response at the observation location of the network of Figure 7.38. (From [7.5]. Reprinted by permission from Academic Press, Inc.)

7.6.2 Development of the Network BLT Equation

For the case where the responses are desired only at the ends of each of the transmission line elements (i.e., at the junctions or loads of the lines), the BLT equations can be extended to provide the solution in a very compact matrix form [7.34–7.36]. Denoting (arbitrarily) the ends of each of the transmission line segments of the network by 1 and 2, the current flowing at each of these locations for *all* the lines in the network can be represented by two *n*-dimensional vectors $[I_1]$ and $]I_2]$, where *n* represents the number of TL segments in the network. To use a shorthand notation for the entire network, it is convenient to represent all of the TL currents as a single supervector containing 2*n* scalar elements:

$$[I_L] = \begin{bmatrix} [I_1] \\ [I_2] \end{bmatrix}$$

At each end of the TL segments, an expression for the reflected *voltage* waves in terms of the incident waves can be developed. For a simple impedance termination of the line segment, this relationship is the scalar voltage reflection coefficient. For a more complex junction involving m TL segments, this relationship becomes an $m \times m$ voltage scattering matrix, computed from a knowledge of the characteristic impedances of each line. For the entire network, these scattering parameters can be put into a $2n \times 2n$ scattering supermatrix $[\rho]$ which has the formal definition

$$[V]_{\text{ref}} = [\rho][V]_{\text{inc}}$$

where $[V]_{\text{inc}}$ represents all of the $2n$ incident voltage waves at all nodes of the network, and $[V]_{\text{ref}}$ represents all of the reflected (or scattered) voltage waves. In constructing this supermatrix, care must be used to ensure that the elements of the matrix correspond to the ordering of the current elements in the vector $[I_L]$.

Between each node of the TL segments, the voltage and current waves propagate according to the exponential function $e^{-\gamma \ell}$. For the entire network, a $2n \times 2n$ propagation supermatrix may be defined as

$$[\Gamma] = \begin{bmatrix} [0] & \text{diag}[e^{+\gamma_i \ell_i}] \\ \text{diag}[e^{+\gamma_i \ell_i}] & [0] \end{bmatrix}$$

where the individual matrix $\text{diag}[e^{+\gamma_i \ell_i}]$ is an $n \times n$ diagonal matrix containing the individual propagation terms for each of the TL segments and involve the propagation constants and segment lengths. In this expression, ℓ_i denotes the length of the ith transmission line section and the matrix $[0]$ is an $n \times n$ null matrix.

With these expressions defined, the BLT equation for the currents on the network can be written as

$$[I_L] = [Y_c]\{[U] - [\rho]\}\{[-\rho] + [\Gamma]\}^{-1}[S] \tag{7.85}$$

where $[U]$ is the $2n \times 2n$ identity matrix, $[Y_c]$ is a $2n \times 2n$ diagonal supermatrix whose elements are the characteristic admittances of each of the TL segments of the network, and $[S]$ is the $2n$ excitation supervector. Note that this equation has the same form as the BLT equation (7.35a) for the single line, except that now the entire network is being considered. A similar equation for the voltages of the network at the nodes can be written as

$$[V_L] = \{[U] + [\rho]\}\{[-\rho] + [\Gamma]\}^{-1}[S] \tag{7.86}$$

which corresponds to Eq. (7.35b) for the single-line case.

For an incident plane-wave excitation, the terms in $[S]$ in Eqs. (7.85) and (7.86) contain expressions similar to those of Eq. (7.36) or (7.37) for each TL segment. For discrete voltage or current excitations of the source, terms given in Eq. (6.90) are to be used.

A typical application of the network BLT equation is found in the study of an electromagnetic field coupling to the fuselage, the wings, and the internal circuits of an aircraft [7.36]. Measurements in the frequency domain on a 1/10-scale model of an aircraft have validated the computed results.

7.7 TRANSMISSION LINES WITH NONLINEAR LOADS

Although it is usually stated that a time-harmonic analysis is not capable of being used for problems involving nonlinearities, there is a method that permits determination of a nonlinear system response in an indirect manner. This method was illustrated for an antenna connected to a nonlinear load [7.38] and has been used subsequently by other researchers for a transmission line [7.39,7.40]. This method uses a frequency-domain analysis of the *linear* portion of the problem to develop a Thévenin or Norton equivalent circuit at the location where the nonlinear element is located. This equivalent circuit is then converted to a time-domain equivalent circuit using a Fourier transform, and with the specified nature of the nonlinear device specified, a Volterra integral equation for the load response is found. This equation is solved by a time-marching procedure.

7.7.1 Volterra Integral Equation

As an example of this procedure, consider the simple transmission line shown in Figure 7.40. The transmission line is assumed to have a nonlinear impedance load located at the $x = \mathcal{L}$ end of the line. For this nonlinear element, the $v-i$ relationship is assumed to be of the form $v_t(t) = F[i_t(t)]$,

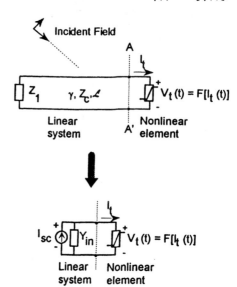

FIGURE 7.40 Transmission line with a nonlinear load.

where F is a nonlinear function. To the left of the load (at cut $A-A'$ in the figure) the remaining portion of the line is a linear system and may be described by the transmission line relations for a line having length $\mathcal{L}$, propagation constant γ, and characteristic impedance Z_c. Although the line is assumed to be excited by an incident field, the following development will also be applicable for a line having a lumped excitation source along the line.

As noted in Figure 7.40, the linear portion of the line can be represented by an equivalent Norton circuit. In the frequency domain, the input admittance of this circuit is given by $Y_{in} = 1/Z_{in}$, where Z_{in} is the input impedance of the line given in Eq. (7.82). The short-circuit current is given by $I_{sc} = V_{oc}/Z_{in}$, where V_{oc} is the open-circuit voltage in Eq. (7.84).

The $v-i$ description of the nonlinear element is provided in the time domain. To include in the analysis the excitation provided by the linear portion of the problem, the time-harmonic functions I_{sc} and Y_{in} may be converted into the time domain by a numerical Fourier transform. In this manner, the equivalent Norton circuit in Figure 7.40 can be viewed as a circuit in the *time domain*, with a time-varying surge admittance $y_{in}(t)$. Applying Kirchhoff's current law at the load and noting that the current through the internal admittance $i_y(t)$ is defined as a convolution operation,

$$i_y(t) = y_{in}(t) * v_t(t) = y_{in}(t) * F[i_t(t)]$$

the transient current through the load can be written as a nonlinear Volterra equation,

$$i_t(t) = i_{sc}(t) - \int_0^t y_{in}(t-\tau) F[i_t(\tau)] \, d\tau \qquad (7.87)$$

The solution for $i_t(t)$ from this equation is described in [7.38, 7.40] and involves a time-marching solution in which a nonlinear root-finding procedure must be employed in each time step. Assuming that $i_{sc}(t)$ and $y_{in}(t)$ are defined at temporal sample points $t_m = m \, \Delta t$, where $m = 0, 1, \ldots, m_{max}$, the integral equation (7.87) at a time t_m can be written as

$$i_t(m \, \Delta t) + y_{in}(0) F[i_t(m \, \Delta t)] \, \Delta t = i_{sc}(m \, \Delta t) - \sum_{k=0}^{m-1} y_{in}(m \, \Delta t - k \, \Delta t) F[i_t(k \, \Delta t)] \, \Delta t$$

$$(7.88)$$

If the line is lossless and has a load impedance equal to the characteristic impedance at the $x = 0$ end, the frequency-domain input admittance at $x = \mathcal{L}$ is $Y_{in} = 1/Z_c$, which is a purely resistive quantity. In the time domain, this admittance corresponds to an impulse function $y_{in}(t) = (1/Z_c)\delta(t)$. Other representations for a more general function $y_{in}(t)$ will usually contain an impulse function at $t = 0$ plus late-time contributions. In the discrete representation for the function $y_{in}(t)$, the term $y_{in}(0)\,\Delta t$ corresponds to the delta-function term.

Noting that the right-hand side of Eq. (7.88) depends only on *past* values of t_i, this equation is of the form

$$i_t(t_i) + y_{in}(0)F[i_t(t_i)]\,\Delta t + K = 0$$

where K is a known function of previous values of i_t. This equation may be solved for the unknown current i_t using a standard root-finding algorithm [7.41].

7.7.2 Example of a Single Transmission Line with a Nonlinear Load Impedance

As an example of such a calculation, consider the case of a field-excited transmission line over a perfectly conducting ground, as shown in Figure 7.19. The line length is $\mathcal{L} = 30$ m, the height over ground is $h = 2$ m, and the radius is $a = 1$ cm. The termination impedance at $x = 0$ is a *linear* load of $Z_1 = 50\,\Omega$. At the $x = \mathcal{L}$ end, the v–i relationship for Z_2 is assumed to be a *nonlinear* function, modeled by a simple piecewise linear relationship: $Z_2 = 50\,\Omega$ for $i(t) > 0$ and $Z_2 = 5000\,\Omega$ for $i(t) < 0$. Other more complicated representations of nonlinear elements are possible in this method, but this simple nonlinearity will serve to illustrate the response.

The line is illuminated by a vertically polarized plane wave-waveform of the form $E^{inc}(t) = 52{,}500 \times (e^{-4\times 10^6 t} - e^{-4.76\times 10^8 t})$, with angles of incidence $\psi = 45°$ and $\phi = 0°$. Thus the excitation is typical of an electromagnetic pulse, with a peak E-field amplitude of 50 kV/m.

Using the frequency-domain analysis presented in this chapter, the spectra of the short-circuit current $I_{sc}(\omega)$ and the input admittance $Y_{in}(\omega)$ at the location $x = \mathcal{L}$ can be calculated. Figure 7.41 illustrates the magnitudes of these spectra. Note that in the spectrum for $I_{sc}(\omega)$, the effects of the excitation field has already been included.

The transient responses corresponding to the spectra of Figure 7.41 are illustrated in Figure 7.42. With these transient responses defined, Eq. (7.88) can be solved for the current through the nonlinear element, $i_t(t)$, at each time step, $i\,\Delta t$, by marching on in time. Figure 7.43a illustrates the calculated load current. The corresponding load voltage $v_t(t)$ can then be easily determined by evaluating the nonlinear equation $v_t(t) = F(i_t(t))$. This is illustrated in Figure 7.43b. Notice that a large current flows through the nonlinear load in the positive direction when the load impedance is equal to $50\,\Omega$. However, when the sign of the current changes, the load impedance becomes $5000\,\Omega$ and the current is considerably reduced. The load voltage, on the other hand, becomes large when the current is blocked.

This treatment of nonlinear loads can be extended to the case of a general linear N-port network with different nonlinear loads at each output port.

7.7 TRANSMISSION LINES WITH NONLINEAR LOADS

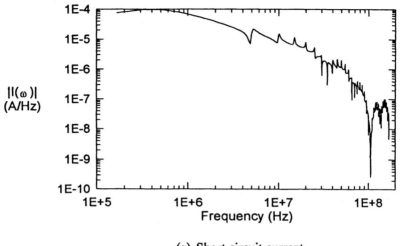

(a) Short circuit current

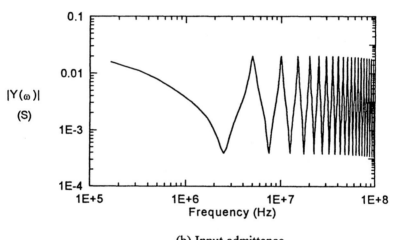

(b) Input admittance

FIGURE 7.41 Plots of the short-circuit current spectrum (*a*) and the input admittance (*b*) at $x = \mathcal{L}$ for a field-excited transmission line as illustrated in Figure 7.19.

This network could be considered to be a transmission line network as in [7.40], or any other system that is described by the short-circuit admittance parameters y_{ij}. The relevant Volterra equation for the n-vector of load currents $[i(t)]$ is given by the matrix integral equation

$$[i(t)] = [i_{sc}(t)] - \int_0^t [y(t-\tau)][F(i_i(\tau))]\,d\tau \qquad (7.89)$$

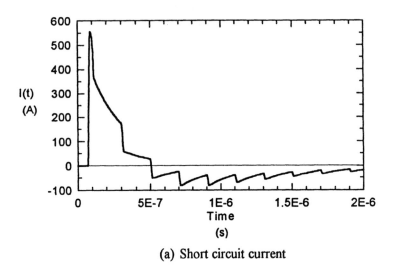

(a) Short circuit current

(b) Surge admittance

FIGURE 7.42 Plots of the transient short circuit current (a) and the surge admittance (b) at $x = \mathcal{L}$.

where $[y(t - \tau)]$ is an $n \times n$ matrix whose elements are the inverse Fourier transforms of the corresponding elements of the y matrix of the frequency domain and $[i_{sc}(t)]$ is an n-vector containing each transient Norton current source for the network. This matrix equation is solved by a time-marching procedure in the same manner as was done for the scalar equation.

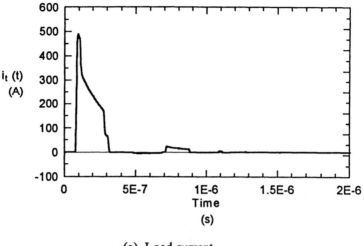

(a) Load current

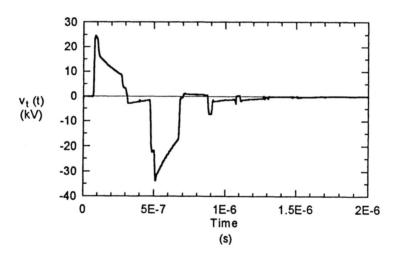

(b) Load voltage

FIGURE 7.43 Plots of the transient load current across the nonlinear load (a) and the corresponding transient load voltage (b), as computed from the frequency-domain data.

As noted in [7.40], a dual equation can be developed for the nonlinearly loaded system involving the open-circuit voltage $v_{oc}(t)$ and the input impedance $z_{in}(t)$ of the system. This alternative expression involves the nonlinear i–v relationship for the load as $i_t(t) = G(v_t(t))$, where G is a nonlinear admittance function. In this manner the Volterra equation for the scalar case is

$$v_t(t) = v_{oc}(t) - \int_0^t z_{in}(t - \tau)G(v_t(\tau)) \, d\tau \qquad (7.90)$$

with a similar matrix equation for the N-port case.

REFERENCES

7.1. Taylor, C. D., R. S. Satterwhite, and C. W. Harrison, "The Response of a Terminated Two-Wire Transmission Line Excited by a Nonuniform Electromagnetic Field," *IEEE Trans. Antennas Propag.*, Vol. AP-13, No. 6, November 1987, pp. 987–989.

7.2. Agrawal, A. K., et al., "Transient Response of Multiconductor Transmission Lines Excited by a Nonuniform Electromagnetic Field," *IEEE Trans. Electromagn. Compat.*, Vol. EMC-22, No. 2, May 1980.

7.3. Rashidi, F., "Formulation of Field-to-Transmission Line Coupling Equations in Terms of Magnetic Excitation Field," *IEEE Trans. Electromagn. Compat.*, Vol. EMC-35, No. 3, August 1993.

7.4. Paul, C. R., *Analysis of Multiconductor Transmission Lines*, Wiley, New York, 1994.

7.5. Tesche, F. M., "Plane Wave Coupling to Cables, Part II," Chapter 2 in *Handbook of Electromagnetic Compatibility*, R. Perez, ed., Academic Press, New York, 1995.

7.6. Ushida, H., *Fundamentals of Coupled Lines and Multiwire Antennas*, Sasaki Press, Sendi, Japan, 1967.

7.7. Smith, A. A., Jr., *Coupling of External Electromagnetic Fields to Transmission Lines*, Wiley, New York, 1977.

7.8. Burke, G. J., and A. J. Poggio, "Numerical Electromagnetic Code (NEC): Method of Moments," *NOSC Technical Document 116*, Naval Ocean Systems Center, San Diego, CA, January 1980.

7.9. Lee, K. S. H., "Two Parallel Terminated Conductors in External Fields," *IEEE Trans. Electromagn. Compat.*, Vol. EMC-20, No. 2, May 1978, pp. 288–297.

7.10. Plonsey, R., and R. E. Collin, *Principles and Applications of Electromagnetic Fields*, McGraw-Hill, New York, 1961.

7.11. Van Bladel, J., *Electromagnetic Fields*, McGraw-Hill, New York, 1964.

7.12. Ramo, S., J. R. Whinnery, and T. Van Duser, *Fields and Waves in Communication Electronics*, 2nd Ed., Wiley, New York, 1989.

7.13. Vance, E. F., *Coupling to Shielded Cables*, R. E. Krieger, Melbourne, FL, 1987.

7.14. Hildebrand, F. B., *Advanced Calculus for Applications*, Prentice Hall, Englewood Cliffs, NJ, 1963.

7.15. Nucci, C. A., and F. Rachidi, "On Field-to-Transmission Line Coupling Models," *Proceedings of the Progress in Electromagnetic Research Symposium*, ESA, Noordwijk, The Netherlands, July 11–15, 1995.

7.16. Ari, N., and W. Blumer, "Analytic Formulation of the Response of a Two-Wire Transmission Line Excited by a Plane Wave", *IEEE Trans. Electromagn. Compat.*, Vol. EMC-30, No. 1, February 1988.

7.17. Bell Laboratories, *EMP Engineering and Design Principles*, Technical Publications Department, Bell Laboratories, Whippany, NJ, 1975.

7.18. Lee, K. S. H., ed., *EMP Interaction: Principles, Techniques and Reference Data*, Hemisphere, New York, 1989.

7.19. Stratton, J. A., *Electromagnetic Theory*, McGraw-Hill, New York, 1941.

7.20. Jones, D. S., *The Theory of Electromagnetism*, Pergamon Press, Oxford, 1964.

7.21. Tai, C.-T., *Dyadic Green's Functions in Electromagnetic Theory*, Intext Publishers, Scranton, PA, 1971.

7.22. Tesche, F. M., and A. R. Neureuther, "The Analysis of Monopole Antennas Located on a Spherical Vehicle, Part 1, Theory," *IEEE Trans. Electromagn. Compat.*, Vol. EMC-18, No. 1, February 1976, pp. 2–8.

7.23. Nucci, C. A., F. Rachidi, M. Ianoz, and C. Mazzetti, "Comparison of Two Coupling Models for Lightning-Induced Overvoltage Calculations," *IEEE Trans. Power Deliv.*, Vol. PD-10, No. 1, January 1995, pp. 330–336.

7.24. Rubinstein, M., A. Y. Tzeng, M. A. Uman, P. J. Medelius, and E. W. Thomson, "An Experimental Test of a Theory of Lightning-Induced Voltages on an Overhead Wire," *IEEE Trans. Electromagn. Compat.*, Vol. EMC-31, No. 4, November 1989, pp. 376-383.

7.25. Rubinstein, M., M. A. Uman, P. J. Medelius, and E. W. Thompson, "Measurements of the Voltage Induced on an Overhead Power Line 20 m from Triggered Lightning," *IEEE Trans. on Electromagn. Compat.*, Vol. EMC-36, No. 2, May 1994, pp. 134–140.

7.26. Uman, M. A., *The Lightning Discharge*, Academic Press, Inc., New York, 1987.

7.27. Master, M. J., and M. A. Uman, "Lightning Induced Voltages on Power Lines: Theory," *IEEE Trans. Power Appar. Syst.*, Vol. PAS-103, No. 9, September 1984, pp. 2502–2518.

7.28. Zeddam, A., and P. Degauque, "Current and Voltage Induced on Telecommunications Cable by a Lightning Return Stroke," in *Lightning Electromagnetics*, R. L. Gardner, ed., Hemisphere, New York, 1990, pp. 377–400.

7.29. Diendorfer, G., "Induced Voltage on an Overhead Line Due to Nearby Lightning," *IEEE Trans. Electromagn. Compat.*, Vol. EMC-32, November 1990, pp. 292-299.

7.30. Chowdhuri, P., "Lightning-Induced Voltages on Multiconductor Overhead Lines," *IEEE Trans. Power Deliv.*, Vol. PD-5, April 1990, pp. 658-667.

7.31. Nucci, C. A., et al., "Lightning-Induced Voltages on Overhead Lines," *IEEE Trans. Electromagn. Compat.*, Vol. EMC-35, No. 1, February 1993.

7.32. Nucci, C. A., and F. Rachidi, "Experimental Validation of a Modification to the Transmission Line Model for LEMP Calculations", *Proceedings of the 8th International Symposium and Technical Exhibition on EMC*, Zurich, March 10–12, 1989, pp. 389–394.

7.33. Monteath, G. D., *Applications of the Electromagnetic Reciprocity Principle*, Pergamon Press, Oxford, 1973.

7.34. Tesche, F. M., and T. K. Liu, "On the Development of a General Transmission Line Model for EMC and EMP Applications," *Proceedings of the 2nd*

Symposium and Technical Exhibition on EMC, Montreaux, Switzerland, June 1977.

7.35. Baum, C. E., T. K. Liu, and F. M. Tesche, "One the Analysis of General Multiconductor Transmission Line Networks," *AFWL Interaction Note 350*, Air Force Weapons Laboratory, Kirtland AFB, NM, 1978.

7.36. Tesche, F. M., and T. K. Liu, "Application of Multiconductor Transmission Line Network Analysis to Internal Problems," *Electromagnetics*, Vol. 6, No. 1, 1987.

7.37. Besnier, Ph., "Etude des couplages électromagnétiques sur des réseaux de lignes de transmission non-uniformes à l'aide d'une approche topologique," Ph.D. thesis, 1060, Université des Sciences et Techniques de Lille, France, 1993.

7.38. Liu, T. K., and F. M. Tesche, "Analysis of Antennas and Scatterers with Nonlinear Loads," *IEEE Trans. Antennas Propag.*, Vol. AP-24, No. 2, March 1977.

7.39. Djordjevic, A. R., T. K. Sarkar, and R. F. Harrington, "Analysis of Lossy Transmission Lines with Arbitrary Nonlinear Terminal Networks," *IEEE Trans. Microwave Theory Tech.*, Vol. MTT-34, No. 6, June 1987.

7.40. D'Amore, M., and M. S. Sarto, "EMP Coupling to Multiconductor Dissipative Lines with Nonlinear Loads Above a Lossy Ground," *Proceedings of the 10th International Symposium and Technical Exhibition on EMC*, Zurich, March 9–11, 1993, pp. 451–456.

7.41. Press, W. H. et al., *Numerical Recipes*, Cambridge University Press, Cambridge, 1987.

PROBLEMS

7.1 Summarize the important distinctions between the total-voltage (Taylor) and scattered-voltage (Agrawal) methods for determining the interaction of an incident plane wave with a transmission line. Is one approach preferred over another?

7.2 A vertically polarized time-harmonic field of frequency ω, denoted as $E^{inc}(\omega)$, is incident on an infinite line at a height h above a lossless ground, as shown in Figure 7.15.

(a) Compute the distributed voltage sources along the line which excite the transmission line currents.

(b) Assuming that the per-unit-length impedance Z' and admittance Y' are known, compute the line characteristic impedance and propagation constant in terms of these quantities.

(c) Determine the line voltage at a particular point x, due to the incident field. (*Hint:* Find the voltage at x due to a lumped voltage source V'_s at another location x' and then integrate from $x' = -\infty$ to x and from x to ∞.)

7.3 In Chapter 6 the transmission line solution for the line voltage and current at a point x due to a lumped voltage and current sources at a point x' on the line is described [see Eqs. (6.38) and (6.39)]. Consider now the excitation of the line by an incident plane wave. Using these equations:

(a) Determine the expressions for the field-induced line current and voltage at an arbitrary point x on the line.

(b) How does this solution differ from that provided by the BLT equation of Eq. (7.35)?

7.4 Show that for an *infinitely long* transmission line with an E-field incident parallel to the wire (see Figure P7.4 for the direction of the excitation field), there is an induced current on the line, but the corresponding charge density is zero. That this is true can be noted most easily by using the current continuity equation and arguments of symmetry in the induced current.

7.5 Verify, either through an analytical proof or by using the NULINE code in Appendix F, that for a finite line such as that shown in Figure P7.4, the line voltage at point P in the middle of the line is identically zero, up to the time $t = \mathcal{L}/2c$, which corresponds to the time that it takes an observer at P to realize that the line is not infinite in length.

7.6 Given the fact that there is a particular excitation of an infinite line that provides an induced current but no charge (line voltage), there must be a dual excitation that provides a line charge but *no current*. Describe this excitation.

7.7 Explain how a transmission line can produce a radiated field, given the fact that the voltage and current traveling waves on the line are composed of transverse EM (TEM) fields, which are nonradiating.

7.8 The NULINE code described in Appendix F also permits analysis of an incident plane-wave excitation of an aboveground transmission line as pictured in Figure 7.19. Either a time-harmonic excitation or various types of transient waveforms can be analyzed using this program. Consider a line 10 m high over a *perfectly conducting* ground, with a length of 100 m and excited by a vertically polarized

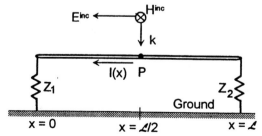

FIGURE P7.4 Finite line excited by a broadside E-field.

($\alpha_p = 0$) transient incident field with angles of incidence $\psi = 45°$ and $\phi = 0°$. Take the termination impedances to be 100 Ω at each end of the line. Using NULINE, compute the transient voltage and current in load 2 for a double-exponential incident field waveform of the form $E^{inc}(t) = A(e^{-\alpha_1 t} - e^{-\beta_1 t})$, where waveform parameters are $\alpha_1 = 4.0 \times 10^6 \, s^{-1}$ and $\beta_1 = 4.76 \times 10^8 \, s^{-1}$, and the term $A = 1.05$, so that the peak value of $E^{inc} = 1.0$. Examine the resulting waveforms and discuss the origins of the various features of the response. (*Hint:* Data set EX12.CMD contains the necessary input data for this calculation.)

7.9 Repeat the calculation for the line in Problem 7.8 with a unit-amplitude time-harmonic incident field over the frequency range 0.1 to 100 MHz. (File EX12A.CMD contains the data for this calculation.) What happens to the spectrum if both load responses are matched to the characteristic impedance of the line?

7.10 An overhead power line is fed by a transformer at the midpoint of the line, as shown in Figure P7.10. It is desired to obtain a Thévenin equivalent circuit of the line at the terminals of the transformer connection to predict the transient behavior when the line is illuminated by an incident transient plane wave. Assuming that the line has the dimensions shown in the figure and that the incident field is the unit-amplitude double exponential waveform discussed in Problem 7.8, compute the following:

(a) The input impedance $Z_{in}(\omega)$ at the terminals between the line and the ground.

(b) The transient open-circuit voltage $V_{oc}(t)$ at these terminals.

(c) The transient input impedance $z_{in}(t) = \text{IFFT}(Z_{in}(\omega))$. What is the meaning of *transient input impedance*?

(*Hint:* The NULINE code can be used to perform this analysis to check your answer. The data file EX14.CMD contains the data for the transient portion of the analysis.)

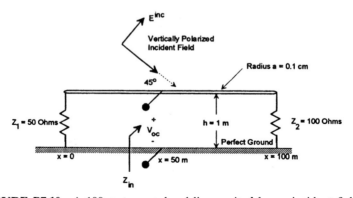

FIGURE P7.10 A 100-meter overhead line excited by an incident field.

7.11 Consider the transmission line shown in Figure P6.15a. Assume that the impedance $Z_1 = 0\,\Omega$ (i.e., a short circuit).

(a) Determine the transient behavior of the open-circuit voltage at the end of the line at $x = 9$ m by tracing the various voltage waves traveling back and forth on the line.

(b) Using NULINE, compute the transient open-circuit voltage response for the line. (*Hint:* Data file EX6.CMD contains the data.

(c) Discuss why the numerical result does not agree with the results of part (a).

(d) The NULINE code allows for an expansion of the resonance term of the BLT equation, as discussed in Section 7.4, to analyze highly resonant transmission lines in the time domain using the FFT. Run NULINE again, changing the *resonance treatment* parameter in the *Excitation* menu to 5 and observe the differences in the solution.

7.12 The transmission line coupling models developed in Chapter 7 and implemented in NULINE assume that the excitation E-field on the ends of the line can be represented by a single, lumped voltage source as shown in Figure 7.8. In many cases this is a good approximation to the actual excitation, which involves a *distributed* set of voltage sources along the risers as well as along the length of the line (see Figure P7.12a). As discussed in [7.26] and developed further by Tesche,† this line can be analyzed by considering the vertical risers to be additional field-excited transmission lines connected to the primary line, as shown in Figure P7.12b. For this particular model of the line:

(a) Discuss the cases when this more elaborate model is needed.

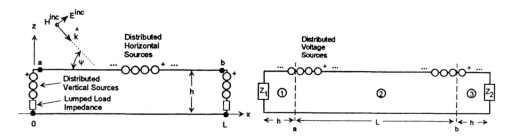

a. Physical model   b. Electrical model

FIGURE P7.12. Transmission line with distributed excitation at the ends of the line.

---

† F. M. Tesche and B. Brändli, "Observations on the Adequacy of Transmission Line Coupling Models for Long Overhead Cables", *Digest of EMC'94 ROMA International Symposium on EMC*, Rome, September 13–16, 1994.

(b) Outline the analysis procedure required for determining the induced line current due to an incident plane-wave field.

(c) Qualitatively describe the difference between responses arising from the lumped source excitation and the more complete distributed field excitation.

7.13 Consider the single-wire transmission line illustrated in Figure 7.19 with the following parameters: $\mathcal{L} = 5$ m, conductor radius $a = 0.15$ cm, $h = 0.8$ m, $Z_1 = Z_2 = 20\ \Omega$. The excitation of the line is assumed to be a transient, vertically polarized double-exponential waveform of the form given in Problem 7.8 with the angles of incidence $\psi = 45°$ and $\phi = 0°$. For this excitation we desire to calculate the voltage induced in load 2 at $x = \mathcal{L}$.

(a) Use the NULINE code to compute the transient response, using a FFT time window of 1.5 μs with 2048 sample points.

(b) The computer program RISER described in Appendix F is designed to calculate the response of a plane-wave excited line, taking into account the distributed field excitation on the vertical risers. Compute the response of the line using this code using the same FFT time window and number of points. Compare and contrast the results obtained with those in part (a).

7.14 The computer code LTLINE described in Appendix F is a program designed to analyze the load responses of the lightning-excited transmission line shown in Figure 7.22. For the transmission line parameters $\mathcal{L} = 1$ km, $h = 10$ m, $a = 5$ mm, and $Z_1 = Z_2 =$ matched, use this code to explore the line responses as the lightning channel parameters and the strike point are varied.

CHAPTER 8

Effects of a Lossy Ground on Transmission Lines

This chapter extends the transmission line models discussed in Chapters 6 and 7 to include the effects of a lossy earth serving as a return conductor. The telegrapher's equations are redefined here to include the lossy ground effects, and this results in additional impedance and admittance elements in the per-unit-length model for the transmission line. Various examples of aboveground cable responses are given both for continuous-wave (CW) and transient excitation. Transmission line models may also be applied to the analysis of buried cables. The theory behind a rigorous analysis is discussed and a suitable transmission line approximation for analyzing such problems is suggested.

8.1 INTRODUCTION

In Chapter 7 the behavior of a single conductor near a perfectly conducting ground was examined. For a lossless conductor of radius a at a height h over the ground, the transmission line propagation constant was equal to that of a plane wave in free space, $\gamma = j\omega/c$, and the characteristic impedance of the line was a real-valued quantity given by

$$Z_c = \sqrt{\frac{Z'}{Y'}} = \sqrt{\frac{j\omega L'}{j\omega C'}} \approx 60 \ln \frac{2h}{a}$$

As a consequence of there being no loss, disturbances on the line propagate without attenuation. If the ground plane is lossy, the impedance and propagation constants become complex quantities and there is an exponential attenuation of the voltage and current waves as they propagate along the line.

For an incident plane-wave field exciting of the line, we have seen that the perfectly conducting ground reflects the incident field and produces the net excitation field on the line. The reflection in the ground is "perfect" in that the amplitude and waveshape of the reflected field are identical to those of the incident field. In the case of a lossy ground, the reflected field has an amplitude less than the incident field and is broader, or dispersed, in time. Furthermore, there is a field transmitted into the ground. In the frequency domain the reflected and transmitted plane-wave fields are described by reflection and transmission coefficients which depend on the frequency and angle of incidence of the incident plane-wave field and on the electrical properties of the ground.

In this chapter we examine more closely the effect that a lossy ground plane has on a transmission line when the ground serves as the return conductor. For the first part of this chapter, the geometry of the problem is the same as in Figure 7.19, except that the ground is treated as a lossy half-space (of infinite extent in the $-z$ direction). Later, the case of a buried conductor will be treated. For both problems, the earth is modeled by a frequency-independent electrical conductivity σ_g and a dielectric constant $\epsilon_g = \epsilon_r \epsilon_0$, where ϵ_r is the relative dielectric constant of the ground.

8.2 DERIVATION OF THE TELEGRAPHER'S EQUATIONS

The analysis of the propagation characteristics of a wire over the earth has been examined in detail in the literature. The early work of Sommerfeld [8.1], Carson [8.2], and Sunde [8.3] has laid the foundation for the treatment of transmission line problems involving a lossy earth. A more recent investigation of multiconductor lines examines in detail the characteristics of the transmission and antenna modes on a multiconductor line over the earth [8.4] and concludes that the transmission line models provide reasonable results for engineering purposes for lines over the earth, except for cases when the incident field arrives at a grazing angle of incidence (i.e., for $\phi = 0$ and $\psi \to 0$ in Figure 7.19). This conclusion was also noted in [8.5].

As an alternative to a rigorous analysis, it is possible to derive the telegrapher's equations for the transmission line mode component of the total current using the approach developed in Chapter 7 for a wire over a perfect ground. Figure 8.1 illustrates an infinitely long conductor over a lossy ground. Once the telegrapher's equations and the appropriate source terms are derived for this line, the solution to a finite line with arbitrary termination impedances can be determined using the BLT equation (7.35) in the same way as for the perfect ground.

One conceptual difficulty in treating the problem shown in Figure 8.1 arises in the definition of the line voltage. For the perfect ground case, this voltage was defined as the line integral of the E-field from the wire at $z = h$ to the ground surface at $z = 0$. For the case of the lossy half-space, the E-field in the earth can be nonzero, and this will contribute to the voltage if the zero-voltage reference point is within the earth. The basic issue, therefore, is where is the location of the zero-voltage point? One possibility

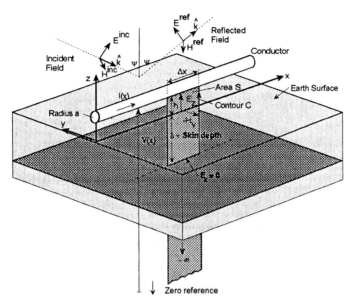

FIGURE 8.1 Geometry of the single-conductor line over a lossy half-space.

is to consider a voltage still between the wire and the earth surface. Another is to consider the voltage defined from the integral from wire to a location at $z = -\infty$, which is considered to be the zero voltage reference point for the entire problem. Regardless of how this voltage is defined, the main requirement is that the *current* and the per-unit-length *charge density* induced in the wire must be correct (to within the limits of the transmission line approximation). As we noted previously, these are the primary responses of the line, and the voltage is a secondary, or derived, response. In the discussion that follows we consider the "generalized voltage" of the line through the integral of the E_z-field from $-\infty$ to h.

8.2.1 Total Voltage Formulation

8.2.1.1 First Telegrapher's Equation. With the foregoing comments in mind, we can derive the transmission line equations for the line in Figure 8.1 using the total voltage formulation. This derivation parallels the development in Chapter 7 and results in equations for the line current and total voltage in terms of the excitation E and H-fields. Using Stokes' theorem, the first Maxwell curl equation (7.1) is integrated over the shaded area S of the figure extending from the wire at $z = h$ to the zero reference point at

$z = -\infty$. This results in the following equation for the total E and H fields:

$$\int_C \mathbf{E} \cdot d\mathbf{l} = -j\omega\mu_0 \iint_S \mathbf{H} \cdot d\mathbf{S} \tag{8.1}$$

which, as before, can be expressed in terms of the field components as

$$\int_{-\infty}^{h} [E_z(x + \Delta x, z) - E_z(x, z)] \, dz - \int_{x}^{x+\Delta x} [E_x(x, h) - E_x(x, -\infty)] \, dx$$

$$= -j\omega\mu_0 \int_{-\infty}^{h} \int_{x}^{x+\Delta x} -H_y \, dx \, dz \tag{8.2}$$

The component $E_x(x, h)$ is zero if the wire is a perfect conductor. If the wire has finite conductivity, $E_x(x, h)$ is related to the current through the impedance relationship of Eq. (7.17). When the ground is lossy, the electromagnetic fields are able to penetrate the ground. This E-field component is usually much smaller than the field in the air. This is because a significant amount of the incident field reflects from the ground surface and does not penetrate the soil. Thus any contribution to the total line response by the E-field in the soil is usually small compared with the contributions of the aboveground fields.

The fields in the earth attenuate exponentially with the depth. At the voltage reference location at $z = -\infty$, the field $E_x(x, -\infty) = 0$. For a finite depth, the fields in the earth never fall completely to zero. However, for practical purposes the skin depth in the ground is a good measure of the penetration depth of the fields and is given by the formula

$$\delta = \sqrt{\frac{2}{\omega\mu_0\sigma_g}} \tag{8.3}$$

Typical penetration depths at 1 MHz range from about 0.1 mm for metallic ground planes to 5 m for a lossy earth with a conductivity of 0.01 S/m. As a result of the low field strengths in the ground and the exponential decay of the fields, the $-\infty$ limits in the integrals in Eq. (8.2) can be replaced by $-\delta$ with little consequence.

The H-field in Eq. (8.1) again can be decomposed into an *excitation* component H^{ex}, which exists when the wire is removed from the problem, and a *scattered* component H^{sca}, due to the current flowing on the wire and returning in the ground. Above ground $(z > 0)$, the excitation field is the sum of the incident field H^{inc} and the earth-reflected field H^{ref}. Below ground $(z < 0)$, the excitation H-field is the transmitted field H^t. With these field components and noting that the line-to-reference voltage V is given by

$$V = -\int_{-\infty}^{h} E_z \, dz \tag{8.4}$$

the limit as $\Delta x \to 0$ in Eq. (8.2) can be taken, resulting in the equation

$$\frac{dV(x)}{dx} + Z_w I(x)$$

$$= -j\omega\mu_0 \int_{-\infty}^{h} H_y(x,z)\,dz = -j\omega\mu_0 \int_{-\infty}^{h} H_y^{ex}(x,z)\,dz - j\omega\mu_0 \int_{-\infty}^{h} H_y^{sca}(x,z)\,dz$$

$$= -j\omega\mu_0 \int_{0}^{h} [H_y^{inc}(x,z) + H_y^{ref}(x,z)]\,dz - j\omega\mu_0 \int_{-\infty}^{0} H_y^t(x,z)\,dz$$

$$- j\omega\mu_0 \int_{-\infty}^{h} H_y^{sca}(x,z)\,dz \qquad (8.5)$$

The last term in Eq. (8.5) is proportional to the flux arising from the wire current and the return current in the ground. If the ground is a perfect conductor, this term is simply equal to $j\omega L'$, where L' is the inductance of the conductor over the ground, given by Eq. (7.42). For the lossy ground case, it is convenient to consider lumping the contributions to the scattered flux from the ground currents into an additional per-unit-length impedance term Z'_g. In this manner, this last integral can be expressed in terms of the line current $I(x)$ as

$$j\omega\mu_0 \int_{-\infty}^{h} H_y^{sca}(x,z)\,dz = (j\omega L' + Z'_g)I(x) \qquad (8.6)$$

Later in this chapter we summarize several different forms for the ground impedance term Z'_g.

The integral of the excitation H-field in Eq. (8.5) has the dimensions of a per-unit-length voltage source and constitutes the voltage excitation source for the equation:

$$V'_{s1}(x) = -j\omega\mu_0 \int_{-\infty}^{h} H_y^{ex}(x,z)\,dz$$

$$= -j\omega\mu_0 \int_{0}^{h} [H_y^{inc}(x,z) + H_y^{ref}(x,z)]\,dz - j\omega\mu_0 \int_{-\infty}^{0} H_y^t(x,z)\,dz$$

$$(8.7)$$

Note that this voltage source involves the incident plus ground-reflected H-fields above the ground as in the case of the single-conductor line, plus an additional component due to the transmitted H-field into the ground. As mentioned previously, the lower limit of $-\infty$ can be replaced by the skin depth δ without changing the result drastically.

With the definitions in Eqs. (8.6) and (8.7), Eq. (8.5) becomes the first telegrapher's equation,

$$\frac{dV(x)}{dx} + Z'(\omega)I(x) = V'_{s1}(x) \qquad (8.8a)$$

with the per-unit-length impedance parameter being defined as

$$Z'(\omega) = j\omega L' + Z_g + Z_w \qquad (8.8b)$$

where the additional impedance term Z_w accounts for the internal impedance of the conductor. Note that this equation is the same form as the previous telegrapher's equation for the line over the perfect earth but has different expressions for distributed impedance and voltage source.

An alternative form for the source term $V'_{S1}(x)$ in Eq. (8.7) is possible, using Stokes' theorem to relate the integral of the transmitted H^t-field to the tangential excitation E-field at the earth surface. This is done by applying Eq. (8.1) to the transmitted E- and H-field over the shaded area in Figure 8.1 for $z < 0$ (i.e., within the earth only). This gives

$$\int_{-\infty}^{0} [E_z^t(x + \Delta x, z) - E_z^t(x, z)] \, dz - \int_{x}^{x+\Delta x} E_x^t(x, 0) \, dx =$$
$$-j\omega\mu_0 \int_{-\infty}^{0} \int_{x}^{x+\Delta x} -H_y^t(x, z) \, dx \, dy \qquad (8.9)$$

Again taking the limit as Δx approaches zero and using the approximation that the vertical component of the transmitted E-field $E_z^t(x, z)$ in the earth is very small [8.6], the following expression results:

$$E_x^t(x, 0) \approx -j\omega\mu_0 \int_{-\infty}^{0} H_y^t(x, z) \, dz$$

Using the fact that the tangential electric fields are continuous at the earth–air interface, we have $E_x^t(x, 0) = E_x^{ex}(x, 0)$. Thus the source term for the first telegrapher's equation (8.8a) can be expressed in terms of only the excitation fields above the ground as

$$V'_{S1}(x) = -j\omega\mu_0 \int_{0}^{h} H_y^{ex}(x, z) \, dz + E_x^{ex}(x, 0) \qquad (8.10)$$

In this manner, the source terms for this alternative form of the telegrapher's equation are related to the excitation (incident plus ground-reflected) magnetic flux linking the area between the line and earth surface *and* the electric field on the earth's surface arising from the tangential components of the incident and reflected E-fields.

8.2.1.2 Second Telegrapher's Equation

The derivation of the second telegrapher's equation proceeds in the same manner as in Section 7.2.1.2. Since this equation is simply the equation of continuity for the current $I(x)$

and the linear charge density $q'(x)$ on the line, we can start from Eq. (7.13),

$$\frac{dI(x)}{dx} + j\omega q'(x) = 0$$

The quantity $j\omega q'$ can be related to a scattered line voltage through a per-unit-length admittance Y'. Thus the continuity equation may be written as

$$\frac{dI(x)}{dx} + Y'V^{\text{sca}}(x) = 0 \tag{8.11}$$

This line admittance Y' consists of a *series* combination of a line-to-ground capacitance and an earth admittance Y'_g. Consequently, Y' can be represented as

$$Y' = \left(\frac{1}{j\omega C'} + \frac{1}{Y'_g}\right)^{-1} \tag{8.12}$$

where C' is the capacitance of the line over a perfectly conducting ground. As discussed by Vance [8.6], the term Y'_g is related to the ground impedance term Z'_g through the expression

$$Z'_g Y'_g \approx \gamma_g^2 = j\omega\mu_0(\sigma_g + j\omega\epsilon_g) \tag{8.13}$$

where σ_g and ϵ_g are the ground parameters and γ_g is the wave propagation constant in the earth. The justification for the form of the line admittance in Eq. (8.12) is not immediately evident from the development of the telegrapher's equations presented here. One must look more closely at the EM scattering solution for the same problem as discussed in [8.5] to verify that this form is correct.

By adding the contribution from the excitation field to Eq. (8.11) to obtain a total voltage response, we get the second telegrapher's equation similar to Eq. (7.45):

$$\frac{dI(x)}{dx} + Y'V(x) = I'_{S1}(x) \tag{8.14}$$

where the distributed current source is now given by

$$I'_{S1}(x) = -Y'\left[\int_0^h [E_z^{\text{inc}}(x,z) + E_z^{\text{ref}}(x,z)]\,dz\right] - Y'\int_{-\infty}^0 E_z^t(x,z)\,dz$$

$$\approx -Y'\left[\int_0^h [E_z^{\text{inc}}(x,z) + E_z^{\text{ref}}(x,z)]\,dz\right] \tag{8.15}$$

As noted previously, the vertical E-field in the conducting ground is generally small compared with the aboveground fields, so the integral from

$-\infty$ to 0 can usually be neglected. The limitations of this approximation are discussed in more detail in Section 8.3, and in Section 8.4.1.2 after Eq. (8.47).

8.2.2 Scattered Voltage Formulation

The telegrapher's equations for the line over the earth can also be developed for the line current and the scattered voltage on the line. This development parallels that of Section 7.2.2.2. The total E_z and H_y fields in the integral relationships of Eq. (8.2) are separated into excitation and scattered components. This equation becomes

$$\int_{-\infty}^{h} [E_z^{ex}(x + \Delta x, z) - E_z^{ex}(x, z)] \, dz + \int_{-\infty}^{h} [E_z^{sca}(x + \Delta x, z) - E_z^{sca}(x, z)] \, dz$$

$$- \int_{x}^{x+\Delta x} [E_x(x, h) - E_x(x, -\infty)] \, dx$$

$$= -j\omega\mu_0 \int_{-\infty}^{h} \int_{x}^{x+\Delta x} -[H_y^{ex}(x, z) + H_y^{sca}(x, z)] \, dx \, dz \qquad (8.16)$$

Note that the E_x components are still the total fields. Again taking the limit as $\Delta x \to 0$, using Eq. (8.6), and observing that $E_x(x, h) = Z'_w I(x)$, and $E_x(x, -\infty) = 0$, Eq. (8.16) becomes

$$\frac{dV^{sca}(x)}{dx} + Z'I(x) = -j\omega\mu_0 \int_{-\infty}^{h} H_y^{ex}(x, z) \, dz$$

$$+ \lim_{\Delta x \to 0} \int_{-\infty}^{h} [E_z^{ex}(x + \Delta x, z) - E_z^{ex}(x, z)] \, dz \qquad (8.17)$$

Applying Eq. (8.2) to the excitation field components and letting $\Delta x \to 0$ allows the integral terms on the right-hand side of Eq. (8.17) to be written in terms of the excitation E-field on the wire. In this manner we obtain

$$-j\omega\mu_0 \int_{-\infty}^{h} H_y^{ex}(x, z) \, dz + \lim_{\Delta x \to 0} \int_{-\infty}^{h} [E_z^{ex}(x + \Delta x, z) - E_z^{ex}(x, z)] \, dz$$

$$= E_x^{ex}(x, h)$$

$$= E_x^{inc}(x, h) + E_x^{ref}(x, h)$$

$$\equiv V'_{S2}(x) \qquad (8.18)$$

and the telegrapher's equation becomes

$$\frac{dV^{sca}(x)}{dx} + Z'I(x) = V'_{S2}(x) \qquad (8.19)$$

The second telegrapher's equation for the case of the scattered voltage formulation is given by Eq. (8.11).

It is important to note that the source term in Eq. (8.19) involves only the tangential excitation E-field on the conductor. It does not contain the tangential E-field on the ground, as was found for the Taylor formulation in Eq. (8.10). This is consistent with the view of this problem from the EM scattering viewpoint, as discussed in Section 7.3.1.3.

In using either the scattering formulation of this section or the total field formulation of previous sections, caution should be exercised in interpreting what is meant by the term *line voltage*. This voltage is the potential difference between the line and a reference point at infinity in the soil. For an observer located on the earth's surface, the potential difference between the wire and top of the ground may also be of interest. This quantity, however, is not what is computed directly from the transmission line model.

8.2.3 Termination Conditions

8.2.3.1 V–I Relationships The telegrapher's equations for the line over the lossy ground can be put into the matrix notation of Eq. (7.22) or (7.29), depending on whether the voltage response is the total or the scattered response. In either case, for a finite line with loads at $x = 0$ and $= \mathcal{L}$, the boundary conditions for the load current and voltage must be enforced. For the total voltage formulation, the boundary conditions are given simply by Eq. (7.23):

$$V(0) = -Z_1 I(0) \quad \text{and} \quad V(\mathcal{L}) = Z_2 I(\mathcal{L})$$

For the case of the scattered voltage formulation, the end conditions are of the form of Eq. (7.30), but with the integration extending to $-\infty$:

$$V^{\text{sca}}(0) = -Z_1 I(0) + \int_0^h [E_z^{\text{inc}}(0, z) + E_z^{\text{ref}}(0, z)] \, dz + \int_{-\infty}^0 E_z^t(0, z) \, dz$$

$$\approx -Z_1 I(0) + \int_0^h [E_z^{\text{inc}}(0, z) + E_z^{\text{ref}}(0, z)] \, dz \quad (8.20a)$$

$$V^{\text{sca}}(\mathcal{L}) = Z_1 I(\mathcal{L}) + \int_0^h [E_z^{\text{inc}}(\mathcal{L}, z) + E_z^{\text{ref}}(\mathcal{L}, z)] \, dz + \int_{-\infty}^0 E_z^t(\mathcal{L}, z) \, dz$$

$$\approx Z_1 I(\mathcal{L}) + \int_0^h [E_z^{\text{inc}}(\mathcal{L}, z) + E_z^{\text{ref}}(\mathcal{L}, z)] \, dz \quad (8.20b)$$

where again the integral of the vertical E-field in the earth can be neglected if the earth is a good conductor.

As before, the presence of the integrals of the vertical excitation E-fields in the end conditions can be interpreted as two lumped voltage sources at each end of the line when the scattering formulation is used. Thus the view of the line excitation provided by Figure 7.17 is also valid for the case of the line over an imperfectly conducting ground.

8.2.3.2 Grounding Impedance.
For both formulations, care must be used in defining the termination impedances Z_1 and Z_2. An impedance element Z_L connecting the transmission line to the ground, as shown in Figure 8.2, will be in series with a grounding (or "footing") impedance Z_f, with the total load impedance seen by the line being $Z_1 = Z_L + Z_f$.

The value of the footing impedance depends on the nature of the connection of physical load Z_L to the earth as well as on the electrical characteristics of the soil. Generally, this is a complex-valued quantity at high frequencies and little information is available about the behavior of this footing impedance for specific electrode shapes. However, Sunde [8.3] presents an analysis of the dc footing resistance of different electrodes in the earth, and these results can be used as an approximation for the footing impedance.

Footing resistances for several grounding electrodes are summarized in Appendix C. As seen in the appendix and as noted in Figure 8.2, the footing resistance is calculated assuming that the reference conductor for the problem is situated an infinite distance from the earth surface. The resistances presented in this table are for the electrode conductors only and neglect any additional resistance contribution due to the wires connecting the electrodes. In realistic grounding installations, care is taken to have this grounding resistance be as small as possible, usually for safety reasons. Typical footing resistances for power system and communications applications can range from several ohms to several tens of ohms.

8.2.4 Solution of the Telegrapher Equations

The telegrapher's equations for the line over the lossy earth have the same form as the equations for the two-conductor line given in Eq. (7.22). As a

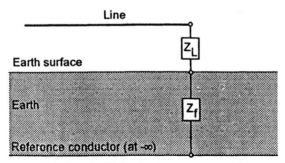

FIGURE 8.2 Connection of a load impedance Z_L to the reference conductor through a footing impedance, Z_f.

result, the solution for the load responses are given by the BLT equation (7.35). The propagation constant and per-unit-length impedance of the line are now given in terms of the per-unit-length line parameters Z' and Y' of Eqs. (8.8b) and (8.12) as $\gamma = \sqrt{Z'Y'}$ and $Z_c = \sqrt{Z'/Y'}$. Since γ now has a real part, the wave propagation term $e^{-\gamma x}$ on the line has an attenuation as well as a propagation component.

For the case of the total voltage formulation, the source term for the BLT equation is given by Eq. (7.36) with the distributed voltage and current sources given by Eqs. (8.10) and (8.15). For the scattered voltage formulation, the source term is given by Eq. (7.37), with the distributed field source given by Eq. (8.18), and the lumped voltage sources V_1 and V_2 are the integrals of Eq. (8.20). An alternative, but equivalent solution to the telegrapher's equations for a single line over a lossy ground excited by a plane wave with vertical polarization had been developed by Vance [8.6].

8.3 PER-UNIT-LENGTH LINE PARAMETERS

8.3.1 Equivalent Circuit for the Line

As noted above, the presence of the lossy return conductor causes a modification of the per-unit-length impedance and admittance parameters of the line. Figure 8.3 illustrates the equivalent circuit for a differential section of line in this case, together with distributed voltage and current sources arising from the incident field using the Taylor formulation. As noted in Section 8.2.1, the line inductance L' and capacitance C' are the same as for a line over a perfect earth:

$$L' = \frac{\mu_0}{2\pi} \ln \frac{2h}{a} \quad \text{and} \quad C' = \frac{2\pi\epsilon}{\ln(2h/a)}$$

where h is the height of the line over the earth and a is the conductor radius. If the line is assumed to be finitely conducting, the internal wire impedance Z'_w is given by Eq. (7.17). The shunt admittance Y'_g representing the ground

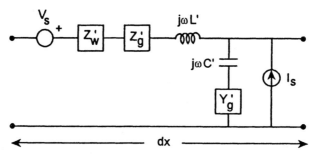

FIGURE 8.3 Differential section of a line over a lossy earth.

is approximately given from Eq. (8.13) as

$$Y'_g = \frac{\gamma_g^2}{Z'_g}$$

where γ_g is the propagation constant in the ground given by Eq. (8.13).

8.3.2 Frequency-Domain Representations for the Ground Impedance

Several different expressions are available for the earth impedance term Z'_g. Starting from a scattering viewpoint, it has been shown to be possible to identify individual terms in the scattering solution that correspond to the per-unit-length line parameters [8.4, 8.5, 8.7]. The term corresponding to the lossy ground can be identified in the external impedance term $Z^e(k_x)$ of a wire at a height h over the ground, which is discussed in [8.4]. For the present application, the axial propagation constant is $k_x = k \cos \psi \cos \phi$ [8.5], $k = \omega/c$ is the free-space wavenumber, and the angles ψ and ϕ are defined in Fig. 7.11.

We start with the expression for $Z_e(k_x)$ given as

$$Z^e(k_x) = \frac{j\omega\mu_0}{2\pi k^2} \frac{1}{(\tau_e a) K_1(\tau_e a)} \{\tau_e^2 [K_0(\tau_e a) - I_0(\tau_e a) K_0(\tau_e 2h)]$$

$$- I_0(\tau_e a)[k^2 J(\tau_e 2h) - k_x^2 G(\tau_e 2h)]\}$$

(8.21)

where the terms J and G are integrals defined as

$$J(\tau_e, 2h) = \int_{-\infty}^{\infty} \frac{1}{\sqrt{\lambda^2 + \tau_e^2} + \sqrt{\lambda^2 + \tau_g^2}} e^{-2h\sqrt{\lambda^2 + \tau_e^2}} \, d\lambda \qquad (8.22)$$

$$G(\tau_e, 2h) = \int_{-\infty}^{\infty} \frac{1}{n^2 \sqrt{\lambda^2 + \tau_e^2} + \sqrt{\lambda^2 + \tau_g^2}} e^{-2h\sqrt{\lambda^2 + \tau_e^2}} \, d\lambda \qquad (8.23)$$

where

$$\tau_e = \sqrt{k_x^2 - k^2} = jk\sqrt{1 - \cos^2\psi \cos^2\phi} \qquad (8.24)$$

denotes the transverse propagation constant, and

$$\tau_g = \sqrt{k_x^2 - k_g^2} \qquad (8.25)$$

with $jk_g \equiv \gamma_g$ denoting the complex propagation constant in the ground defined by Eq. (8.13). The symbol k_g has been chosen here to be consistent with [8.4]. In Eq. (8.21) the terms $I_0(\tau_e a)$, $K_0(\tau_e a)$ and $K_1(\tau_e a)$ are modified Bessel functions of complex argument.

The transverse propagation constant is always less than the free-space value k (i.e., $|\tau|_e \ll k$). For small arguments the Bessel functions $I_0(\tau_e a) \approx 1$ and $(\tau_e a) K_1(\tau_e a) \approx 1$ [8.8], which gives

$$Z^e(k_x) = \frac{j\omega\mu_0}{2\pi k^2} \{\tau_e^2 [K_0(\tau_e a) - K_0(\tau_e 2h)] - [k^2 J(\tau_e, 2h) - k_x^2 G(\tau_e, 2h)]\} \quad (8.26)$$

As noted in [8.5], the term corresponding to the ground impedance is $(j\omega\mu/2\pi) J(\tau_e, 2h)$. By defining $k_g/k = \sqrt{\epsilon_r + \sigma_g/j\omega\epsilon_0} \equiv n$ to be the complex index of refraction of the ground, the term τ_g becomes $\tau_g = \sqrt{k_x^2 - n^2 k^2}$, and the expression for the integral $J(\tau_e, 2h)$ may be written as

$$J(\tau_e, 2h) = 2 \int_0^\infty \frac{1}{\sqrt{\lambda^2 + k_x^2 - k^2} + \sqrt{\lambda^2 + k_x^2 - n^2 k^2}} e^{-2h\sqrt{\lambda^2 + k_x^2 - k^2}} d\lambda \quad (8.27)$$

Defining $\lambda^2 + k_x^2 = \xi^2$, the ground impedance can be written as [8.5]

$$Z_g'(\psi, \phi) = \frac{j\omega\mu_0}{\pi} \int_{k_x}^\infty \left(\frac{1}{\sqrt{\xi^2 - k^2} + \sqrt{\xi^2 - n^2 k^2}} \right) \frac{e^{-2h\sqrt{\xi^2 - k^2}}}{\sqrt{\xi^2 - k_x^2}} \xi \, d\xi \quad (8.28)$$

This last expression is a generalization of Sunde's formula for the ground impedance (see [8.3], p. 112), for the case of an incident wave illuminating a wire with angles of incidence ψ and ϕ. By setting $\psi = 0$ and $\phi = 0$, in (8.28), $k_x = k$ and with a change of variable $u = (\xi^2 - k^2)^{1/2}$, one arrives at Sunde's expression,

$$Z_g' \approx \frac{j\omega\mu_0}{\pi} \int_0^\infty \frac{e^{-2hu}}{u + \sqrt{u^2 + \gamma_g^2}} du \quad (8.29)$$

One approximation of Sunde's formula for low frequencies when $\sigma_g \gg \omega\epsilon_0\epsilon_r$ has been proposed by Carson in 1926 and is given in [8.2] as

$$Z_g' \approx \frac{\omega\mu_o}{\pi} \int_0^\infty \left(\sqrt{j + v^2} - v\right) e^{-2h'v} dv \quad (8.30)$$

where $v = u/\sqrt{\omega\mu_0\sigma_g}$ and $h' = h\sqrt{\omega\mu_0\sigma_g}$.

These expressions for Z_g' are not convenient for numerical calculations, since they involve an integral over an infinite interval. Chen [8.9] has found an approximation of the ground impedance expression that is independent of the angles (ψ, ϕ) and involves an integral over a finite interval as

$$Z'_g \approx \frac{j\omega\mu_0}{2\pi}\left[\frac{-1}{2(\gamma_g h)^2} - \frac{K_1(j2\gamma_g h)}{j\gamma_g h} - 2j\int_0^1 \sqrt{(1-\xi^2)}e^{-j2\gamma_g h\xi}\,d\xi\right] \quad (8.31)$$

The term $K_1(\cdot)$ is a modified Bessel's function. As noted in [8.5], this and the other approximate forms for Z'_g are accurate for nongrazing angles of incidence. However, as $\psi \to 0$, the transmission line solution can have significant error.

Another approximation to Z'_g is offered by Vance [8.6] as

$$Z'_g \approx \frac{-j\gamma_g}{2\pi h\sigma_g}\frac{H_0^{(1)}(j\gamma_g h)}{H_1^{(1)}(j\gamma_g h)} \quad (8.32a)$$

where $H^{(1)}(\cdot)$ are cylindrical Hankel functions. For high frequencies such that the skin depth $\delta = \sqrt{2}/\sqrt{\omega\mu_0\sigma_g} \ll 2h$, but with $\sigma_g \gg \omega\epsilon_g$, this impedance expression can be approximated further as

$$Z'_g \approx \frac{1+j}{2\pi\sigma_g\delta h} \quad (8.32b)$$

A very simple but accurate expression for Sunde's expression (8.29) is quoted in [8.3] as

$$Z'_g \approx \frac{j\omega\mu_0}{2\pi}\ln\frac{1+\gamma_g h}{\gamma_g h} \quad (8.33)$$

Other approximations have been proposed by Fontaine et al. [8.10] and by Gary [8.11]. The latter introduces the concept of the complex plane by replacing the real soil surface by a perfect conducting ground plane at a depth $d = 1/\sqrt{j\omega\mu_0\sigma_g}$, which as seen, is a complex quantity. It can be shown that in this case the expression of $Z'_g = (j\omega\mu_0/2\pi)\ln(1 + d/h)$ represents the low-frequency approximation of (8.32).

In examining the voltage or current responses of a transmission line using these various approximations, it has been concluded that expression (8.33) for the ground impedance provides a very good and easily calculable representation for the lossy ground effects [8.4, 8.7].

As an example of a comparison of these terms, Figure 8.4 illustrates the magnitude of the per-unit-length ground impedance term for a line with a height $h = 10$ m over a lossy ground with conductivity $\sigma_g = 0.01$ S/m and $\epsilon_r = 10$. The expression of Vance (8.32a) and the approximate expression of Sunde (8.33) are almost identical for frequencies from 10 kHz to 100 MHz. The more complicated expression of Sunde (8.29) is very close to these results and shows only a hint of a deviation at high frequencies. A calculation of the results of Chen (8.31) are virtually identical to these

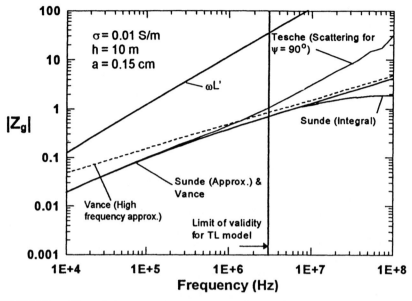

FIGURE 8.4 Plot of the per-unit-length impedance Z_g' due to a lossy earth using different expressions.

curves and are not shown in this figure. The dashed curve shows the high-frequency approximation of Eq. (8.32b). The result of the presumably more accurate scattering solution of [8.4] is shown for the angle of incidence $\psi = 90°$. For frequencies below 1 MHz, this curve is identical with the result of Sunde and Vance, but begins to deviate from these values at high frequencies. Note that it can be shown that for high frequencies, the ground impedance value tends asymptotically to a frequency-independent value [8.12].

This figure also shows a plot of the inductive component of the per-unit-length impedance of the transmission line (assuming a wire radius of $a = 0.15$ cm). It is seen that this term is larger than the earth impedance for all frequencies. However, because the earth impedance contains a real part, there is a frequency-dependent loss that causes distortions of the traveling waves on the line. Also shown in this figure is the frequency at which the line height is equal to $\lambda/10$, a point where the transmission line model is no longer strictly valid. Frequently, however, transmission line models are used for these higher frequencies, especially when a transient response is to be calculated using Fourier inversion. Typically, the excitation spectrum for the problem will decrease in amplitude at the higher frequencies, so that the effects of the errors in the transmission line coupling model will not be significant in the resulting transient response.

8.3.3 Time-Domain Representation of the Ground Impedance

For time-domain calculations of the field-to-transmission line coupling, Eqs. (8.19) and (8.11) can be written as

$$\frac{\partial v^{sca}(x,t)}{\partial x} + L'\frac{\partial i(x,t)}{\partial t} + \int_0^t z'_g(t-\tau)i(\tau)\,d\tau = V'_{s2}(x) \qquad (8.34a)$$

$$\frac{\partial i(x,t)}{\partial x} + G'v^{sca}(x,t) + C'\frac{\partial v^{sca}(x,t)}{\partial t} = 0 \qquad (8.34b)$$

where the transient term $z'_g(t)$ denotes the inverse Fourier transform of the per-unit-length impedance due to the earth, $Z'_g(\omega)$. Notice that this equation neglects the contribution from the per-unit-length admittance of the lossy ground.

The numerical calculation of the convolution integral in (8.34a) requires a considerable time and computer memory storage. To avoid this difficulty, one approach suggested by [8.13] is to calculate the ground impedance at a fixed frequency. The choice of the appropriate frequency can be difficult to determine and it cannot easily be related to physical quantities in a straightforward manner.

Attempts have been made to determine the function $z'_g(t)$ in closed form. Indeed, it is possible to find analytical inverse Fourier transforms for approximate expressions of the frequency-dependent ground impedance. In [8.14] it was shown that if the simplified approximation of Vance (8.32b) is written as $Z'_g \approx (1/2\pi h)\sqrt{j\omega\mu_0/\sigma_g}$, it has the following inverse Fourier transform

$$z'_g(t) = -\frac{1}{2\pi h}\sqrt{\frac{\mu_0}{4\pi\sigma_g t^3}} \qquad (8.35)$$

However, as observed in [8.15], at high frequencies the Vance simplified formula does not correctly reproduce the magnitude and phase of the ground impedance. A more accurate expression has been proposed by Timotin [8.16] in a paper published in the late 1960s. Using Sunde's expression (8.29) with the assumption $\sigma_g \gg \omega\epsilon_0\epsilon_r$, he derived an analytical inverse Fourier transform for the term $Z'_g/j\omega$, which is

$$\zeta'(t) = F^{-1}\left\{\frac{Z'_g}{j\omega}\right\} = \frac{\mu_0}{\pi\tau_g}\left[\frac{1}{2\sqrt{\pi}}\sqrt{\frac{\tau_g}{t}} + \frac{1}{4}e^{\tau_g/t}\,\text{erfc}\left(\sqrt{\frac{\tau_g}{t}}\right) - \frac{1}{4}\right] \qquad (8.36)$$

where

$$\text{erfc}(x) = \frac{2}{\sqrt{\pi}}\int_x^\infty e^{-t^2}\,dt \qquad (8.37)$$

is the complementary error function and $\tau_g = h^2 \mu_0 \sigma_g$ is the time constant of the ground. Note that to be consistent with the notations of the referred papers, we use τ_g to denote a time constant, which is different from the τ_g defined by Eq. (8.25), which represents a transverse propagation constant in the ground.

It should be noted that as Timotin's expression is a low-frequency approximation, with the validity limits of relation (8.36) given by $f \leq 10$ MHz for ground conductivities of about 10^{-3} S/m. To use this approximation at higher frequencies, the ground conductivity must be larger than 10^{-3} S/m.

The inverse Fourier transform function $z'_g(t)$ can be determined simply by taking the time derivative of Eq. (8.36). However, as suggested in [8.14], it is more convenient to use (8.36) directly in the coupling equation (8.34a) by noting that

$$Z'I(x) = \frac{Z'}{j\omega} j\omega I(x) \tag{8.38}$$

In the time domain, Eq. (8.34a) becomes

$$\frac{\partial v^{sca}(x,t)}{\partial x} + L' \frac{\partial i(x,t)}{\partial t} + \int_0^t \zeta'(t-\tau) \frac{\partial i(x,t)}{\partial t} d\tau = V'_{s2}(x) \tag{8.39}$$

The term $\zeta'(t)$ is the time integral of the impedance term $z'_g(t)$ and as noted in [8.15], this has a singular behavior at $t = 0$. This singularity, however, is integrable. From a numerical point of view, it can be modeled as a distribution function (a Dirac delta function) plus an additional continuous function.

8.4 REFLECTED AND TRANSMITTED PLANE-WAVE FIELDS

In the derivation of the excitation terms of the telegrapher's equations, we see that it is necessary to know certain components of the excitation (i.e., the incident plus ground-reflected) fields to compute the induced transmission line response. Summarizing these results, we see the following:

- For the total voltage formulation, we require the E_z^{ex} and H_y^{ex} field components for $h \geq z > -\infty$, or we need to know E_z^{ex} for $h \geq z > -\infty$, H_y^{ex} for $h \geq z > 0$, and E_x^{ex} on the earth surface for $z = 0$.
- For the scattered voltage formulation, we need to know E_x^{ex} along the wire at $z = h$, and E_z^{ex} for $h \geq z > -\infty$. As we will see, in many instances, the ground is sufficiently highly conducting so that the vertical E-field within the earth is small. Consequently, the E_z^{ex} field is needed only for $h \geq z > 0$.

8.4.1 Plane-Wave Reflection and Transmission from the Earth

8.4.1.1 General Expressions for the Fields.
The electric and magnetic fields produced by an incident plane wave striking an imperfectly conducting half-space is described in terms of the Fresnel reflection coefficients [8.5, 8.17]. With reference to Figure 8.5, an arbitrarily polarized incident plane wave of magnitude E_0 is divided into a vertically polarized component and a horizontally polarized component. If the incident E-field vector makes an angle α with the plane of incidence (measured clockwise as seen by an observer looking towards the source in Figure 7.11), the vertically polarized component is given by $E^{inc} = E_0 \cos \alpha$ and the horizontally polarized component is $E^{inc} = E_0 \sin \alpha$.

By Snell's law, the reflected field has the vertical angle $\psi_r = \psi$, and the transmitted angle ψ_t is given by the expression

$$\cos \psi_t = \frac{jk}{\gamma_g} \cos \psi \tag{8.40}$$

where γ_g is the propagation constant in the soil given by Eq. (8.13). Because γ_g is a complex quantity, the transmitted angle ψ_t is also complex, which implies that there is an attenuation of the fields propagating into the ground.

The incident fields shown in Figure 8.5 have angles of incidence (ψ and ϕ) and are expressed as follows:

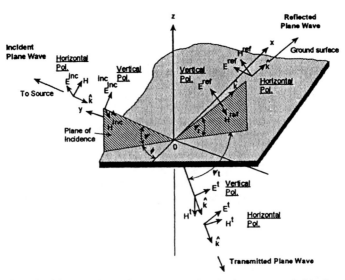

FIGURE 8.5 Incident, reflected, and transmitted plane-wave fields from an air–earth interface.

8.4 REFLECTED AND TRANSMITTED PLANE-WAVE FIELDS

VERTICAL POLARIZATION

$$\mathbf{E}^{inc} = E_0 \cos\alpha(\hat{x}\sin\psi\cos\phi - \hat{y}\sin\psi\sin\phi + \hat{z}\cos\psi)\,e^{-jkx\cos\psi\cos\phi}\,e^{jky\cos\psi\sin\phi}\,e^{jkz\sin\psi}$$

$$\mathbf{H}^{inc} = \frac{E_0}{Z_0} \cos\alpha(-\hat{x}\sin\phi - \hat{y}\cos\phi + \hat{z}0)\,e^{-jkx\cos\psi\cos\phi}\,e^{jky\cos\psi\sin\phi}\,e^{jkz\sin\psi}$$

HORIZONTAL POLARIZATION

$$\mathbf{E}^{inc} = E_0 \sin\alpha(\hat{x}\sin\phi + \hat{y}\cos\phi + \hat{z}0)\,e^{-jkx\cos\psi\cos\phi}\,e^{jky\cos\psi\sin\phi}\,e^{jkz\sin\psi}$$

$$\mathbf{H}^{inc} = \frac{E_0}{Z_0} \sin\alpha(\hat{x}\sin\psi\cos\phi - \hat{y}\sin\psi\sin\phi + \hat{z}\cos\psi)\,e^{-jkx\cos\psi\cos\phi}\,e^{jky\cos\psi\sin\phi}\,e^{jkz\sin\psi}$$

where $Z_0 \approx 377\,\Omega$ is the free-space wave impedance.

The reflected fields are related to the incident fields through the Fresnel refection coefficients, which are determined by matching the appropriate field components across the air–earth interface [8.17]. For the two polarizations, these reflected fields are expressed as follows:

VERTICAL POLARIZATION

$$\mathbf{E}^{ref} = E_0 \cos\alpha\, R_v(-\hat{x}\sin\psi\cos\phi + \hat{y}\sin\psi\sin\phi + \hat{z}\cos\psi)\,e^{-jkx\cos\psi\cos\phi}\,e^{jky\cos\psi\sin\phi}\,e^{-jkz\sin\psi}$$

$$\mathbf{H}^{ref} = \frac{E_0}{Z_0} \cos\alpha\, R_v(-\hat{x}\sin\phi - \hat{y}\cos\phi + \hat{z}0)\,e^{-jkx\cos\psi\cos\phi}\,e^{jky\cos\psi\sin\phi}\,e^{-jkz\sin\psi}$$

HORIZONTAL POLARIZATION

$$\mathbf{E}^{ref} = E_0 \sin\alpha\, R_h(\hat{x}\sin\phi + \hat{y}\cos\phi + \hat{z}0)\,e^{-jkx\cos\psi\cos\phi}\,e^{jky\cos\psi\sin\phi}\,e^{-jkz\sin\psi}$$

$$\mathbf{H}^{ref} = \frac{E_0}{Z_0} \sin\alpha\, R_h(\hat{x}\sin\psi\cos\phi - \hat{y}\sin\psi\sin\phi - \hat{z}\cos\psi)\,e^{-jkx\cos\psi\cos\phi}\,e^{jky\cos\psi\sin\phi}\,e^{-jkz\sin\psi}$$

where R_v and R_h are the Fresnel reflection coefficients for the vertically polarized and horizontally polarized fields, respectively. These coefficients are given in [8.6] as

$$R_v = \frac{\epsilon_r(1 + \sigma_g/j\omega\epsilon_r\epsilon_0)\sin\psi - [\epsilon_r(1 + \sigma_g/j\omega\epsilon_r\epsilon_0) - \cos^2\psi]^{1/2}}{\epsilon_r(1 + \sigma_g/j\omega\epsilon_r\epsilon_0)\sin\psi + [\epsilon_r(1 + \sigma_g/j\omega\epsilon_r\epsilon_0) - \cos^2\psi]^{1/2}} \qquad (8.41a)$$

$$R_h = \frac{\sin\psi - [\epsilon_r(1 + \sigma_g/j\omega\epsilon_r\epsilon_0) - \cos^2\psi]^{1/2}}{\sin\psi + [\epsilon_r(1 + \sigma_g/j\omega\epsilon_r\epsilon_0) - \cos^2\psi]^{1/2}} \qquad (8.41b)$$

These plane-wave reflection coefficients are complex functions of the earth parameters and the incidence angle ψ. Noting that a common factor in these expressions is $\sigma_g/\omega\epsilon_0$, it is possible to plot the magnitudes of the reflection coefficients as a surface plot, with the independent variables being ψ and $\sigma_g/\omega\epsilon_0$. In doing this, the relative permittivity of the ground is a parameter, so that a family of surfaces can be generated, one for each value of ϵ_r.

Figure 8.6 illustrates plots of the magnitude of the reflection coefficients for the vertical and horizontal reflection coefficients for $\epsilon_r = 1$ and 10. The vertical reflection coefficient exhibits a more complicated behavior than does the horizontal coefficient. For low values of conductivity, the ground appears as a perfect dielectric, and a "trough" or null in the vertical polarization coefficient surface appears for angles $\psi \approx 20$ to $30°$, depending on ϵ_r. This

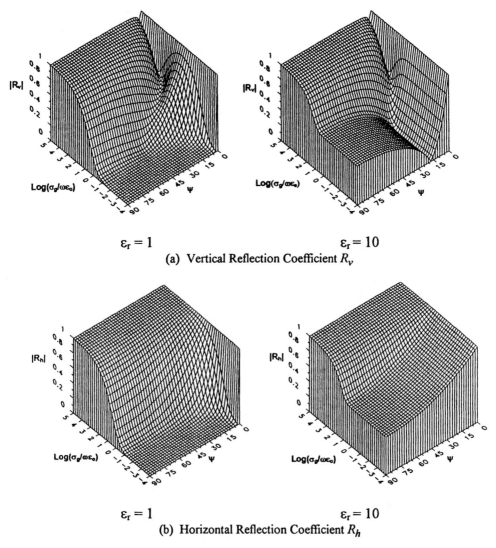

FIGURE 8.6 Surface plots of the magnitudes of the vertical reflection coefficient (*a*) and the horizontal reflection coefficient (*b*) for two different values of ϵ_r.

corresponds to the Brewster angle, where there is no reflected field. Notice that as the ground becomes perfectly conducting, $\sigma \to \infty$, and the reflection coefficients in Eqs. (8.27) have the limits $R_v \to 1$ and $R_h \to -1$, as long as $\psi \neq 0$. For $\psi = 0$ both R_v and $R_h \to -1$. Thus there is a rapid change of sign in the real part of the vertical reflection coefficient, which appears as the trough in part (a) of the figure.

For observer locations in the ground, the transmitted fields can be expressed in terms of a similar plane-wave function, but using the Fresnel transmission coefficients, T_v and T_h. These fields are given as

VERTICAL POLARIZATION

$$\mathbf{E}^t = E_0 \cos \alpha \, T_v (\hat{x} \sin \psi_t \cos \phi - \hat{y} \sin \psi_t \sin \phi + \hat{z} \cos \psi_t) \, e^{-\gamma_g x \cos \psi_t \cos \phi} e^{\gamma_g y \cos \psi_t \sin \phi} e^{\gamma_g z \sin \psi_t}$$

$$\mathbf{H}^t = \frac{E_0}{Z_{0g}} \cos \alpha \, T_v (-\hat{x} \sin \phi - \hat{y} \cos \phi + \hat{z} 0) \, e^{-\gamma_g x \cos \psi_t \cos \phi} e^{\gamma_g y \cos \psi_t \sin \phi} e^{\gamma_g z \sin \psi_t}$$

HORIZONTAL POLARIZATION

$$\mathbf{E}^t = E_0 \sin \alpha \, T_h (\hat{x} \sin \phi + \hat{y} \cos \phi + \hat{z} 0) \, e^{-\gamma_g x \cos \psi_t \cos \phi} e^{\gamma_g y \cos \psi_t \sin \phi} e^{\gamma_g z \sin \psi_t}$$

$$\mathbf{H}^t = \frac{E_0}{Z_{0g}} \sin \alpha \, T_h (\hat{x} \sin \psi_t \cos \phi - \hat{y} \sin \psi_t \sin \phi + \hat{z} \cos \psi_t) \, e^{-\gamma_g x \cos \psi_t \cos \phi} e^{\gamma_g y \cos \psi_t \sin \phi} e^{\gamma_g z \sin \psi_t}$$

In these expressions for the transmitted fields, the term Z_{0g} is the characteristic wave impedance in the soil, given by $Z_{0g} = \sqrt{j\omega\mu_0/(\sigma_g + j\omega\epsilon_r\epsilon_0)}$, and the transmission coefficients are given by

$$T_v = \frac{2Z_{0g} \sin \psi}{Z_0 \sin \psi + Z_{0g} \sin \psi_t} \tag{8.42a}$$

$$T_h = \frac{2Z_{0g} \sin \psi}{Z_{0g} \sin \psi + Z_0 \sin \psi_t} \tag{8.42b}$$

The complex transmission angle ψ_t is a solution to Eq. (8.40), and the corresponding function $\sin \psi_t$ is

$$\sin \psi_t = \sqrt{1 - \cos^2 \psi_t} = \sqrt{1 + \left(\frac{k \cos \psi}{\gamma_g}\right)^2} \tag{8.43}$$

8.4.1.2 Excitation Fields for a Transmission Line.
From these general expressions for the plane-wave fields incident, reflected, and transmitted from the air–earth interface, it is possible to write the excitation fields for the transmission line. We will consider only the fields required by the scattering formulation, where only the tangential E-fields along the conduc-

416 EFFECTS OF A LOSSY GROUND ON TRANSMISSION LINES

tors are needed. For the line shown in Figure 7.17, this amounts to the E_x and E_z components.

The tangential excitation E field along the line at height $z = h$ and $y = 0$ is given by the sum of the incident and reflected fields for both polarizations. This is expressed as

$$E_x^{ex}(x, 0, h) = E_x^{inc} + E_x^{ref}$$
$$= E_0[\cos \alpha \sin \psi \cos \phi (e^{jkz \sin \psi} - R_v e^{-jkz \sin \psi})$$
$$+ \sin \alpha \sin \phi (e^{jkz \sin \psi} + R_h e^{-jkz \sin \psi})] e^{-jkx \cos \psi \cos \phi} \quad (8.44)$$

Above the ground, the vertical E-field at a location $(x, 0, z)$ is expressed as

$$E_z^{ex}(x, 0, z) = E_z^{inc} + E_z^{ref}$$
$$= E_0 \cos \alpha \cos \psi (e^{jkz \sin \psi} + R_v e^{-jkz \sin \psi}) e^{-jkx \cos \psi \cos \phi} \quad (8.45a)$$

and for $z < 0$, the excitation field at $(x, 0, z)$ is

$$E_z^{ex}(x, 0, z) = E_z^{t}(x, 0, z)$$
$$= E_0 \cos \alpha \cos \psi (1 - R_v) e^{\gamma_g z \sin \psi_t} e^{-\gamma_g x \cos \psi_t \cos \phi} \quad (8.45b)$$

Notice that for a perfectly conducting ground $R_v = 1$ and $R_h = -1$, and inserting these values into Eqs. (8.44) and (8.45a) we obtain the expressions used in Eqs. (7.58) and (7.60) for the line over a perfectly conducting earth in Chapter 7.

For conductivities in the range of $\sigma_g = 0.1$ to 0.003 S/m and a relative dielectric constant of $\epsilon_r = 10$ (typical values for a conducting earth), $|\gamma_g| \gg k$ for frequencies over 1 MHz. Thus the transmission angle ψ_t becomes

$$\psi_t = \arccos\left(\frac{jk}{\gamma_g} \cos \psi\right) = \arccos\left[\frac{jk}{\sqrt{j\omega\mu_0(\sigma_g + j\omega\epsilon_g)}} \cos \psi\right]$$
$$= \arccos\left[\frac{1}{\sqrt{\epsilon_r(1 + \sigma_g/j\omega\epsilon_g)}} \cos \psi\right] \approx 90° \quad \text{for } \sigma_g \gg \omega\epsilon_g \quad (8.46)$$

For this case, the term $1 - R_v$ in Eq. (8.45b) becomes

$$1 - R_v \approx \frac{2}{\sin \psi} \sqrt{\frac{j\omega\epsilon_0}{\sigma_g}}$$

and the propagation constant γ_g is approximately

$$\gamma_g \approx \frac{1 + j}{\delta}$$

where $\delta = \sqrt{2/(\omega\mu_0\sigma_g)}$ is the skin depth in the ground. With these approximate expressions, the expression for the z-component of the E-field in the ground in Eq. (8.45b) becomes

$$E_z^{ex}(x, 0, z) \approx E_0 \cos\alpha \cos\psi \frac{2}{\sin\psi} \sqrt{\frac{j\omega\epsilon_0}{\sigma_g}} e^{(1+j)z/\delta} e^{-jkx\cos\psi\cos\phi} \quad (8.47)$$

Under the assumption that $\sigma_g \gg \omega\epsilon_0$ and for nongrazing angles of incidence ($\psi \neq 0$), we see that this E-field in the earth is negligible compared with the field exciting the aboveground conductor. Thus for these cases the portion of the excitation of the line arising from the integral from 0 to $-\infty$ in Eq. (8.15) can be neglected, and only the excitation of the vertical field given by Eq. (8.45a) is needed.

8.4.2 Transient Field Reflected from the Ground

In problems relating to lightning or the nuclear EMP, it is often necessary to find the fields above the earth in the time domain. This is especially important for problems involving a direct time-domain solution for field coupling to lines or for problems involving nonlinear responses. One way of evaluating the transient fields is by the use of Fourier transform methods. For a specified incident transient field, the frequency-domain spectrum $E_0(j\omega)$ can be determined either analytically or numerically. Then the E- or H-field spectra for the reflected or total fields are evaluated at each frequency in the spectrum and the inverse Fourier transform is evaluated. Generally, this last step is done by a numerical FFT.

8.4.2.1 Transient Fields Evaluated by the FFT. As an example of the results provided by this procedure, consider the case of a vertically polarized incident field as shown in Figure 8.5, with the angles of incidence $\psi = 30°$ and $\phi = 0°$. This particular E-field will provide excitation field components in both the $\hat{x}$ and $\hat{z}$ directions. If a transmission line is located along the x-axis as shown in Figure 7.19, this particular incident field contributes to the distributed sources and to the two lumped sources at the end of the line. Using Eqs. (8.44) and (8.45a), the E_x- and E_z-field components at different heights over an earth with $\sigma_g = 0.01$ S/m and $\epsilon_r = 10$ can be evaluated as a function of frequency, and the resulting magnitudes of the normalized responses are shown in Figures 8.7a and 8.8a.

For illustrating the transient responses of the E-fields, the double-exponential incident waveform, defined as $E^{inc}(t) = 1.05 \times (e^{-4\times 10^6 t} - e^{-4.76\times 10^8 t})$, can be used. Figures 8.7b and 8.8b illustrate the behavior of the calculated transient excitation E_x and E_z components, along with the vertically polarized incident field. In plots for the horizontal field it is noticed that the pulse reflected from the ground plane tends to cancel the incident field, while for the vertical component, the reflected field adds to the

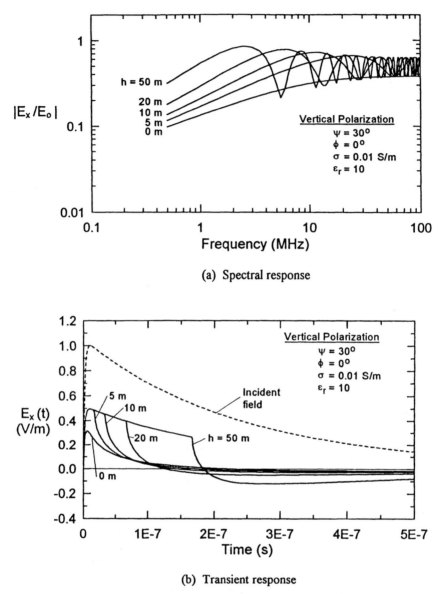

FIGURE 8.7 Total horizontal (E_x) field at different heights over a lossy ground for a vertically polarized incident field.

incident field. The corresponding case for a horizontal polarization of the incident field is shown in Figure 8.9. In this case, there is no vertical field component.

The use of Fourier transforms to obtain the transient fields is a relatively standard approach, and [8.6] contains additional results of a parametric

8.4 REFLECTED AND TRANSMITTED PLANE-WAVE FIELDS 419

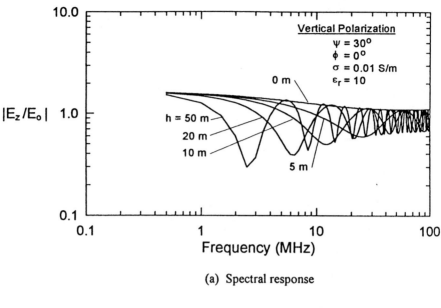

(a) Spectral response

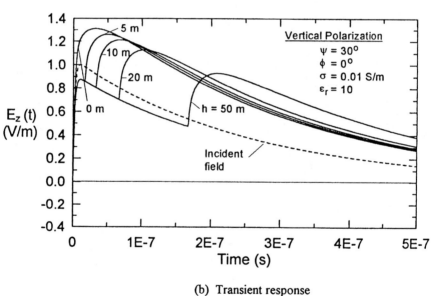

(b) Transient response

FIGURE 8.8 Total vertical (E_z) field at different heights over a lossy ground for a vertically polarized incident field.

study showing different E-field components above and in the earth, for variations in the observation height, angle of incidence, and ground conductivity. In addition, variations in the frequency domain of the E-fields are illustrated.

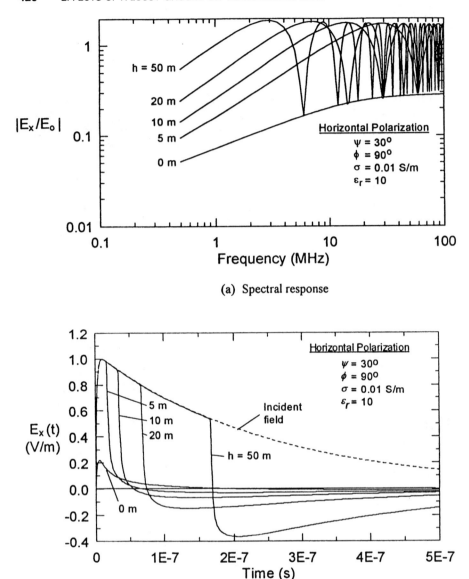

(a) Spectral response

(b) Transient response

FIGURE 8.9 Total horizontal (E_x) field at different heights over a lossy ground for a horizontally polarized incident field.

8.4.2.2 Direct Evaluation of the Transient Reflected E-Field.
As an alternative to the frequency-domain approach, it is possible to obtain approximate, yet accurate expressions for the earth-reflected fields directly in the time domain without resorting to numerical Fourier transforms [8.18]

8.4 REFLECTED AND TRANSMITTED PLANE-WAVE FIELDS

This method can be useful in cases when a transient calculation of EM coupling is being performed directly in the time domain.

By omitting the various sine and cosine terms relating to vector coordinates in Eq. (8.44) and the exponential terms which correspond to time shifts in the transient response, we see that reflected field has the form

$$E^{\text{ref}}(j\omega) = R(j\omega)E_0(j\omega) \qquad (8.48a)$$

where $R(j\omega)$ represents either the vertical or horizontal reflection coefficient. In this expression the dependence on frequency of the incident field spectrum E_0 is shown explicitly.

In the time domain, this corresponds to the convolution between the transient field $E_0(t)$ and the impulse response of the ground, $r(t)$ as

$$E^{\text{ref}}(t) = E^{\text{inc}}(t) * r(t) = \int_{-\infty}^{t} E^{\text{inc}}(t-\tau)r(\tau)\,d\tau \qquad (8.48b)$$

The impulse response of the ground $r(t)$ is the inverse Fourier transform of the Fresnel reflection coefficients in Eq. (8.41). By replacing the frequency $j\omega$ by the Laplace transform variable s, and defining the following parameters:

$$\tau = \frac{\sigma_g}{\epsilon_r \epsilon_0} \qquad \beta = \frac{\sqrt{\epsilon_r - \cos^2\psi}}{\epsilon_r \sin\psi} \qquad \nu = \tau/(1 - \cos^2\psi/\epsilon_r)$$

the vertical reflection coefficient in Eq. (8.41a) can be written as

$$R_v(\psi, s) = \frac{s + \tau - \beta\sqrt{s(s+\nu)}}{s + \tau + \beta\sqrt{s(s+\nu)}}$$

For grazing angles ($\psi \approx 0°$), and metallic ground planes ($\epsilon_r = 1$), the maximum value of $\cos^2\psi/\epsilon_r$ is 1. However, in many practical applications, the ground plane is the earth, for which $\epsilon_r \approx 10$. Thus $\cos^2\psi/\epsilon_r \leq 0.1$, and this term can be neglected with respect to unity. This implies that $\nu = \tau$ and the expression of R_v becomes

$$R_v(\psi, s) \approx \frac{\sqrt{s+\tau} - \beta\sqrt{s}}{\sqrt{s+\tau} + \beta\sqrt{s}} \qquad (8.49)$$

Similarly, the horizontally polarized reflection coefficient can be written as

$$R_h(\psi, s) = \frac{\sqrt{s} - \epsilon_r\beta\sqrt{s+\nu}}{\sqrt{s} + \epsilon_r\beta\sqrt{s+\nu}} \approx -\frac{\sqrt{s+\tau} + (\epsilon_r\beta)^{-1}\sqrt{s}}{\sqrt{s+\tau} + (\epsilon_r\beta)^{-1}\sqrt{s}} \qquad (8.50)$$

Both the vertical and horizontal reflection coefficients in Eqs. (8.49) and

(8.50) now have the same functional form:

$$R(\psi, s) \approx \pm \frac{\sqrt{s+2a} - \kappa\sqrt{s}}{\sqrt{s+2a} + \kappa\sqrt{s}} \quad \left(\text{for } \frac{\cos^2\psi}{\epsilon_r} \ll 1\right) \quad (8.51)$$

where $a = \tau/2$ and $\kappa = \beta$ for vertical polarization and $\kappa = (\epsilon_r\beta)^{-1}$ for horizontal polarization. The leading + sign is used for vertical polarization and the − sign is used for horizontal polarization. Thus, to determine the transient response for either polarization, the same inverse Laplace transform can be used.

The inverse Laplace transform of Eq. (8.51) can be performed analytically as described in [8.18] to obtain the following transient impulse function response:

$$r(t) \approx \pm \left[K\delta(t) + \frac{4\kappa}{1-\kappa^2} \frac{e^{-at}}{t} \sum_{n=1}^{\infty} (-1)^{n+1} n K^n I_n(at) \right] \quad (8.52)$$

In this expression, $K = (1-\kappa)/(1+\kappa)$, $\delta(t)$ is the Dirac delta function, and $I_n(\cdot)$ is the modified Bessel function of order n.

Figure 8.10 illustrates the transient fields reflected from the earth as calculated by this approximate method. Shown in this figure are the fields at angles $\psi = 45°$ and $10°$ for the unit amplitude double-exponential incident field defined earlier. The ground has the same electrical parameters used earlier: $\sigma_g = 0.01$ S/m and $\epsilon_r = 10$. In these plots, all of the sine and cosine factors multiplying the reflected field components in Eq. (8.44) have been neglected. The solid lines labeled "approximation" result from the evaluation of the reflected field using Eq. (8.48b), with the impulse response function $r(t)$ being given by Eq. (8.52). The short-dashed lines labeled "Fourier transform" arise from the calculation of the reflected fields in the frequency domain using the reflection coefficients as in Eq. (8.48a), and subsequent inversion into the time domain. As can be seen from these curves, the analytical approximation for the impulse response of the reflected fields is quite good.

8.5 EXAMPLES OF ABOVEGROUND TRANSMISSION LINE RESPONSES

Using the preceding lossy coupling model for the line over the ground, a number of cases have been calculated to illustrate the behavior of the induced responses by an incident field. As indicated in Figure 8.11, the current at the $x = \mathcal{L}$ end of the line is selected as the observable quantity. In all cases the incident field is assumed to be a vertically polarized plane wave with the angle of incidence $\phi = 0°$. For this excitation, the required integrals

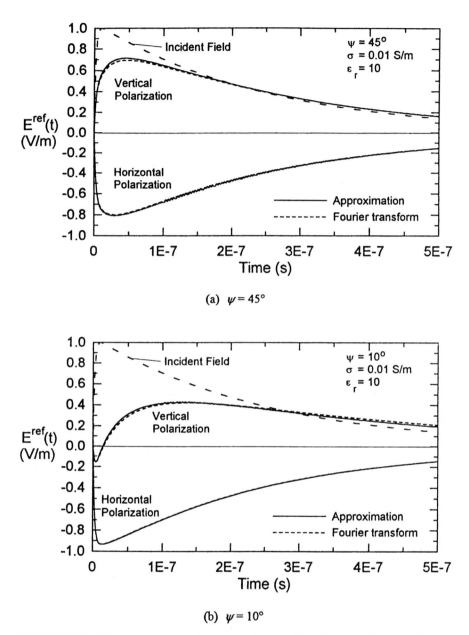

FIGURE 8.10 Plots of the approximate transient and Fourier-transformed reflected E-field for two angles of incidence.

for the BLT source term can be evaluated analytically, as in Eq. (7.64). For the case of transient responses, the excitation waveform is assumed to be the unit amplitude double exponential $E^{inc}(t) = 1.05 \times (e^{-4\times 10^6 t} - e^{-4.76 \times 10^8 t})$.

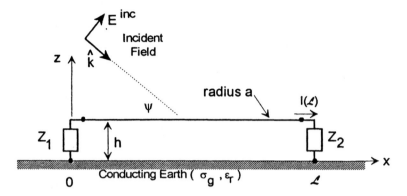

FIGURE 8.11 Geometry of the single line over the lossy earth.

For the frequency-domain responses, the quantities calculated are the normalized current spectrum magnitude, $|I(j\omega)/E^{\text{inc}}(j\omega)|$.

8.5.1 Variations in Earth Conductivity

Figure 8.12 presents the frequency-domain and transient load current responses for the line for different values of the earth conductivity. The line has the parameters $\mathcal{L} = 5$ m, $h = 0.8$ m, $a = 0.15$ cm, and the termination impedances are resistive and are $Z_1 = Z_2 = 20\,\Omega$. The incident field strikes the line with an angle $\psi = 45°$. The frequency-domain response in part (*a*) of the figure exhibits the periodic resonance structure due to the reflections from the end of the line. The transient response in part (*b*) also shows the effects of the resonances in the form of the fast oscillations superimposed on the slower waveform. Note that in this calculation, the incident field strikes the end of the line at $x = 0$ at $t = 0$. Consequently, there is no response at the $x = \mathcal{L}$ end of the line until time $t = (L \cos \psi)/c \approx 0.12 \times 10^{-7}$ s.

The effect of the lossy ground *increases* the amplitude of current induced in the line. One often hears that a "perfectly conducting ground is used to give a worst-case response," presumably because there is no attenuation of the traveling waves on the line. Here we see that just the opposite is true. The fact that the response is smaller is due to the fact that the tangential excitation *E*-field along the horizontal section of the line is smaller for the case of the perfectly conducting ground.

8.5.2 Variations with Angle of Incidence

Figure 8.13 shows the load current response for the same line, as the vertical angle of incidence ψ varies from 0° to 90°, with $\phi = 0°$. For this calculation the earth has parameters $\sigma_g = 0.01$ S/m and $\epsilon_r = 10$. For such a short line we see that the induced current response tends to a maximum as the angle ψ

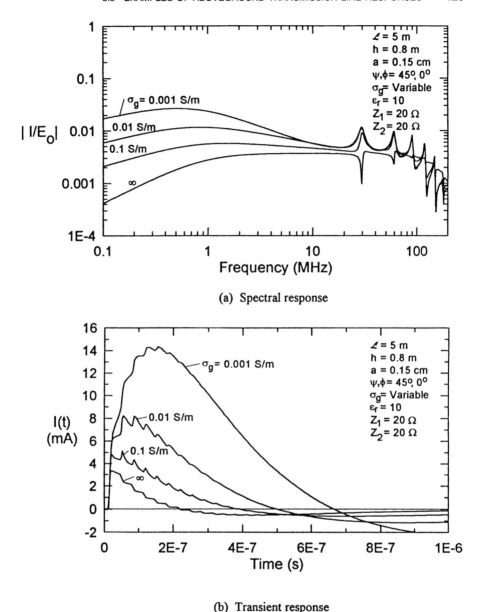

FIGURE 8.12 Variations of the load current at $x = \mathcal{L}$ for different ground conductivities.

approaches 90°. A long line behaves differently, however. Figure 8.14 shows the load current response for a line of length $\mathcal{L} = 500$ m. For this line the waveform turn-on time $t = (\mathcal{L} \cos \psi)/c$ varies from 0 to 1.666 μs, depending on the angle of incidence. At grazing angles of incidence, the response of the

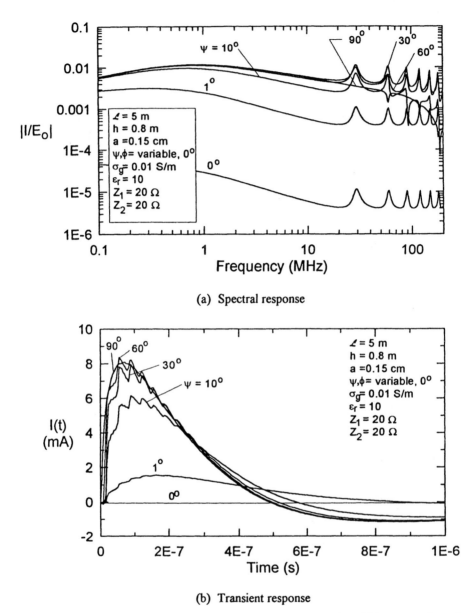

FIGURE 8.13 Variations of the load current at $x = \mathcal{L}$ for different angles of incidence ψ.

line tends to increase if the ground is lossy. Thus the maximum line response does not occur for broadside ($\psi = 90°$) incidence but for an angle ψ in the range 10 to 20°, depending on the conductivity. This increase is due to the

8.5 EXAMPLES OF ABOVEGROUND TRANSMISSION LINE RESPONSES

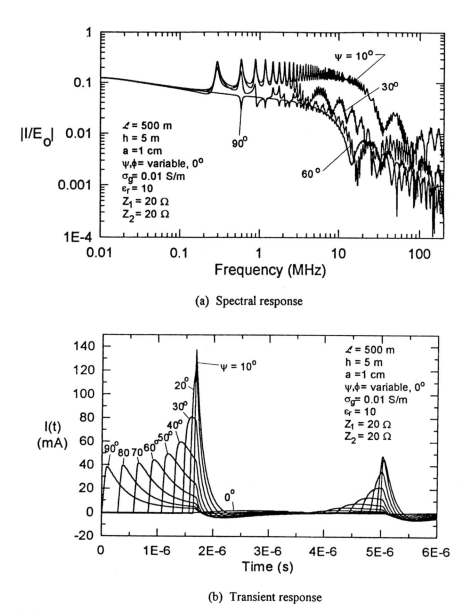

FIGURE 8.14 Variations of the load current at $x = \mathcal{L}$ on a long line for different angles of incidence ψ.

denominator $(\gamma - jk \cos \psi \cos \phi)$ in Eq. (7.64) becoming small for a certain set of angles ψ and ϕ.

Figure 8.15 illustrates the transient response for the same line discussed in

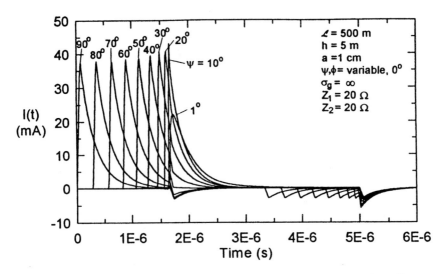

FIGURE 8.15 Vartiations of the transient load current at $x = \mathcal{L}$ on a long line over a perfectly conducting earth for different angles of incidence ψ.

Figure 8.14, but for a perfectly conducting ground plane. As noted here, there is no buildup of the response for grazing angles of incidence. Because the propagation constant on the line is now $\gamma = jk$, the denominator term in Eq. (7.64) is given by $jk(1 - \cos\psi \cos\phi)$, which approaches 0 as $\psi \to 0$. However, this equation has a $\sin\psi$ term in the numerator that cancels the denominator at small angles of incidence.

8.5.3 Variations with Line Height

As a final example, Figure 8.16 shows variations of the line responses as the height of the line h is varied from 0.01 m to 1.5 m. For this example, the shorter line ($\mathcal{L} = 5$ m) is considered. These figures indicate that increasing the height of the line over the earth will tend to increase its induced response. However, because the excitation E-field does not go to zero right on the surface of the earth as in the case of a perfectly conducting ground, a cable located on the surface of the earth will still have an induced response.

8.6 BURIED CABLES†

Cables that are buried in the earth may also be analyzed using transmission line models. The presence of the conducting ground near the cable will

---

† Note that in contrast with the other parts of this chapter, in this section the wave propagation along the line is described by the function, $e^{-jk_x x}$, where k_x is complex. This is done to be consistent with the references cited.

8.6 BURIED CABLES 429

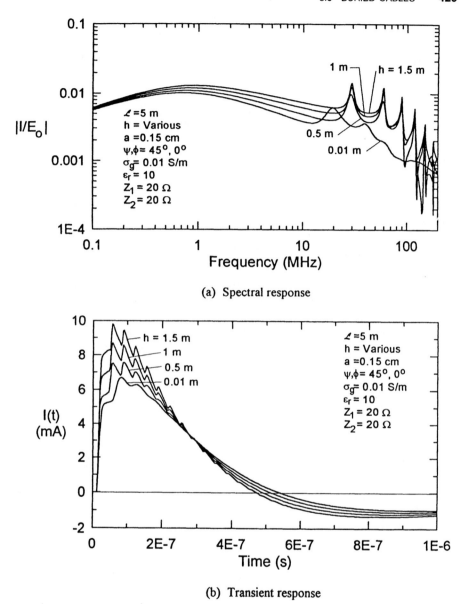

(a) Spectral response

(b) Transient response

FIGURE 8.16 Variations of the load current at $x = \mathcal{L}$ for different line heights.

affect the propagation of induced current and charge on the cable more than on an aboveground line. Furthermore, the excitation field at the cable will be significantly reduced from that of the aboveground line, due to the reflection of part of the incident field at the earth's surface and to subsequent attenuation of the transmitted field as it propagates from the surface down to the location of the cable in the ground. Consequently, it is expected that the responses of buried cables will be significantly lower than for aboveground lines.

Because of the close proximity of the ground to the cable and the ill-defined nature of the voltage in this case, the usual starting point for developing transmission line equations for the buried cable problem is EM scattering theory. The geometry under consideration is illustrated in Figure 8.17. The buried cable of infinite length has an inner conductor of radius a and electrical parameters μ_0, ϵ_w, and σ_w. Surrounding this wire is a dielectric jacket of radius b and electrical parameters μ_0, ϵ_d, and σ_d. The cable is buried at a depth d in the ground, which is described by μ_0, ϵ_g, and σ_g. The incident field providing the excitation for this problem is the same as in the aboveground problem of Figure 8.5. The incident field is a plane wave having vertically polarized and horizontally polarized components and angles of incidence ψ and ϕ.

For this problem it is desired to determine a rigorous solution for the current induced on the line due to the incident field. Once this is accomplished, the solution can be approximated and be put into a form that is similar to the transmission line solution. This provides approximate expressions for per-unit-length impedance and admittance parameters of the line. With these line parameters, it is then possible to develop transmission line solutions for lines of finite length and suitable termination impedances at the ends.

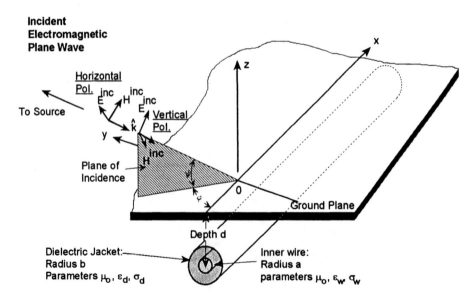

FIGURE 8.17 Geometry of the infinite buried cable.

8.6.1 Summary of Rigorous Solution

The development of the scattering solution for the buried cable has been discussed in [8.19] and involves the derivation of a frequency-domain integral equation for cable current. This requires a Green's function for an infinitesimal x-directed current element in the earth and the surface impedance of a circular dielectric-coated conductor. The excitation field for the integral equation is the tangential incident E-field in the earth at the cable location, and a useful approximation made at this stage in the analysis is that the cable is electrically small, so that the variation of the incident E-field across the cable is negligible.

The solution of the integral equation for the induced cable current is obtained by using a spatial Fourier transform in the direction of the line to transform the spatial domain variable x to the spectral domain k_x. In doing this the spatial equation involving an integral from $x = -\infty$ to ∞ is converted into an algebraic equation in k_x, which may be solved algebraically. Then an inverse Fourier transform is applied to obtain the desired spatial solution for the line current. This resulting solution contains the usual traveling-wave (transmission line) mode, plus other surface-wave and radiated field modes, which are typically neglected in the approximate transmission line solution.

8.6.1.1 The Integral Equation

Outside the cable ($r > b$) the E-field consists of the portion of the incident field transmitted into the ground, E^t, plus the scattered field from the currents induced on the cable in the soil). Inside the cable ($r < b$) the E-field is described by a cylindrically symmetric field arising from the cable current and described by a per-unit-length wire impedance. At $r = b$, the tangential E-field components must be continuous, hence

$$E_x^t + E_x^{sca} = E_x^{int} \tag{8.53}$$

The total internal and scattered fields are determined by integrals over the induced currents on the cable, from $x = -\infty$ to $x = +\infty$. The differential fields at a point x are related to the current at x' as

$$dE_x^{sca}(x) = Z_s'(x - x')I(x') \tag{8.54a}$$

$$dE_x^{int}(x) = Z_i'(x - x')I(x') \tag{8.54b}$$

The integral equation for the cable current is obtained by integrating Eqs. (8.54) and substituting into Eq. (8.53) to yield

$$\int_{-\infty}^{\infty} [Z_i'(x - x') - Z_s'(x - x')]I(x')\,dx' = E_x^t(x') \tag{8.55}$$

With the spatial Fourier transform pair

$$\mathcal{F}(k_x) = \int_{-\infty}^{\infty} F(x)e^{jk_x x}\, dx \quad \text{and} \quad F(x) = \frac{1}{2\pi}\int_{-\infty}^{\infty} \mathcal{F}(k_x)\, e^{-jk_x x}\, dk_x$$

the integral equation in Eq. (8.55) can be transformed to the k_x domain as

$$[Z'_i(k_x) - Z'_s(k_x)]I(k_x) \equiv Z'(k_x)I(k_x) = E'_x(k_x) \tag{8.56}$$

This equation has a simple spectral solution given by

$$I(k_x) = \frac{E'_x(k_x)}{Z'(k_x)} \tag{8.57}$$

In the spatial domain, the solution for the line current is obtained by taking the inverse transform of Eq. (8.57) as

$$I(x) = \frac{1}{2\pi}\int_{-\infty}^{\infty} \frac{E'_x(k_x)}{Z'(k_x)} e^{-jk_x x}\, dk_x \tag{8.58}$$

8.6.1.2 Soil Impedance. The term Z'_s in the overall coupling impedance $Z'(k_x)$ in Eq. (8.56) is the per-unit-length soil impedance term, representing the E-field on the wire produced by an infinitely long traveling-wave current having the form $e^{-jk_x x}$ at a depth d in the ground. This impedance may be determined from solutions of the Hertz potentials π in the air and in the ground [8.20]. As discussed in [8.21], this expression is

$$Z'_s(k_x) = \frac{j\omega\mu_0}{2\pi k_g^2} \frac{1}{\tau_g b K_1(\tau_g b)} \left\{ \begin{matrix} \tau_g^2 K_0(\tau_g b) \\ -I_0(\tau_g b)[\tau_g^2 K_0(\tau_g 2d) + k_g^2\eta_1 - k_x^2\eta_2] \end{matrix} \right\} \tag{8.59}$$

where I_0, K_0, and K_1 are modified Bessel functions. In this expression the following definitions are used:

$$\tau_0 = \sqrt{k_x^2 - k_0^2} \qquad \tau_g = \sqrt{k_x^2 - k_g^2}$$

$$k_0 = \frac{\omega}{c} = \sqrt{\omega^2 \mu_0 \epsilon_0} \qquad k_g = \sqrt{\omega^2 \mu_0 \epsilon_g - j\omega\mu_0 \sigma_g} \tag{8.60a}$$

$$\eta_1 = \int_{-\infty}^{\infty} \frac{e^{-2d\sqrt{\xi^2 + \tau_g^2}}}{\sqrt{\xi^2 + \tau_0^2} + \sqrt{\xi^2 + \tau_g^2}}\, d\xi \tag{8.60b}$$

$$\eta_2 = \int_{-\infty}^{\infty} \frac{n^2 e^{-2d\sqrt{\xi^2 + \tau_g^2}}}{n^2\sqrt{\xi^2 + \tau_0^2} + \sqrt{\xi^2 + \tau_g^2}}\, d\xi \tag{8.60c}$$

8.6 BURIED CABLES

In the integral of Eq. (8.60c), the term n is the complex index of refraction of the ground, given by

$$n = \sqrt{\frac{\epsilon_g}{\epsilon_0} + \frac{\sigma_g}{j\omega\epsilon_0}} \qquad (8.61)$$

Note that the form of the impedance term $Z'_s(k_x)$ for the buried line in Eqs. (8.59) and (8.60) is the same as that for the aboveground line in Eqs. (8.21) to (8.25).

8.6.1.3 Cable Impedance.

The term Z'_i is the per-unit-length internal cable impedance, which contains the effects of the lossy inner core of the cable as well as the dielectric jacket. It has the form [8.22]

$$Z'_i(k_x) = \frac{-j\omega\mu_0\tau_d^2}{2\pi k_d^2} \frac{I_0(\tau_d b)K_0(\tau_d a) - I_0(\tau_d a)K_0(\tau_d b)}{(\tau_d a)I_0(\tau_d a)K_1(\tau_d a) + (\tau_d a)I_1(\tau_d a)K_0(\tau_d a)}$$
$$+ Z'_w(k_x) \frac{I_0(\tau_d b)K_1(\tau_d a) + I_1(\tau_d a)K_0(\tau_d b)}{I_0(\tau_d a)K_1(\tau_d a) + I_1(\tau_d a)K_0(\tau_d a)} \qquad (8.62a)$$

where Z'_w corresponds to the impedance of the inner conductor of the cable:

$$Z'_w(k_x) = \frac{-j\omega\mu_0}{2\pi k_w^2} \frac{\tau_w^2 I_0(\tau_w a)}{(\tau_w a)I_1(\tau_w a)} \qquad (8.62b)$$

In these expressions, the following terms are defined

$$\begin{aligned}\tau_d = \sqrt{k_x^2 - k_d^2}, \quad & k_d = \sqrt{\omega^2\mu_0\epsilon_d - j\omega\mu_0\sigma_d} \\ \tau_w = \sqrt{k_x^2 - k_w^2} \quad & k_w = \sqrt{\omega^2\mu_0\epsilon_w - j\omega\mu_0\sigma_w}\end{aligned} \qquad (8.63)$$

For typical cables, $|\tau_g b|$ and $|\tau_d b| \ll 1$, and for a highly conducting conductor, $k_x \ll k_w \approx \sqrt{-j\omega\mu_0\sigma_w}$. In this case the cable internal impedance term can be approximated as

$$Z'_i(k_x) \approx j\omega \frac{\mu_0}{2\pi} \ln\frac{b}{a} + Z'_w(k_x) + k_x^2 \left[j\omega 2\pi\epsilon_d \frac{1}{\ln(a/b)} \right]^{-1}$$
$$= j\omega L' + Z'_w(k_x) + k_x^2(j\omega C')^{-1} \qquad (8.64a)$$

with

$$Z'_w(k_x) \approx \frac{1}{2\pi a} \sqrt{-j\omega\mu_0/\sigma_w} \frac{J_0(k_w a)}{J_1(k_w a)} \qquad (8.64b)$$

where J_0 and J_1 are cylindrical Bessel functions and L' and C' represent the inductance and capacitance contributions from the dielectric jacket surrounding the cable. This is equivalent to Eq. (7.17).

8.6.1.4 Solution for the Current.
The induced cable current is determined through evaluation of the integral of Eq. (8.58). This is accomplished by complex variable theory, where it is noted that the integrand has both pole and branch cut contributions to the response. The pole term corresponds to the transmission line mode, and the branch cuts gives the radiated modes and the surface-wave mode.

For the special case of excitation by a plane wave, the solution for the current is simple to evaluate. The excitation E-field on the wire in the earth is of the form

$$E_x^t(x) = E_0(\psi, \phi) e^{-jk_0 x \cos \psi \cos \phi} \quad (8.65a)$$

where the term

$$E_0(\psi, \phi) = (E_v T_v \sin \psi_t \cos \phi + E_h T_h \sin \phi) e^{-jk_g d \sin \psi_t} \quad (8.65b)$$

has two polarization components, one for the vertically polarized component of the incident field and another for the horizontally polarized component. The terms T_v and T_h are the Fresnel transmission coefficients introduced in Section 8.4.1.1 and have the form

$$T_v = \frac{2n \sin \psi}{n^2 \sin \psi + \sqrt{n^2 - \cos^2 \psi}} \quad (8.66a)$$

$$T_h = \frac{2 \sin \psi}{\sin \psi + \sqrt{n^2 - \cos^2 \psi}} \quad (8.66b)$$

with n being the index of refraction given in Eq. (8.61). The angle of transmission into the ground ψ_t is pictured in Figure 8.5 and has the form $n \cos \psi_t = \cos \psi$.

For the incident plane wave, the spectral transform of $E_x^t(x)$ is

$$E_x^t(k_x) = \int_{-\infty}^{\infty} E_0(\psi, \phi) e^{-jk_0 x \cos \psi \cos \phi} e^{jk_x x} dx$$

$$= 2\pi E_0(\psi, \phi) \delta(k_x - k_0 \cos \psi \cos \phi)$$

As a consequence of this simple expression, the solution for current induced on the infinite cable becomes

$$I(x) = \frac{1}{2\pi} \int_{-\infty}^{\infty} \frac{E'_x(k_x)}{Z'(k_x)} e^{-jk_x x} \, dk_x$$

$$= \int_{-\infty}^{\infty} \frac{E_0(\psi, \phi)}{Z'(k_x)} e^{-jk_x x} \, \delta(k_x - k_0 \cos \psi \cos \phi) \, dk_x$$

$$= \frac{E_0(\psi, \phi)}{Z'(k_0 \cos \psi \cos \phi)} e^{-jk_0 x \cos \psi \cos \phi}$$

or

$$I(x) = \frac{(E_v T_v \sin \psi_t \cos \phi + E_h T_h \sin \phi) e^{-jk_g d \sin \psi_t}}{Z'_i(k_0 \cos \psi \cos \phi) - Z'_s(k_0 \cos \psi \cos \phi)} e^{-jk_0 x \cos \psi \cos \phi}$$

(8.67)

8.6.2 Transmission Line Approximation

The transmission line solution for the induced cable current corresponds to the response at the pole of the integrand of Eq. (8.58). This occurs when $Z'(k_p) = 0$, where k_p is the characteristic propagation constant of the line [8.19]. The general expression for the coupling impedance $Z'(k_x)$ can be put into the form

$$Z'(k_x) = Z'_i(k_x) - Z'_s(k_x) = Z'_{\text{line}}(k_x) + k_x^2 [Y'_{\text{line}}(k_x)]^{-1}$$

which is described by the per-unit-length equivalent circuit shown in Figure 8.3 with

$$Z'_{\text{line}} = Z'_w + Z'_g + j\omega L' \qquad (8.68)$$

where

Z'_w = internal wire impedance

$$= \frac{-j\omega\mu_0}{2\pi k_w^2} \frac{\tau_w^2 I_0(\tau_w a)}{(\tau_w a) I_1(\tau_w a)} \qquad (8.69a)$$

$$\approx \frac{1}{\pi a^2 \sigma_w} + j\omega \frac{\mu_0}{8\pi} \qquad (\tau_w a \ll 1)$$

$Z'_g(k_x)$ = ground impedance contribution

$$\approx \frac{-j\omega\mu_0}{2\pi} \left[\ln\left(\frac{\Gamma}{2} \tau_g b\right) + K_0(\tau_g 2d) - \eta_1(k_x) \right] \qquad (\Gamma = 1.7811\ldots)$$

(8.69b)

and

$$j\omega L' = \text{dielectric jacket contribution}$$
$$= j\omega \frac{\mu_0}{2\pi} \ln \frac{b}{a} \tag{8.69c}$$

The total per-unit-length line admittance is given by

$$Y'_{\text{line}} = Y'_\ell \| Y'_g = [(j\omega C')^{-1} + Y'^{-1}_g]^{-1} \tag{8.70}$$

with

$$j\omega C' = \text{dielectric jacket contribution}$$
$$= \frac{j\omega 2\pi\epsilon_d}{\ln(b/a)} \tag{8.71a}$$

$Y'_g = $ ground admittance contribution

$$\approx -j\omega 2\pi\epsilon_g \left[\ln\left(\frac{\Gamma}{2}\tau_g b\right) + K_0(\tau_g 2d) - \eta_2(k_x) \right]^{-1}$$
$$\tag{8.71b}$$

Note that the definitions of Z'_g and Y'_g involve the infinite integrals defined in Eqs. (8.60b) and (8.60c).

The propagation constant k_p is the solution to the modal equation

$$Z'(k_p) = Z'_{\text{line}}(k_p) + (k_p^2)[Y'_{\text{line}}(k_p)]^{-1} = 0 \tag{8.72}$$

For cables in air, the cable propagation constant is $k_p \approx k_0$. For cables on the surface,

$$k_p \approx \sqrt{(k_0^2 + k_g^2)/2}.$$

A buried cable has $k_p \approx k_g$. The transmission line approximation for buried cables assumes that $k_p = k_g$ for the line impedance and admittance terms in Eq. (8.72) and thus permits the approximate evaluation of the line propagation constant as

$$k_p \approx \sqrt{-Z'_{\text{line}}(k_g) \cdot Y'_{\text{line}}(k_g)} \tag{8.73}$$

The characteristic impedance of the line is then defined as

$$Z_c(k_p) = \sqrt{\frac{Z'_{\text{line}}(k_g)}{Y'_{\text{line}}(k_g)}} \tag{8.74}$$

For the transmission line model, a single discrete voltage source at a position x' on an infinite line induces a current at a point x on the line of the form

$$I(x) \approx \frac{V_0}{2Z_c} e^{-jk_p|x-x'|} \tag{8.75}$$

Considering the excitation of the buried line to be due to a distribution of voltage sources arising from the tangential E-field incident on the cable, the solution for the current is given by the integral over the source location x' in Eq. (8.75):

$$I(x) = \int_{-\infty}^{\infty} \frac{E_0(\Psi, \phi) e^{-jk_0 x' \cos \psi \cos \phi}}{2Z_c} e^{-jk_p|x-x'|} dx'$$

$$= \frac{E_0(\psi, \phi) e^{-jk_0 x \cos \psi \cos \phi}}{Z_c} \frac{-jk_p}{k_p^2 - (k_0 \cos \psi \cos \phi)^2} \tag{8.76}$$

8.6.3 Additional Simplifications to the TL Solution

The approximate expression for the induced current in Eq. (8.76) can be simplified further for cases when the wave velocity in the ground is much slower than c. In this case, $k_p \gg k_0 \cos \psi \cos \phi$, and for *nongrazing angles of incidence* the current becomes

$$I(x) \approx \frac{E_0(\psi, \phi)}{Z'_{\text{line}}(k_p)} e^{-jk_0 x \cos \psi \cos \phi} \tag{8.77}$$

Neglecting internal wire impedance Z'_w and approximating the ground impedance Z'_g in Eq. (8.69b) yields

$$Z'_{\text{line}} \approx \frac{j\omega\mu_0}{2\pi} \ln \frac{b}{a} - \frac{j\omega\mu_0}{2\pi} \ln\left(\frac{\Gamma}{2} \tau_g b\right) \tag{8.78}$$

The transverse propagation constant is approximated by $\tau_g = \sqrt{k_x^2 - k_g^2} \approx jk_g$, and Eq. (8.78) becomes

$$Z'_{\text{line}} \approx -\frac{j\omega\mu_0}{2\pi} \ln\left(jk_g \frac{\Gamma}{2} a\right) \tag{8.79}$$

By approximating the line propagation constant as $k_p \approx k_g$, the line current is given by

$$I(x) \approx \frac{(E_v T_v \sin \psi_t \cos \phi + E_h T_h \sin \phi) e^{-jk_g d \sin \psi_t}}{-(j\omega\mu_0/2\pi) \ln[jk_g(\Gamma/2)a]} e^{-jk_0 x \cos \psi \cos \phi} \tag{8.80}$$

Further approximations are possible as discussed by Vance [8.6], who assumes that the displacement current in the ground is negligible. This occurs at low frequencies or for high conductivities, when $\sigma_g \gg \omega\epsilon_g$. Under this assumption we have

$$jk_g \approx (1+j)/\delta_g \quad \left(\delta_g = \text{skin depth} = \frac{1}{\sqrt{\pi f \mu_0 \sigma_g}}\right)$$

$$|n| \gg 1: \psi_t \approx 90°: T_v \approx \frac{2}{n}; T_h \approx \frac{2\sin\psi}{n}$$

and the tangential E-field on the wire becomes

$$E_0(\psi, \phi) \approx 2(E_v \cos\phi + E_h \sin\psi \sin\phi)\sqrt{\frac{j\omega\epsilon_0}{\sigma_g}}\, e^{-(1+j)d/\delta_g} \qquad (8.81)$$

The resulting line current in this case is then given as

$$I(x) \approx -10^7 \frac{E_v \cos\phi + E_h \sin\psi \sin\phi}{\sqrt{j\omega\sigma_g/\epsilon_0}[(1+j)a/\delta_g]} e^{-(1+j)d/\delta_g} e^{-jk_0 \times \cos\psi \cos\phi} \qquad (8.82)$$

8.6.4 Example of Current Responses on an Infinite Buried Cable.
Bridges [8.19] has calculated several examples of comparing the rigorous solution for the induced cable current of Eq. (8.67) with the approximate solutions of Eqs. (8.80) and (8.82). Considering a single exponential transient incident E-field of the form

$$E^{\text{inc}}(t) = 1000\, e^{-t/8.854\times 10^{-8}} \qquad (8.83)$$

with polarization angle $\alpha = 45°$ ($E_v = E_h$) and angles of incidence $\phi = 0°$, $\psi = 90°$, $10°$, and $2°$, Figure 8.18 presents the calculated current for a cable at a depth of $d = 1$ m and earth parameters $\sigma_g = 0.001$ S/m and $\epsilon_{rg} = 10$. Each of the equations indicated was used to generate the current response spectrum, which was then converted to the time domain using the inverse Fourier transform. Notice that the agreement between the exact and approximate transmission line solution provided by Eq. (8.80) is quite acceptable. The approximation of Eq. (8.82), however, yields a result that is less accurate. In fact, for the field incident with the angle $\phi = 0°$, the response is seen to be independent of the angle ψ.

In summary, we see that the exact (i.e., scattering) model can be used for finding the current induced on an infinite buried line due to an incident plane wave. This solution, however, requires the evaluation of two infinite integrals, which can be time consuming. Good results are obtained for the same response using a transmission line model in which the excitation E-field along the line in the ground is evaluated using the Fresnel transmission coefficients. The accuracy of the TL solution decreases, however, when the E-field in the ground is approximated by neglecting the displacement

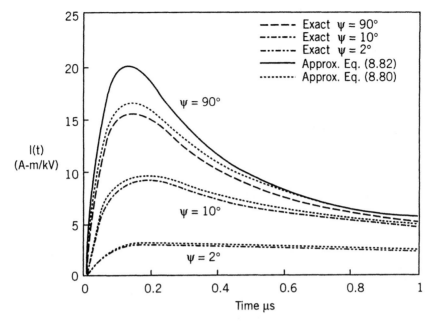

FIGURE 8.18 Plot of the induced current in an infinite buried cable, using the exact expression [Eq. (8.67)] and the two approximate expressions of Eqs. (8.80) and (8.82). (From [8.19]). © 1995 Institute of Electrical and Electronics Engineers.)

current in the ground. As discussed in [8.19], this final approximation is not really needed, as Eq. (8.80) can easily be evaluated numerically.

8.6.5 Application to Buried Lines of Finite Length. The development of the transmission line solution for the infinite cable suggests the possibility of using the same model for a line of finite length. Figure 8.19 illustrates a

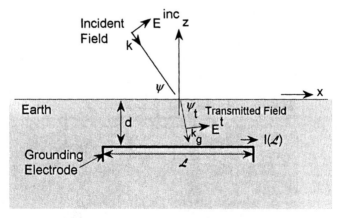

FIGURE 8.19 Buried line of length $\mathcal{L}$.

cable of length $\mathcal{L}$ buried in the earth and excited by an incident field. For this line, the per-unit-length section shown in Figure 8.3 is used, together with the impedance and admittance parameters given by Eqs. (8.67) to (8.71) with the approximation that $\tau_g \approx jk_g$.

Once the propagation constant of the line and the characteristic impedance are determined from Eqs. (8.73) and (8.74), the solution for the induced current at the ends of a line of length $\mathcal{L}$ may be determined from the BLT equation (7.35a). Noting that the line propagation constant is $\gamma = jk_g$, the BLT equation for the load currents becomes

$$\begin{bmatrix} I(0) \\ I(\mathcal{L}) \end{bmatrix} = \frac{1}{Z_c} \begin{bmatrix} 1-\rho_1 & 0 \\ 0 & 1-\rho_2 \end{bmatrix} \begin{bmatrix} -\rho_1 & e^{jk_g\mathcal{L}} \\ e^{jk_g\mathcal{L}} & -\rho_2 \end{bmatrix}^{-1} \begin{bmatrix} \frac{1}{2}\int_0^{\mathcal{L}} e^{jk_g x_s} V'_{s2}(x_s)\,dx_s \\ -\frac{1}{2}\int_0^{\mathcal{L}} e^{jk_g(\mathcal{L}-x_s)} V'_{s2}(x_s)\,dx_s \end{bmatrix}$$

(8.84)

In developing this model, there is uncertainty as to the proper termination conditions at the ends of the line. The reflection coefficients ρ_1 and ρ_2 in Eq. (8.84) are formally defined through Eq. (6.37) and require the specification of the effective "load" impedances at each end of the buried line. These quantities are difficult to define, as it is necessary to introduce the concept of a voltage into the problem. As was done for the aboveground line, we can reference the line voltage to a point at infinity and use the footing resistances for deeply buried electrodes in Appendix D as an approximation to these impedances.

For a real buried cable, the details of the termination are usually unknown. In the present model we assume that there is a vertical ground "stake" at the end of the line, as shown in Figure 8.19, and compute an effective resistance of the stake into the ground. From this grounding resistance, ρ_1 and ρ_2 can be calculated.

Notice that the riser contributions found in the excitation vector of Eq. (7.37) for the BLT equation for an aboveground line are absent for the buried cable, due to the fact that the E_z component of the E-field in the earth is small.

The excitation waveform of the cable for this example is given by Eq. (8.83). The incident field is assumed to be vertically polarized ($\alpha = 0°$) with angles of incidence $\psi = 45°$ and $\phi = 0°$, and this provides the distributed voltage source V'_{s2} in Eq. (8.65). The integration of this source term along the line can again be performed analytically, with the details being left to the reader.

As an example of calculated responses, Figure 8.20 illustrates the calculated current in the line at the $x = \mathcal{L}$ end for different values of earth conductivity. Part (a) of the figure shows the spectral response of the current, and part (b) illustrates the transient waveforms. The details of the line geometry and other parameters are displayed in the figures. It is evident that as the earth conductivity decreases, the oscillations of the current waves

8.6 BURIED CABLES   441

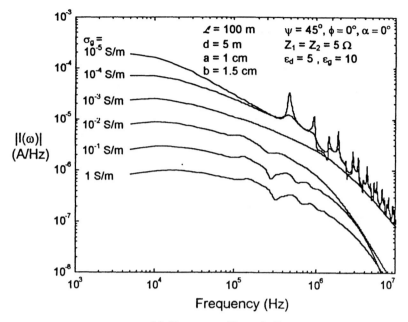

(a) Frequency Domain Response

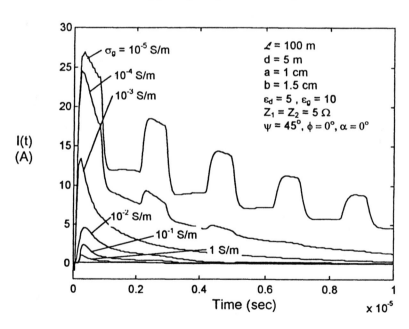

(b) Transient Response

FIGURE 8.20 Frequency-domain spectrum (*a*) and transient response (*b*) for the field-induced current at $x = \mathscr{L}$ of a buried cable, for different values of earth conductivity. Double exponential excitation as defined in section 7.2.6.2.

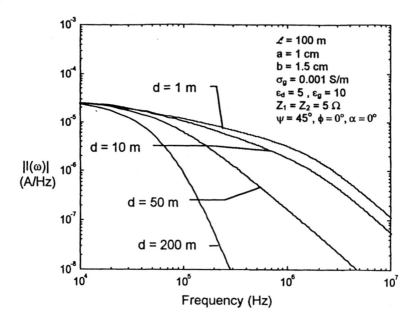

(a) Frequency Domain Response

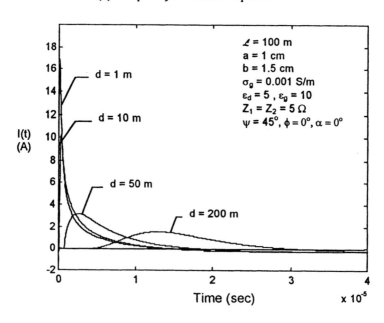

(b) Transient Response

FIGURE 8.21 Frequency-domain spectrum (*a*) and transient response (*b*) for the field-induced current at $x = \mathcal{L}$ of a buried cable, for different values of burial depth and $\sigma_g = 0.001$ S/m. Double exponential excitation as defined in section 7.2.6.2.

along the line become more evident, due to the decreasing attenuation of the traveling waves. In addition, the excitation field is not attenuated as much for low conductivities, so the overall response amplitude is larger with a smaller conductivity.

It should be noted that the terminating impedances in this example have been fixed at a constant value of 5 Ω. In the real case of a line in a conducting earth, these resistances are also functions of the soil conductivity. As seen from Appendix D this resistance increases as the conductivity decreases.

Another example of buried cable responses is offered in Figure 8.21, where the frequency domain spectrum and the transient response for the field-induced current at $x = \mathcal{L}$ of a buried cable are shown for different values of burial depth. For this case the earth conductivity was assumed to be $\sigma_g = 0.001$ S/m. For this conductivity the line resonances are highly damped, and the increasing attenuation of the high-frequency components of the incident field width of the burial depth is clearly evident.

REFERENCES

8.1. Sommerfeld, A., "Uber die ausbreitung der Wellen in der drahtlosen Telegraphie," *Ann. Phys.*, Vol. 28, 1909, p. 665.

8.2. Carson, J. R., "Wave Propagation in Overhead Wires with Ground Return," *Bell Syst. Tech. J.*, Vol. 5, 1926, pp. 539–554.

8.3. Sunde, E. D., *Earth Conduction Effects in Transmission Systems*, Van Nostrand, New York, 1949.

8.4. Bridges, G. J., and L. Shafai, "Plane Wave Coupling to Multiple Conductor Transmission Lines Above a Lossy Earth," *IEEE Trans. Electromagn. Compat.*, Vol. EMC-31, No. 1, February 1989.

8.5. Tesche, F. M., "Comparison of the Transmission Line and Scattering Models for Computing the HEMP Response of Overhead Cables", *IEEE Trans. Electromagn. Compat.*, Vol. EMC-34, No. 2, May 1992.

8.6. Vance E. F., *Coupling to Shielded Cables*, R. E. Krieger, Melbourne, FL, 1988.

8.7. Neff, H. P., and D. A. Reed, "The Effects of Secondary Scattering on the Induced Current in an Infinite Wire over an Imperfect Ground from an Incident Electromagnetic Pulse," *IEEE Trans. Antennas Propag.*, Vol. AP-37, December 1989.

8.8. Janke, E., F. Emde, and F. Loesch, *Tafeln Hoeherer Funktionen*, Teubner, Stuttgart, 1960.

8.9. Chen, K. C., and K. M. Damrau, "Accuracy of Approximate Transmission Line Formulas for Overhead Wires," *IEEE Trans. Electromagn. Compat.*, Vol. EMC-31, No. 4, November 1989.

8.10. Fontaine J. M., A. Umbert, A. Djebari, and J. Hamelin, "Ground Effects in the Response of a Single Wire Transmission Line Illuminated by an EMP,"

Proceedings of the 4th International Symposium on EMC, Zurich, March 10–12, 1981, paper 20E2.

8.11. Gary, C., "Approche complète de la propagation multifilaire en haute fréquence par l'utilisation des matrices complexes," *Bull. EDF*, Ser. B, No. 3/4, 1976, pp. 5–20.

8.12. Rachidi, F., private communication, Ecole Polytechnique Fédérale de Lausanne, 1994.

8.13. Rubinstein, M., et al., "Lightning-Induced Voltages on an Overhead Wire," *IEEE Trans. Electromagn. Compat.*, Vol. EMC-31, No. 4, November 1989, pp. 376–383.

8.14. Tesche, F. M., "On the Inclusion of Loss in Time-Domain Solutions of Electromagnetic Interaction Problems," *IEEE Trans. Electromagn. Compat.*, Vol. EMC-32, No. 1, February 1990, pp. 1–4.

8.15. Rachidi, F., C. A. Nucci, M. Ianoz, and C. Mazzetti, "The Inclusion of a Lossy Ground in the Calculation of Lightning-Induced Voltages on an Overhead Line," accepted for publication for *IEEE Trans. on Electromagn. Compat.*

8.16. Timotin, A., "Longitudinal Transient Parameters of a Unifilar Line with Ground Return," *Rev. Roum. Sci. Tech. Electrotech. Energ.*, Vol. 12, No. 4, 1967, pp. 523–535.

8.17. Jordan, E. C., and K. G. Balmain, *Electromagnetic Waves and Radiating Systems*, Prentice Hall, Englewood Cliffs, NJ, 1968.

8.18. Barnes, P. R., and F. M. Tesche, "On the Direct Calculation of a Transient Plane Wave Reflected from a Finitely Conducting Half-Space," *IEEE Trans. Electromagn. Compat.*, Vol. EMC-33, No. 2, May 1991.

8.19. Bridges, G. J., "Transient Plane Wave Coupling to Bare and Insulated Cables Buried in a Lossy Half-Space," *IEEE Trans. Electromagn. Compat.*, Vol. EMC-35, No. 1, February 1995.

8.20. Olsen, R. G., and D. C. Chang, "Current Induced by a Plane Wave on a Thin Infinite Wire near the Earth," *IEEE Trans. Antennas Propag.*, Vol. AP-22, No. 4, July 1974, pp. 586–589.

8.21. Bridges, G., "Fields Generated by Bare and Insulated Cables Buried in a Lossy Half-Space," *IEEE Trans. Geosci. Remote Sens.*, Vol. GRS-30, No. 1, January 1992, pp. 720–725.

8.22. Stratton, J. A., *Electromagnetic Theory*, McGraw-Hill, New York, 1941.

PROBLEMS

8.1 Compute and plot the skin depth δ as a function of frequency from $f = 10\,\text{kHz}$ to $10\,\text{GHz}$ for the following materials: aluminum, copper, silver, iron, and three different lossy grounds (with $\epsilon_r = 10$ and $\sigma_g = 0.01$, 0.001, and $0.0001\,\text{S/m}$). From these plots draw a conclusion about the accuracy of treating a lossy ground as a perfect conductor for power system problems.

8.2 The dc footing resistance of a vertical ground rod of length L and

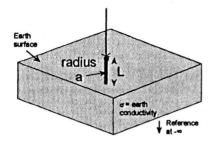

FIGURE P8.2 Vertical grounding electrode in a lossy earth.

radius a embedded in earth having a conductivity σ is given in Appendix D. Figure P8.2 shows this electrode configuration. At higher frequencies there can be a reactive component to this impedance. One way to model such a ground rod is to consider the rod as a buried transmission line, open circuited at the bottom end. Develop the appropriate formulas for the frequency-dependent input impedance for such an electrode.

8.3 Compute the peak current in the vertical riser shown in Figure P8.3, neglecting excitation on the horizontal portion of the power line, for an incident electromagnetic plane wave with vertical polarization coming from the left at a grazing angle ($\psi = 0°$). The electric peak field strength is 10 kV/m, the rise time is 1 µs, and the decay time is 100 µs. The height of the line is $h = 10$ m, the cable radius $a = 2.5$ cm, and the soil has a conductivity of $\sigma_g = 0.01$ S/m and a relative permeability of $\epsilon_r = 5$.

8.4 For the transmission line shown in Figure P6.15a, let the ground conductivity be $\sigma_g = 0.01$ S/m. Using the NULINE code of Appendix F:

(a) Compute the transient open-circuit voltage $V_2(t)$ and compare the results with those computed in Problem 6.15 for a perfect ground. Explain the differences between the two results.

(b) Repeat the calculation above for $\sigma_g = 0.1$ S/m and 0.001 S/m. Qualitatively summarize the effects that σ_g has on traveling waves on the line.

(*Hint:* The data file EX2.CMD can be used for this calculation.)

8.5 For the line of Problem 7.8 excited by an incident plane wave,

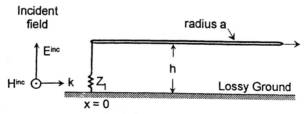

FIGURE P8.3 Transmission line excited by a grazing incident plane wave.

compare the responses computed previously for the perfectly conducting ground plane, with results obtained for the ground conductivity $\sigma_g = 0.001, 0.01,$ and 0.1 S/m.

8.6 Consider a long aboveground power line (Figure P8.6b) that is excited by a transient, vertically polarized incident plane wave with angles of incidence ψ and ϕ, and a temporal behavior given by the waveform in Problem 7.8. For this excitation, the current $I_d(t)$ induced in load Z_2 is of interest. It is desired to *simulate* the effects of this distributed field excitation by using a transient pulse generator located somewhere along the line, as shown in Figure P8.6b. This type of *current injection testing* causes a different current $I_s(t)$ to flow in the load and is frequently used when it is difficult to test large systems electrically.
  (a) Discuss qualitatively the differences that will arise in the responses for the two load currents with these two types of excitations.
  (b) Assuming that the transmission line has $\mathscr{L} = 100$ m, line height $h = 10$ m, radius $a = 0.5$ cm, $Z_1 = Z_2 = 10\,\Omega$, and that the earth parameters are $\sigma_g = 0.01$ S/m and $\epsilon_r = 10$, explore the behavior of the load current response $I_s(t)$ for the simulation case (Figure P8.6b) as a function of the pulser parameters (pulser current wave shape, pulser admittance Y_p and pulser location x_s), and compare with the desired current response $I_d(t)$. (Assume that $Y_p = 0.1$ S.)
  (c) Discuss the dual excitation of a *voltage* source located in series with the transmission line. Which source (series voltage or shunt current) is more practical for experimental purposes?

8.7 The double-exponential transient waveform given in Problem 7.8 is vertically incident on a lossy ground.
  (a) Compute and plot the total transient E-field at a height of 10 m for earth conductivities ranging from ∞ to 0.001 S/m and $\epsilon_r = 10$.
  (b) Repeat the calculations for σ_g fixed at 0.01 S/m and ϵ_r varying

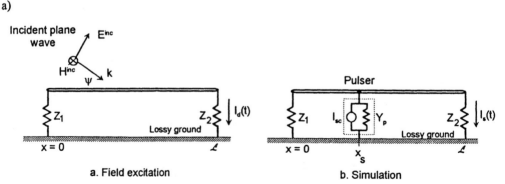

FIGURE P8.4 Transmission line excited by an incident field or a lumped current source.

from 1 to 100. Which parameter (σ_g or ϵ_r) is most important in determining the transient field behavior?
(*Hint:* Consider using the program TOTALFLD in Appendix F.)

8.8 Repeat Problem 8.7 for an observation point located at a burial depth of 5 m in the earth. (*Hint:* Be very careful to select the FFT time window so as to avoid the late-time fold-through of the transient response.)

8.9 In Figure 8.6a, it is apparent that for very low conductivities of the earth, there is a *null* in the vertically polarized reflection coefficient for a particular angle ($\psi \approx 20°$). This is known as Brewster's angle. Assuming that $\sigma_g = 0$, derive and plot the expression for the Brewster's angle as a function of the relative dielectric constant.

8.10 A realistic ground is seldom modeled adequately by a single, homogeneous lossy half-space. A better model is provided by two or more layers each having a different conductivity and permittivity. For the two-layer earth shown in Figure P8.10, derive the reflection coefficients for vertical and horizontal polarization as a function of the angle of incidence ψ. (*Hint:* Consult W.C. Chew, *Waves and Fields in Inhomogeneous Media*, Van Nostrand Reinhold, New York, 1990, p. 49.)

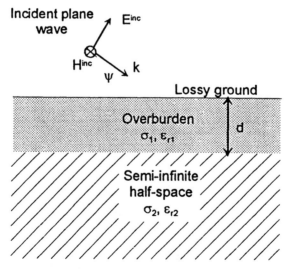

FIGURE P8.10 Two-layer model for a lossy earth.

PART V
SHIELDING MODELS

CHAPTER 9

Shielded Cables

Shielded cables are frequently used to transmit information between equipment contained within two protective enclosures. As such, one can define two distinct transmission lines: an *external* line having currents and charges flowing on the exterior of the cable, together with a possible ground-plane return, and an *internal* line consisting of the conductors inside the shield. EMI fields can excite the external transmission line and if the cable shield is imperfect, some of the external currents and charges can penetrate through the shield and excite the internal line. This leads to an unwanted response at the "protected" equipment. In this chapter this coupling mechanism is examined for both solid and braided cables, and several examples are given to illustrate the resulting calculational models.

9.1 INTRODUCTION

In its simplest configuration, a shielded transmission line system consists of a cable with the shield connected to each enclosure. The enclosures and the cable are usually located over a conducting ground plane with the cable at a height h. As shown in Figure 9.1, the two enclosures are connected to the ground plane by external impedances $Z^{(e)}$, so that either a grounded or open-circuit configuration can be modeled.

For a shielded cable illuminated by an electromagnetic field, the external electric and magnetic fields can penetrate through imperfections in the cable sheath and give rise to disturbing currents and voltages on the internal conductors. The coupling between the external electromagnetic field and the inner conductors occurs through three basic phenomena [9.1]:

1. Diffusion of the E and H fields through the sheath material.
2. Penetration of the fields through the small apertures of the braided shields.
3. A more complicated induction phenomenon due to overlapping of the individual strands (or carriers) of the shield.

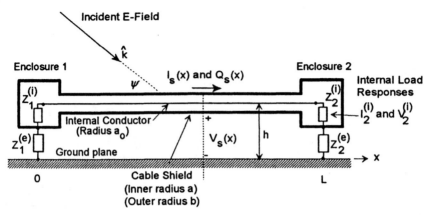

FIGURE 9.1 Geometry of a shielded coaxial line with an internal circuit. (From [9.20]. Reprinted by permission from Academic Press, Inc.)

The last two phenomena occur only for braided shields.

The behavior of the induced response on the inner conductors of a shielded cable can be described in terms of a transfer impedance Z_t and a transfer admittance Y_t of the shield. To understand the transfer impedance and admittance concepts, we begin with a qualitative discussion describing the physical phenomena occurring for homogeneous tubular shields. The issues of field penetration through apertures and induction, which are specific to the braided sheaths, are considered later.

9.2 FUNDAMENTALS OF CABLE SHIELD COUPLING

A time-varying external EM field will induce both a cable sheath current I_s and a sheath-to-ground voltage V_s (or equivalently, a sheath charge density), as discussed in earlier chapters. Portions of the external electric and magnetic fields are able to penetrate the shield, and these induce internal voltage and current responses, V_i and I_i, respectively. These quantities and their assumed polarities are shown in Figure 9.2. For this conductor system, a return current I_g flows in the ground which is the sum of I_s and I_i.

The current flow in the cable sheath will create an axial electric field inside the sheath. Due to the skin effect, the current distribution and the associated electric field distribution in the sheath cross section are not uniform. If I_s is the total current flowing in the sheath, the electric field E_i on the inner surface of the shield is produced by an attenuated current density, with the reduction being determined approximately by δ, the skin depth in the shield material, given by

$$\delta = \sqrt{\frac{1}{\pi f \sigma \mu}} \quad \text{m} \tag{9.1}$$

9.2 FUNDAMENTALS OF CABLE SHIELD COUPLING

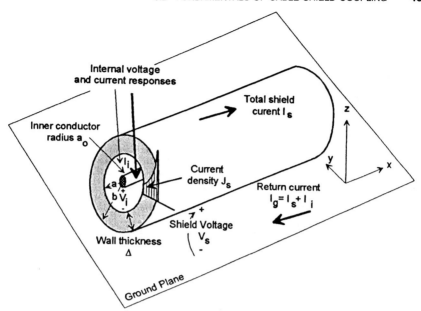

FIGURE 9.2 Coaxial cable located over a conducting ground plane.

where σ is the conductivity of the sheath material, f is the frequency of the induced current, and μ is the permeability of the sheath material.

This axial electric field component on the inner surface of the sheath will create a voltage between the internal conductors and the sheath, and depending on the termination impedances of the inner signal conductor, a current may flow. The protection capability of the cable shield is reflected by the amount of reduction of the electric field component due to the skin-depth attenuation. This reduction capability will be determined more precisely in the next section through the concept of a transfer impedance, which is defined as the ratio between the inner electric field E_i and the total shield current I_s in the frequency domain.

The dual of the transfer impedance is the transfer admittance, which describes the process by which a portion of the induced charge on the cable sheath finds its way onto the internal wire inside the shield. This induction of charge on the inner conductor amounts to a current injected on the internal cable. This effect can be related to the external shield-to-ground voltage V_s by a transfer admittance.

9.2.1 Definitions of Transfer Impedance and Transfer Admittance

The cable sheath with its external return (generally, a ground plane) and the internal conductor form two coupled circuits, as shown in Figure 9.3. Note that in this development it is assumed that the external circuit is independent of the behavior of the internal circuit, and that the internal circuit has both a

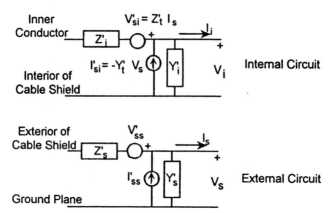

FIGURE 9.3 Two per-unit-length circuits formed by the sheath and its ground return and the sheath and the internal conductor.

voltage- and a current-controlled source, providing the excitation from the external circuit. This circuit configuration results from the assumption that the cable performs as a good shield.

For these two coupled lines, the current and voltage I_s and V_s of the outer shield and the current and voltage I_i and V_i of the inner conductor are described by the following set of differential equations [9.1]:

EXTERNAL CIRCUIT

$$\frac{dV_s}{dx} + Z'_s I_s = V'_{ss} \quad \text{V/m} \qquad (9.2a)$$

$$\frac{dI_s}{dx} + Y'_s V_s = I'_{ss} \quad \text{A/m} \qquad (9.2b)$$

INTERNAL CIRCUIT

$$\frac{dV_i}{dx} + Z'_i I_i = V'_{si} \quad \text{V/m} \qquad (9.3a)$$

$$\frac{dI_i}{dx} + Y'_i V_i = I'_{si} \quad \text{A/m} \qquad (9.3b)$$

The primary sources in this problem are the distributed voltage and current sources V'_{ss} and I'_{ss} on the external circuit. The excitation sources for the internal circuit are the voltage and current sources shown in Figure 9.3 and are denoted by V'_{si} and I'_{si}, respectively. These internal sources are related to the external line responses by

$$V'_{si} = Z'_t I_s \quad \text{(V/m)} \tag{9.4}$$

$$I'_{si} = -Y'_t V_s \quad \text{(A/m)} \tag{9.5}$$

where Z'_t and Y'_t are the transfer impedance and transfer admittance of the shield. These quantities may be formally defined from Eqs. (9.3a) and (9.3b) by setting I_i and V_i to zero, as

$$Z'_t = \frac{1}{I_s} \frac{dV_i}{dx}\bigg|_{I_i=0} \quad \Omega/\text{m}, \tag{9.6}$$

$$Y'_t = -\frac{1}{V_s} \frac{dI_i}{dx}\bigg|_{V_i=0} \quad \text{S/m} \tag{9.7}$$

9.2.2 Relative Importance of Z'_t and Y'_t

The physical behavior of a shielded cable as described above is valid for both tubular and braided cables. For a solid tubular shield, the electrostatic shielding is much greater than the magnetostatic shielding, and as a result, the transfer impedance term dominates at low frequencies. This fact has led many investigators to neglect the transfer admittance term in EMC coupling problems.

For braided cables, the coupling mechanism giving rise to the transfer impedance and admittance are enhanced, due to the field penetration through the shield apertures. Again, at low frequencies, the electrostatic shielding of the braid is much better than the magnetic field shielding, and Y'_t is usually small compared with Z'_t. However, as the frequency increases, both the E and H fields are able to penetrate the braid apertures, and the induced effects on the inner conductor from both field components can be on the same order of magnitude. In this case, neglecting the transfer admittance can lead to roughly twofold errors in internal response.

The magnetic and electric field coupling to shielded cables can be controlled somewhat by a procedure called *braid optimization*, in which the braid material and weave characteristics are tailored to obtain certain relationships between Y'_t and Z'_t.

9.3 EM COUPLING THROUGH A SOLID TUBULAR SHIELD

9.3.1 Transfer Impedance

The tubular shield consists of a thin-walled metallic tube of uniform cross section and uniform wall thickness, as shown in Figure 9.2. Coupling of external disturbances through such a shield occurs only by diffusion. An analytical expression for the transfer impedance of tubular shields has been derived by Schelkunoff [9.3]. If a and b are the inner and outer radii of the tubular shield, respectively, and σ is the conductivity of the shield material, the general transfer impedance expression is

$$Z'_t = \frac{\gamma}{2\pi\sigma b} \frac{J_1(\gamma a)Y_0(\gamma a) - Y_1(\gamma a)J_0(\gamma a)}{J_1(\gamma a)Y_1(\gamma b) - Y_1(\gamma a)J_1(\gamma b)} \quad \Omega/m \quad (9.8a)$$

where

$$\gamma = \sqrt{j\omega\mu(\sigma + j\omega\epsilon)} \quad m^{-1} \quad (9.8b)$$

with $J_i(\cdot)$ and $Y_i(\cdot)$ representing the cylindrical Bessel functions of first and second kind and order i.

For most practical cable shields, the wall thickness $\Delta = b - a$ is small compared to radius a of the tube, and the radii are small compared to the wavelength of interest. In this case, the exact analytical expression in Eq. (9.8) can be approximated by

$$Z'_t \approx R'_0 \frac{(1+j)\Delta/\delta}{\sinh[(1+j)\Delta/\delta]} \quad \Omega/m \quad (9.9a)$$

where

$$R'_0 = \frac{1}{\pi\sigma(b+a)(b-a)} \approx \frac{1}{2\pi\sigma a\Delta} \quad \Omega/m \quad (9.9b)$$

is the dc per-unit-length resistance of the sheath and δ the skin depth defined by Eq. (9.1).

Low- and high-frequency simplifications to the approximation of the transfer impedance in Eq. (9.9) are developed in Vance [9.2]. These are:

LOW-FREQUENCY APPROXIMATION $(\Delta/\delta \ll 1)$

$$Z'_t \approx R'_0 \quad \Omega/m \quad (9.10a)$$

HIGH FREQUENCY APPROXIMATION $(\Delta/\delta \gg 1)$

$$Z'_t = 2\sqrt{2} R'_{hf} e^{-(1+j)\Delta/\delta} e^{j\pi/4} \quad \Omega/m \quad (9.10b)$$

where $R'_{hf} = 1/(2\pi a\delta\sigma)$ represents the high-frequency approximation of the per-unit-length tube resistance.

As an example of the shielding provided by a practical cable, the case of a semirigid UT-141 cable has been considered in [9.4]. Figure 9.4 presents the magnitude of the transfer impedance $|Z'_t|$ as calculated from Eq. (9.9a), as well as measured values. At low frequencies below about 2 to 3 MHz, good agreement between these two results is noted. However, at higher frequencies, leakage through cable connectors begins to cause an increase in

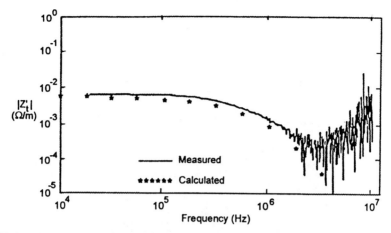

FIGURE 9.4 Comparison between measurements and calculation for thin-walled semirigid UT-141 cable. (From [9.4]. Reprinted by permission.)

the transfer impedance. Such an increase of Z'_t with increasing frequency is also typical of cables with braided shields, as discussed in the next section.

9.3.2 Transfer Admittance

Although it is possible to define a transfer admittance Y'_t formally for a solid tubular shield, it is rare to find this discussed in the literature. This is because the contribution of Y'_t to the total response of a cable is very small: the solid sheath is a very good shield for the E-field. For practical problems involving solid shields, this term is usually neglected. However, when the shield contains apertures, such as in the case of a braided shield, the E-field is able to penetrate more effectively and the contribution of Y'_t can be important. This is discussed in the next section.

9.4 MODELS FOR BRAIDED SHIELDS

Many practical cables are shielded by a flexible, woven braid made of conducting filaments. Figure 9.5a shows a developed (unwrapped) view of the braided shield of a cable with an outer radius b. The shield consists of many bands of parallel wire filaments called *carriers* which are woven over and under each other. The braid is characterized by the following parameters: $\mathcal{C}$, the number of carriers in the shield; N, the number of filaments in each carrier; d, the diameter of the filaments; σ, the electrical conductivity of the shield material; and ψ, the weave angle of the shield. Such a woven construction provides a series of periodic apertures along the cable and around its circumference. Figure 9.5b shows additional detail of the apertures and their dimensions.

458 SHIELDED CABLES

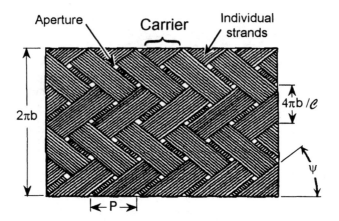

a. View of the braided cable showing the weave parameters (adapted from [9.2]).

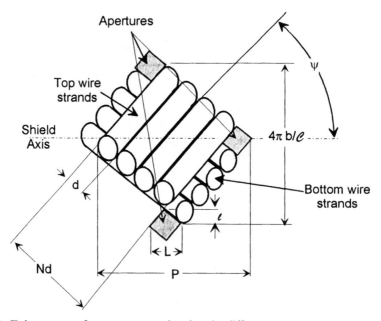

b. Enlargement of weave pattern showing the different weave parameters

FIGURE 9.5 Geometry of a braided shield. (a) View of the braided cable showing the weave parameters (adapted from [9.2]. Copyright © 1978 John Wiley & Sons, Inc., reprinted by permission of the publisher.); (b) Enlargement of weave pattern showing the different weave parameters.

This inhomogeneous construction of braided shields and the complexity in the way the external electromagnetic field penetrates the braid has presented difficulties in the attempts at modeling the transfer impedance and admittance of such cables. As a consequence, the most reliable method for determining these transfer quantities is by measurement. Many contributions in the literature have been devoted to an improvement of the measurement methods (see, e.g., [9.1–9.9]). However, as this book is aimed toward EMC modeling techniques, we will discuss the state-of-the-art in braided cable modeling while stressing the point that much work remains to be done in this field.

Although not describing the entire complex nature of a sheath response to an external electromagnetic environment, it can be assumed that the three penetration mechanisms listed in Section 9.1 occur. Diffusion for a braided shield is basically the same phenomenon as for tubular shields. However, due to the weave pattern of the shield, the presence of several carriers in the braid, and the individual filaments in each carrier, the expression of the transfer impedance must be modified to take into account the braided shield construction. Vance [9.2] has extended the approximate transfer impedance expression (9.9) to obtain the diffusion term in the expression of the transfer impedance for braided cables as

$$Z'_d = \frac{4}{\pi d^2 N \ell \sigma \cos \psi} \frac{(1+j)d/\delta}{\sinh(1+j)d/\delta} \quad \Omega/m \qquad (9.11)$$

where δ is the skin depth given in Eq. (9.1), and N, ℓ, ψ, d, and σ are the shield parameters defined previously.

9.4.1 EM Field Penetration and Diffraction into Braided Shields

As the overlapping of the braided cables is not perfect, the external magnetic and electric fields will penetrate the apertures in the same way that a field penetrates through small holes of a solid metallic plate. The magnetic and electric field lines penetrating through one aperture are illustrated in Figure 9.6, and it is these fields that induce an additional contribution to the internal response of the inner conductor.

The magnetic and electric field penetration into braided cables has been discussed by Lee and Baum [9.10]. To solve the interaction of braided shielded cables with external sources, they divided the problem into three separate problems:

1. The exterior scattering problem of determining the induced currents and charges on the outer surface of the shield with all shield apertures short-circuited.
2. The problem of calculating the transmission coefficients of the aper-

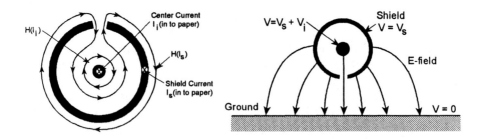

FIGURE 9.6 Penetration of the magnetic and electric field inside the shield of a coaxial cable through a single aperture. (Adapted from [9.2]. Copyright © 1978 John Wiley & Sons, Inc., reprinted by permission of the publisher.)

tures (i.e., the static electric and magnetic polarizabilities α_e and $\overset{\leftrightarrow}{\alpha}_m$) using a low-frequency approximation.
3. The interior problem of determining the magnitudes and the propagation characteristics of the voltage and current induced by the penetration of the EM fields through an aperture.

In this chapter, only item 3 is treated, as problems 1 and 2 have been discussed in Chapter 5.

9.4.2 Single Aperture Excitation

Consider first the case of a single aperture in an otherwise solid coaxial shield. Let V_i and I_i be the voltage and current induced on the internal conductor by the field penetration through a single aperture in the braided shield located at a position $x = x_a$. The assumed polarities of these internal responses are indicated in Figure 9.2. Because the transverse dimension of the braided shield is assumed to be smaller than the wavelength λ, we assume that there is only a TEM mode propagating within the coaxial region and the transmission line equations for V_i and I_i are given by Eqs. (9.3a) and (9.3b). With the assumption that the loss in the line is negligible, these equations become

$$\frac{dV_i}{dx} + j\omega L' I_i = V'_{eq} \quad \text{V/m} \tag{9.12a}$$

$$\frac{dI_i}{dx} + j\omega C' V_i = I'_{eq} \quad \text{A/m} \tag{9.12b}$$

where L' and C' are the per-unit-length inductance and capacitance coefficients of the inner region of the coaxial cable given by

$$L' = \frac{\mu_0}{2\pi} \ln \frac{a}{a_0} \quad \text{H/m} \tag{9.13}$$

$$C' = 2\pi\epsilon_0 \frac{1}{\ln(a/a_0)} \quad \text{F/m} \tag{9.14}$$

In this discussion it is assumed that the coaxial line has no dielectric filling; that is, the dielectric constant inside the line is ϵ_0. Later, the results developed here will be generalized for different internal and external dielectric material. The case of different magnetic material in the cable is not considered.

The excitation sources V'_{eq} and I'_{eq} take into account the localized excitation of the line by the aperture at position $x = x_a$ and are given in [9.10] as

$$V'_{eq} = -\frac{j\omega\mu_0}{2\pi a} m_\phi \delta(x - x_a) \quad \text{V/m} \tag{9.15}$$

$$I'_{eq} = -\frac{1}{\eta} \frac{j\omega}{2\pi a} p_r \delta(x - x_a) \quad \text{A/m} \tag{9.16}$$

where m_ϕ and p_r are the azimuthal and radial components of the magnetic and electric dipole moments of the aperture, a is the inner sheath radius, a_0 is the radius of the inner conductor, $\delta(x - x_a)$ is the Dirac delta function, and η is a geometrical factor defined as

$$\eta = \frac{1}{2\pi} \ln \frac{a}{a_0} \tag{9.17}$$

As described in [9.10], the dipole moment components m_ϕ and p_r can be evaluated using the short-circuited magnetic field and electric field in the vicinity of the aperture and the corresponding aperture polarizabilities. This implies that m_ϕ and p_r depend on the external shield current I'_s and on the external per-unit-length charge Q'_s. (In consulting this reference, note that the time dependence is $e^{-i\omega t}$ and that the definitions of the directions of positive current flow and voltage differ from those used here.)

Because the presence of the aperture in the shield will affect the propagation characteristics inside the cable, we expect that the aperture will also modify the capacitance and inductance of the internal line. Reference

[9.10] illustrates how Eq. (9.12) can be written as

$$\frac{dV_i}{dx} + j\omega L'[1 + \Delta_L \delta(x - x_a)]I_i = j\omega L_a I_s \delta(x - x_a) \quad \text{V/m} \quad (9.18a)$$

$$\frac{dI_i}{dx} + j\omega C'[1 - \Delta_C \delta(x - x_a)]V_i = j\omega \frac{C_a}{C'} Q'_s \delta(x - x_1) \quad \text{A/m}$$

$$(9.18b)$$

where L_a is an effective aperture inductance given by

$$L_a = \frac{\mu_0 \alpha_m}{(2\pi a)^2} \quad \text{H} \quad (9.19)$$

with α_m representing the ϕ–ϕ component of the magnetic polarizability tensor $\overleftrightarrow{\alpha}_m$. Similarly, an aperture capacitance is defined as

$$C_a = \frac{\epsilon_0 \alpha_e}{(2\pi a \eta)^2} \quad \text{F} \quad (9.20)$$

where α_e is the electrical polarizability of the aperture. The electric and magnetic polarizabilities are discussed in more detail in Section 9.4.4. In Eq. (9.18), the terms $\Delta_L = L_a/L'$ (m) and $\Delta_C = C_a/C'$ (m) represent small changes to the line inductance and capacitance arising from the aperture. These can be thought of as coming from small positive lumped inductance and negative lumped capacitance elements on the internal transmission line at the location of the aperture.

9.4.3 Multiple Apertures

The development of the transmission line equations for a single aperture can be extended to the case of a braided-shielded cable having n apertures per unit length with a distance between the centers of these apertures given by $P = 1/n$, as shown in Figure 9.7. For such a line, the response should consist

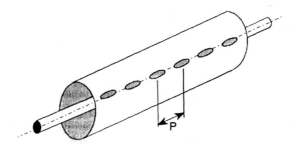

FIGURE 9.7 Cable shield containing multiple apertures.

of a sum over all of the aperture contributions, which yields a sum of delta-function contributions in the telegrapher's equations (9.18), as $\sum_{n=-\infty}^{\infty} \delta(x - nP)$. This sum may be transformed into a cosine Fourier series in the interval $-P/2$ to $+P/2$ as

$$\sum_{n=-\infty}^{\infty} \delta(x - nP) = \frac{1}{P} + \frac{2}{P} \sum_{k=1}^{\infty} \cos \frac{2k\pi x}{P} \qquad (9.21)$$

As observed previously, the shielded cable can be regarded as a waveguide, which means that a dominant TEM mode in the cable can be assumed. Consequently, the series expansion (9.21) can be approximated by the average value term $1/P$ [9.10]. Thus the transmission line equations for a braided cable with n apertures per unit length can be written simply as

$$\frac{dV_i}{dx} + j\omega L'\left(1 + \frac{\Delta_L}{P}\right)I_i = j\omega \frac{L_a}{P} I_s \qquad \text{V/m} \qquad (9.22a)$$

$$\frac{dI_i}{dx} + j\omega C'\left(1 - \frac{\Delta_C}{P}\right)V_i = -j\omega \frac{C_a}{PC'} Q'_s \qquad \text{A/m} \qquad (9.22b)$$

Replacing $1/P = n$, it is possible to write the expressions for the equivalent aperture inductance and capacitance terms for a braided cable with n apertures per unit length as

$$L'_a \equiv \frac{L_a}{P} = nL_a = \frac{n\mu_0 \alpha_m}{(2\pi a)^2} \qquad \text{H/m} \qquad (9.23a)$$

$$C'_a \equiv \frac{C_a}{P} = nC_a = \frac{n\epsilon_0 \alpha_e}{(2\pi a\eta)^2} \qquad \text{F/m} \qquad (9.23b)$$

$$\Delta^*_L \equiv \frac{L'_a}{L'} \qquad (9.23c)$$

$$\Delta^*_C \equiv \frac{C'_a}{C'} \qquad (9.23d)$$

Noting that the per-unit-length capacitance of the shield relative to the external ground plane C'_s can be used to express the external charge on the shield in terms of the external voltage as $Q'_s = C'_s V_s$, the source terms in Eq. (9.22) can be written in terms of a transfer impedance Z'_t and transfer admittance Y'_t as

$$\frac{dV_i}{dx} + j\omega L'(1 + \Delta^*_L)I_i = Z'_t I_s \qquad \text{V/m} \qquad (9.24a)$$

$$\frac{dI_i}{dx} + j\omega C'(1 - \Delta^*_C)V_i = -Y'_t V_s \qquad \text{A/m} \qquad (9.24b)$$

464 SHIELDED CABLES

where

$$Z'_t = j\omega L'_a \qquad \Omega/\text{m} \qquad (9.25)$$

$$Y'_t = j\omega \frac{C'_a C'_s}{C'} \qquad \text{S/m} \qquad (9.26)$$

Notice in Eq. (9.25) that the transfer impedance depends on L'_a, which in turn, depends only on the properties of the shield. However, the transfer admittance in Eq. (9.26) depends on *both* the cable shield properties and the external cable capacitance. Thus, as discussed in [9.2], the transfer admittance does not uniquely characterize the behavior of the shield.

9.4.4 Expressions for the Aperture Polarizabilities

Details of aperture polarizabilities $\overset{\leftrightarrow}{\alpha}_m$ and α_e that are suitable for treating the EM penetration through the holes in the braided cable have been discussed in Section 5.4.2. Most braided cables have diamond-shaped apertures, as suggested in Figure 9.5. More complicated aperture shapes can be treated with numerical methods.

As illustrated in Figure 9.8, the diamond-shaped aperture can be approximated by an elliptical hole, for which analytical expressions for the polarizabilities have been given in Table 5.1. For the diamond-shaped aperture with dimensions L and l shown in the figure, the major axis length of the equivalent elliptical aperture is denoted by L_{eq}, where $L_{eq} = L$. The relation between the large and small effective ellipse axes depends on the weave angle ψ of the shield and is given by [9.11]:

$$l_{eq} = L_{eq} \tan \psi \qquad \text{for } \psi < 45° \qquad (9.27a)$$

$$l_{eq} = L_{eq} \cot \psi \qquad \text{for } \psi > 45° \qquad (9.27b)$$

As noted in Section 5.4.2, the tangential magnetic field can be oriented in an arbitrary manner over the elliptical aperture. Consequently, it is necessary to express $\overset{\leftrightarrow}{\alpha}_m$ using two different components, one for the external magnetic field parallel to the principal axis and another for the magnetic

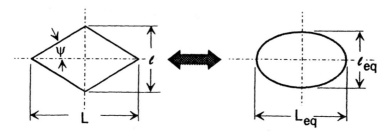

FIGURE 9.8 Equivalence between the diamond-shaped aperture and an ellipse.

field perpendicular to the principal axis. From Table 5.1, these polarizabilities can be put into a different, but equivalent form as [9.11]:

FOR H PARALLEL TO THE PRINCIPAL AXIS

$$\alpha_m = \frac{\pi L_{eq}^3}{24} \frac{e^2}{K(e) - E(e)} \quad m^3 \quad (9.28a)$$

FOR H PERPENDICULAR TO THE PRINCIPAL AXIS

$$\alpha_m = \frac{\pi L_{eq}^3}{24} \frac{e^2(1-e^2)}{K(e) - (1-e^2)E(e)} \quad m^3 \quad (9.28b)$$

with the corresponding electric polarizability α_e given by

$$\alpha_e = \frac{\pi L_{eq}^3}{24} \frac{1-e^2}{E(e)} \quad m^3 \quad (9.29)$$

Here L_{eq} is the effective length of the principal axis of the ellipse, l_{eq} the minor axis length, and the parameter e is the eccentricity of the ellipse given by

$$e = \sqrt{1 - \left(\frac{l_{eq}}{L_{eq}}\right)^2} \quad (9.30)$$

The terms $E(\cdot)$ and $K(\cdot)$ denote the complete elliptic integrals of the first and second kind, which are defined in Table 5.1. It is important to note that the aperture dipole moments provided by the polarizabilities in Eqs. (9.28) and (9.29) must be imaged in a conducting screen, as discussed in Section 5.4.2, to calculate the total field leaking through the aperture.

9.4.5 Shield Transfer Characteristics in Terms of Braid Weave Parameters

9.4.5.1 Transfer Impedance. The shielding characteristics of a braided shield can be expressed in terms of shield radius b, the number of carriers in the braid $\mathcal{C}$, the braid wire diameter d, the number of wires in one carrier N, and weave angle ψ (see Figure 9.5) [9.11]. The derived parameters describing the braid are the fill factor F,

$$F = \frac{Nd\mathcal{C}}{4\pi b \cos \psi} \quad (9.31)$$

and the optical coverage $\mathcal{K}$,

$$\mathcal{K} = 2F - F^2 \quad (9.32)$$

The number of apertures per unit length n can be expressed in terms of weave parameters as

$$n = \frac{e^2 \tan \psi}{4\pi b} \quad (9.33)$$

For $\psi < 45°$, the transfer impedance of Eq. (9.25) can be expressed in terms of the braid parameters by an aperture inductance L_a' as

$$Z_t' = j\omega L_a' = j\omega \left[\frac{\pi \mu_0}{6e} \frac{e^2}{E(e) - (1-e^2)K(e)} \left(\frac{el}{4\pi b}\right)^3 \right] \quad \Omega/m \quad (9.34)$$

where it has been assumed that $l \approx l_{eq}$. The small axis of the diamond-shaped aperture in Figure 9.5 is

$$l = \frac{4\pi b}{e} - \frac{Nd}{\cos \psi} \quad m \quad (9.35)$$

and using the definitions of the fill in Eq. (9.31) and of the optical coverage in Eq. (9.32), we obtain the expression for the aperture inductance to be

$$L_a' = \frac{\pi \mu_0}{6e} (1-\mathscr{K})^{3/2} \frac{e^2}{E(e) - (1-e^2)K(e)} \quad H/m \quad (9.36a)$$

where $\psi < 45°$. Similarly, for the braid angle $\psi > 45°$ the aperture inductance is given by

$$L_a' = \frac{\pi \mu_0}{6e} (1-\mathscr{K})^{3/2} \frac{e^2/\sqrt{1-e^2}}{E(e) - (1-e^2)K(e)} \quad H/m \quad (9.36b)$$

where $\psi > 45°$.

Using this approach for calculating the transfer inductance of the braided shield, Vance [9.2] has added another term to Z_d' to account for the effects of diffusion through the braid. In this manner, the complete transfer impedance of the braided shield is

$$Z_t' = Z_d' + j\omega L_a' \quad \Omega/m \quad (9.37)$$

where Z_d' is the diffusion term given by Eq. (9.11). According to Vance, this model gives very accurate results at low frequency ($d/\delta \ll 1$) and is accurate within a factor or 3 or less at high frequencies ($\omega L_a' \gg |Z_d|$). The validity of these expressions has also been checked by comparison with measured data by Degauque and Hamelin [9.1].

9.4.5.2 Transfer Admittance.

Using the electric polarizability expression, the transfer admittance term in Eq. (9.26) arising from the per-unit-length braid aperture capacitance can be written as [9.2]:

$$Y'_t = j\omega \frac{\pi C' C'_s}{6\ell\epsilon_0}(1-\mathcal{K})^{3/2}\frac{1}{E(e)} \quad \text{S/m} \qquad (9.38a)$$

where $\psi < 45°$, and

$$Y'_t = j\omega \frac{\pi C' C'_s}{6\ell\epsilon_0}(1-\mathcal{K})^{3/2}\frac{1}{(1-e^2)E(e)} \quad \text{S/m} \qquad (9.38b)$$

where $\psi > 45°$.

9.4.5.3 Dielectric Filling in the Cable.
It is common to encounter braided cables that contain a dielectric filling within the coaxial region. In such a case it is necessary to modify Eq. (9.38) to account for the presence of the dielectric. Consider the case when the interior of the coaxial cable has a dielectric constant $\epsilon_{int} = \epsilon_{r_{int}}\epsilon_0$ and the region outside the cable has a dielectric constant $\epsilon_{ext} = \epsilon_{r_{ext}}\epsilon_0$, where the terms ϵ_r are relative dielectric constants. The effects of this change in dielectric constant are that the capacitance terms C' and C'_s must change to reflect the change in the value of ϵ, and that the aperture polarizability must also change [9.11]. By defining an equivalent permittivity of a composite medium of two different dielectric materials as [9.12]

$$\epsilon_{eq} = \frac{2\epsilon_{ext}\epsilon_{int}}{\epsilon_{ext}+\epsilon_{int}} \quad \text{F/m} \qquad (9.39)$$

the transfer admittance terms can be written as follows:

$$Y'_t = j\omega \frac{\pi\epsilon_{eq} C'' C''_s}{6\ell\epsilon_{int}\epsilon_{ext}}(1-\mathcal{K})^{3/2}\frac{1}{E(e)} \quad \text{S/m} \qquad (9.40a)$$

where $\psi < 45°$, and

$$Y'_t = j\omega \frac{\pi\epsilon_{eq} C'' C''_s}{6\ell\epsilon_{int}\epsilon_{ext}}(1-\mathcal{K})^{3/2}\frac{1}{(1-e^2)E(e)} \quad \text{S/m} \qquad (9.40b)$$

where $\psi > 45°$.

In evaluating this expression, the term C'' is the capacitance of the coaxial region, as given by Eq. (9.14), but with the dielectric constant ϵ_{int} used in place of ϵ_0. Similarly, C''_s is the external capacitance of the line to the ground, assuming that ϵ_{ext} is used in place of ϵ_0.

9.4.5.4 Comparison with Measurements.
The measurement of the transfer impedance is not trivial. As mentioned at the beginning of Section 9.4, several measurement techniques can be used but their discussion is

beyond the scope of this book. A useful method based on the injection of a transient current pulse has been developed [9.11] and used to obtain experimental values of the inductance component L'_a and the net transfer capacitance term $(C'C'_s/C'_a)$ for different types of cables (different diameters, weave angles, and optical coverage). Table 9.1 compares experimental results with values calculated using Eqs. (9.36) and (9.38).

For L'_a, the disagreement between measurement and calculation ranges from 16% up to a factor of 6. For the aperture transfer capacitance, the results of the modeling seem even more discouraging: Differences up to a factor 11 between measurement and calculation can be observed! Two causes have been proposed to explain these discrepancies [9.1]. The first is the possibility of errors in the determination of the aperture dimensions. These are evaluated from photographs taking an average value from about 10 photos. As the aperture length L_{eq} enters into the expressions to the third power in the formulas for L'_a and $(C'C'_s/C'_a)$, a small error in this quantity has an important influence on the overall result. Another reason is that the diffraction theory applied to apertures in the braided shields assumes that the aperture dimensions are much larger than the thickness of the sheath, which is not the case for real cables (0.6 mm thickness). The assumption of a negligible thickness increases the polarizability value, and this tendency can clearly be seen in the values calculated for the transfer capacitance.

9.4.6 Improved Expressions for the Transfer Impedance of Braided Shields

It is important to note that although rather high discrepancies can be observed for the aperture capacitance for all the braid samples in Table 9.1, this is not the case for the aperture inductance. Sample 2 shows a reasonable

TABLE 9.1 Measured and Calculated Aperture Inductance and Transfer Capacitance Values for Different Braided Cables

Sample	$2b$ (mm)	ψ (deg)	$\varkappa$	L_{eq} (mm)	L'_a (nH/m) Measured	L'_a (nH/m) Calculated	$C'C'_s/C'_a$ (pF/m) Measured	$C'C'_s/C'_a$ (pF/m) Calculated
1	6.5	37	0.91	0.70	0.45	0.98	0.071	0.375
2	7.3	43	0.90	0.66	1.27	1.06	0.168	0.456
3	11	28	0.97	0.90	0.26	0.16	0.181	0.129
4	9.2	43.6	0.94	0.64	0.25	0.46	0.023	0.257
5	8.8	40	0.96	0.57	0.047	0.292	0.015	0.143

Source: Reference [9.1] © Oxford University Press, 1993, reprinted by permission.

agreement between the measured and calculated values. This sample has a weave angle of 43° and an optical coverage of 0.90. It is noted that in the other samples for which poor agreement between calculations and measurements is observed, the weave angles are substantially smaller than 45° (i.e., for samples 1 and 3), or there is a high optical coverage (samples 3, 4, and 5). This suggests that the reasons for the errors in the aperture capacitance proposed in [9.1] may also be valid for the aperture inductance.

The experimental evidence suggests that for the transfer inductance, Eq. (9.34) can be used for cables having a weave angle $\psi \approx 45°$ and a rather poor optical coverage. However, for small or large ψ values or for a high optical coverage, another phenomenon should be considered. Demoulin has shown experimentally that for cables with an optical coverage very near 1:

- The absolute transfer impedance value is not linearly increasing with frequency and that this disagreement can be corrected by introducing an additional term proportional to $\sqrt{f}$.

- The phase does not vary between $+\pi/2$ and $-\pi/2$ as theory predicts for the diffraction model, but between zero and about $-3\pi/4$.

Figure 9.9 shows the measured magnitude and phase of four braided shield cables with a high optical coverage, and these observations are evident. Part (*a*) of the figure shows the magnitude of the transfer impedance and it is clear that at higher frequencies, the transfer impedance is proportional to $\sqrt{f}$. Part (*b*) of the figure illustrates the phase behavior.

Based on these and other experimental results, other authors have proposed alternative models for the transfer impedance to take into account additional factors, such as the presence of an induction phenomenon called *porpoising*, which adds to the diffusion and penetration terms. The porpoising effect is more important when the optical coverage is high, because less field penetration occurs. This additional inductive term results from the layer construction of the braid. The alternation of wires between the outer and inner layer distributes half of the current in each layer. Due to the skin effect, the current is always concentrated in the upper part of each wire. This means that the outer layer will contribute very little to create a field in the space between the two layers. This field is due essentially to the current flowing in the inner layer.

In the following sections we summarize results from some of the other calculational models which improve upon the accuracy of the basic transfer impedance model of braided shields, as developed by Vance.

9.4.6.1 Tyni's Model.
The first model to take into account the magnetic coupling between the inner and outer braid layers at the crossovers was

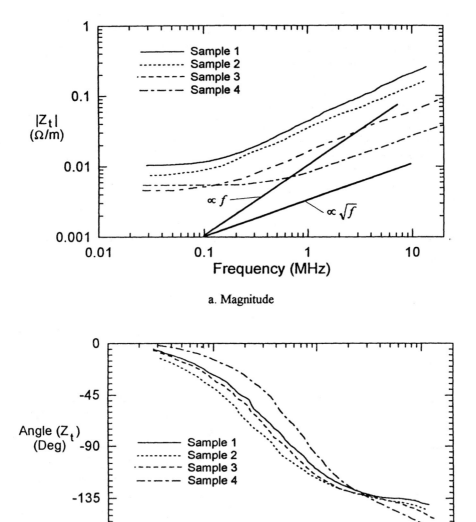

FIGURE 9.9 Measured magnitude and phase of four braided shield cables with very good optical coverage. (From [9.11].)

proposed by Tyni [9.13]. The transfer impedance expression in his model has three terms, one due to diffusion, a second due to the penetration through the diamond-shaped holes in the braid, and an additional porpoising term given by the mutual inductance between the two layers:

$$Z'_t = Z'_d + j\omega L'_a + j\omega L'_p \quad \Omega/m \qquad (9.41)$$

where Z'_d and $j\omega L'_a$ are given by Eqs. (9.11) and (9.34), respectively. In this model the porpoising inductance L'_p is given by

$$L'_p = \frac{\mu_0 h_1}{4\pi D}(1 - \tan^2\psi) \quad H/m \qquad (9.42)$$

where $D = 2b$ is the outer diameter of the braided shield and h_1 is the effective distance between the two layers of the braid.

As discussed by Tyni, when the mean distance between two carriers in the braid approaches zero, the effective layer separation of the braid is twice the diameter of the individual conductors of the braid. However, as the separation of the carriers increases and the apertures begin to appear, the effective separation of the braid decreases. He models this effect by using an effective braid thickness given approximately as

$$h_1 \approx \frac{2d}{1 + \frac{w}{d}} \quad m \qquad (9.43)$$

where w denotes the mean distance between two carriers and d is the diameter braid filaments. Sali has found that Eq. (9.43) overestimates the flux area between the braid layers and has proposed new expressions to correct this approximation [9.14]. A comparison if calculation results using Vance, Tyni, and Sali expressions and measurements is presented in Figure 9.10.

As seen in Figure 9.10, the coincidence between the transfer impedance calculated using Sali's expression and experimental values is very good at low frequencies. At medium frequencies, the two curves begin to diverge and it is clear that the linear behavior which can be observed for the calculated curve will not follow the trend of the results measured at low frequencies.

In his early work, Tyni [9.13] observed that a loss term caused by the currents induced in the outer layer (which tend to oppose the magnetic field between the two layers) must be included in the model. Based on Kaden's work [9.15], he adds a term inversely proportional to $\sqrt{f}$ to the aperture inductance L'_p, thereby providing a frequency-dependent aperture inductance

$$L'_{pl} = L'_p\left(1 + \frac{a}{\sqrt{j\omega}}\right) \quad H/m \qquad (9.44)$$

In this approximation, a is a factor that Tyni has not determined explicitly. With such a corrected porpoising inductance, the reactance term $\omega L'_{pl}$ will

472 SHIELDED CABLES

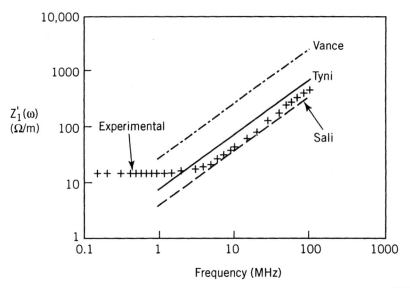

FIGURE 9.10 Measured and calculated transfer impedance magnitude for a URM-43-0 cable. (From [9.14]. © 1991 Institute of Electrical and Electronics Engineers.)

be proportional to $\sqrt{f}$ at low frequencies. Note that this correction has not been taken into account in Tyni's curve in Figure 9.10.

9.4.6.2 Demoulin's Model.
The problem of induction, or porpoising, has been reconsidered by Demoulin in [9.11] and [9.16]. Like Tyni, he has also defined an additional porpoising term proportional to $\sqrt{\omega}$. Physically, this term is justified by the flow of eddy currents in the braid, which produce a tangential electric field component. This electric field is proportional to the shield current I_s and to the resistance of the circuit of the eddy currents. This resistance is dependent on the surface conductivity of the inner carriers, and as such, it is proportional to the square root of the frequency. It means that the three phenomena which explain the transfer impedance are diffusion, penetration, and induction, but that *two* porpoising (or induction) terms must be introduced.

Using Tyni's and Sali's work and combining them with a careful comparison between measured and calculated results for cables with different weave angles and optical coverages, Demoulin [9.17] proposes a transfer impedance expression consisting of *four* terms:

$$Z_t' = Z_d' + j\omega L_a' + k'\sqrt{\omega}\,e^{j\pi/4} \pm j\omega L_p' \quad \Omega/\text{m} \qquad (9.45)$$

where L_p' is positive for $\psi > 45°$ and negative for $\psi < 45°$, and where Z_d', L_a', and L_p' have the same expressions as in Tyni's model. Demoulin [9.11] has proposed the following expression for k':

$$k' = -\frac{1.16}{eNd} \arctan \frac{N}{3} \sin\left(\frac{\pi}{2} - 2\psi\right)\sqrt{\frac{\mu}{\sigma}} \qquad (9.46)$$

The terms e, N, d, and ψ, are as defined previously, and σ is the conductivity of the braid.

For angles ψ near 45°, $k' \approx 0$ and there is no porpoising effect. In this case the transfer impedance can be represented as

$$Z'_t = Z'_d + j\omega L'_a \qquad \Omega/m \qquad (9.47)$$

For weave angles whose values differ significantly from 45°, the diffusion term Z'_d can be replaced by the dc resistance R'_0.

If the weave angle is smaller than 45°, it has been found that the field penetration is very small and can be neglected with respect to the porpoising (i.e., $L'_a \ll L'_p$), and the transfer impedance becomes

$$Z'_t = R'_0 + k'\sqrt{\omega}e^{+j\pi/4} - j\omega L'_p \qquad \Omega/m \qquad (9.48a)$$

where $\psi < 45°$. If the weave angle is larger than 45°, the penetration is dominant with respect to the porpoising and L'_p can be neglected, giving

$$Z'_t = R'_0 + k'\sqrt{\omega}e^{+j\pi/4} + j\omega L'_a \qquad \Omega/m \qquad (9.48b)$$

where $\psi > 45°$.

A comparison of calculations using the model described by Eq. (9.48) and measurements is given in Table 9.2 for frequencies of 20 kHz and 10 MHz. It can be seen that for both frequencies, the calculated and measured phase are the same. This result shows that the discrepancy between the transfer impedance model using only two terms (the diffusion term Z'_d and the aperture penetration term L'_a), together with the phase behavior shown in Figure 9.9, can be corrected by adding the porpoising terms. The agreement between the calculated and measured magnitudes is remarkably good at low frequency. At high frequencies it can be noted that the calculated values are in general lower than the measured values and that the disagreement is sometimes rather large.

9.4.6.3 Kley's Model.
The transfer impedance model proposed by Kley [9.18], based on the work of Tyni [9.13] and Halme [9.19], relies on "tuning" the model using measured cable data. The transfer impedance has the expression

$$Z'_t = Z'_d + j\omega L'_t + (1+j)\omega L'_s \qquad \Omega/m \qquad (9.49)$$

Here Z'_d is the transfer impedance of the equivalent solid tube with holes, as

TABLE 9.2 Comparison of Measured and Calculated Transfer Impedances

Sample	$2b$ (mm)	ψ (deg)	κ (Optical Coverage)	L_{eq} (mm)	$|Z'_t|$(mΩ/m), Phase (rad) $f = 20$ kHz		$|Z'_t|$(mΩ/m), Phase (rad) $f = 10$ MHz	
					Meas.	Calc.	Meas.	Calc.
1	3.0	27	1.0	0	7.5∠0	8.8∠0	151 ∠−3π/4	44∠−3π/4
2	3.0	19	1.0	0	9.3∠0	9.2∠0	171 ∠−3π/4	59∠−3π/4
3	9.2	23	0.99	0.5	5.3∠0	4.0∠0	24 ∠−3π/4	31∠−3π/4
4	6.5	1.7	1.0	0	4.6∠0	4.0∠0	57 ∠−3π/4	37∠−3π/4
5	7.3	31	0.98	0.4	6.5∠0	6.8∠0	33 ∠−3π/4	29∠−3π/4
6	7.7	35	0.97	0.52	7.0∠0	7.2∠0	20 ∠−3π/4	20∠−3π/4
7	20.0	28	0.99	0.30	2.3∠0	1.9∠0	11 ∠−3π/4	10∠−3π/4
8	8.0	38	0.97	0.47	7.5∠0	7.5∠0	9.0∠−3π/4	14∠−3π/4

Source: [9.1]. © Oxford University Press, 1993, reprinted by permission.

defined by Eq. (9.11). The penetration inductance L'_t is expressed using two terms,

$$L'_t = M'_L + M'_G \qquad \text{H/m} \qquad (9.50)$$

The term M'_L is the aperture inductance L'_a suitably corrected to take into account the curvature of the braid and the "chimney effect" (extra attenuation due to the thickness of the braid apertures), as proposed by Halme [9.19]. For the chimney effect, a magnetic field attenuation factor $e^{-\tau_H}$ is proposed where

$$\tau_H = 9.6 \, F \left(\frac{\mathcal{K}^2 d}{2b} \right)^{1/3} \qquad \text{s} \qquad (9.51)$$

and $\mathcal{K}$ is the optical coverage defined in Eq. (9.32), F is the shield fill factor of Eq. (9.31), and b is the average braid radius. For the braid curvature, a reduction in the term L'_a by a factor of 0.875 is proposed by Kley, based on his comparisons between calculations and measurements. The final expression for the term M'_L becomes

$$M'_L = 0.875 L'_a \, e^{-\tau_H} \qquad \text{H/m} \qquad (9.52)$$

with L'_a being given by Eq. (9.34).

The term M'_G in Eq. (9.50) accounts for the porpoising effect and is expressed as a mutual inductance between the braid carriers. It must be maximum when the carriers are parallel ($\psi = 0°$) and zero when the carriers are orthogonal ($2\psi = 90°$). One approximation to M'_G is

$$M'_G \sim \mu_0 \cos 2\psi \qquad (\text{H/m}). \qquad (9.53)$$

An empirical expression for M'_G using the total inductance term L'_T of Eq. (9.49) and M'_L from (9.52) and the proportionality relation (9.53), but with coefficients based on measurements, is proposed by Kley. He suggests the following expression

$$M'_G \approx -\mu_0 \frac{0.11 d}{4\pi b F_0} \cos 2 k_1 \psi \qquad \text{H/m} \qquad (9.54)$$

where $F_0 = F \cos \psi$ is the minimal filling factor and

$$k_1 = \frac{\pi}{4} \left(\frac{2}{3} F_0 + \frac{\pi}{10} \right)^{-1} \qquad (9.55)$$

As discussed in [9.18], the braid angle for which the term M'_G in Eq. (9.54) is zero does not occur at an angle of $\psi_0 = \pi/4$, as suggested by Eq. (9.53), but depends on the term F_0:

$$2k_1\psi_0 = \frac{\pi}{2} \Rightarrow \psi_0 = \frac{2}{3}F_0 + \frac{\pi}{10} \qquad (9.56)$$

Kley explains this dependence by the mutual influence of the braid's elements. In reality, the proposed expression (9.54) shows a slightly different behavior from the measurements as the weave angle is changed. The computed value of ψ_0 for the zero of the inductance term M'_G in Eq. (9.56) has an error of less than $\pm 3°$. For braid angles Ψ quite different from ψ_0, an accuracy of ± 1.5 dB in the value of M'_G has been found. Kley has also experimentally verified that M'_G is independent of the contact resistance between the braids.

As done by Demoulin [9.11,9.17], Kley [9.18] assumes that the magnetic field that penetrates the shield causes eddy currents in the walls of the elliptic holes, giving an ohmic term of the form ωL_s. Other eddy currents are induced by the magnetic field that exists between the outer and the inner carriers of the braid, giving a quadrature component of $j\omega L_s$. These two effects account for the additional term $(1+j)\omega L_s$ in Eq. (9.49) of the form

$$\omega L_s = \frac{1}{\pi\sigma\delta}\left(\frac{1}{D_L} + \frac{1}{D_G}\right) \quad \Omega/\text{m} \qquad (9.57)$$

Notice that this impedance term is proportional to $\sqrt{f}$. The terms D_L and D_G are fictitious diameters that model the corresponding skin effect, whose values are a function of F the filling factor and Ψ the weave angle

$$D_L^{-1} \approx 10\pi F_0^2 \frac{\cos\psi}{2b}(1-F)e^{-\tau_E} \quad \text{m}^{-1} \qquad (9.58)$$

$$D_G^{-1} \approx -\frac{3.3}{4\pi b F_0}\cos(2k_2\psi) \quad \text{m}^{-1} \qquad (9.59)$$

with

$$\tau_E = 12F\left(\frac{\varkappa^2 d}{2b}\right)^{1/3} \qquad (9.60)$$

and $k_2 = (\pi/4)(\frac{2}{3}F_0 + \frac{3}{8})^{-1}$. Contrary to Demoulin, in Kley's model the term $\omega L'_s$ adds to the diffusion and penetration terms in a way that is independent of the value of the weave angle. This means that Eq. (9.49) is valid for all weave angles ψ.

Using Eq. (9.49), Kley has compared measurements and calculations of the transfer impedance and has found very good agreement at low frequencies and differences of 3 to 6 dB (i.e., a factor 1.4 to 2) for $f > 1$ MHz. Figure 9.11 presents a comparison of these data.

9.4 MODELS FOR BRAIDED SHIELDS

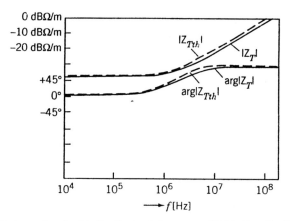

a. Braid with parameters $\ell = 24$, $N = 8$, $d = 0.1$ mm, $\psi = 25.7°$, $2b = D = 7.5$ mm, $\mathcal{K} = 0.70$

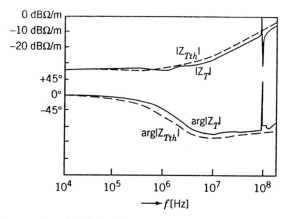

b. Braid with parameters $\ell = 16$, $N = 8$, $d = 0.1$ mm, $\psi = 24.7°$, $2b = D = 3.2$ mm, $\mathcal{K} = 0.91$

FIGURE 9.11 Comparison of calculated and measured transfer impedances of two braided cables. (From [9.18]. © 1993, Institute of Electrical and Electronics Engineers.)

9.4.6.4 Comparison of Demoulin's and Kley's Models.
Two approaches have been presented for calculation of the transfer impedance of braided cables. The differences between them are:

- In the transfer inductance, due to the diffraction through the apertures of the braid, two more parameters are added by Kley with respect to Demoulin's model: an attenuation factor which accounts for the "chimney effect" of the aperture, and a coefficient to take the braid curvature into consideration.
- The terms used in Kley's model are independent of the weave angle and filling factor; he assumes that the induction phenomenon is always present.

However, if the complete expressions of the terms from Demoulin and Kley relations are written, it can be seen that the two approaches are very similar.

DEMOULIN'S TRANSFER IMPEDANCE

$$Z_t' = \frac{4}{\pi d^2 N e \sigma \cos \psi} - \sqrt{\frac{\omega \mu}{\sigma}} \frac{1.16}{eNd} \arctan \frac{N}{3} \sin\left(\frac{\pi}{2} - 2\psi\right) e^{+j\pi/4}$$

$$+ j\omega L_t' \pm j\omega \frac{\mu_0 h_1}{4\pi b} (1 - \tan^2 \psi) \quad \Omega/m \tag{9.61}$$

KLEY'S TRANSFER IMPEDANCE WITH THE ORDER OF TERMS REARRANGED

$$Z_t' = \frac{4}{\pi d^2 N e \sigma \cos \psi} \frac{(1+j)d/\delta}{\sinh[(1+j)d/\delta]}$$

$$+ \sqrt{\frac{\omega \mu}{\sigma}} (1+j) \frac{1}{2\pi b} \left[7\pi F^2 \cos^2 \psi (1-F) e^{-\tau_E} - \frac{2.34 \cos(k_2 \psi)}{2\pi \cos \psi} \right]$$

$$+ j\omega L_t' \cdot 0.875 e^{-\tau_H} + j\omega \frac{\mu_0}{4\pi} \frac{0.11d}{b} \frac{\cos(k_1 \psi)}{F \cos \psi} \quad \Omega/m \tag{9.62}$$

Taking into account the facts that

$$1 + j = 2 \cos \frac{\pi}{4} e^{j\pi/4} \qquad \sin\left(\frac{\pi}{2} - \Psi\right) = \cos \Psi \qquad h_1 \approx d$$

as well as neglecting the term containing $(1 - F)$ in Eq. (9.62) since $F \approx 1$, and replacing the filling factor F by its expression (9.31), the expression by Kley becomes

$$Z_t' = \frac{4}{\pi d^2 N e \sigma \cos \psi} \frac{(1+j)d/\delta}{\sinh[(1+j)d/\delta]} - \sqrt{\frac{\omega \mu}{\sigma}} \frac{0.53 \cos(k_2 \psi)}{eNd} e^{j\pi/4}$$

$$+ j\omega L_t' \cdot 0.875 e^{-\tau_H} + j\omega \frac{\mu_0}{4\pi} \frac{0.11d}{b} \frac{\cos(k_1 \psi)}{F \cos \psi} \quad \Omega/m \tag{9.63}$$

Comparing the different terms of Eqs. (9.61) and (9.63), it can be seen that there is a similarity in the expressions for the two models. In looking at the comparisons between measurements and calculations, it is noted that for a frequency of 10 MHz, the calculated results in Table 9.2 are systematically smaller than the measured values. A straightforward conclusion could be that these smaller values are explained by the transfer inductance term, which is neglected for $\psi < 45°$ in Demoulin's model. However, the comparisons presented by Kley in Figure 9.11 are only for two weave angles, both very near 25° and for a very poor optical coverage. This makes evaluation of the accuracy of his model difficult for general cables.

This comparison of the models illustrates the point that the modeling of the braided shields cannot yet take into account the entire complexity of the field penetration through the sheath. Presently, measurements are still the only completely reliable way to qualify braided cables. However, the 20 years of effort devoted by different research groups to try to build reliable models has permitted a better understanding of the physics of field penetration into the braid, and in some cases, as seen in Table 9.2 and Figure 9.11, a remarkable agreement between calculations and measurements can be reached.

9.4.7 Effect of an Axial Magnetic Field Component

By postulating that the sheath current has the direction shown in Figure 9.2, it has been assumed that the magnetic field which induces this current has only an azimuthal component (see also Figure 9.6a). The effect due to a magnetic field oriented along the cable axis has been considered by Broyde and Clavelier [9.20]. Theoretical and experimental evaluations [9.21] seem to confirm that this effect is about one to two orders of magnitude lower than the effect of the azimuthal magnetic field component, which gives rise to the transfer impedance discussed in this chapter. Investigations into this coupling mode are continuing.

9.4.8 Alternative Expression for the Transfer Admittance of Braided Shields

For good cables and at low frequencies, the contribution of the transfer impedance is generally much more important than that from the transfer admittance. Consequently, not as much effort has gone into developing alternative models for Y_t'. One exception is found in Kley [9.18], where experimental data for Y_t' are examined and incorporated into a modified transfer admittance expression.

According to Kley's model, another representation for Y_t' is given by

$$Y_t' = j\omega C_0' C_{s0}' \frac{\epsilon_{r_{ext}} \epsilon_{r_{int}}}{\sqrt{\epsilon_{r_{ext}} + \epsilon_{r_{int}}}} 0.875 \frac{\pi \cos \psi}{6\epsilon_0 \ell} (1 - F)^3 e^{-\tau_E} \quad \text{S/m} \quad (9.64)$$

where C_0' and C_{s0}' are the internal and external per-unit-length capacitances of the cable with the internal and external relative dielectric constants $\epsilon_{r_{int}}$ and $\epsilon_{r_{ext}}$ set to unity. The exponential term involving τ_E from Eq. (9.60) again takes into account the thickness of the apertures in the braid. Notice that this expression assumes a different correction factor for the differences between the inner and outer dielectric constants.

480 SHIELDED CABLES

9.5 CALCULATED RESPONSES OF A BRAIDED CABLE

As an example of the use of the braided shield cable models developed in this chapter, let us consider the problem of evaluating the internal load responses in Figure 9.1. The coaxial cable of length $\mathcal{L}$ and the two enclosures are located over a ground plane at a height h, and the system is assumed to be excited by an incident plane wave. This incident field is defined by its spectral content $E_0(\omega)$, angles of incidence ψ and ϕ, and the polarization angle α, all of which were discussed in Chapter 7 and shown in Figure 7.11.

As noted earlier, there are two separate transmission line problems in this example. One is an external problem in which the incident plane wave serves as the excitation and the response is the current and charge (or voltage) along the outside of the coaxial cable shield. The other problem involves the internal transmission line geometry, in which the excitation arises from the penetration of the external coaxial shield current and charge into the cable interior. The internal responses of interest are $I_1^{(i)}$ and $V_1^{(i)}$ within enclosure 1 and $I_2^{(i)}$ and $V_2^{(i)}$ in enclosure 2, as shown in Figure 9.1. In this discussion, the superscript $^{(e)}$ is used to denote a quantity for the external transmission line and $^{(i)}$ is for the internal line.

The approach to determining the internal load responses is first to obtain an expression for the distributed current and charge on the exterior of the cable shield using the expressions developed in Chapter 7. Then by using the transfer impedance and admittance terms in Eqs. (9.37) and (9.38), the internal responses are determined using the BLT equation.

9.5.1 External Transmission Line

The first part of the analysis involves determining the current and charge density on the external portion of the coaxial cable. The exterior of the shielded cable, together with the ground plane, form a transmission line structure similar to that shown in Figure 7.19. The conductor radius of this line is the outer cable shield radius b, and the two termination impedances connecting each enclosure to the ground plane are given by $Z_1^{(e)}$ and $Z_2^{(e)}$. As a result of the ground-plane symmetry of the structure and the excitation, there is no antenna mode induced on the cable shield, and the transmission line model can be used to determine the external responses.

For an incident field excitation, a general solution for the line current at any location x can be expressed as a superposition of the Green's function solutions of Eq. (7.32a) as

$$I_s(x) = \int_0^{\mathcal{L}} G_I^{(V)}(x; x_s) V_s'(x_s)\, dx_s + G_I^{(V)}(x; 0)V_1 - G_I^{(V)}(x; \mathcal{L})V_2 \quad \text{A}$$

(9.65)

9.5 CALCULATED RESPONSES OF A BRAIDED CABLE

In this expression $V'_s(x_s)$ is the tangential E-field excitation (i.e., incident plus ground-reflected fields) along the line, and V_1 and V_2 represent the lumped voltage sources at each end of the line.

As discussed in Section 7.3.3, these sources are related to the incident E field for the special case of a plane wave. Assuming a perfectly conducting ground, Eqs. (7.59), (7.61), and (7.62) provide analytic expressions for these source terms as

$$V'_s = E_x^{\text{inc}}(x, h) + E_x^{\text{ref}}(x, 0) = E_x(e^{jk_z h} - e^{-jk_z h}) e^{-jk_x x}$$
$$= V'_x e^{-jk_x x} \qquad \text{V/m} \qquad (9.66a)$$

where $k_x = k_0 \cos\psi \cos\phi$, $k_z = k_0 \sin\psi$ and $k_0 = \omega/c$. The term V'_x is defined as

$$V'_x \approx jk2h \sin\psi E_0(\omega)[\cos\alpha \sin\psi \cos\phi + \sin\alpha \sin\phi] \qquad \text{V/m} \quad (9.66b)$$

The two lumped sources at the ends of the line are

$$V_1 = -\int_0^h [E_z^{\text{inc}}(0, z) + E_z^{\text{ref}}(0, z)] \, dz \approx -2h E_z^{\text{inc}}(0, 0) \equiv V_z \qquad \text{V} \qquad (9.67a)$$

$$V_2 = -\int_0^h [E_z^{\text{inc}}(\mathcal{L}, z) + E_z^{\text{ref}}(\mathcal{L}, z)] \, dz \approx -2h E_z^{\text{inc}}(\mathcal{L}, 0) = V_z \, e^{-jk_x \mathcal{L}} \qquad \text{V} \qquad (9.67b)$$

where

$$V_z = -2h E_0(\omega) \cos\alpha \cos\psi \qquad \text{V} \qquad (9.67c)$$

For this type of exponential phase variation in the excitation field, the required integrals in Eq. (9.65) can be evaluated analytically to yield the following expression for the line current $I(x)$:

$$I_s^{(e)}(x) = K_1 e^{\gamma^{(e)} x} + K_2 e^{-\gamma^{(e)} x} + K_3 e^{-jk_x x} \qquad \text{A} \qquad (9.68)$$

where $\gamma^{(e)} = \sqrt{Z'^{(e)} Y'^{(e)}}$ (m^{-1}) is the propagation constant for exterior transmission line in terms of its per-unit-length impedance and admittance parameters. The coefficients K_1, K_2, and K_3 are determined to be

$$K_1 = \frac{1}{2Z_c^{(e)}[1 - \rho_1^{(e)}\rho_2^{(e)} e^{-2\gamma^{(e)}\mathcal{L}}]} \left\{ V_x' \left[-e^{-\gamma_-^{(e)}\mathcal{L}} \left(\frac{1}{\gamma_+^{(e)}} + \frac{\rho_2^{(e)}}{\gamma_-^{(e)}} \right) + \rho_2^{(e)} e^{-2\gamma^{(e)}\mathcal{L}} \left(\frac{1}{\gamma_-^{(e)}} + \frac{\rho_1^{(e)}}{\gamma_+^{(e)}} \right) \right] \right.$$

$$\left. + V_z[(1 - \rho_1^{(e)})\rho_2^{(e)} e^{-2\gamma^{(e)}\mathcal{L}} + (1 - \rho_2^{(e)}) e^{-\gamma_-^{(e)}\mathcal{L}}] \right\} \quad (9.69a)$$

$$K_2 = \frac{1}{2Z_c^{(e)}[1 - \rho_1^{(e)}\rho_2^{(e)} e^{-2\gamma^{(e)}\mathcal{L}}]} \left\{ V_x' \left[\rho_1^{(e)} e^{-\gamma_-^{(e)}\mathcal{L}} \left(\frac{1}{\gamma_+^{(e)}} + \frac{\rho_2^{(e)}}{\gamma_-^{(e)}} \right) - \left(\frac{1}{\gamma_-^{(e)}} + \frac{\rho_1^{(e)}}{\gamma_+^{(e)}} \right) \right] \right.$$

$$\left. - V_z[(1 - \rho_1^{(e)}) + (1 - \rho_2^{(e)})\rho_1^{(e)} e^{-\gamma_-^{(e)}\mathcal{L}}] \right\} \quad (9.69b)$$

$$K_3 = \frac{V_x'}{2Z_c^{(e)}} \left(\frac{1}{\gamma_-^{(e)}} + \frac{1}{\gamma_+^{(e)}} \right) \quad (9.69c)$$

In these expressions, the terms $\rho_1^{(e)}$ and $\rho_2^{(e)}$ are the load voltage reflection coefficients for the external circuit, and the propagation constants $\gamma_+^{(e)}$ and $\gamma_-^{(e)}$ are

$$\gamma_+^{(e)} = \gamma^{(e)} + jk_x$$

$$\gamma_-^{(e)} = \gamma^{(e)} - jk_x$$

The term $Z_c^{(e)} = \sqrt{Z'^{(e)}/Y'^{(e)}}$ (Ω) is the characteristic impedance of the external line, defined in terms of the external per-unit-length line parameters.

Notice that although these coefficients are rather complicated functions of the external line parameters, the solution for the line current is very simple, consisting of forward and backward propagating waves and a third contribution propagating along the line with the incident field. These correspond to the homogeneous and particular solutions of the wave equation for the line current.

The second external response needed for the cable penetration calculation is the charge on the cable exterior. This is computed most easily by using the continuity equation

$$\frac{dI_s(x)}{dx} + j\omega Q_s'(x) = 0 \quad (9.70)$$

Applying this expression to Eq. (9.68) provides the following expression for the external charge

$$Q_s'(x) = \frac{-1}{j\omega} [\gamma^{(e)} K_1 e^{\gamma^{(e)}x} - \gamma^{(e)} K_2 e^{-\gamma^{(e)}x} - jk_x K_3 e^{-jk_x x}] \quad (9.71)$$

9.5.2 Internal Excitation Sources

The excitation of the internal line of the cable is determined by the distributed voltage and current sources $V'_{si} = Z'_t I_s$ and $I'_{si} = -Y'_t V_s$, as shown in Figure 9.3. For a braided shield, Eqs. (9.45) and (9.38) provide the expressions for the transfer impedance and admittance in terms of various cable parameters. Because Y'_t depends on both the shield parameters and the external circuit capacitance, it is more convenient to express the current source in terms of the external charge, as done in Eq. (9.22b). Thus, using Eq. (9.38), the internal current source is expressed as

$$I'_{si} = -j\omega S_s C' Q'_s \tag{9.72}$$

where C' is inner coax capacitance given in Eq. (9.14) and S_s is an electrostatic shield leakage parameter described in [9.2]. From Eq. (9.38) this parameter is seen to have the following values:

$$S_s = \frac{\pi}{6\ell\epsilon_0}(1-\mathcal{K})^{3/2}\frac{1}{E(e)} \qquad \psi < 45° \tag{9.73a}$$

$$S_s = \frac{\pi}{6\ell\epsilon_0}(1-\mathcal{K})^{3/2}\frac{1}{(1-e^2)E(e)} \qquad \psi < 45° \tag{9.73b}$$

Reference [9.2] provides data computed for the dc shield resistance R_0, given by

$$R_0 \approx \frac{4}{\pi d^2 N \ell \sigma \cos\psi} \quad \Omega/m$$

the aperture inductance L'_a, the electrostatic shield leakage parameter S_s, and various other cable shield parameters for several different types of common coaxial cables. Appendix E summarizes some of these data for common cables.

9.5.3 Internal Load Responses

The internal load current and voltage responses can be determined using the BLT equations with the appropriate parameters for the internal line by integrating over the internal source distributions V'_s and I'_s. In this manner, the source terms S_1 and S_2 in Eq. (7.36) for the internal line are

$$S_1 = \frac{1}{2}\int_0^{\mathscr{L}} e^{\gamma^{(i)}x_s}[V'_{si}(x_s) + Z_c^{(i)}I'_{si}(x_s)]\,dx_s \tag{9.74a}$$

$$S_2 = -\frac{1}{2}\int_0^{\mathscr{L}} e^{\gamma^{(i)}(\mathscr{L}-x_s)}[V'_{si}(x_s) - Z_c^{(i)}I'_{si}(x_s)]\,dx_s \tag{9.74b}$$

These integrals also can be evaluated analytically, since V'_{si} and I'_{si} have only an exponential dependence on the line position x. The resulting expressions for the internal BLT sources are given in terms of the constants K of Eq. (9.69) as

$$S_1 = \frac{1}{2} K_1(Z'_t - Z'_{\text{eff}}) \frac{e^{(\gamma^{(i)}+\gamma^{(e)})\mathcal{L}} - 1}{\gamma^{(i)} + \gamma^{(e)}} + \frac{1}{2} K_2(Z'_t + Z'_{\text{eff}}) \frac{e^{(\gamma^{(i)}-\gamma^{(e)})\mathcal{L}} - 1}{\gamma^{(i)} - \gamma^{(e)}}$$

$$+ \frac{1}{2} K_3 \left(Z'_t + Z'_{\text{eff}} \frac{jk_x}{\gamma^{(e)}} \right) \frac{e^{(\gamma^{(i)}-jk_x)\mathcal{L}} - 1}{\gamma^{(i)} - jk_x} \quad (9.75a)$$

$$S_2 = \frac{e^{\gamma^{(i)}\mathcal{L}}}{2} K_1(Z'_t + Z'_{\text{eff}}) \frac{e^{(\gamma^{(e)}-\gamma^{(i)})\mathcal{L}} - 1}{\gamma^{(e)} - \gamma^{(i)}}$$

$$- \frac{e^{\gamma^{(i)}\mathcal{L}}}{2} K_2(Z'_t - Z'_{\text{eff}}) \frac{e^{-(\gamma^{(e)}+\gamma^{(i)})\mathcal{L}} - 1}{\gamma^{(e)} + \gamma^{(i)}}$$

$$- \frac{e^{\gamma^{(i)}\mathcal{L}}}{2} K_3 \left(Z'_t - Z'_{\text{eff}} \frac{jk_x}{\gamma^{(e)}} \right) \frac{e^{-(\gamma^{(i)}+jk_x)\mathcal{L}} - 1}{\gamma^{(i)} + jk_x} \quad (9.75b)$$

where the term Z'_{eff} is defined as

$$Z'_{\text{eff}} = j \frac{\gamma^{(e)}\gamma^{(i)}}{\omega} S_s \quad (9.76)$$

This term accounts for the charge coupling in the shield and is not completely independent of the external transmission line, since the external propagation constant is contained in the expression. However, for cases when the propagation losses in the external and internal transmission line are negligible, $\gamma^{(e)} = \gamma^{(i)} = jk$ and this term becomes $Z'_{\text{eff}} = -j\omega S_s/c^2$.

With the BLT source terms of Eq. (9.75) determined, the internal transmission line responses can be obtained by direct solution of the BLT equation (7.35), with the propagation constant γ, reflection coefficients ρ_1 and ρ_2, and characteristic impedance Z_c all corresponding to the values for the inner transmission line.

9.5.4 Numerical Example of a Shielded Cable System

As a numerical example of the responses for a shielded cable system using the formalism above, consider the case of a RG-58 cable as described in [9.22]. This line configuration is shown in Figure 9.1, with a length of $\mathcal{L} = 30$ m at a height $h = 1$ m over an imperfectly conducting earth and with parameters $\sigma_g = 0.01$ S/m and $\epsilon_r = 10$. For this example the expressions for the excitation fields in Eqs. (9.66) and (9.67) must be replaced by those developed in Section 8.4 for a lossy earth. Moreover, the expressions for the per-unit-length impedance and admittance parameters for the external line should use the values given in Section 8.3, which contain the effects of the loss in the ground. The external load impedances for this example are taken to be $Z_1^{(e)} = Z_2^{(e)} = 100 \, \Omega$, and the internal impedances are assumed to be matched to the 50-Ω internal characteristic impedance of the RG-58 cable.

Reference [9.2] provides additional data on this particular cable, including the cable shield radius of $a = 0.152$ cm, the shield wall thickness of $\Delta = 0.127$ mm, the dc resistance of the shield of $R_0 = 14.2$ mΩ/m, the aperture leakage inductance of $L'_a = 1.0$ nH/m, and the electrostatic shield leakage parameter of $S_s = 6.6 \times 10^7$ m/F. For this example, only the diffusion and aperture penetration effects will be taken into account for the transfer impedance expression. This means that Eq. (9.37) will be used instead of Eq. (9.45) or (9.49) for Z'_t. The transfer impedance Z'_t and the effective charge coupling impedance parameter Z'_{eff} of Eq. (9.76) are plotted in Figure 9.12 as a function of frequency. From this plot is clear that at low frequencies below 1 MHz the resistive coupling through the cable shield dominates. At higher frequencies, both the inductive and the capacitive coupling through the cable shield is important. Furthermore, it is noted that the inductive and capacitive couplings are of the same order of magnitude, and if one component were neglected, the resulting internal cable responses would be in error.

To obtain an indication of the cable responses, consider the case of the incident plane wave of Figure 7.11 representing a high-amplitude transient E-field, typical of an electromagnetic pulse (EMP) [9.23]. The assumed waveform has a double exponential expression of the form $E(t) = E_0(e^{-\beta_1 t} - e^{-\beta_2 t})$, with $E_0 = 52.5$ kV/m, $\beta_1 = 4.0 \times 10^6$ s^{-1}, and $\beta_2 = 4.76 \times 10^8$ s^{-1}. The angle of incidence parameters are $\psi = 30°$ and $\phi = 0°$, and a polarization angle $\alpha = 0°$ is used.

For this excitation field, the spectrum $E_0(\omega)$ used in Eqs. (9.66) and (9.67)

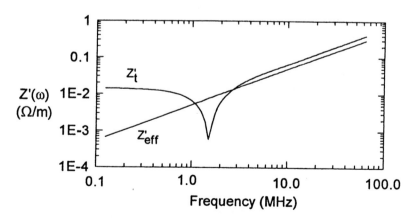

FIGURE 9.12 Plots of the magnitude of the transfer impedance Z'_t and the effective charge coupling impedance Z'_{eff}. (From [9.20]. Reprinted by permission from Academic Press, Inc.)

is given as $E_0(\omega) = E_0[1/(j\omega + \beta_1) - 1/(j\omega + \beta_2)]$. An evaluation of Eq. (9.68) for the line current provides the spectral response at any point along the line. Figure 9.13a shows the shield current response spectrum for the location $x = 15$ m, and Figure 9.13b shows the corresponding transient response. Note that this waveform has several discontinuities due to wave reflections at the loads of the exterior transmission line.

The computed spectrum of the internal voltage response at the $x = \mathcal{L}$ load is shown in Figure 9.14a, and the corresponding transient response is shown in Figure 9.14b. In the latter plot, the contributions to the response from only diffusion through the shield [i.e., considering only the Z'_d term in Eq.

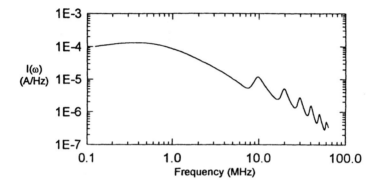

a Spectral magnitude

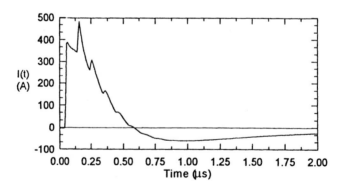

b Transient response

FIGURE 9.13 Plot of the induced cable shield current at $x = 15$ m due to a double exponential incident plane-wave field. (From [9.22]. Reprinted by permission from Academic Press, Inc.)

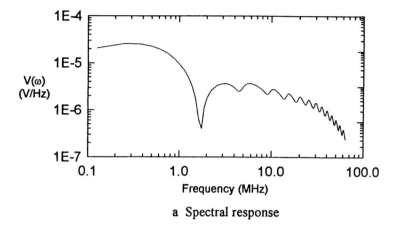

a Spectral response

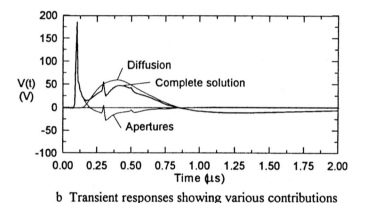

b Transient responses showing various contributions

FIGURE 9.14 Internal voltage response at the $x = \mathcal{L}$ load impedance. (From [9.22]. Reprinted by permission from Academic Press, Inc.)

(9.37)] and from only the aperture penetrations (containing both the electric and magnetic field coupling effects) are shown, together with the total solution.

It is seen that even though the internal cable is matched, there are oscillations on the line due to the resonances on the external line. The shield is seen to do a relatively good job in protecting the internal electronics from this external pulsed environment. In the present case, a peak current of $I_2^{(i)} \approx 200/50 = 4$ A flows through the internal load. If the line were not shielded, an induced current on the exterior of the shield ($I_2^{(e)} \approx 500$ A) would be expected to flow into the equipment loads, giving rise to peak voltages on the order of $500 \times 50 = 25$ kV. We see that the shield thus provides a shielding effectiveness value for the peak current of $SE = 20 \log(500/4) \approx 42$ dB.†

9.6 CABLES WITH SHIELD INTERRUPTIONS

9.6.1 Introduction

From a strict EMC point of view, the correct interconnection between two shielded enclosures using a shielded transmission line should have no EM penetrations. The shielded cable and the shielded boxes should form a single, continuous enclosure with no apertures, seams, or other penetrations. As discussed in previous sections, braided cables represent the majority of the shielded connections between enclosures, and these have apertures that limit their shielding effectiveness.

Moreover, practical cable shields are rarely uniform and homogeneous. There are usually connectors or other shield interruptions, and these can seriously degrade the EMI protection provided by the braid of an otherwise well-shielded cable system. It has been observed experimentally that cable connectors often play the major role in determining the internal responses of a shielded system, with the braid-shield transfer impedance and admittance being less important [9.24].

Other kinds of discontinuities can represent even more severe violations of the shielding continuity. These occur when the shield is interrupted in one or several places along the cable length, or the cable is not terminated with connectors. These situations are due primarily to lack of EMC knowledge on the part of the electrical engineer designing the installation or the worker performing the cabling, and for economical reasons. In some cases the effect of these discontinuities can be disastrous and may result in cancellation of the effect of the shield of the rest of the cable. This is discussed in this section.

Modeling cable connectors and other shield interruptions from first principles is very difficult, if not impossible, due to the physical complexity of the connector geometry. Furthermore, terminations of the cable shield, such as a pigtail, can differ from cable to cable, and indeed, from day to day on a particular cable installation. Consequently, measurements of these effects are usually the best way of determining the shielding behavior of the shield interruptions. Modeling concepts can be useful on a limited basis,

---

† This example provides a good illustration of why the characterization of a shield in terms of "shielding effectiveness" is not desired. The internal current depends not only on the shield properties but also on the internal impedance level and the terminating impedances. A very high shielding effectiveness could be obtained simply by open-circuiting the load impedance, causing the current to vanish. Of course, the voltage at this point will be large. The shield is best characterized, therefore, by parameters that depend only on the shield, such as the transfer impedance Z'_t and the electrostatic shield factor S_s.

however, in determining how to characterize cable shield interruptions and in interpreting measured data.

The shielding discontinuities can be of two types: (1) one having an electrical contact (such as a connector or pigtail connection) or (2) one without an electrical contact (such as a completely open shield). Figure 9.15 illustrates a shielded cable system with a general discontinuity in the shield. The width of the discontinuity Δ is assumed to be electrically small, so that it appears as a lumped imperfection in the shield. The equipment enclosures are located to the left and right of the discontinuity, at distances $\mathcal{L}_1$ and $\mathcal{L}_2$ along lines 1 and 2, respectively. The excitation of this cable system is from some unspecified external source that induces current on the outer cable shield. At present we are not concerned with the nature of this excitation. What is desired is to develop a methodology for analyzing this system to determine the interior voltage or current at the interior loads.

The procedure for determining the response at the interior of the cable is similar to that described in Section 7.6.1, which uses a series of Thévenin circuit transformations to collapse the transmission lines into a lumped circuit. Considering the geometry of Figure 9.15, point d located on the bottom of enclosure 1 can be thought of as being electrically translated through length $\mathcal{L}_1$ to location a at the left side of the shield discontinuity. In this manner, the external transmission line *to the left* is represented at the terminal pair a–c by its lumped Thévenin equivalent circuit, as shown in Figure 9.16a. The translated voltage source $V_{oc1}^{(e')}$ represents the effects of the

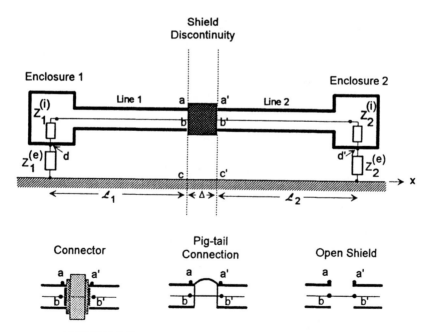

FIGURE 9.15 Cable with an interruption in the shield.

490 SHIELDED CABLES

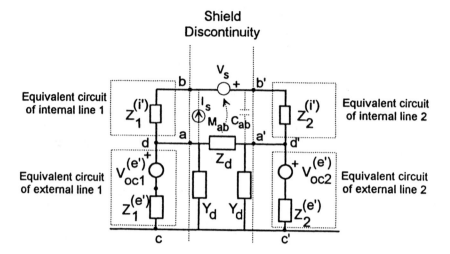

a. High frequency model

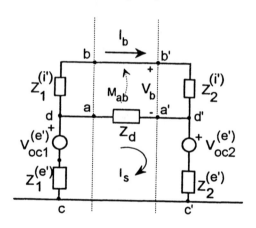

b. Low frequency approximation

FIGURE 9.16 Equivalent circuits of the cable system at the point of interruption.

external excitation of this segment of the shield and is given by Eq. (7.84) as

$$V_{oc1}^{(e')} = \frac{2}{1 - \rho_1^{(e)} e^{-2\gamma^{(e)} \mathcal{L}_1}} (e^{-\gamma^{(e)} \mathcal{L}_1} S_1 + \rho_1^{(e)} e^{-2\gamma^{(e)} \mathcal{L}_1} S_2) \qquad (9.77)$$

where the BLT source terms S_1 and S_2 for this equation are given in Eq. (7.37) and depend on the particular details of the excitation of the external

9.6 CABLES WITH SHIELD INTERRUPTIONS

line (i.e., on the incident field, angles of incidence, etc.) The term $\rho_1^{(e)}$ is the reflection coefficient for the outer transmission line at the load in enclosure 1 and is given as

$$\rho_1^{(e)} = \frac{Z_1^{(e)} - Z_c^{(e)}}{Z_1^{(e)} + Z_c^{(e)}} \qquad (9.78)$$

and the terms $Z_c^{(e)}$ and $\gamma^{(e)}$ denote the characteristic impedance and propagation constant on the outer transmission line.

The impedance of the Thévenin source, and $Z_1^{(e')}$, is determined by translating the terminating impedance of the enclosure $Z_1^{(e)}$ through the length $\mathcal{L}_1$. Equation (7.82) provides an expression for this transformed impedance as

$$Z_1^{(e')} = Z_c^{(e)} \frac{1 + \rho_1^{(e)} e^{-2\gamma^{(e)}\mathcal{L}_1}}{1 - \rho_1^{(e)} e^{-2\gamma^{(e)}\mathcal{L}_1}} \qquad (9.79)$$

In a similar manner, the internal transmission line connected between terminal pairs a–b can be represented by a lumped, transformed impedance $Z_1^{(i')}$. This impedance again is given by Eq. (9.79), but using the characteristic impedance, propagation constant, and voltage reflection coefficient of the inner coaxial line, $Z_c^{(i)}$, $\gamma^{(i)}$, and $\rho_1^{(i)}$, respectively. There may also be a voltage source in this part of the equivalent circuit arising from field diffusion and penetration through the braided shield of the cable. This is neglected in the present model.

Line 2 *to the right* of the discontinuity at terminals a'–b' and a'–c' is transformed in the identical manner through line length $\mathcal{L}_2$. This provides the additional source and impedance elements to the right of the discontinuity, $V_{0c2}^{(e')}$, $Z_2^{(e')}$, and $Z_2^{(i')}$, as shown in Figure 9.16a.

At the shield discontinuity, there are other lumped-circuit elements needed to complete the equivalent circuit. The interruption of the shield provides an impedance discontinuity in the outer circuit, Z_d, which can be thought of as arising from any resistance offered by the discontinuity and by a line impedance change due to the change of radius of the conductor. At the same time, there are shunt admittance elements Y_d, usually in the form of capacitance elements, which also arise from the change in outer conductor radius.

The inner circuit contains a mutual inductance element M_s which accounts for the inductive coupling from the external shield current to an induced voltage on the inner conductor (b–b'). In addition, the coupling capacitance C_s relates the external charge on the shield at the discontinuity to an injected current on the inner conductor. Two additional sources are also present in the circuit: V_s and I_s. V_s is a lumped voltage source arising from the linkage of the excitation B-field (incident plus ground reflected) between the inner circuit comprised of conductors $b-b'$ and $a-a'$. The current source I_s arises from the charge deposited on the inner conductor by the excitation E-field. These sources were not needed in the previous model for the braided shield, since the braid provided a very good barrier against the incident fields. For the braid, the only excitation was that due to the external current and charge leaking through the shield. In the present case with the shield being partially or completely removed over the discontinuity, this additional mode of excitation may need to be included.

The model in Figure 9.16a is a general high-frequency model, with parameters that must be determined experimentally. It is possible to omit several of these circuit elements, however, if we are willing to restrict the range of frequencies over which the model is applied. Frequently, such models are used at low frequencies, where the capacitance effects in the model are neglected. Furthermore, the effects of the direct field illumination of the internal shielded wire are usually neglected by again assuming that the main excitation comes from the coupling of the external cable response onto the inner wire. Making these simplifications leads to the equivalent circuit of Figure 9.16b.

The behavior of the exterior shield current I_s and the inner wire current I_b in the circuit of Figure 9.16b can be analyzed using conventional circuit theory. A mesh analysis yields the following matrix equation for these quantities:

$$\begin{bmatrix} Z_1^{(e')} + Z_2^{(e')} + Z_d & -(Z_d + j\omega M_{ab}) \\ -(Z_d + j\omega M_{ab}) & Z_1^{(i')} + Z_2^{(i')} + Z_d \end{bmatrix} \begin{bmatrix} I_s \\ I_b \end{bmatrix} = \begin{bmatrix} V_{oc1}^{(e')} - V_{oc2}^{(e')} \\ 0 \end{bmatrix} \quad (9.80)$$

which can be evaluated either numerically or symbolically for both currents. Note that this is a coupled set of equations, in that the current I_b influences the behavior of the shield current.

Once the inner current I_b has been determined, the voltage V_b between the inner conductor and the shield, to the right of the discontinuity, can be determined as

$$V_b = Z_2^{(i')} I_b \quad (9.81)$$

Using the compensation theorem [9.25], the voltage at the termination impedance within enclosure 2 can now be determined. As shown in Figure 9.17, this amounts to placing an ideal voltage source having the same strength as V_b at terminals $a'-b'$ and then removing the network. In doing this, the current I_b remains the same, as does the response at the load impedance $Z_2^{(i)}$ in enclosure 2. The voltage source V_b can be translated through the length $\mathcal{L}_2$ using Eq. (7.83), with $\rho_1 = -1$ to account for the fact that the voltage source is ideal; that is, it has a zero source impedance. This provides the voltage source V_{oc} at the load. Similarly, the input impedance as seen by the load $Z_2^{(i)}$ looking into the ideal voltage generator is given by Eq.

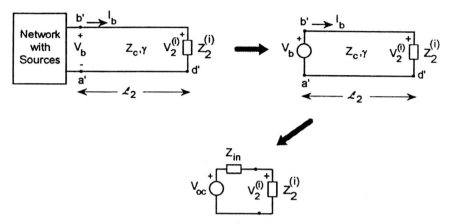

FIGURE 9.17 Compensation theorem applied to a transmission line load on a network containing sources.

(7.82), again with $\rho_1 = -1$. These two transformations yield the final lumped circuit at the load in enclosure 2, as shown in the figure. Combining these circuit elements and using Eq. (9.81), the internal voltage response in enclosure 2 can be written in terms of the computed inner current at the interruption I_b as

$$V_2^{(i)} = \frac{2Z_2^{(i)} Z_c^{(i)} e^{-\gamma^{(i)} \ell_2}}{Z_c^{(i)}(1 + e^{-2\gamma^{(i)} \ell_2}) + Z_2^{(i)}(1 - e^{-2\gamma^{(i)} \ell_2})} I_b \qquad (9.82)$$

A similar expression results for the load voltage within enclosure 1 simply by replacing the subscripts 2 in Eq. (9.82) by 1.

From Eq. (9.82), it is seen that the key in determining the inner load responses is a knowledge of the current I_b, as determined from a solution of Eq. (9.80). Depending on the nature of the cable interruption, this current can take on different values, as discussed in the next sections.

9.6.2 Cable Connectors

For the case of a cable connector, Eq. (9.80) can be simplified further by assuming that the connector sufficiently isolates the internal region from the external region so that there is no reaction of the internal current back onto the shield current. Furthermore, the effect of the connector impedance discontinuity on the inner and outer circuit is usually negligible compared with the Thévenin impedances. Under this assumption, the shield current is given by

$$I_s \approx \frac{V_{oc1}^{(e')} - V_{oc2}^{(e')}}{Z_1^{(e')} + Z_2^{(e')}} \qquad (9.83a)$$

and the internal current is

$$I_b \approx \frac{Z_t I_s}{Z_1^{(i')} + Z_2^{(i')}} \qquad (9.83b)$$

where $Z_t = Z_d + j\omega M_{ab}$ is a lumped transfer impedance for the connector. Eq. (9.82) can then be used to estimate the internal response for this excitation.

According to [9.24], measurements suggest that for a new, tightly connected cable connector, the direct magnetic field penetration is minimal, so that the inductance term $j\omega M_{ab}$ is small compared with the series impedance element Z_d. Furthermore, the latter impedance is primarily resistive. However, as the cable ages and the connector flexes, the connection between the braid and the connector becomes loose and M_{ab} increases. Vance [9.2] presents the results of measurements of the transfer impedance Z_t for several different cable connectors. Appendix E summarizes these measured data.

9.6.3 Pigtail Terminations

For the pigtail connection of Figure 9.15, the same formalism as used for the cable connector is used, but with a modification of the circuit parameters. The pigtail in the shield provides a discontinuity in the inductance of the shield due to the large change in the radius of the conductor. As a result, the external circuit may be affected by this discontinuity more than by the cable connector. Consequently, the effect of the pigtail is retained in the solution for the external circuit.

Neglecting the resistance of the pigtail connection, the coupling element between the two circuits of Figure 9.16 is primarily inductive: $Z_d = j\omega L_d$. This leads to the shield current

$$I_s \approx \frac{V_{oc1}^{(e')} - V_{oc2}^{(e')}}{Z_1^{(e')} + Z_2^{(e')} + j\omega L_d} \qquad (9.84a)$$

and the internal current

$$I_b \approx \frac{Z_t I_s}{Z_1^{(i')} + Z_2^{(i')} + j\omega L_d} \qquad (9.84b)$$

where the transfer impedance of the pigtail is $Z_t = j\omega(L_d + M_{ab})$. As in the case of the cable connector, it is difficult to obtain the inductance terms L_s and M_{ab} for a pigtail from first principles, and they are best found through experiments.

A specific example of coupling to a shielded cable with pigtail terminations has been discussed in [9.26]. The interference source was another shielded cable parallel to the victim. In this study the current induced in the pigtail was completely neglected and, over the pigtail length, only the direct coupling between the generator circuit and the internal conductor was considered. The good agreement obtained by comparison with measurements indicated that for frequencies up to about 1 MHz, this kind of approach is justified.

If the shielded cable with pigtails is excited by an incident EM field, the expressions for the induced currents in Eq. (9.84) can be used, with the excitation $V_{oc1}^{(e')}$ and $V_{oc2}^{(e')}$ being calculated using Eq. (9.77).

9.6.4 Discontinuous Shields

The third configuration for the cable shield interruption is when the shield is completely open at the discontinuity. In this case the impedance element Z_d becomes a capacitance which for interruptions not smaller than 0.1 mm is less than 1 pF [9.27], and is neglected at low frequencies. For this configuration all of the induced cable shield current is forced to flow onto the inner "protected" cable. Setting the impedance element Z_d to infinity and neglecting the mutual coupling term M_{ab} yields the following solution for the currents:

$$I_b = I_s \approx \frac{V_{oc1}^{(e')} - V_{oc2}^{(e')}}{Z_1^{(e')} + Z_1^{(i')} + Z_2^{(e')} + Z_2^{(i')}} \tag{9.85}$$

With this expression, the internal load voltage at Z_2 can be calculated directly from Eq. (9.82) as

$$V_2^{(i)} = \frac{2 Z_2^{(i)} Z_c^{(i)} e^{-\gamma^{(i)} \mathcal{L}_2}}{Z_c^{(i)}(1 + e^{-2\gamma^{(i)} \mathcal{L}_2}) + Z_2^{(i)}(1 - e^{-2\gamma^{(i)} \mathcal{L}_2})} \times \frac{V_{oc1}^{(e')} - V_{oc2}^{(e')}}{Z_1^{(e')} + Z_1^{(i')} + Z_2^{(e')} + Z_2^{(i')}}, \tag{9.86}$$

This configuration of the cable shield gives rise to the largest internal response, as the excitation term $V_{oc1}^{(e')} - V_{oc2}^{(e')}$ is not multiplied by any sort of a transfer impedance, as in Eq. (9.83b) or (9.84b). It amounts to a direct injection of the exterior braid current onto the inner electronics and clearly is a case to avoid.

9.6.5 Example of an Interrupted Cable Shield

As an example consider a shielded cable of length $\mathcal{L}$ with the shield interrupted at the middle of the cable. The exterior of the cable shield is represented by a transmission line over a ground plane as denoted by Figure 7.19, and it is assumed to be short-circuited to ground at the two ends ($Z_1 = Z_2 = 0\,\Omega$). The excitation field is taken to be a vertically polarized E-field illuminating the line end-on with angles of incidence $\psi = 0°$ and $\phi = 0°$.

The solution for this line configuration can be validated easily by an experiment using a bounded-wave EMP simulator. It permits the illustration of the fact that the induced voltage can behave in different ways, depending on the termination of the internal conductor. Using a low-frequency approximation for the present example, it is seen that the frequency dependence of the internal response is quite different if, with all other conditions remaining unchanged, the internal conductor is terminated at one end on the characteristic impedance or is open-circuited. As discussed in [9.1], the variation of the induced voltage at low frequencies for the case of the characteristic impedance load is proportional to the frequency, while in the second case (the open-circuit load), it is proportional to the square of the frequency.

Using expression (9.87), we first introduce the general solution valid for any frequency and for both terminations. We then discuss the low-frequency approximation. If the excitation field is a vertically polarized plane wave having only the vertical component of the electric field, the only excitation sources for the line are at the two terminations and are given by

$$V_0 = \int_0^h E_z^i \, dx \tag{9.87}$$

The values of the other parameters needed to apply Eq. (9.86) for this particular case are

$$\mathcal{L}_1 = \mathcal{L}_2 = \frac{\mathcal{L}}{2} \tag{9.88}$$

$$Z_1^{(e)} = Z_2^{(e)} = 0 \to \rho_1^{(e)} = \rho_2^{(e)} = -1 \tag{9.89}$$

Applying (9.79), we obtain

$$Z_1^{(e')} = Z_2^{(e')} = Z_c^{(e)} \frac{1 - e^{-\gamma^{(e)}\mathcal{L}}}{1 + e^{-\gamma^{(e)}\mathcal{L}}} \tag{9.90}$$

Let us first assume that the internal conductor is matched at the two ends, that is,

$$Z_1^{(i)} = Z_2^{(i)} = Z_c^{(i)} \to \rho_1^{(i)} = \rho_2^{(i)} = 0 \tag{9.91}$$

9.6 CABLES WITH SHIELD INTERRUPTIONS

and by applying Eq. (9.79) gives for the internal transmission line

$$Z_1^{(i')} = Z_2^{(i')} = Z_c^{(i)} \qquad (9.92)$$

Introducing Eqs. (9.90) and (9.92) into (9.86), the voltage response can be written as

$$V_2^{(i)} = \frac{2(Z_c^{(i)})^2 e^{-\gamma^{(i)}\mathcal{L}/2}}{Z_c^{(i)}(1+e^{-\gamma^{(e)}\mathcal{L}}+1-e^{-\gamma^{(e)}\mathcal{L}})} \frac{V_{oc1}^{(e')} - V_{oc2}^{(e')}}{2\{Z_c^{(e)}[(1-e^{-\gamma^{(e)}\mathcal{L}})/(1+e^{-\gamma^{(e)}\mathcal{L}})] + Z_c^{(i)}\}}$$

$$= \frac{Z_c^{(i)} e^{-\gamma^{(i)}\mathcal{L}/2}(1+e^{-\gamma^{(e)}\mathcal{L}})(V_{oc1}^{(e')} - V_{oc2}^{(e')})}{2[Z_c^{(e)}(1-e^{-\gamma^{(e)}\mathcal{L}}) + Z_c^{(i)}(1+e^{-\gamma^{(e)}\mathcal{L}})]} \qquad (9.93)$$

The voltage at the far end of the shield $V_{oc2}^{(e')}$ can be written as a function of the voltage $V_{oc1}^{(e')}$ at the near end as

$$V_{oc2}^{(e')} = V_{oc1}^{(e')} e^{-\gamma^{(e)}\mathcal{L}} \qquad (9.94)$$

From Eq. (9.87) with $\rho_1^{(e)} = -1$ and $\mathcal{L}_1 = \mathcal{L}/2$

$$V_{oc1}^{(e')} = \frac{2}{1+e^{-\gamma^{(e)}\mathcal{L}}} (e^{-\gamma^{(e)}\mathcal{L}/2} S_1 - e^{-2\gamma^{(e)}\mathcal{L}/2} S_2)$$

$$= \frac{2e^{-\gamma^{(e)}\mathcal{L}/2}}{1+e^{-\gamma^{(e)}\mathcal{L}}} (S_1 - e^{-\gamma^{(e)}\mathcal{L}/2} S_2) \qquad (9.95)$$

The BLT source functions S_1 and S_2 are given by Eq. (7.37). For the particular excitation of the shielded cable considered in this example, we see that (1) there is no horizontal component of the electric field, and (2) of the two possible vertical sources at the end of the line to the left of the shield interruption, only the one at the far left end at $x = 0$ excited the line, and this excitation source is $V_1 = V_0$, which is given by Eq. (9.87). This implies that $S_1 = -V_1/2$ and $S_2 = (V_1/2) e^{\gamma^{(e)}\mathcal{L}/2}$. Introducing these expressions for S_1 and S_2 into Eq. (9.95) gives

$$V_{oc1}^{(e')} = \frac{2e^{-\gamma^{(e)}\mathcal{L}/2}}{1+e^{-\gamma^{(e)}\mathcal{L}}} \left(-\frac{V_0}{2} - e^{-\gamma^{(e)}\mathcal{L}/2} \frac{V_0}{2} e^{\gamma^{(e)}\mathcal{L}/2} \right) = -\frac{2V_0 e^{-\gamma^{(e)}\mathcal{L}/2}}{1+e^{-\gamma^{(e)}\mathcal{L}}}$$

$$(9.96)$$

Substituting Eqs. (9.94) and (9.96) into (9.93) gives

$$V_2^{(i)} = \frac{-Z_c^{(i)} e^{-\gamma^{(i)}\mathcal{L}/2}(1+e^{-\gamma^{(e)}\mathcal{L}})(1-e^{-\gamma^{(e)}\mathcal{L}})}{2[Z_c^{(e)}(1-e^{-\gamma^{(e)}\mathcal{L}}) + Z_c^{(i)}(1+e^{-\gamma^{(e)}\mathcal{L}})]} \frac{2V_0 e^{-\gamma^{(i)}\mathcal{L}/2}}{1+e^{-\gamma^{(e)}\mathcal{L}}}$$

$$= \frac{-Z_c^{(i)} e^{-\gamma^{(i)}\mathcal{L}/2}(1-e^{-\gamma^{(e)}\mathcal{L}})}{Z_c^{(e)}(1-e^{-\gamma^{(e)}\mathcal{L}}) + Z_c^{(i)}(1+e^{-\gamma^{(e)}\mathcal{L}})} V_0 e^{-\gamma^{(e)}\mathcal{L}/2} \quad (9.97)$$

With this expression, the voltage at the interruption without the effects of propagation along the half-length of the internal conductor (i.e. $e^{-\gamma^{(i)}\mathcal{L}/2} = 1$) is given by

$$V_{\text{inter}}^{(i)} = \frac{-Z_c^{(i)}(1-e^{-\gamma^{(e)}\mathcal{L}})}{Z_c^{(e)}(1-e^{-\gamma^{(e)}\mathcal{L}}) + Z_c^{(i)}(1+e^{-\gamma^{(e)}\mathcal{L}})} V_0 e^{-\gamma^{(e)}\mathcal{L}/2} \quad (9.98)$$

This expression is the same as that arising from an alternative analysis discussed in [9.28].

For low frequencies, for which $\gamma^{(e)}\mathcal{L} \ll 1$, the relation above can be approximated as

$$V_{\text{inter}}^{(i)} = \frac{-V_0 Z_c^{(i)} \gamma^{(e)} \mathcal{L}}{Z_c^{(e)} \gamma^{(e)} \mathcal{L} + 2Z_c^{(i)}} \approx \frac{j\omega V_0 \mathcal{L}\sqrt{\epsilon_0 \mu_0}}{2} \quad (9.99)$$

where we have replaced

$$\gamma^{(e)} \approx j\beta_e = j\omega\sqrt{\epsilon_0 \mu_0} \quad (9.100)$$

Equation (9.99) shows that for low frequencies and lossless lines, the induced voltage is proportional to the frequency. This low-frequency solution has also been found using a different approach in [9.1].

Figure 9.18 shows a comparison between the general solution of Eq. (9.98), the low-frequency approximation (9.99), and an experimental result obtained by illuminating a cable with an interrupted shield in an EMP simulator [9.28]. The calculated values are determined using a measured value for E_z. It can be seen that the low-frequency approximation is valid up to about 1 MHz but that the disagreement with the measurements is 15 dB or higher for frequencies above 10 MHz. The solution of Eq. (9.98), which takes propagation into account, shows an improved agreement, although some differences still exist. These differences could be attributed to factors such as presence of losses and variation of the cable termination resistance as a function of frequency.

A different behavior of the internal conductor response is noted if the internal load is open-circuited at the left end and matched at the right end, with all the other parameters of the previous example remaining unchanged.

9.6 CABLES WITH SHIELD INTERRUPTIONS

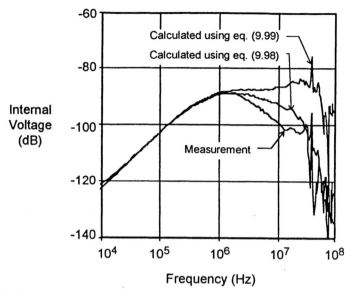

FIGURE 9.18 Internal voltage induced at the far end of a shielded cable with an aperture at half-length of the cable (in dB with respect to 1 V). The internal cable is terminated with the characteristic impedance at both ends. (From ref. [9.28].

For this case, the open-circuit impedance of the left end $Z_1^{(i)}$ can be approximated as

$$Z_1^{(i)} = \frac{-j}{\omega C^{(i)} \mathcal{L}/2} \qquad (9.101)$$

where $C^{(i)}$ is the internal distributed capacitance of the cable. This approximation is valid for $\mathcal{L} \ll \lambda$.

According to Eq. (9.79), the lumped transformed impedance of the internal transmission line formed by the half length $\mathcal{L}/2$ can be expressed as

$$Z_1^{(i')} = Z_c^{(i)} \frac{1 + \rho_1^{(i)} e^{-2\gamma^{(i)} \mathcal{L}/2}}{1 - \rho_1^{(i)} e^{-2\gamma^{(i)} \mathcal{L}/2}} \qquad (9.102)$$

with

$$\rho_1^{(i)} = \frac{Z_1^{(i)} - Z_c^{(i)}}{Z_1^{(i)} + Z_c^{(i)}} \qquad (9.103)$$

so that the lumped transformed impedance is expressed as

$$Z_1^{(i')} = Z_c^{(i)} \frac{1 + \dfrac{Z_1^{(i)} - Z_c^{(i)}}{Z_1^{(i)} + Z_c^{(i)}} e^{-2\gamma^{(i)} \mathcal{L}/2}}{1 - \dfrac{Z_1^{(i)} - Z_c^{(i)}}{Z_1^{(i)} + Z_c^{(i)}} e^{-2\gamma^{(i)} \mathcal{L}/2}}$$

$$= Z_c^{(i)} \frac{Z_1^{(i)}(1 + e^{-\gamma^{(i)} \mathcal{L}/2}) + Z_c^{(i)}(1 - e^{-\gamma^{(i)} \mathcal{L}/2})}{Z_1^{(i)}(1 - e^{-\gamma^{(i)} \mathcal{L}/2}) + Z_c^{(i)}(1 + e^{-\gamma^{(i)} \mathcal{L}/2})}$$

(9.104)

As the right end of the internal conductor is matched, we have

$$Z_2^{(i')} = Z_c^{(i)} \quad (9.105)$$

Introducing Eqs. (9.104) and (9.105) into (9.86), using (9.94), with $V_{\text{ocl}}^{(e')}$ given by Eq. (9.96), with the external lumped transformed impedances $Z_1^{(e')}$ and $Z_2^{(e')}$ given by Eq. (9.90), and assuming as in the previous case that $M \approx 0$, the internal voltage at the right-end termination becomes, after rearrangement and simplification,

$$V_2^{(i)} = \frac{-Z_c^{(i)} e^{-\gamma^{(i)} \mathcal{L}/2} V_0 e^{-\gamma^{(i)} \mathcal{L}/2}(1 - e^{-\gamma^{(i)} \mathcal{L}})[Z_1^{(i)}(1 - e^{-\gamma^{(i)} \mathcal{L}}) + Z_c^{(i)}(1 + e^{-\gamma^{(i)} \mathcal{L}})]}{Z_c^{(e)}(1 - e^{-\gamma^{(i)} \mathcal{L}})[Z_1^{(i)}(1 - e^{-\gamma^{(i)} \mathcal{L}}) + Z_c^{(i)}(1 + e^{-\gamma^{(i)} \mathcal{L}})] + Z_c^{(i)}(1 + e^{-\gamma^{(i)} \mathcal{L}})(Z_1^{(i)} + Z_c^{(i)})}$$

(9.106)

For low frequencies, $\gamma^{(e)} \mathcal{L} \ll 1$ and $\gamma^{(i)} \mathcal{L} \ll 1$. It can also be assumed that $Z_1^{(i)} \gamma^{(i)} \mathcal{L} \ll 2 Z_c^{(i)}$ but that $Z_1^{(i)} \gg Z_c^{(i)}$. In this case the voltage across the interruption simplifies to

$$V_{\text{inter}}^{(i)} = \frac{-Z_c^{(i)} V_0 \gamma^{(e)} \mathcal{L}}{Z_c^{(e)} \gamma^{(e)} \mathcal{L} + Z_1^{(i)}} \approx \frac{-Z_c^{(i)} V_0 \gamma^{(e)} \mathcal{L}}{Z_1^{(i)}} \quad (9.107)$$

Introducing the value of the left-end open-circuit impedance $Z_1^{(i)}$ given by Eq. (9.101) and replacing $\gamma^{(e)}$ by its expression in Eq. (9.100) gives

$$|V_{\text{inter}}^{(i)}| = \frac{\omega^2 Z_c^{(i)} C^{(i)} \mathcal{L}^2 \sqrt{\epsilon_0 \mu_0}}{2} \quad (9.108)$$

This result has be found using a different approach in [9.1], where it is also supported by experimental measurements. Figure 9.19 illustrates a comparison of the measured and calculated internal voltage responses of the shielded cable.

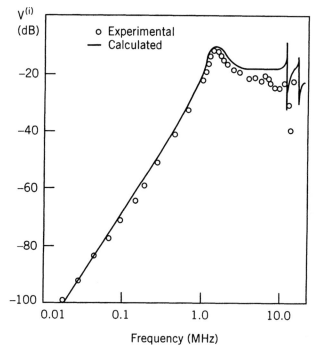

FIGURE 9.19 Internal voltage induced at the far end of a shielded cable with a break aperture at the midpoint of the cable (in dB with respect to 1 V). The cable is terminated in the characteristic impedance at the far end and open-circuited at the near end. (From [9.1]. © Oxford University Press, 1993.)

REFERENCES

9.1. Degauque, P. and J. Hamelin, *Compatibilité Electromagnétique*, Dunod, Paris, 1990.

9.2. Vance, E. F., *Coupling to Shielded Cables*, R. E. Krieger, Melbourne, FL, 1987.

9.3. Schelkunoff, S. A., "Theory of Lines and Shields," *Bell Syst. Tech. J.*, Vol. 13, (1934), pp. 522–579.

9.4. Bech, Ph., Y. Dijamatovic, and M. Ianoz, "Mesure de l'impédance de transfert en régime impulsionnel," *Bulletin ASC/USC 78(1987)9*, May 9, 1987.

9.5. Fowler, E. P., and L. K. Halme, "State of Art in Cable Screening Measurements," *Proceedings of the 9th International Symposium on EMC*, Zurich, March 12–14, 1991, paper 30F1.

9.6. Fowler, E. P., "Test Methods for Cable Screening Effectiveness: A Review," *Proceedings of the IERE Symposium on EMC*, York, 1988, IERE Publication 81.

9.7. Hoeft, L. O., and J. S. Hostra, "Measured Electromagnetic Shielding Performance of Commonly-Used Cables and Connectors," *IEEE Trans. Electromagn. Compat.*, Vol. EMC-30, No. 3, August 1988, pp. 260–275.

9.8. Goldstein, C., and P. Mani, "CW and Pulsed Mode Transfer Impedance

Measurements in Coaxial Cables," *IEEE Trans. Electromagn. Compat.*, Vol. EMC-34, No. 1, February 1992, pp. 50–57.

9.9. "Radio-Frequency Cables, Part 1, General Requirements and Measuring Methods," *International Electrotechnical Commission Standard 96-1*, Fourth ed., 1986.

9.10. Lee, K. S. H. and C. Baum, "Application of Modal Analysis to Braided-Shielded Cables," *IEEE Trans. Electromagn. Compat.*, Vol. EMC-17, No. 3, August 1975, pp. 159–169.

9.11. Demoulin, B., "Etude de la pénétration des ondes électromagnétiques à travers des blindages homogènes ou des tresses à structure coaxiale," Ph.D. thesis, Université des Sciences et Techniques de Lille, June 17, 1981.

9.12. Aguet, M., and M. Ianoz, *Haute Tension*, Traité d'Electricité, Presses Polytechniques, Romandes, Switzerland, 1987.

9.13. Tyni, M., "The Transfer Impedance of Coaxial Cables with Braided Outer Conductor," *Digest of the 10th International Wroclaw Symposium on EMC*, 1976, pp. 410–419.

9.14. Sali, S., "An Improved Model for the Transfer Impedance Calculations of Braided Coaxial Cables," *IEEE Trans. on Electromagn. Compat.*, Vol. EMC-33, No. 2, May 1991, pp. 139–143.

9.15. Kaden, H., *Wirbelstroeme und Schirmung in der Nachrichtentechnik*, Springer-Verlag, Berlin, 1959.

9.16. Demoulin, B., P. Degauque, and M. Cauterman, "Shielding Effectiveness of Braids with High Optical Coverage," *Proceedings of the International Symposium on EMC*, Zurich, 1981, pp. 491–495.

9.17. Demoulin, B., private communication.

9.18. Kley, T., "Optimized Single-Braided Cable Shields," *IEEE Trans. Electromagn. Compat.*, Vol. EMC-35, No. 1, February 1993.

9.19. Halme, L., *Transmission-Line and Electromagnetic Screening*, Parts 1 and 2, Kyriiri Oy, Helsinki, ISBN 951-672-088-9 and 951-672-089-7, 1989.

9.20. Broyde, F., and E. Clavelier, "Comparison of Coupling Mechanism on Multiconductor Cables," *IEEE Trans. Electromagn. Compat.*, Vol. EMC-35, No. 4, November 1993, pp. 409–416.

9.21. Mohamudally, N., "Etude des fuites électromagnétiques introduites sur le raccordement des câbles blindes et du couplage produit par une composante de champ magnétique perpendiculaire a la section des câbles coaxiaux," Ph.D. thesis, Université des Sciences et Techniques de Lille, January 11, 1996.

9.22. Tesche, F. M., "Plane Wave Coupling to Cables, Part II," Chapter 2 in *Handbook of Electromagnetic Compatibility*, R. Perez, ed., Academic Press, New York, 1995.

9.23. Bell Laboratories, *EMP Engineering and Design Principles*, Technical Publications Department, Bell Laboratories, Whippany, NJ. 1975.

9.24. Hoeft, L. O., EM Consultant, Albuquerque, NM, private communication with the authors, December 1994.

9.25. Balabanian, N., *Fundamentals of Circuit Theory*, Allyn and Bacon, Boston, 1962.

9.26. Paul, C. R. "Effect of Pigtails on Crosstalk to Braided Cables," *IEEE Trans. Electromagn. Compat.*, Vol. EMC-22, No. 3, August 1980, pp. 161–172.

9.27. Duvinage, P., "Etude et caractérisation électromagnétique des discontinuités de blindage. Application à la mesure des paramètres de transfert de câbles coaxiaux aux fréquences élevées," Thèse de 3e cycle 1203, Université des Sciences et Techniques de Lille, October 5, 1984.

9.28. Ianoz, M., F. Rachidi, and P. Zweiacker, "Electromagnetic Field Coupling to Imperfect Shielded Cables," *Proceedings of the International Symposium on EMC*, Sao Paolo, Brazil, December 2–5, 1994.

PROBLEMS

9.1 It is desired to perform measurements on a piece of coaxial cable to determine its transfer impedance and transfer admittance parameters. Design a suitable experimental test setup for both of these measurements and outline detailed test procedures to be followed. (*Hint:* Consider the basic definitions of Z'_t and Y'_t in Eqs. (9.6) and (9.7). In addition, [9.1] may be useful.)

9.2 Show that the thin-walled approximation for the transfer impedance in Eq. (9.9) can be derived from the more general expression of Eq. (9.8).

9.3* A braided cable shield has the following parameters: optical coverage $\mathcal{K} = 0.9$; number of carriers $\mathcal{C} = 12$; weave angle $\psi = 45°$; radius $b = 5$ mm; major axis length of the aperture $L_{eq} = 1$ mm; thickness (braid wire diameter) $d = 0.25$ mm; wire conductivity $\sigma = 5.7 \times 10^7$ S/m. For this cable shield:

(a) Calculate the dc resistance per meter,

(b) Calculate the transfer inductance of the shield.

9.4* For the shielded cable of Problem 9.3, an external current waveform given by $i(t) = 100(1 - e^{-t/\tau})u(t)$, with $\tau = 2 \times 10^{-9}$ s and $u(t)$ the unit step function, propagates at the speed of light down the shield. Assume that the external line is matched at the far end of the shield. For an *internal* transmission line impedance of 50-Ω, which is also assumed to be matched at both ends, calculate the voltage across the internal load at each end: (a) for early time; (b) for late time.

9.5* Critical braided cables in some installations may be shielded further by using solid conduits, as illustrated in Figure P9.5. In this instance there can be diffusion of electromagnetic energy through the imperfectly conducting conduit (i.e., a distributed penetration) and subsequent coupling with the braided cable. Discuss this phenomenon and formulate an appropriate transmission line model for treating this problem.

---

*Problems denoted by an asterisk were used in the EMP short courses conducted by the Summa Foundation in 1983 through 1989. These contributions by C. Baum and the other lecturers are gratefully acknowledged.

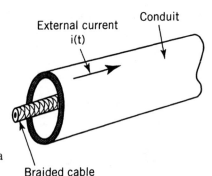

FIGURE P9.5 Braided cable shielded by a solid, imperfectly conducting conduit.

9.6* In the context of Problem 9.5 there can be aperture diffusion through the conduit (a local penetration) when two sections of conduits are joined together improperly. Discuss this phenomenon of local penetration and suggest a suitable model for evaluating the internal cable response.

9.7* A braided wire shield contains apertures through which both electric and magnetic fields can penetrate. Discuss the magnetic and electric coupling of an isolated aperture in terms of the transfer impedance and a circuit model for an incremental length of the shield.

9.8 The coaxial line of Figure 9.1 is excited by a distant lightning stroke, as illustrated in Figure 7.22.
  (a) Develop expressions for the external charge and currents on the exterior cable sheath for this non-plane-wave excitation.
  (b) Formulate the internal load responses for this excitation.
  (c) Discuss the computational difficulty of these solutions, and suggest an alternative, approximate solution for the internal responses.

CHAPTER 10
Shielding

In most practical cases, the metallic shield protecting a facility or an equipment enclosure is imperfectly conducting, thereby permitting EMI to penetrate the interior. In Chapter 5 we discussed the penetration of such signals through discrete imperfections in a shield, such as an aperture or a seam. It is possible, however, for the disturbing signals to penetrate the shield material directly. In this chapter we discuss the principles of EM shielding and illustrate results from some of the simple models that can be used for predictions of shielding.

10.1 INTRODUCTION

Shielding is an effective way of preventing undesirable electromagnetic interference. However, as a form of system or component protection, it is only one element of many that must be considered together to create a complete and effective protection concept. A correct choice of shields and other elements of an EMI protection scheme must take into account not only the surrounding environment and its electromagnetic properties, but the user requirements and mission of the system.

The limitations and shortcomings in EM shielding are the openings of the shield. For real systems, such openings are unavoidable. Wires and pipes must penetrate the shield, and people, air, and other media have to pass in and out of the shielded region. Generally, these penetrations provide the major degradations in the shielding: the EM field diffusing through the shield material is only a small fraction of the total response. The simple models presented in this chapter are sufficiently accurate for practical needs. It must be remembered, however, that to obtain adequate protection against interference, it is often necessary to use a combination of shielding and other protective measures, such as filters or overvoltage supressors (nonlinear components such as zener diodes or spark gaps).

In this chapter a presentation is given of the general principles of shielding, including a discussion of the properties of shielding material and the impact that openings in the shield have on shielding behavior. We begin with a development of an overall shielding concept by defining a generalized shield. This is followed by a description of the electromagnetic mechanism that gives rise to shielding and a description of the various qualitative properties of ordinary shielding material by developing simple shield models that can be used in practical applications.

The shielding that is provided by a material can vary significantly from case to case, depending on the particular application in question. In this context, the question of how the shielding depends on distance, shield size, frequency-time function, shield thickness, density (net shields), and material properties is treated. Connections of cable shields, entry panels, and openings in the shield, important aspects to be considered when designing the protective shielding, are dealt with separately.

10.2 GENERALIZED SHIELD CONCEPT

As discussed in Chapter 2, a boundary surface in the electromagnetic topology of a system does not necessarily have to correspond to a solid shield in the practical case. The important conceptual aspect in defining such a boundary surface is that there should be no coupling between circuits located on different sides of the closed boundary. The term *generalized shield*, which refers to an extension of the traditional shield concept, is the designation of the boundary surface that *represents* the actual limiting coupling—or attenuation—between circuits on the various sides [10.1–10.3]. The generalized shield is the practical equivalent of the theoretical, topological boundary surface. Attenuation does not have to take place in an identifiable metallic shield; the generalized shield simply indicates that neither side communicates with the other. The coupling between circuits on either side of the shield is less than a fixed value, the maximum permissible coupling.

As a simple example of a generalized shield that limits coupling without a metal shield, consider two antennas polarized perpendicularly to each other. The electrical field strength generated by one of the antennas is perpendicular to that of the other. If one antenna is rotated or if a reflecting object comes near them, the field configuration changes and the generalized shield disappears. This example also shows that the generalized shield describes the coupling conditions and is *not* some form of physical shield. In many cases the generalized shield coincides with a constructed metallic shield, which is often necessary for the shield to function as envisioned.

At times, a sufficient degree of coupling reduction can be achieved through a combination of different coupling-reducing factors, such as shielding, distance, proximity, and shadowing. With the help of the generalized shielding concept, it is possible to formulate the requirements for a sufficiently effective shield. Sometimes, the optimum solution to the shielding design may be to use a compact metal shield made of thin netting. In other occasions, when the distance between disturbed and disturbing circuits

10.2 GENERALIZED SHIELD CONCEPT

is sufficiently large, it may be sufficient simply to locate the equipment needing protection near a flat metal surface that reduces the primary field components. Thus a well-designed installation can be viewed as serving as a generalized shield.

One simple way of reducing the EM field coupling to wires is to install them close to a conductive surface or ground plane as shown in Figure 10.1. In this case the electrical field is largely perpendicular to the ground plane, since the component parallel to the plane drops to zero right on the metal surface. Circuits that are parallel to the ground plane are perpendicular to the electrical field, and the field coupling to these circuits will be weak. The same conclusion can be drawn by studying the magnetic coupling, which is weak to loops that lie on a plane parallel to the ground plane since the component of the magnetic field perpendicular to the plane drops toward zero next to the metal surface. This effect depends on the proximity to the metal plane and decreases rapidly with increasing distance from the plane. This is the same physical effect that has been discussed in the context of field coupling to transmission lines in Chapter 7.

A ground plane limits the coupling in the same way as a traditional metal shield, despite the fact that it does not enclose the circuits. It is clearly unnecessary for the containment surface to consist of an enclosing shield; it may be sufficient for certain critical parts of the generalized shield to consist of conductive material. Once the attenuation requirement of the shield is determined by the EM field coupling of the particular case in question, it is possible to determine what parts of the containment surface need to be conductive.

All practical shields can be viewed as generalized shields, since they always have some form of imperfection: windows, cracks around doors, ventilation grills, sheet-metal joints, and all the wires that are needed in

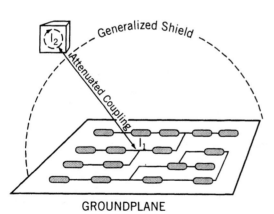

The generalized shield can be illustrated with the help of a ground plane. The coupling between a wire that is positioned close to the ground plane and a circuit, which is situated further away, is limited by the proximity of the wire to the plane. The physical explanation is that the electrical field is perpendicular to the metal surface and has a weak coupling to wires parallel with the ground plane. In other words, the coupling is limited although there is no closed metal shield between the circuits. We call this a generalized shield.

FIGURE 10.1 Generalized shield.

order for the system to function. Despite these shortcomings, the shields function properly, provided that the connections between the circuits inside and outside the screen are sufficiently small. One important factor is the *distance* from the disturbing or disturbed circuits to the imperfections—holes and cracks—in the shield. If the distance is sufficiently large, the coupling will be weak and the imperfections may be accepted.

Often, it is sufficient for circuits belonging to different electromagnetic environment zones to be situated far away from each other. No metal screening is needed to lower the interference level. The generalized screen that surrounds such zones represents the attenuation that takes place as a result of distance, and no other limitation of the coupling is needed.

This generalized shield concept is very important for a definition of grounding, because it always contains the potential references in those cases where one exists. Ground is equivalent to the reference point for the generalized shield, which is often coincident with the cable penetration panel, frequently referred to as an *intake point*.

Experienced electromagnetics engineers will sometimes use their intuitive feeling for precisely this property in the generalized shield as the potential reference, in order to locate it. Even in a maze of electrical wires and metal building components there is generally some part that can serve as a natural potential reference.

From the definition it can be clearly seen how generalized shields—all shields in practice—function within a certain frequency range since all connections are dependent on frequency. Although the topological appearance of the shield may vary with frequency, it will always remain a closed surface.

10.3 SHIELDING MECHANISMS

In the field of electromagnetism, shielding can mean different things in different circumstances. In this section a description is given of the different types of shielding and the electromagnetic mechanisms that give rise to shielding. Shielding can be viewed as being either electrical or magnetic in nature. It can also be viewed as arising from eddy currents that are induced in the shield material, which in practice is the only shielding mechanism that is active when counteracting electromagnetic interference.

10.3.1 Shielding of Static Fields

10.3.1.1 Electrical Shielding. It is possible to shield an electrostatic field using fairly simple means. The classical designation for an electrostatic shield is a Faraday's cage, named after the English experimental physicist who discovered electromagnetic induction in the 1830s. As indicated in Figure 10.2, electrostatic shielding is created by the fact that electrical charges on

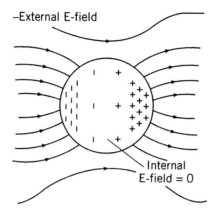

−External E-field

Internal E-field = 0

The electrical charges in a metal body, free electrons, distribute themselves over the metal surface so that the electrical field inside the body is extinguished. The field lines on the outside terminate in charges. Since the charges are easy to move, they will be applied in exactly the right position to make the interior free of fields. The shielding is extremely effective for electrostatic fields. Since the charges can be moved in a totally unrestricted manner on all metallic conductors, even on very thin wires, netting was judged to function very well as electrostatic shielding.

FIGURE 10.2 Electrostatic shielding.

metal surfaces try to distribute themselves such as to cancel the electrical field inside the metal. Consequently, if there is a void, or volume, enclosed by the metal, it is shielded from external electrical fields. The electrical charges are fully mobile in all conductive materials. Even in extremely thin metal foils the mobile charges are present in sufficient quantities to cause an effective shielding of external fields. This is why the optical coverage of the shield material, rather than its thickness, is important in connection with electrostatic shielding.

If there is a hole in the shield, some of the external electrical influence will penetrate the shield (Figure 10.3). To understand the mechanism, it is a good idea to remember that the electrical field lines run between charges of different polarity and symbolize a force between the charges. The field lines

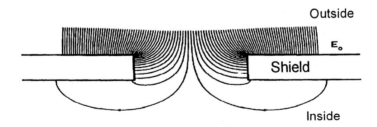

Outside

Shield E_o

Inside

The electrical field forces its way through a hole in the shield. The field lines terminate in charges on the inside of the shield. At a distance of several hole diameters from the hole, the shielding functions anyway, despite the hole. At a distance of two hole diameters in, the field has weakened by approximately 44 dB.

FIGURE 10.3 Behavior of the E-field near a hole in an electrostatic shield.

that enter through holes in the shield will, in other words, have an effect on charges on the inside. But if no conductive objects are situated in the vicinity of the hole, the electrical field lines will bend and terminate on the inside of the shield, which means that the shield is still serving to protect the internal components.

The shielding will be impaired in a certain volume around the hole, but at a distance larger than several hole diameters, the field weakens by the cube of the distance [10.4]. At a large distance r from a circular hole with a diameter of d, the E-field magnitude is given by Eqs. (4.13a), (5.63) and Table 5.1 as

$$|E| \approx \frac{E_0}{6\pi}\left(\frac{d}{r}\right)^3 \quad \text{V/m} \tag{10.1}$$

where E_0 is the strength of the unperturbed, or incident, normal E-field component.

This equation implies that the field attenuation is already more than 40 dB at a distance of only two hole diameters from the opening. Thus, even if a shield has one or more apertures in it, by locating sensitive equipment away from the holes, an effective electrostatic shielding can be obtained.

10.3.1.2 Magnetostatic Shielding. A magnetostatic field is extremely difficult to shield. A basic physical reason for this is described by a nonsymmetry in Maxwell's equations, which shows that there is a fundamental difference between electrical and magnetic fields. Maxwell's fourth equation, $\nabla \cdot \mathbf{B} = 0$, says that the magnetic field lacks sources equivalent to the electrical charges, which generate the electrical field in accordance with the third equation, $\nabla \cdot \mathbf{D} = \rho$. There are, in other words, no magnetic charges. Since it is the electrical charge that produces the effective electrical shielding, it is only natural that in the absence of magnetic charge, the magnetic shielding is weaker.

It is only possible to obtain magnetostatic shielding from ferromagnetic material, the permeability of which is considerably greater than that of vacuum. There are five elements that are ferromagnetic at room temperature: iron, nickel, cobalt, and the rare earth elements gadolinium and terbium. The oxides and alloys of these elements are also ferromagnetic. Other nonferromagnetic metals, such as manganese, copper, and aluminum, can also form ferromagnetic alloys (Hustler's alloys). Alloys of manganese chrome and chromium platinum are other well-known ferromagnetic materials.

As an example of magnetostatic shielding, consider the case of a spherical ferromagnetic shell of inner and outer radii a and b, respectively, and relative permeability μ_r, located in a homogeneous magnetic field H^{inc} as shown in Figure 10.4. The figure shows how the magnetic field is strengthened above and below the points where the field meets the surface

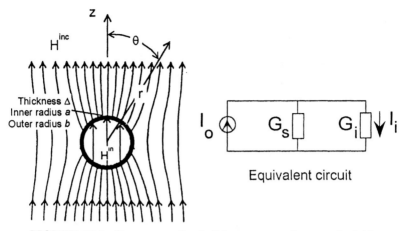

FIGURE 10.4 Ferromagnetic shell in an external magnetic field.

of the sphere, whereas it is weakened at the sides around the sphere. This is due to the fact that magnetic material attracts the magnetic field.

Inside the sphere the field H^{in} is homogeneous and only slightly attenuated. To calculate the internal field, Van Bladel [10.5] introduces a scalar magnetic potential† ϕ which is a solution to Laplace's equation and from which the H-field may be calculated as $\mathbf{H} = -\nabla\phi$. Outside the sphere, the potential has the form

$$\phi_e = -H^{inc} r \cos\theta + \frac{A}{r}\cos\theta \qquad (10.2a)$$

where r and θ are the spherical coordinates illustrated in Figure 10.4. Within the spherical shell material (for $a \geq r \geq b$) the potential is of the form

$$\phi_s = \left(Br + \frac{C}{r^2}\right)\cos\theta \qquad (10.2b)$$

and within the center hollow region of the shell, it is

$$\phi_i = Dr\cos\theta \qquad (10.2c)$$

The unknown coefficients A, B, C, and D must be determined from the boundary conditions that both ϕ and $(\mu\,\partial\phi/\partial r)$ be continuous functions at $r = a$ and $r = b$. Imposing these conditions on Eq. (10.2) results in a system of four equations in four unknowns that can be solved algebraically.

As a result of this development, the static H-field within the sphere can

---

† In the more general case with the region of interest containing a volume current density, the H-field is described by a vector potential. However, in the present case there is no current and $\nabla \times \mathbf{H} = 0$, which permits the representation of H by a scalar potential.

be expressed as

$$H^{in} = \frac{H^{inc}}{1 + (2(\mu_r - 1)^2/9\mu_r)(1 - a^3/b^3)} \quad (10.3)$$

Notice that if the shell material is nonmagnetic, $\mu_r = 1$ with the result that $H^{in} = H^{inc}$. For a thin shell of thickness $\Delta = b - a$ made of a highly magnetic material ($\mu_r \gg 1$), the internal field is approximately

$$H^{in} = \frac{H^{inc}}{1 + \mu_r(2\Delta/3a)} \quad \Delta \ll a, \quad \mu_r \gg 1 \quad (10.4)$$

A common way of representing the shielding behavior of an enclosure is through a magnetic field transfer function, T_H, which is the ratio of the inner field to the excitation field, as

$$T_H = \frac{H^{in}}{H^{inc}} \quad (10.5)$$

An alternative description is the magnetic field shielding capacity, or shielding effectiveness, denoted by S_H, which is the inverse of T_H:

$$S_H = \frac{1}{T_H} = \frac{H^{inc}}{H^{in}} \quad (10.6)$$

The term S_H, referred to as the *attenuation* of the shield, is usually expressed in decibels as

$$(S_H)_{dB} = 20 \log \frac{H^{inc}}{H^{in}}$$

A formal analogy with Ohm's law applies for magnetostatic shielding, whereby the strength of a current I is equivalent to the magnetic flux Φ, conductivity σ is equivalent to permeability μ, and conductance G is equivalent to permeance G_m. In other words, the magnetic flux prefers a route that represents a high permeance. Table 10.1 summarizes these relationships.

Magnetic shields can be analyzed in much the same way as electrical dc circuits, using a circuit model [10.6]. For the shield illustrated in Figure 10.4, the shield diverts the flux from the inner part of the shield in the same proportion as the relationship between the permeance of the shield and that of the medium inside the shield. The permeance of the spherical shield is proportional to its cross-sectional area and permeability and conversely proportional to the extent it has in the "direction of the magnetic flux," all

10.3 SHIELDING MECHANISMS

TABLE 10.1 Formal Correspondence Between the Electrical and Magnetic Circuits

Electricity	Magnetism
Ohm's law: $I = GV$	Hopkinson's law: $\Phi = G_m IN$
$j = \sigma E$ (A/m^2)	$B = \mu H$ (T)
I = current intensity (A)	Φ = magnetic flux (Wb)
σ = conductivity (S/m)	μ = permeability (H/m)
$1/R = G$ = conductance (S)	G_m = permeance (Wb/A)[a] (or giorgi)
R = resistance (Ω)	R_m = reluctance (A/Wb)[a]
V = electromotive force (V)	$I \cdot N$ = magnetomotive force (A)[a]

[a] In these units, A is actually ampere-*turn*, with "turn" being dimensionless.

in analogy with Ohm's law. Thus the internal flux can be expressed as

$$I_i = \text{interior flux} = \frac{G_i I_0}{G_s + G_i} \tag{10.7}$$

Consequently, the internal H-field has the same strength as that given in Eq. (10.4).

One of the difficulties in calculating the shielding efficiency of practical shields is determining a suitable value for μ_r, since it varies with the magnetic field strength. Certain materials, such as mu-metal and supermalloy, have a very high permeability for relatively low magnetic field strength, but become rapidly saturated, with the consequence that the permeability decreases with increasing magnetic field strength. Table 10.2 presents the conductivity and relative permeability of selected materials.

It is evident from Eq. (10.7) that the shielding of large volumes is particularly difficult, since the permeance of the internal medium increases with the cross-sectional area of the shielded area, whereas the permeance of the shield increases with the circumference. An example of calculating the shielding capacity of the spherical iron shell shown in Figure 10.4 illustrates the difficulty in obtaining a good magnetostatic shield. A typical value for the relative permeability that is used fairly often in engineering assessments for ordinary sheet iron is $\mu_r = 400$. For a spherical shell with 1 m in diameter and 1 cm thick, Eq. (10.3) gives an H-field attenuation factor S_H of about 16 dB. Figure 10.5 presents the behavior of the shielding as a function of the shield radius for two different shell thicknesses.

To improve the shielding, the shell can be made thicker, but it is impossible to achieve a shielding efficiency higher than $2\mu_r/9$, which applies for an iron sphere with a very small void. Since for static shielding the shape of the shield is of little importance to the permeance, the example with a sphere can give a good approximation of the magnetic field shielding efficiency of other shield configurations.

The shielding is further impaired if a hole is made in the shield. In the

TABLE 10.2 Conductivity and Relative Permeability of Selected Materials

Material	σ (S/m)	μ_r
Selenium	83×10^6	1
Silver	62×10^6	1
Copper	58×10^6	1
Gold	41×10^6	1
Aluminum	38×10^6	1
Chromium	38×10^6	1
Brass[a]	26×10^6	1
Tungsten	18×10^6	1
Zinc	17×10^6	1
Nickel	14×10^6	max 600
Cobalt	10×10^6	max 250
Iron[b]	10×10^6	max 4000–8000
Platinum	9.5×10^6	1
Tin	8.8×10^6	1
Lead	4.6×10^6	1
Lead dioxide (PbO_2)	1.1×10^6	1
Mu-metal[c]	$2 \times 10^6 - 4 \times 10^6$	max 100,000
Supermalloy[d]	1.7×10^6	max 1,000,000
Stainless steel[e]	1.1×10^6	1
Mercury	1.0×10^6	1
Graphite	71×10^3	1
Seawater	3	1

[a] 66% Cu, 34% Zn.

[b] The figures in the table refer to technically pure iron. Iron of the highest possible purity can have $\mu_r = 25{,}000{-}350{,}000$. The conductivity of steel is somewhat lower than that of pure iron. In the case of hard steel it is significantly lower: $\approx 1{-}6$ MS/m. It is meaningless to give permeability values for different types of steel here, as they vary within wide limits and with the field strength. When performing a shielding calculation, you should therefore try to find out what the magnetization curve is for the type of iron or steel concerned. A value often used for "unknown" iron is $\mu_r = 400$, which should provide a fairly accurate result. ($B_{max} \approx 2$ T.)

[c] 71–78% Ni, 4.3–6% Cu, 0–2%; $B_{max} = 0.72$ T.

[d] 79% Ni, 5% Mo; $B_{max} = 0.8$ T.

[e] 0.1% C, 18% Cr, 8% Ni.

inner part of the shield a long way from all the holes, the impairment in the shielding is proportional to the decrease in the permeance, which is a result of the fact that there will be a smaller amount of magnetic material to divert the magnetic flux. Near openings, it is more suitable to describe the impairment as a result of the magnetic field concentration that is formed in the actual hole and which decreases rapidly with increasing distance from the opening; one hole diameter inward, the leakage is normally negligible. It can be compared, for example, with the field configuration around the gap in the recording head on a tape recorder. Holes in a magnetic shield do not

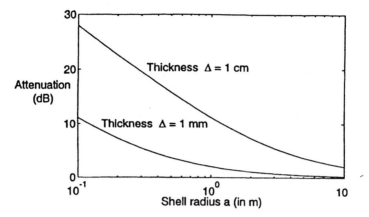

FIGURE 10.5 Static magnetic field attenuation factor S_H of a thin spherical shell of iron with $\mu_r = 400$.

involve too much of an impairment because the shielding effect is weak, even with a homogeneous shield.

10.3.2 Shielding of Time-Varying Fields: Eddy Current Shielding

10.3.2.1 General Concepts. What we have discussed in the preceding section concerns the shielding of static E and H fields. These static results are also applicable for very slow time variations, but as the frequency increases, another shielding mechanism becomes important, even for frequencies as low as a few hertz. This mechanism is the effective shielding of the EM fields by the eddy currents that are induced to flow in the conducting shield. These currents themselves produce electric and magnetic fields that oppose the incident fields, thereby resulting in a decrease of the total field strength within the shielded region.

A fundamental physical principle described by Maxwell's equations is that a varying electrical field gives rise to a magnetic field, and vice versa. This implies that for this more general case of nonstatic fields, the shielding cannot be described separately as being "magnetic" or "electric," but instead, must be explained as a more complex electromagnetic phenomenon. However, in dealing with dynamic shielding problems, it is usual to characterize the "shielding" of an enclosure by a simple ratio of fields, as in Eq. (10.6). Depending on whether the E-fields or the H-fields are used, the same enclosure will have two different measures of shielding effectiveness. Consequently, the statement that "a shielded room has an attenuation of 50 dB at 1 kHz" is meaningless, unless additional information is provided as to exactly what field quantities are being considered.

For dynamic shielding in a general enclosure, the behavior of the eddy currents is very complex. Each individual case requires complete solution of an electromagnetic boundary value problem. However, to better understand the shielding mechanism and to find approximate solutions to shielding problems, it is common to study a number of generic shielding enclosures.

Consider again the shielding provided by a thin, spherical shell of magnetic material, but this time for a time-harmonic EM field excitation. The quasistatic magnetic shielding, given by Eq. (10.3), dominates up to a frequency f, where the size of the shield enclosure (as measured by its radius a) is approximately equal to $\mu_r \delta$ where δ is the skin depth given by the expression

$$\delta = \sqrt{\frac{1}{\pi f \mu_r \mu_0 \sigma}} \quad \text{m} \tag{10.8}$$

Here $\mu_0 = 4\pi \times 10^{-7}$ H/m is the magnetic permeability of vacuum and σ is the conductivity of the shield material. Figure 10.6 presents the skin depth as a function of frequency for common shielding materials.

For a nearly spherical shield (the exact shape is of no particular relevance) made of iron with $\mu_r = 400$ and having a radius of a meters, the eddy current shielding begins to dominate at frequencies above $f \approx 10/a^2$ Hz. This means that eddy current shielding in a sphere with a 1-m diameter (i.e., like a small shielded room) is already the most important shielding mechanism at power frequencies of 50 to 60 Hz.

As an example of the shielding efficiency of a spherical iron shell of different radii at a frequency of 50 Hz, Figure 10.7 plots the behavior of S_H (in dB) for two different shield thicknesses. For small spheres, the shielding behaves like that of Figure 10.5, with the process being dominated by the

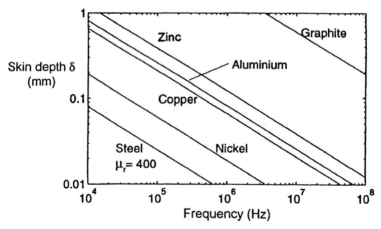

FIGURE 10.6 Skin depth as a function of frequency for common shielding materials.

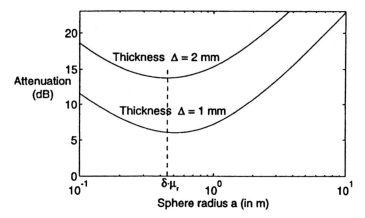

FIGURE 10.7 Magnetic field shielding efficiency S_H for a spherical shell of iron with $\mu_r = 400$ at a frequency of 50 Hz.

static shielding process. As the radius increases, the shielding begins to increase again, due to the eddy currents flowing within the shell.

Figure 10.8 illustrates the frequency dependence of the shielding provided by a 1-mm-thick spherical shell with a 1-m radius made of copper, and by iron with $\mu_r = 400$. For very low frequencies the iron shell has better shielding than the copper shell. But even at 5 Hz, the shielding of the copper shell will be as good as that of the iron shell. Copper has better conductivity, and consequently, it provides better shielding in the important frequency range where the shield is "electrically thin." At high frequencies the iron shell is better, since the skin depth is less in iron than in copper.

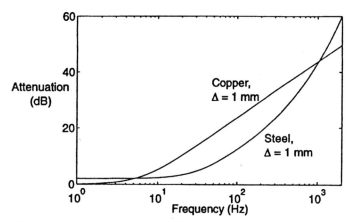

FIGURE 10.8 Frequency dependence of the magnetic field shielding factor of copper and iron spherical shells with a radius of $a = 1$ m.

10.3.2.2 Skin Effect and Skin Depth.

When a direct current is carried in a highly conducting wire, the charges move uniformly over the entire cross section of the conductor. However, if the amplitude of the current varies over time, the charges are forced out toward the surface of the conductor. This effect, which is called *current displacement* or the *skin effect*, will become stronger as the frequency increases. However, even at frequencies as low as 50 to 60 Hz, this effect can be important. One consequence of this is that electricity distributors can reduce the weight of power lines by making them tubular-shaped without significantly impairing the resistance, since most of the current is carried close to the outer surface.

In a metal plate, the induced current density decreases exponentially inward from the surface as $j(x) = j_0 e^{-x/\delta}$, where δ is the skin depth of the material, given by Eq. (10.8). Figure 10.9 illustrates the behavior of this current density. Under many circumstances it is useful to approximate the exponential distribution with a constant current density, extending down to the depth δ, as shown by the shaded rectangle in the figure. In this manner, the total current in the conductor is the same and only the distribution is changed slightly. As noted in Eq. (10.8), when the frequency increases, the skin depth decreases. The skin depth is less in materials having a higher conductivity and a higher permeability.

For planar shields that are many skin depths thick, the thickness expressed in the number of skin depths gives an indication of the shielding capacity of the material, due to the exponentially decreasing current density. In the case of thinner shields, the situation is more complicated. Two shields that are both one skin depth thick can give very different shielding. However, one must be careful of thinking of shielding only in terms of a skin-depth attenuation of the current. Significant field reduction can occur from a *reflection* of the external field on the shield surface. Moreover, the actual shape of the shield can have an effect on the current distribution, so that the simple exponential form is not always appropriate. More will be said about these issues later in the chapter.

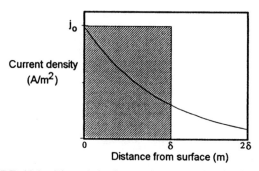

FIGURE 10.9 Plot of the internal current density in a shield.

10.3.2.3 Plane-Wave Shielding by an Infinite Metal Plate.

To further understand the shielding mechanisms, it is useful to study the simple example in which a plane electromagnetic wave produced by a distant source impinges on an infinite metal plane of finite thickness.† This example is well suited for illustrating the dependence of the shielding on the thickness of the metal and on the frequency of the incident field. Figure 8.5 can be used to illustrate the geometry for this problem, assuming that the earth is replaced by a plate, that the plate has a thickness Δ and that there is free space on each side of it. The incident field arrives with an elevation angle ψ, measured between the incident k-vector to the shield as shown in the figure, passes through the plate, and departs with the same angle on the shielded side.

Some of the energy in the incident field will be reflected back to the volume containing the source, some will penetrate through to the other side of the shield, and the rest will be transformed into heat inside the metal. We define the shielding factor S as the ratio between the field strength of the incident field (E^{inc} or H^{inc}) and the penetrating field (E^t or H^t) on the screened side, as in Eq. (10.6). Because the fields on the source and the shielded sides of the plate are both plane waves, the E and H fields in these regions are related by the impedance of free space Z_0 as $|\mathbf{E}| = Z_0 |\mathbf{H}|$. This implies that for this special case, the shielding factor S_E computed using the E-fields will be the same as S_H computed for the H-fields.

The shielding factor does depend on the polarization of the incident field, however. As shown in Figure 8.5, the incident plane wave can be decomposed into two components, one with the E-field vector parallel to the ground plane (denoted as horizontal polarization in the figure), and another with the H-field parallel to the plane (vertical polarization). For each polarization component, the reflection and transmission through the slab can be analyzed using the transmission line analogy as described in [10.7]. For the slab, the shielding factor can be expressed in a standard form as

$$S_H = \cosh(\sqrt{j\omega\tau_d}) + F \sinh(\sqrt{j\omega\tau_d}) \tag{10.9}$$

where τ_d is a characteristic diffusion constant for the shield wall having the dimensions of time and is given by

$$\tau_d = \mu\sigma\Delta^2 \quad \text{s} \tag{10.10}$$

The term in parentheses in Eq. (10.9) can also be written as $\sqrt{j\omega\tau_d} = \sqrt{j\omega\mu\sigma\Delta^2} = \gamma\Delta$, where γ is the propagation constant of the EM field through the shield wall under the assumption that $\sigma \gg \omega\epsilon$. Figure 10.10 plots this relationship for the diffusion time versus thickness for various shielding materials.

---

† Note that this is a non-physical example, since all realistic shields are of finite extent. As a result, some of the shielding attenuation numbers that will be obtained in this section - at times more than 100 dB - are unrealistically high. Actual (i.e., physical) shields typically have much smaller field attenuation values.

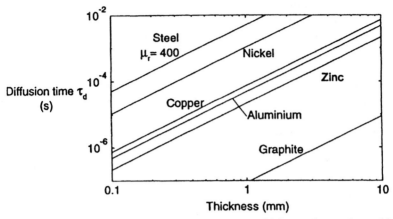

FIGURE 10.10 Plot of the diffusion time versus thickness for various shielding materials.

The factor F depends on the frequency, the shield material, the angle of incidence, and the polarization. For a horizontally polarized incident field F is expressed as

HORIZONTAL POLARIZATION

$$F = \frac{1}{2}\left(\frac{f_s}{\sin\psi} + \frac{\sin\psi}{f_s}\right) \qquad (10.11a)$$

Similarly, for a vertically polarized incident field, the function F is:

VERTICAL POLARIZATION

$$F = \frac{1}{2}\left(f_s \sin\psi + \frac{1}{f_s \sin\psi}\right) \qquad (10.11b)$$

where

$$f_s = \frac{Z_0}{Z_s} \approx \frac{1}{\sqrt{j\omega\epsilon_0\mu_r/\sigma}} = \frac{(1-j)(\lambda_0/2\pi)}{\mu_r\sigma} \qquad (10.12)$$

The term Z_s is the wave impedance in the shield material, given by $Z_s = \sqrt{j\omega\mu_r\mu_0/(j\omega\epsilon_r\epsilon_0 + \sigma)}$ and Z_0 is the wave impedance of free space $\sqrt{\mu_0/\epsilon_0} \approx 377\,\Omega$. The term $\lambda_0 = c/f$ is the wavelength in free space.

The formulas above require that both δ and Δ be much larger than the parameter $1/(Z_0\sigma)$. For good electrical conductors at low frequencies, the wavelength λ is much larger than the skin depth δ, and consequently, the parameter f_s is very large. Under this assumption, Eq. (10.11) becomes

10.3 SHIELDING MECHANISMS

HORIZONTAL POLARIZATION

$$F \approx \frac{1}{2}\frac{f_s}{\sin \psi} \qquad \lambda \gg \delta \qquad (10.13a)$$

VERTICAL POLARIZATION

$$F \approx \frac{1}{2} f_s \sin \psi \qquad \lambda \gg \delta \qquad (10.13b)$$

These formulas for S_H look deceptively simple, but on closer inspection we see that they contain hyperbolic functions of complex arguments. A bit of algebra can put Eq. (10.9) into a different form:

$$S_H = \left[\frac{1+F}{2}\right]\left[1 + \frac{1-F}{1+F}e^{-2\sqrt{j\omega\tau_d}}\right](e^{-\sqrt{j\omega\tau_d}}) \qquad (10.14)$$

in which the first term corresponds to an attenuation due to the reflection at the front surface of the slab, the middle term corresponds to multiple reflection loss within the material, and the last term corresponds to the exponential propagation loss in the material (i.e., the skin-depth attenuation).

Since we are usually interested in only the absolute value of S_H, we can derive more applicable formulas by approximating the hyperbolic functions with simpler functions. We presuppose that $\lambda/2\pi \gg \mu_r\delta$ and assume a perpendicular incidence ($\theta = 0°$) so that all field components are parallel to the shield. In the case of thick metal plates, the formula for the shielding is simplified as follows:

$$|S| \approx \frac{\sqrt{2}}{4}\frac{(\lambda/2\pi)}{\mu_r\delta} e^{\Delta/\delta} \qquad \text{for } \Delta > 1.3\delta \qquad (10.15a)$$

and for thin plates or foils:

$$|S| \approx \frac{(\lambda/2\pi)\Delta}{\mu_r\delta^2} = \frac{Z_c\sigma\Delta}{2} \qquad \text{for } \frac{\pi\mu_r\delta^2}{\lambda} = \frac{1}{Z_0\sigma} \ll \Delta < 1.3\delta \qquad (10.15b)$$

It is also of interest to illustrate how the shielding depends on the frequency. From Eq. (10.14) we can see that the shielding is constant from low frequencies up to frequencies at which the skin depth approximately equals the shield thickness. When the frequency increases further, Δ becomes larger than 1.3δ, and according to Eq. (10.15a), the attenuation increases exponentially with the frequency. It should be noted, however, that this line of argument presupposes that the shield and radius of curvature of the incident wave is larger than $\lambda/2\pi$.

Figure 10.11 illustrates the frequency dependence of a large, thin copper foil. The figure shows the calculated shielding curves according to the

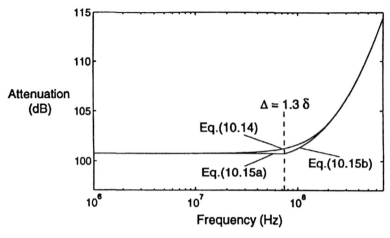

FIGURE 10.11 EM shielding of a plane wave normally incident on an infinite copper plate with $\Delta = 0.01$ mm.

different formulas. The upper graph is calculated in accordance with the exact formula (10.14), whereas the two lower graphs are taken from the approximate formulas as indicated in (10.15). As can be seen, the deviation is less than 1 dB.

In the case of thin-metal foils, according to Eq. (10.15b) the shielding increases proportionally to the thickness of the metal, up to a thickness of about one skin depth. It is a common misconception that the shielding of a metal foil should be approximately $e^{\Delta/\delta}$ when the thickness is on the order of the skin depth. As mentioned earlier, this misconception neglects the effects of reflection on the field reduction. To illustrate the importance of this reflection contribution, consider the shielding factor for a copper foil at 1 MHz with a thickness of one skin depth ($\delta = 47$ μm). Evaluating Eq. (10.14) gives a shielding factor of $e^{13.5} \approx 10^6$. This should be compared with $e^1 \approx 2.7$, which would have been obtained from considering only the attenuation of the eddy currents in the shield material. Put in another way, due to reflections from the surface, a 0.05-mm-thick foil gives a reduction that is equivalent to an exponential attenuation in 13.5 skin depths, which is equal to the current attenuation through 1 mm of copper. Such large values of the shielding factor show that in most cases it is sufficient to have thin metal foil shields. This is illustrated in Figure 10.12, which shows the shielding efficiency of a large, flat copper shield of several different thicknesses.

Equation (10.15b) shows that a high conductivity is the most important property of thin metal shields. As long as reflection is the most important attenuation mechanism, the shielding is independent of μ_r. At high frequencies, when absorption (attenuation) in the shield material becomes dominant, a large μ_r contributes to a rapidly increasing shielding effective-

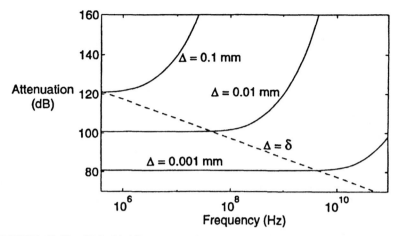

FIGURE 10.12 EM shielding of a plane wave normally incident on an infinite copper plate of different thicknesses.

ness. This can be seen clearly from a comparison between shields made of different materials, as illustrated in Figure 10.13.

If some other material with a considerably lower conductivity than metal is used as a shield, the shielding from the first skin depth will be of less importance. Seawater with conductivity $\sigma \approx 3$ S/m has a skin depth at 1 MHz of 3 dm. The shielding at a depth of 3 dm is only a factor of 169, which

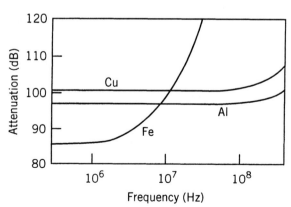

FIGURE 10.13 EM shielding of a plane wave normally incident on an infinite plate of different materials ($\Delta = 0.01$ mm).

corresponds to an exponential attenuation in a layer just over 5 skin depths thick, which can be compared with the 13.5 skin depths for the copper foil in the preceding example.

In these shielding examples, we have considered only a perpendicular incidence of the field. Equation (10.13) permits the consideration of plane waves incident at an arbitrary angle θ. The shielding factor decreases with the angle of incidence for the case where the magnetic field vector is parallel to the shield, and it increases with θ when the electrical field vector is parallel to the shield. Later in this chapter, we shall see that shielding does not change much between different canonical geometries, which is why the simple reasoning presented above is useful in illustrating that the shielding mechanisms must be understood to have a command of the more general shielding problem.

10.3.2.4 Plane Wave Shielding by Two Infinite Parallel Plates.
Another case of shielding by infinite plates is shown in Figure 10.14, in which two infinite plates both of thickness Δ are located a distance $2a$ apart [10.8]. For a normally incident field ($\theta = 0°$) the H-field in the region between the two plates is independent of polarization and is again given by Eq. (10.9), but with the term F being given by

$$F = \frac{1+j}{\mu_r} \frac{a}{\delta} = \frac{a}{\mu_r \Delta} \sqrt{j\omega\tau_d} \qquad (10.16)$$

10.3.2.5 Plane-Wave Shielding by a Conducting Mesh.
At times, shielding is accomplished by using a wire mesh instead of a completely solid shield. At low frequencies, a mesh shield behaves in the same way as a homogeneous metal shield of the same material, but with only one-half the amount of material per unit surface of the mesh. The reason why only half of the mesh material is used when determining the equivalent thickness of the planar sheet representing the mesh is because the surface impedance of the mesh is not affected by the wires in running perpendicularly to the shield

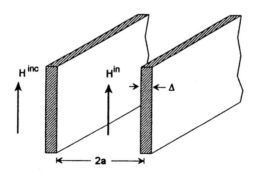

FIGURE 10.14 Two infinite parallel plates.

current. Consequently, one half of the wires in the mesh have no effect on the shielding efficiency.

For increasing frequencies, the inductive behavior of the net shield will give less shielding than the homogeneous shield. The external fields are able to penetrate the holes in the mesh, much as in the case of a braided cable shield, described in Chapter 9. For frequencies that are lower than a fixed frequency f_{min}, which depends on the mesh properties, the attenuation of plane waves by a planar net shield is constant, irrespective of the frequency. As the frequency increases, the mesh attenuation decreases as $1/f$, or in other words, at 20 dB per decade. The high-frequency limit for the shielding is denoted by f_{max}, which occurs when the shield begins to be completely transparent. This frequency is determined purely by the surface inductance of the mesh.

Casey [10.9] develops a theory for mesh shields. As shown in Figure 10.15, the mesh is modeled by two sets of parallel wires, of diameter d and separation a. The wires are assumed to be bonded electrically at the junctions. The mesh surface is characterized by a surface impedance of the form

$$Z_s = R_s + j\omega L_s \quad \Omega \qquad (10.17)$$

where the surface resistance and inductance terms are given by

$$R_s = \frac{4a}{\pi d^2 \sigma} \frac{\sqrt{j\omega \tau_w} I_0(\sqrt{j\omega \tau_w})}{2 I_1(\sqrt{j\omega \tau_w})} \qquad (10.18a)$$

$$\approx \frac{4a}{\pi d^2 \sigma} \quad \text{for } \omega \tau_w < 1 \quad \Omega$$

$$L_s = \frac{\mu_0 a}{2\pi} \ln \frac{1}{1 - e^{-\pi d/a}} \quad H \qquad (10.18b)$$

In these expressions, the term $\mu_0 = 4\pi \times 10^{-7}$ H/m, τ_w is the diffusion time in the wire material, given by $\tau_w = \mu_r \sigma d^2/4$, and $I_0(\cdot)$ and $I_1(\cdot)$ are modified Bessel functions.

For the case of a horizontally polarized field incident on a horizontally

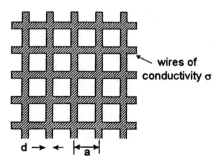

FIGURE 10.15 Geometry of the wire mesh surface.

orientated planar net (see Figure 8.5), Casey provides an expression for the transfer function of the mesh, $T = |H^{in}|/|H^{inc}|$. In our notation of this chapter, we are interested in the shielding function $S_H \equiv 1/T$, which in this case has the simple expression

$$S_H = 1 + \frac{Z_0}{2Z_{sh} \cos \theta} \tag{10.19}$$

where Z_{sh} is the effective mesh impedance for a horizontally polarized field $Z_{sh} = Z_s$ and Z_0 is the free-space wave impedance. (Recall that the angle $\theta = \pi/2 - \psi$, where ψ is shown in the Figure 8.5.)

For the vertically polarized incident field, the mesh impedance is given by

$$Z_{sv} = Z_s - \frac{j\omega L_s}{2} \sin^2 \theta$$

$$S_v = 1 + \frac{Z_0 \cos \theta}{2Z_{sv}} \tag{10.20}$$

Notice that Eqs. (10.19) and (10.20) are identical for normal angles of incidence ($\theta = 0°$ or $\psi = 90°$). For nearly normal angles of incidence the low break frequency f_{min} is determined by the ratio between the surface resistance of the net shield and the surface inductance:

$$f_{min} \approx \frac{R_s}{2\pi L_s} = \frac{2}{\tau_w} \left[\ln(1 - e^{-\pi d/a})^{-1}\right]^{-1} \quad \text{Hz} \tag{10.21}$$

Furthermore, the high-frequency limit f_{max} is defined where the asymptote of the logarithmic attenuation curve has the value 0 dB. This is the frequency where $2Z_s = Z_0$, and by neglecting the wire loss, the frequency is given by

$$f_{max} \approx \frac{Z_0}{4\pi L_s} = \frac{Z_0}{2\mu_0 a} \left[\ln(1 - e^{-\pi d/a})^{-1}\right]^{-1} \quad \text{Hz} \tag{10.22}$$

Figure 10.16 shows the plane-wave attenuation of a copper mesh shield. The break frequencies are $f_{min} \approx 5 \times 10^5$ Hz and $f_{max} \approx 4 \times 10^9$ Hz. For frequencies below f_{min} the attenuation is constant, irrespective of the frequency. For frequencies between f_{min} and f_{max}, the attenuation decreases inversely with the frequency. Above f_{max} the shielding provided by the mesh is small. It should be remembered that the simplified formulas in this section apply to large volumes in which the shield wall and bending radius are very large compared to the wavelength. Smaller volumes can have a different shielding behavior and are dealt with in Section 10.4.

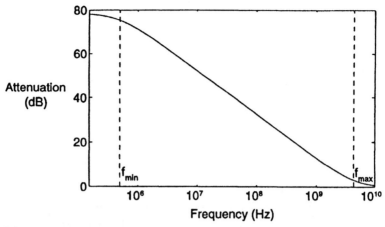

FIGURE 10.16 Shielding at normal incidence provided by a copper wire mesh with the parameters $\sigma = 58 \times 10^6$ S/m. wire diameter $d = 0.1$ mm and wire separation $a = 1$ cm.

10.3.3 Summary of Shielding Dependence on Frequency

For a plane wave that strikes a thin planar shield with a bending radius that is larger than $\lambda/2\pi$, and which is thinner than one skin depth, the attenuation is independent of the frequency up to frequencies where the penetrating field from openings begins to dominate. This is what is happening at the frequency f_{min} in the case of the shield mesh. If the shield is thicker, the attenuation is constant at low frequencies and then begins to grow exponentially as the frequency is increased. Finally, the effects of the fields penetrating through the openings become important and the shielding decreases.

A general rule for estimating the frequency dependence of eddy-current shielding for enclosures can now be postulated. At low frequencies, the shielding is proportional to the frequency up to the point at which any of the following situations occur:

- The wavelength is no longer much larger than the bending radius of the shield.
- The thickness of the shield is the same as a skin depth.
- The openings and imperfections that are always present in a shield allow so much field to penetrate that it dominates the field diffused through the metal.

Details of the low-frequency behavior for volume shields now are developed next.

10.4 VOLUMETRIC SHIELDS

The cases of planar shields discussed in previous sections are relatively straightforward to develop and use, but as mentioned, they can provide unrealistic responses for the shielding effectiveness due to their infinite structure. A better model is usually found in one of the canonical shielding problems that have been discussed by King [10.10], and later by Kaden [10.8], Dahlberg [10.11], and Lee [10.12].

10.4.1 Closed, Homogeneous Metal Shield

Many practical shielding problems involving a closed shield can be solved if the shield is approximated by a spherical or cylindrical shape. For such canonical problems, the wavelength of the incident field and distance to the source must be large in relation to the extent of the shield, so that the approaching field can be regarded as being homogeneous. In the vicinity of corners and other parts of the shield with a small bending radius, the approximation will be worse, but usually this is of no practical importance.

Both frequency- and time-domain solutions are needed for shielding problems. As with the other transient problems discussed in this book, the time-domain solution is obtained by performing a Fourier transform on the spectral data.

10.4.1.1 Shielding of Time-Harmonic Fields.
For the time-harmonic shielding problem, we can again define the H-field shielding effectiveness S_H as the ratio between the incident magnetic field and the penetrating magnetic field, as in Eq. (10.6). It is usually true that the penetrating electrical field is acceptably low if the penetrating magnetic field strength is small enough. Consequently, S_H is generally used as the sole indicator as to the shielding behavior of the enclosure.

Should the internal E-field behavior be of concern, the ratio between the electric fields can also be used to define an E-field shielding effectiveness S_E. The terms *H-field shielding effectiveness* and *E-field shielding effectiveness* refer to specific measures of eddy-current shielding and must not be confused with the more general topics of electrical and magnetic shielding. Both S_H and S_E are valid measures of enclosure shielding, the only difference between them being that they show the relationship between two different quantities. Since the magnetic shielding efficiency S_H is usually the more important of the two, we discuss it here in detail.

10.4.1.1.1 Evaluation of the H-Field Shielding Effectiveness.
Considering again a spherical shell with permeability $\mu = \mu_0 \mu_r$, permeability $\epsilon = \epsilon_0 \epsilon_r$, and electrical conductivity σ, as shown in Figure 10.4, the rigorous expression for

the magnetic field transfer function can be expressed as

$$S_H = \frac{b}{a} \left\{ \begin{array}{l} \cosh \gamma\Delta \left\{ \frac{2}{3}\left(1 + \frac{b}{2a} + \frac{1}{2(\gamma a)^2} - \frac{1}{2(\gamma a)(\gamma b)}\right) + \frac{4}{3}\left(\frac{1}{\gamma b} - \frac{1}{\gamma a}\right)\left[\frac{\mu_r}{2\gamma a} + \frac{1}{4\mu_r}\left(\gamma b - \frac{1}{\gamma a}\right)\right] \right\} \\ + \sinh \gamma\Delta \left\{ \frac{1}{3\mu_r}\left[\gamma b + \frac{1}{\gamma a}\left(\frac{b}{a} + \frac{a}{b} - 1\right) + \frac{1}{\gamma b(\gamma a)^2}\right] + \frac{2}{3}\left[\frac{1}{\gamma b} - \frac{1}{2\gamma a}\left(1 + \frac{b}{a}\right) + \frac{1}{2(\gamma a)^2 \gamma b}\right] \\ + 2\frac{\mu_r}{3}\left[\frac{1}{\gamma a} - \frac{1}{(\gamma a)^2 \gamma b}\right] \right\} \end{array} \right\}$$

(10.23a)

where a and b are the outer and inner radii of the sphere, and $\Delta = a - b$ is the shell thickness. In this expression the term γ is the wave propagation constant in the shield material and is given by

$$\gamma = j\omega\sqrt{\mu\epsilon}\left(1 + \frac{\sigma}{j\omega\epsilon}\right)^{1/2}$$

$$\approx \sqrt{j\omega\mu\sigma} \quad \text{for good conductors}$$

(10.23b)

For shields of sufficiently high conductivity, so that the terms of orders $o(1/\gamma a)$ and $o(1/\gamma b)$ are negligible, Eq. (10.23a) becomes

$$S_H \approx \frac{b}{a} \left\{ \begin{array}{l} \cosh \gamma\Delta \left[\frac{2}{3}\left(1 + \frac{1}{2\mu_r} + \frac{b}{2a}\left(1 - \frac{1}{\mu_r}\right)\right)\right] \\ + \sinh \gamma\Delta \left[\frac{\gamma b}{3\mu_r} + \frac{2\mu_r}{3\gamma a}\left[1 + \frac{1}{\mu_r}\left(\frac{a}{b} - \frac{1}{2}\left(1 + \frac{b}{a}\right)\right)\right] + \left[\frac{1}{2\mu_r^2}\left(\frac{b}{a} + \frac{a}{b} - 1\right)\right]\right] \end{array} \right\}$$

(10.24)

This expression can be further simplified for thin shields, when $b \to a$, as discussed by Kaden [10.8]. In this reference, a comprehensive theory of shielding has been developed, and various expressions for the magnetic shielding factor S_H for several canonical shielding shapes are given for the special case of thin shields.

For each of the enclosures shown in Figure 10.17, the shield is described by a characteristic radius a and a shield thickness Δ, as well as by μ_r, ϵ_r, and σ. The general functional form for the shielding factor is given by Eq. (10.9), with the factor F in this equation being a function of the specific shape of the enclosure:

SPHERICAL SHIELD

$$F_s = \frac{1}{3}\left[\frac{(1+j)a}{\mu_r \delta} + \frac{2\mu_r \delta}{(1+j)a}\right]$$

(10.25a)

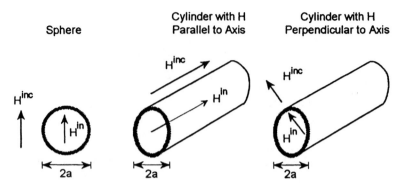

FIGURE 10.17 Simple canonical shapes for volume shielding (Δ is the thickness of the shell).

CYLINDRICAL SHIELD, H PARALLEL

$$F_{c\|} = \frac{1}{2}\left[\frac{(1+j)a}{\mu_r\delta}\right] \qquad (10.25b)$$

CYLINDRICAL SHIELD, H PERPENDICULAR

$$F_{c\perp} = \frac{1}{2}\left[\frac{(1+j)a}{\mu_r\delta} + \frac{\mu_r\delta}{(1+j)a}\right] \qquad (10.25c)$$

As in Eq. (10.16), the term $(1+j)a/\mu_r\delta$ can also be expressed as $(a/\mu_r\Delta)\sqrt{j\omega\tau_d}$ to illustrate the frequency dependence of the factor F explicitly.

As can be seen from the equations above, there is only a slight difference between the shielding efficiency of a spherical shield and that of a cylindrical shield. From this it can be concluded that the actual shape of the shield is of little importance. The formulas that are specified here for spherical and cylindrical shields can be used to calculate the shielding efficiency of a wide variety of differently shaped shields.

For electrically thin shields, the following approximations can be made to the sinh and cosh terms of Eq. (10.9):

$$\cosh\sqrt{j\omega\tau_d} \approx 1 \quad \text{and} \quad \sinh\sqrt{j\omega\tau_d} \approx \sqrt{j\omega\tau_d} \qquad \text{for } \Delta \ll \delta \quad (10.26a)$$

Similarly, for electrically thick shields, we have

$$\cosh\sqrt{j\omega\tau_d} = \sinh\sqrt{j\omega\tau_d} \approx \tfrac{1}{2} e^{\sqrt{j\omega\tau_d}} \qquad \text{for } \Delta \gg \delta \quad (10.26b)$$

Using these expressions, the approximate expressions in Table 10.3 result for the shielding of these canonical shapes.

The expressions for S_H can be modified to be useful for estimating the shielding of enclosures that are not exactly spherical or cylindrical in shape.

TABLE 10.3 Approximate Expressions for S_H of Thin and Thick Shields

Shield Configuration	Thin Shield ($\Delta \ll \delta$)	Thick Shield ($\Delta \gg \delta$)
Spherical shield	$(S_H)_{\text{sphere}} = 1 + \dfrac{j\omega\mu_0\sigma a\Delta}{3} + \dfrac{2\mu_r\Delta}{3a}$	$(S_H)_{\text{sphere}} = \left(\dfrac{1}{6}\dfrac{a}{\mu_r\Delta}\sqrt{j\omega\tau_d} + \dfrac{1}{2} + \dfrac{1}{3}\dfrac{\mu_r\Delta}{a\sqrt{j\omega\tau_d}}\right)e^{\sqrt{j\omega\tau_d}}$
Cylindrical shield, H parallel	$(S_H)_{\text{cyl}\|} = 1 + \dfrac{j\omega\mu_0\sigma a\Delta}{2} + \dfrac{\mu_r\Delta}{2a}$	$(S_H)_{\text{cyl}\|} = \left(\dfrac{1}{4}\dfrac{a}{\mu_r\Delta}\sqrt{j\omega\tau_d} + \dfrac{1}{2} + \dfrac{1}{4}\dfrac{\mu_r\Delta}{a\sqrt{j\omega\tau_d}}\right)e^{\sqrt{j\omega\tau_d}}$
Cylindrical shield, H perpendicular	$(S_H)_{\text{cyl}\perp} = 1 + \dfrac{j\omega\mu_0\sigma a\Delta}{2}$	$(S_H)_{\text{cyl}\perp} = \left(\dfrac{1}{4}\dfrac{a}{\mu_r\Delta}\sqrt{j\omega\tau_d} + \dfrac{1}{2}\right)e^{\sqrt{j\omega\tau_d}}$

This is done by introducing a characteristic number of the shield, determined by the ratio of the shielding volume V to the area of the shield, A. We can replace radius a in the equations for S_H by $3 \times V/A$ for a sphere and by $2 \times V/A$ for a cylinder. If $\mu_r\delta < a$, which is normally the case where shielding is concerned, the formula can be used to calculate the shielding efficiency for an arbitrary, "round" closed metal shield with volume V and area A as:

$$S_H \approx \begin{cases} 1 + \dfrac{j\omega\mu_0\sigma V\Delta}{A} & \text{for thin shields } \Delta \ll \delta \qquad (10.27a) \\ \left(\dfrac{1}{2} + \dfrac{V}{2\mu_r\Delta A}\sqrt{j\omega\tau_d}\right)e^{\sqrt{j\omega\tau_d}} & \text{for thick shields } \Delta \gg \delta \qquad (10.27b) \end{cases}$$

Notice that under the initial assumption that $\mu_r\delta < a$, the shielding efficiency begins at 1 (i.e., no attenuation) at very low frequencies and initially, increases linearly with frequency. For frequencies where the shield is thicker than one skin depth, the attenuation then begins to increase exponentially with the frequency.

10.4.1.1.2 Limitations of the Shielding Expressions.

The formulas for S_H apply to the attenuation of a plane-wave field, which can be understood to be the attenuation of the field from a source at a distance that is much greater than the equivalent radius a of the shield. Using reciprocity, we can also say that the same attenuation applies for interference that can affect a sensitive item of equipment at a distance much greater than a from a source enclosed in the shield. It is always difficult to quantify "much greater than," but a general rule of thumb is that a factor of 3 gives a sufficiently high level of accuracy in practical cases.

For electrical circuits situated near corners or other points where the bending radius of the shield is small, the formulas for S_H give an erroneous attenuation value. This is unimportant in most practical cases, but it is possible to define a "forbidden region" near those points with a small bending radius. This region is defined so that the generalized shield, within which all circuits are to be located, has a bending radius that is larger than $\mu_r\delta$, and preferably with a certain safety margin (e.g., so that the bending radius of the generalized shield is $3\mu_r\delta$).

The formulas for shielding efficiency are low-frequency approximations and apply only to quasistatic cases: They presuppose a field that is uniform around the entire shield. When the wavelength is not much longer than the shield, the situation becomes more complicated. Different types of resonance phenomena appear. Surface currents on the shield form traveling waves and the field inside the shield form a complicated pattern of waves. The approaching field can no longer be regarded as being independent of the shield and the penetrating field.

By far the most important limitation of the shielding formulas is that almost no shields are fully sealed. There is often leakage through cracks and bad joints. The field that forces its way in through such imperfections in the shield generally predominates over the field that diffuses through the shield material. Since leakage at high frequencies is particularly important, there is no real reason to develop other formulas for diffusion through the shield apart from the quasistatic ones.

10.4.1.1.3 Examples of the H-Field Shielding Effectiveness.
As an example of the shielding efficiency of closed metal shields for a plane-wave excitation, Figure 10.18 presents calculated responses of S_H for several different shielding geometries. For this calculation, a spherical copper shell with a fixed thickness $\Delta = 0.1$ mm and varying diameters of 1, 2, 4, 8, and 16 m were considered. Notice that these curves can also apply with a high level of accuracy to arbitrary "round" shield shapes with V/A values of 0.17, 0.33, 0.67, 1.33, and 2.67 m, respectively. Also shown on this plot is S_H for an infinite planar shield having the same thickness Δ. It is again clear from this plot that the infinite plane provides unrealistic shielding results for practical problems.

Since it is assumed that the shield has no openings, the shielding efficiency increases monotonically with frequency. For high frequencies, the attenuation increases exponentially with the frequency and can exceed 200 dB. For this high a shielding value to be achieved in practice, the shield must be extremely tight (i.e., it must have no openings.) There are always openings in practical shields that limit the attenuation at high frequencies. The effect of these openings appear at lower frequencies if their size increases, which means that an increasingly smaller part of the closed shield's attenuation is used.

As an example of the behavior of the magnetic field shielding effectiveness as the shield thickness varies, Figure 10.19 presents plots of S_H for a

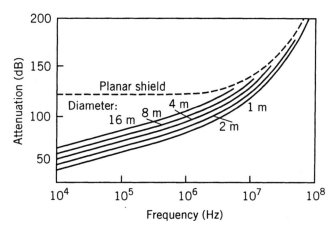

FIGURE 10.18 Example of the H-field shielding effectiveness S_H for closed copper shields of different diameters as a function of frequency ($\Delta = 0.1$ mm).

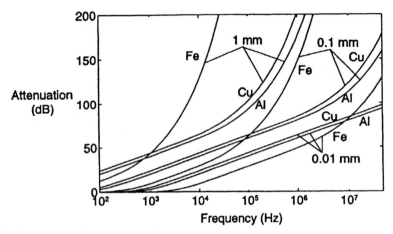

FIGURE 10.19 Example of the H-field shielding effectiveness S_H for closed spherical shields of different material and thickness as a function of frequency (diameter $2a = 3$ m).

spherical shell made of three different materials (Cu, Al, and Fe) and for three distinct shield thickness: $\Delta = 0.01$, 0.1, and 1 mm. The sphere has a diameter of 3 m, which means that these data can be applied to arbitrary "round" shield shapes with V/A value of 0.5 m.

10.4.1.1.4 Determination of the E-Field Shielding Effectiveness. Due to the duality between electrical and magnetic fields, it is possible to infer the electrical shielding efficiency S_E from the formulas for the magnetic shielding efficiency S_H in Eq. (10.9). To do this, it is only necessary to make the following exchanges in the equations:

$$S_H \Rightarrow S_E$$
$$\mu_r \Rightarrow \frac{\sigma}{j\omega\epsilon_0} \qquad (10.28)$$
$$\sigma \Rightarrow j\omega\epsilon_0\mu_r$$

10.4.1.2 Shielding of Transient Electromagnetic Fields. One important type of electromagnetic interference is a short, transient pulse. In the case of transient EMI, it is difficult to understand the shielding capacity of a metal shield by means of its thickness expressed in a number of skin depths, because the skin depth depends on frequency, and therefore, it varies in the course of a transient event. Instead, use is made of the concept of the diffusion time τ_d.

Diffusion through the metal is a relatively slow process. The time it takes for the electromagnetic field to penetrate 1 mm of aluminum corresponds to the time it takes for the field to propagate 12 km in free space. This means that by the time that the internal field has grown to some measurable value, the external field is long gone. Consequently, on the time scale used to represent the internal response, the external field appears as a delta function.

The starting point for obtaining the transient response $h^{in}(t)$ inside a shielded enclosure is the Fourier integral,

$$h^{in}(t) = \frac{1}{2\pi} \int_{-\infty}^{\infty} H^{in}(\omega) e^{j\omega t} \, d\omega \quad \text{A/m} \quad (10.29)$$

where H^{in} is the spectrum of the internal H-field. From Eq. (10.9) this field equation has the form

$$H^{in}(\omega) = \frac{1}{S_H(\omega)} H^{inc}(\omega) = \frac{H^{inc}(\omega)}{\cosh \sqrt{j\omega \tau_d} + F(\omega) \sinh \sqrt{j\omega \tau_d}} \quad \text{A/m/Hz} \quad (10.30)$$

where $F(\omega)$ is the frequency-dependent shape factor given in Eq. (10.25) and the diffusion time parameter τ_d is given by Eq. (10.10).

For most practical shielding problems, the spectrum of $1/S_H$ falls off with frequency much more rapidly than does the spectrum of the incident field, $H^{inc}(\omega)$. This is a consequence of the external field appearing as a delta function on the time scale of the internal response. Thus a good approximation is to neglect the frequency dependence of the incident field, replacing it by its dc value H_0 which is simply the area under the time-domain waveform:†

$$H_0 \equiv H^{inc}(0) = \int_0^{\infty} h^{inc}(t) \, dt \quad \text{A/m/Hz} \quad (10.31)$$

Note that this last relation comes directly from the Fourier transform counterpart to Eq. (10.29), evaluated at $\omega = 0$.

Using Eqs. (10.29) and (10.30), the transient internal field can be approximated by the integral

$$h^{in}(t) \approx \frac{H_0}{2\pi} \int_{-\infty}^{\infty} \frac{e^{j\omega t}}{\cosh \sqrt{j\omega \tau_d} + F(\omega) \sinh \sqrt{j\omega \tau_d}} \, d\omega \quad \text{A/m} \quad (10.32)$$

For practical shielding enclosures where the size is much larger than the shield thickness ($a \gg \Delta$), the function F in Eq. (10.25) may be simplified,

---

† This approach does not work for excitation waveforms that are bipolar and have a zero total impulse, because this implies that $H_0 = 0$. For these waveforms, the integral in Eq. (9.31) must be evaluated using the actual spectrum $H^{inc}(\omega)$. Many practical waveforms have the property $H_0 = 0$. For example, it is well known that an antenna cannot radiate a static field, implying that $H_0 = 0$ for this waveform.

and Eq. (10.32) can be put into the form

$$h^{in}(t) \approx \frac{H_0}{2\pi} \int_{-\infty}^{\infty} \frac{e^{j\omega t}}{\cosh\sqrt{j\omega\tau_d} + \xi\sqrt{j\omega\tau_d}\sinh\sqrt{j\omega\tau_d}} d\omega \qquad (10.33)$$

where the parameter ξ has the value

$$\xi = \ell \frac{a}{\mu_r \Delta} \qquad (10.34)$$

with $\ell = 1$ for two parallel plates, $\ell = \frac{1}{2}$ for the case of the cylindrical shell (either polarization), and $\ell = \frac{1}{3}$ for the spherical shell.

The integral in Eq. (10.33) can be evaluated numerically, or as done in [10.4], approximations to the integral can be found for two different time regimes. For early times, the internal field is approximated by

$$h^{in}(t) \approx \frac{2H_0}{\sqrt{\pi}\,\xi\tau_d} \sqrt{\frac{\tau_d}{t}}\, e^{-\tau_d/4t} \qquad \text{for } t < 0.1\,\tau_d \qquad (10.35a)$$

and for the intermediate and late times, the field is given by

$$h^{in}(t) \approx \frac{H_0}{\xi\tau_d}(e^{-t/\xi\tau_d} - 2e^{-\pi^2 t/\tau_d} + 2e^{-4\pi^2 t/\tau_d}) \qquad \text{for } t > \tau_d \qquad (10.35b)$$

Recalling that this discussion is valid for cases when the pulse length of the incident field is short in relation to the diffusion time, it is possible to calculate important characteristics of the penetrating field into a closed shield in a very simple manner. Inside the shield, the duration of the field depends on how long the induced eddy currents continue to circulate. These currents flow in the shield in the same way as in a coil surrounding the shield. At late times when $t > \tau_d$, the first term of Eq. (10.35b) dominates and the internal field can be written as

$$h^{in}(t) \approx \frac{H_0}{\tau_{fall}} e^{-t/\tau_{fall}} \qquad (10.36)$$

where the time constant for a spherical enclosure is

$$\tau_{fall} = \xi\tau_d = \frac{1}{3}\frac{a}{\mu_r \Delta}\tau_d \quad s \qquad (10.37)$$

Generalizing to an arbitrary enclosure of volume V and area A, the equivalent radius of the enclosure is $a_{eq} = 3V/A$ and the field decay time constant becomes

$$\tau_{fall} = \frac{V}{\mu_r \Delta A}\tau_d = \frac{V}{A}\mu_0 \sigma \Delta \quad s \qquad (10.38)$$

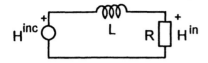

FIGURE 10.20 Equivalent circuit for representing H-field within the enclosure.

By defining an equivalent resistance of the enclosure R as

$$R = \frac{1}{\sigma \Delta} \quad \Omega \tag{10.39a}$$

and an equivalent inductance as

$$L = \mu_0 \frac{a_{eq}}{3} = \frac{\mu_0 V}{A} \quad H \tag{10.39b}$$

Note that this cavity inductance depends only on the geometry of the shield—the material is of no importance.

The field decay time constant is then given as $\tau_{fall} = L/R$, and Eq. (10.36) becomes

$$h^{in}(t) \approx H_0 \frac{R}{L} e^{-(R/L)t} \tag{10.40}$$

which is the solution of the RL circuit analog shown in Figure 10.20. Notice that in the case of metal shields this decay time constant is much greater than the pulse length of the approaching field, the energy in the excitation transient field has been distributed over a much larger time interval. As a consequence, the peak value will be correspondingly lower. It is seen that the permeability of the material is of no importance in determining this decay-time constant and the peak value of the inner field. The most important property of the shield at late times is a high conductivity. Using such circuit representations, models for the SPICE code which permit calculation of the H-field attenuation at different frequencies and in different locations inside the shielded volume have been developed for parallel-plate [10.13,10.14] or cylindrical shields [10.15].

Table 10.4 summarizes several important parameters of the penetrating magnetic field $h^{in}(t)$. For these formulas to be valid, the incident field $h^{inc}(t)$

TABLE 10.4 Parameters Representing the Approximate Internal Transient H-Field[a]

Peak Value (A/m)	Maximum Derivative (A/m/s)	10–90% Rise Time (s)	$1/e$ Fall Time (s)
$\frac{R}{L} H_0$	$6 \frac{R}{L} \frac{H_0}{\tau_d}$	$\frac{\tau_d}{4}$	$\frac{L}{R}$

[a] Note the definition of the term $H_0 = \int_0^\infty h^{inc}(t)\, dt$.

must have a duration shorter than the diffusion time τ_d, and the wall thickness Δ must be small compared to the dimensions of the box. For an example of transient shielding, consider the case of a cube with 1-m sides and walls made of 0.2-mm aluminum foil, as shown in Figure 10.21. Figure 10.10 shows that the diffusion time for a 0.2-mm layer of aluminum foil is approximately 2 ms. The conductivity and relative permeability of aluminum are provided in Table 10.2, and with these values, it is possible to apply to compute the internal waveform characteristics from Table 10.4.

For an incident H-field pulse with a duration of 1 µs, the calculations indicate that even aluminum foil significantly attenuates the peak value and derivative of the internal field. The decay time of the penetrating field is about 1.5 ms. If the pulse length of the approaching field is less than 1 µs, the peak value of the field inside the box will be approximately 1/50,000 of the peak value outside. The rise time inside the box will be 0.5 µs, a value that depends on the diffusion time.

One important parameter needed when assessing the effects of shielding against external transient fields is the time derivative of the magnetic field exciting the electronics within the shield, since the coupling to a circuit increases proportionally with frequency up to the resonance frequency. The peak value of the time derivative of the internal field is seen to be in no way dependent on the derivative of the incident field.

To check the simple formulas and to illustrate in more detail the transient behavior of the shielding, the penetrating field into the aluminum box has been calculated using the formula given in Eq. (10.30). For this example, an incident plane wave with a double-exponential time dependence shown in Figure 10.22 is assumed. The rise time is about 3 ns and the pulse length around 20 ns.

The spectral densities of the external and internal H-fields are presented in Figure 10.23. These do not differ very much for frequencies below 100 Hz. From a few 100 Hz and upwards it can be seen how the shielding efficiency increases linearly with frequency. Also noted in this figure is the fact that the incident H-field spectrum is well represented by a constant value, H_0, which

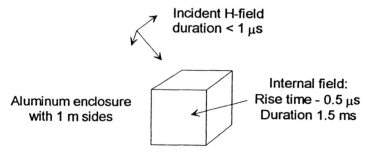

FIGURE 10.21 Shielding example with a cube with 1-m sides and walls made of 0.2 mm aluminum.

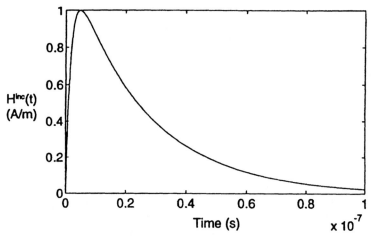

FIGURE 10.22 Time behavior of the incident H-field exciting the aluminum box.

implies that a delta-function representation of the external field is a good approximation.

The computed late-time internal transient H-field is shown in Figure 10.24a. Note that the length of the x-axis in this figure is 80,000 times longer than that of Figure 10.22b. From this plot we see that a transient field can be effectively attenuated without there being any significant field attenuation at very low frequencies. However, because the two spectra in Figure 10.23 have the same value as $\omega \to 0$, the area under this curve is equal to the area under the excitation waveform, H_0. The early-time behavior of the waveform is shown in Figure 10.24b. From these plots it can be seen that the fall time and

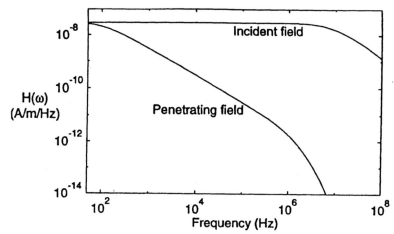

FIGURE 10.23 Internal and external H-field spectral responses for incident and penetrating fields for Figure 10.21.

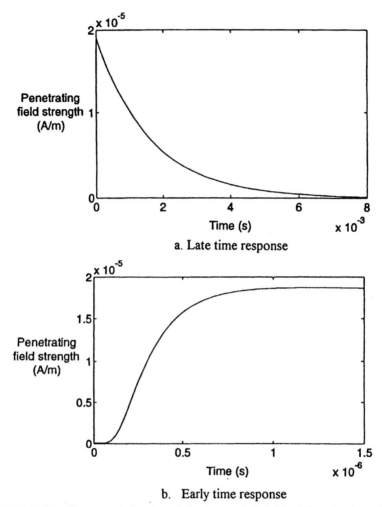

FIGURE 10.24 Transient behavior of the internal H-field of the aluminum box.

the peak value agree with those calculated from the simple formulas in Table 10.3.

10.4.2 Closed Metallic Mesh Shield

In Section 10.3.2.5, the shielding of a plane wave by a large planar mesh shield was discussed. If we deform such a mesh into a finite-volume shield, the shielding efficiency will depend on the overall size of the shield, as is the case with the homogeneous shield. One important difference, however, is that the net shield has openings, and at high frequencies, this restricts the attenuation.

A precondition for describing the shielding efficiency of such a shield with a reasonably simple formula is that the wavelength of the incident field must be large in comparison to the extent of the shield. In the quasistatic case, the field will be uniform around the entire shield. If, at the same time, the mesh size of the shield is assumed to be small compared to the dimensions of the shield, the following is valid [10.9]:

$$S_H = 1 + \frac{j\omega\mu_0 V}{A(R_s + j\omega L_s)} \quad \text{for } \omega < \frac{3c}{2a} \quad (10.41)$$

where V and A are the volume and area of the enclosure, respectively, and the surface resistance R_s and surface inductance L_s of the net shield are determined in accordance with formulas (10.18). This expression is valid for frequencies $\omega < 3c/2a$, where c is the speed of light and a is the radius of the enclosure.

As an example of how the shielding efficiency for closed mesh shields varies with frequency, Figure 10.25 plots the attenuation of a mesh shield and a homogeneous shield (foil), both made of copper and having a spherical shape with a diameter of 3 m (arbitrary shield shape with $V/A = 0.5$ m). The mesh shields have wire diameters and mesh sizes as indicated in the figure, and the homogeneous shield has a thickness of 0.1 mm.

These curves show that a closed net shield has a shielding efficiency that increases proportionally with the frequency for low frequencies. In this frequency regime, the same shielding efficiency is obtained for a homoge-

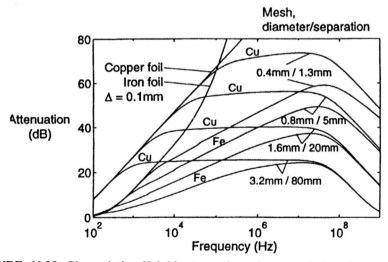

FIGURE 10.25 Plots of the H-field attenuation of a spherical enclosure with diameter = 3 m made of foil or mesh with double the amount of metal. The metals are copper with $\sigma = 58 \times 10^6$ S/m and iron with $\sigma = 10 \times 10^6$ S/m and $\mu_r = 400$.

neous shield, with half the material quantity of the net shield per surface unit (see Section 10.3.2.5). At a frequency determined by the surface inductance and surface resistance of the mesh, the shielding levels off to a maximum value. A low surface impedance (i.e., high conductivity) and a finely meshed net with a large quantity of material per surface unit gives a high level of shielding. The maximum value of shielding has the upper bound of $\mu_0 V/AL_s$ and is dependent primarily on the geometric dimensions of the shield and mesh. In the case of high frequencies when the wavelength is less than the size of the shielded space, the shielding efficiency will decrease as described in Section 10.3.2.5. When the frequency is so high that the wavelength is approximately the same as the size of the net mesh, the shielding efficiency will approach zero with a limit frequency of $f_{max} = Z_0/4\pi L_s$.

For a mesh shield, the shielding formulas apply only at distances sufficiently far from the individual holes so that the influence of an individual hole is negligible. It is obvious that the attenuation in the middle of a mesh opening is poor. However, at a distance from the shield surface of only a few mesh sizes on each side of the shield, the attenuation begins to be in line with what the formulas indicate. In practical applications, a forbidden area should be created inside the shield where no equipment installation is allowed. A general rule of thumb is that the thickness of this forbidden area should be one or two mesh sizes. If, for example, a closed mechanical reinforcement such as rebar is used as a mesh shield, the surrounding wall of concrete can serve as the forbidden area.

Since rebar reinforcement in a building can often be used to create an external shield at a low price, Figure 10.26 shows the shielding efficiency of a

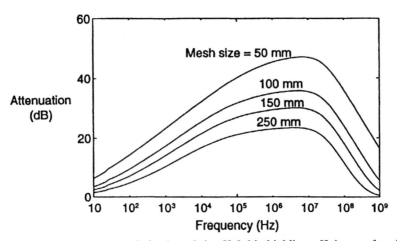

FIGURE 10.26 Frequency behavior of the H-field shielding efficiency of a closed iron mesh shield for varying mesh sizes (enclosure diameter = 6 m, mesh conductor diameter = 16 mm).

closed mesh shield with varying mesh sizes. The shields are spherical in shape with a diameter of 6 m (or arbitrary shield shapes with $V/A = 1$ m). The reinforcement rods are made of iron with a diameter of 16 mm. The mesh sizes are 5, 10, 15, and 25 cm. Measurements made on reinforcement rod cages show that measured shielding values concur well with the foregoing theoretical values.

Since the exterior walls of facilities are often built with several layers of rebar reinforcement, higher values for shielding efficiency than those given by the theoretical expression in Eq. (10.41) for a single layer can be expected. However, using the same attenuation for each of the layers in a multilayer shield provides an exaggeration of the total shielding effect.

10.4.2.1 Induced Responses Within a Mesh-Protected Shield.
CW measurements have been conducted on a reinforced and protected facility where the external layer of reinforcement is interconnected to form an electromagnetic shield [10.16]. Other layers of mechanical reinforcement have been built in the traditional way, with no consideration given to EM shielding. The shield was created by forming the reinforcement rods to give a mesh size of 20 cm. Most of the pipe lead-throughs penetrating the shield were concentrated on one common entry plate.

As an illustration of the shielding effect of such a reinforced facility, Figure 10.27 shows induced current in a rectangular loop with dimensions of 2.5×6 m outside and inside the plant, normalized to a unit incident E-field (i.e., $H^{inc} = 1/377$ A/m). It can be seen from the curves that the current will be approximately 60 dB lower inside the plant.

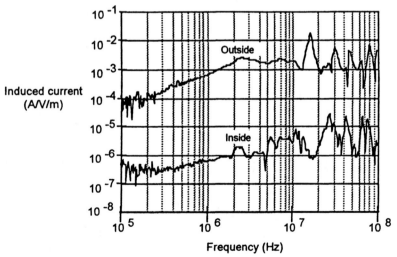

FIGURE 10.27 Measured normalized short-circuit current in a loop (2.5 m $\times$ 6 m) outside and inside the reinforcement structure.

10.5 SHIELDING OF NON-PLANE WAVE FIELDS

10.5.1 Overview of Near-Field Shielding

Shielding can be viewed as an attenuation of the EM coupling between two circuits. The formulas presented in previous sections of this chapter apply for cases when the circuits are widely separated, so that the EM fields produced by the source circuit are uniform over the shield protecting the second circuit. If both of the circuits are close to the shield, with a separation distance that is much smaller than the shield's characteristic length, the significance of the dimensions of the shield in determining the overall shielding becomes less important. In such a case we can replace the shield enclosure with a large, planar shield. The characteristic length of interest thus becomes the distance between the circuits and the shield.

In somewhat simplified terms, it can be viewed that the term $\lambda/2\pi$ is replacing the characteristic length in the formulas relating to the shielding of plane waves by a planar shield in Eqs. (10.9) to (10.12). This means that in the case of thin shields, the shielding efficiency is proportional to the frequency. When the frequency increases, the wavelength decreases, and eventually it becomes comparable to the characteristic length. The field at the shield then increasingly assumes the character of a plane wave, and the shielding efficiency eventually ceases to increase with the frequency.

10.5.2 Shielding Between Two Circular Loops.

As an example of shielding between two adjacent circuits, we have chosen to describe the attenuation a plane shield provides for the connection between two circular loops, as illustrated in Figure 10.28. There is a special reason for choosing this particular example, as some standards for the measurement of shielding efficiency suggest this arrangement with loops on either side of the shield for measuring shielding efficiency of enclosures. In part (*a*) of the figure, the two loops are placed in the same plane on either side of the shield. The loops are equal in size, and the distance from the closest point of the loops to the shield is equal to the loop diameter. This configuration is used in MIL-STD 285, which stipulates that radius *a* shall be 15 cm. In part (*b*), the two loops are placed coaxially on either side of the shield. The latter arrangement is presented in the standards NSA 65-6 and NSA 73-2a for the measurement of shielding efficiency. The loops are equal in size and the distance from the closest point of the loops to the shield is equal to the loop diameter. These standards also stipulate that radius *a* shall be 15 cm.

In this case it is appropriate to define a shielding efficiency S_L as the ratio between coupling with and coupling without the shield present. Here the term *coupling* refers to the voltage induced in one of the loops by a current in the other, in other words, a transfer impedance. The load impedance in the receiving loop is assumed to be sufficiently high that currents induced in

10.5 SHIELDING OF NON-PLANE WAVE FIELDS

it provide a negligible contribution to the field. The voltage generator that drives the current in the transmitting loop should have such a low impedance that the current will remain the same regardless of the presence of the shield. In practice, it may be difficult to achieve such ideal conditions. In connection with practical measurements, anomalous results can sometimes be explained by the fact that these impedances have differed from the ideal values.

Reference [10.11] presents the general results for the shielding efficiency S_L for the loops of Figure 10.28. For a shield with good conductivity so that $\mu_r \delta^2 \ll 4a\Delta$ and $\mu_r \delta \ll 4a$, the following formula can be used:

$$S_L = \frac{\sinh[(1+j)(\Delta/\delta)]}{(1+j)(\Delta/\delta)} \frac{ja\Delta}{\mu_r \delta^2} G = \frac{\sinh \sqrt{j\omega\tau_d}}{\sqrt{j\omega\tau_d}} \frac{j\omega\tau_d}{\mu_r} \frac{a}{\Delta} G \quad (10.42a)$$

where τ_d is the diffusion time in the shield, given by Eq. (10.10). For a ferromagnetic shield with a very high μ_r so that $\mu_r \Delta^2 \gg 4a$ and $\mu_r \Delta \gg 4a$, the shielding expression is

$$S_L = \frac{\sinh[(1+j)(\Delta/\delta)]}{(1+j)(\Delta/\delta)} \frac{\mu_r \Delta}{a} F = \frac{\sinh \sqrt{j\omega\tau_d}}{\sqrt{j\omega\tau_d}} \frac{\mu_r \Delta}{a} F \quad (10.42b)$$

In these expressions the terms G and F depend on the orientation of the loops, and they have the following values:

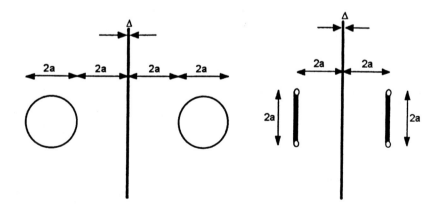

a. Circular loops in one plane   b. Coaxial circular loops

FIGURE 10.28 Geometry of two loops separated by an EM shield.

Loop Orientation	G	F
Coplanar (Figure 10.28a)	1.9	0.17
Coaxial (Figure 10.28b)	1.2	0.23

General formulas for loops of different sizes and different distances to the shield are similar to those above. The only factors that vary are the values for G and F.

For practical shields that are good conductors, Eq. (10.42a) is most applicable. For electrically thick shields the magnitude of the shielding is

$$|S_L| = \frac{1}{2\sqrt{2}} \frac{a}{\mu_r \delta} G e^{\Delta/\delta} \qquad \Delta \gg \delta \qquad (10.43a)$$

For the case of electrically thin shields, the shielding magnitude is developed in [10.11] as

$$|S_L| = 1 + \frac{\omega}{2} \mu_0 \sigma a \, \Delta G \qquad \Delta \ll \delta \qquad (10.43b)$$

We can see from these expressions that the shielding efficiency for thin shields is proportional to the frequency and shield thickness. The attenuation is also proportional to the conductivity of the material but is *independent* of its permeability. When the shield is thicker than a skin depth, attenuation increases exponentially with the frequency and shield thickness, and with *both* increasing conductivity and permeability.

Due to the differences in the terms G and F, shielding values measured in accordance with a standard method with loops in one plane will be somewhat greater than those measured in accordance with a standard method with coaxial loops. The ratio is approximately 1.6 which is equivalent to 4 dB.

Using the formulas above, we can calculate the shielding efficiency between two 30-cm-diameter loops lying in the same plane, in accordance with the test method in MIL-STD-285. For a copper foil 0.1 mm thick, the following applies:

$$|S_L| \approx 1 + 0.00656 f \qquad (10.44a)$$

for frequencies up to 440 kHz, at which point the thickness of the shield is equal to the skin depth. The attenuation is then 70 dB. In the case of higher frequencies, the shielding efficiency increases exponentially in accordance with

$$|S_L| = 1.5 \sqrt{f} e^{\Delta/\delta} \qquad (10.44b)$$

Figure 10.29 presents the computed shielding efficiency of copper and

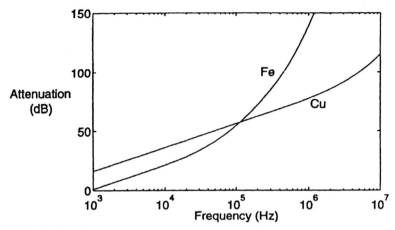

FIGURE 10.29 Shielding efficiency of a copper foil and an iron foil, as calculated from the attenuation between two circular loops in accordance with MIL-STD 285.

iron foils, both 0.1 mm thick, as a function of the frequency. These curves can be seen to coincide with the curves describing the shielding efficiency of an equivalent spherical shield with a diameter of 86 cm. Values measured in accordance with MIL-STD 285 can be interpreted as describing the shielding efficiency the measured material would provide if it formed a sphere with this diameter.

REFERENCES

10.1. Vance, E. F., and W. Graf, "The Role of Shielding in Interference Control," *IEEE Trans. Electromagn. Compat.*, Vol. EMC-30, No. 3, August 1988, pp. 294–297.

10.2. Karlsson, T., "On Grounding: Practical Procedures Based on Electromagnetic Theory," *Proceedings of the 9th International Symposium and Technical Exhibition on EMC*, Zurich, March 3–5, 1987.

10.3. Karlsson, T. "The Topological Concept of a Generalized Shield," *AFWL Interaction Note 461, Air Force Weapons Laboratory*, Kirtland AFB, NM, January 1988.

10.4. Lee, K. S. H., ed., *EMP Interaction: Principles, Techniques and Reference Data*, Hemisphere, New York, 1989.

10.5. Van Bladel, J., *Electromagnetic Fields*, McGraw-Hill, New York, 1964.

10.6. Watson, J. K., *Applications of Magnetism*, published by the author, University of Florida, Gainesville, FL, 1985.

10.7. Tyras, G., *Radiation and Propagation of Electromagnetic Waves*, Academic Press, New York, 1969.

10.8. Kaden, H, *Wirbelstroeme und Schirmung in der Nachrichtentechnik*, Springer-Verlag, Berlin, 1959.

10.9. Casey, K. F., "Electromagnetic Shielding Behavior of Wire-Mesh Screens," *IEEE Trans. Electromagn. Compat.*, Vol. EMC-30, No. 3, August 1988, pp. 298–311.

10.10. King, L. V., "Electromagnetic Shielding at Radio Frequencies," *Philos. Mag. J. Sci.*, Vol. 15, No. 97, February 1933, pp. 201–223.

10.11. Dahlberg, E., "Electromagnetic Shielding, Some Simple Formulae for Closed Uniform Shields," *TRITA-EPP-75-27*, KTH, Stockholm, December 1975.

10.12. Lee, K. S. H., "Electromagnetic Shielding," Chapter 11 in *Recent Advances in Electromagnetic Theory*, H. N. Kritikos and D. L. Jagard, eds., Springer-Verlag, New York, 1990.

10.13. Radu, S., and G. Maxim, "A SPICE Model for the Plane Parallel Shield," *Proceedings of the International Symposium on Signals, Circuits and System*, SCS'95, Iasi, Romania, October 19–21, 1995.

10.14. Radu, S., *Introducere in Compatibilitatea Electromagnetica*, Vol. 1, Ed. Gh. Asachi, Iasi, Romania, 1995 (in Romanian).

10.15. Radu, S., and M. Albulet, "A SPICE Model for the Cylindrical Shields," *Proceedings of the 3rd Electronic Devices and Systems International Conference*, EDS'95, Brno, Czech Republic, June 28, 1995.

10.16. Garmland, S., "Evaluation of the EMP Protection in the Civil Defence Facility RC Åmål," *EMTECH Report MTR-9218*, Linköping, Sweden, February 1992.

PROBLEMS

10.1 EM shielding is often discussed using a model consisting of a flat, infinite slab of conducting material, as in Section 10.3.2.3. Discuss why this model is inappropriate for many practical shielding problems and why it leads to an overestimation of shielding levels provided by real enclosures.

10.2 Consider a shield in the form of a sphere or cylinder (as illustrated in Figure 10.17) made of steel with a thickness of $\Delta = 0.15$ mm.
  (a) Compute and plot the magnetic field shielding effectiveness S_H at a frequency of $f = 10$ kHz, as a function of the enclosure size a for the three different geometries shown in the figure.
  (b) Which configuration provides the best shielding? Why?

10.3 A computer video monitor is located approximately 10 m from a railway line that is energized by a 16.7-Hz power supply. Every time a train passes, the monitor screen is disrupted, with distorted and wavy images.
  (a) Discuss the cause of this interference on the monitor.
  (b) Suggest ways of eliminating this interference.

10.4 A portable shielded enclosure with dimensions $2 \times 1.75 \times 1.4$ m is constructed from a metallized fabric having a surface resistance of 100 Ω/square and a thickness of 100 μm.

   (a) Compute the electrical conductivity of this material. (*Hint:* The concept of surface resistance is discussed in [4.7].)

   (b) Compute and plot the ratio $|H^{in}/H^{out}|$, (where H^{out} is the external H-field with the enclosure removed) for the enclosure as a function of frequency, from 10 kHz to 100 MHz.

   (c) At what frequencies will the calculations in part (*b*) become questionable?

10.5 Suppose that the enclosure in Problem 10.4 is located on a lossy earth having a conductivity and relative permittivity σ and ϵ_r and is illuminated by a vertically polarized plane-wave field with an E-field spectrum $E_0(\omega)$ and an elevation angle of incidence ψ.

   (a) Determine the internal H-field spectrum for this case.

   (b) Repeat (a) for a horizontally polarized field.

   (*Hint:* See J. P. Blanchard, et al., "Electromagnetic Shielding by a Metalized Fabric Enclosure: Theory and Experiment," *IEEE Trans. Electromagn. Compat.*, Vol. EMC-30, No. 3, August 1988, pp. 282–288.

APPENDIX A

Tables of Physical Constants

TABLE A.1 Primary Constants

Quantity	Value
μ_0 permeability of free space	$4\pi \times 10^{-7}$ H/m $\approx 1.257 \times 10^{-6}$ H/m
ϵ_0 permittivity of free space	8.85415×10^{-12} F/m $\approx 1/(36\pi) \times 10^{-9}$ F/m
c speed of light in vacuum	$1/\sqrt{\epsilon_0 \mu_0} = 2.998 \times 10^8$ m/s
Z_0 impedance of free space	$\sqrt{\mu_0/\epsilon_0} \approx 120\pi \,\Omega \approx 377 \,\Omega$

TABLE A.2 Approximate Static Relative Permeabilities of Magnetic Materials

Material	Class	Relative Permeability (μ_r)
Bismuth	Diamagnetic	0.999834
Silver	Diamagnetic	0.99998
Lead	Diamagnetic	0.999983
Copper	Diamagnetic	0.999991
Water	Diamagnetic	0.999991
Vacuum	Nonmagnetic	1.0
Air	Paramagnetic	1.0000004
Aluminum	Paramagnetic	1.00002
Nickel chloride	Paramagnetic	1.00004
Palladium	Paramagnetic	1.0008
Cobalt	Ferromagnetic	250
Nickel	Ferromagnetic	600
Mild steel	Ferromagnetic	2,000
Iron	Ferromagnetic	5,000
Silicon iron	Ferromagnetic	7,000
Mu-metal	Ferromagnetic	100,000
Purified iron	Ferromagnetic	200,000
Supermalloy	Ferromagnetic	1,000,000

Source: C. A. Balanis, *Advanced Engineering Electromagnetics*, Wiley, New York, 1989. (Copyright © 1989 John Wiley & Sons, Inc. Reprinted by permission from the publisher.)

TABLE A.3 Approximate Static Dielectric Constants (Relative Permittivities) of Selected Dielectric Materials

Material	Static Dielectric Constant (ε_r)
Air	1.0006
Styrofoam	1.03
Paraffin	2.1
Teflon	2.1
Plywood	2.1
RT/duroid 5880	2.20
Polyethylene	2.26
RT/duroid 5870	2.35
Glass-reinforced Teflon (microfiber)	2.32–2.40
Teflon quartz (woven)	2.47
Glass-reinforced Teflon (woven)	2.40–2.62
Cross-linked polystyrene (unreinforced)	2.56
Polyphenelene oxide (PPO)	2.55
Glass-reinforced polystyrene	2.62
Amber	3.0
Soil (dry)	3.0
Rubber	3.0
Plexiglas	3.4
Lucite	3.6
Fused silica	3.78
Nylon (solid)	3.8
Quartz	3.8
Sulfur	4.0
Bakelite	4.8
Formica	5.0
Lead glass	6.0
Mica	6.0
Beryllium oxide (BeO)	6.8–7.0
Marble	8.0
Sapphire	$\epsilon_x = \epsilon_y = 9.4, \epsilon_z = 11.6$
Flint glass	10.0
Ferrite (Fe_2O_3)	12.0–16.0
Silicon (Si)	12.0
Gallium arsenide (GaAs)	13.0
Ammonia (liquid)	22.0
Glycerin	50.0
Water	81.0
Rutile (TiO_2)	$\epsilon_x = \epsilon_y = 8.09, \epsilon_z = 173$

Source: C. A. Balanis, *Advanced Engineering Electromagnetics*, Wiley, New York, 1989. (Copyright © 1989 John Wiley & Sons, Inc. Reprinted by permission from the publisher.)

TABLE A.4 Typical Conductivities of Insulators, Semiconductors, and Conductors

Material	Class	Conductivity σ (S/m)
Fused quartz	Insulator	~10^{-17}
Sulfur	Insulator	~10^{-15}
Mica	Insulator	~10^{-15}
Porcelain	Insulator	~10^{-14}
Glass	Insulator	~10^{-12}
Bakelite	Insulator	~10^{-9}
Distilled water	Insulator	~10^{-4}
Fused silica	Semiconductor	~2.1×10^{-4}
Cross-linked polystyrene (unreinforced)	Semiconductor	~3.7×10^{-4}
Beryllium oxide (BeO)[a]	Semiconductor	~3.9×10^{-4}
Intrinsic silicon	Semiconductor	~4.39×10^{-4}
Sapphire[a]	Semiconductor	~5.5×10^{-4}
Glass-reinforced Teflon (microfiber)[a]	Semiconductor	~7.8×10^{-4}
Dry soil	Semiconductor	~10^{-4}–10^{-3}
Ferrite (Fe_2O_3)[a]	Semiconductor	~1.3×10^{-3}
Glass-reinforced polystyrene[a]	Semiconductor	~1.45×10^{-3}
Glass-reinforced Teflon (woven)[a]	Semiconductor	~2.43×10^{-3}
Plexiglas[a]	Semiconductor	~5.1×10^{-3}
Gallium arsenide (GaAs)[a]	Semiconductor	~8×10^{-3}
Wet Soil	Semiconductor	~10^{-3}–10^{-2}
Freshwater	Semiconductor	~10^{-2}
Human and animal tissue	Semiconductor	~0.2–0.7
Intrinsic germanium	Semiconductor	~2.227
Seawater	Semiconductor	~4.0
Tellurium	Conductor	~5.0×10^3
Carbon	Conductor	~3.0×10^4
Graphite	Conductor	~3.0×10^4
Cast iron	Conductor	~10^5
Mercury	Conductor	10^6
Nichrome	Conductor	10^6
Silicon steel	Conductor	~2.0×10^6
German silver	Conductor	2.0×10^6
Lead	Conductor	5.0×10^6
Tin	Conductor	9.0×10^6
Iron	Conductor	1.03×10^7
Nickel	Conductor	1.45×10^7
Zinc	Conductor	1.7×10^7
Tungsten	Conductor	1.83×10^7
Brass	Conductor	2.56×10^7
Aluminum	Conductor	3.96×10^7
Gold	Conductor	4.1×10^7
Copper	Conductor	5.76×10^7
Silver	Conductor	6.1×10^7

Source: C. A. Balanis, *Advanced Engineering Electromagnetics*, Wiley, New York, 1989. (Copyright © John Wiley & Sons, Inc. Reprinted by permission from the publisher.)
[a] For most semiconductors the conductivities are representative for a frequency of about 10 GHz.

APPENDIX B

Vector Analysis and Functions

B.1 MULTIPLICATIVE RELATIONSHIPS

Let **A**, **B**, **C** and **D** denote different vectors in three dimensions. Then the following relations hold:

1. $\mathbf{A} \cdot (\mathbf{B} \times \mathbf{C}) = \mathbf{C} \cdot (\mathbf{A} \times \mathbf{B}) = \mathbf{B} \cdot (\mathbf{C} \times \mathbf{A})$
2. $\mathbf{A} \times (\mathbf{B} \times \mathbf{C}) = \mathbf{B} \cdot (\mathbf{A} \cdot \mathbf{C}) - \mathbf{C} \cdot (\mathbf{A} \cdot \mathbf{B})$
3. $\mathbf{A} \times (\mathbf{B} \times \mathbf{C}) - \mathbf{C} \times (\mathbf{B} \times \mathbf{A}) = \mathbf{B} \times (\mathbf{A} \times \mathbf{C})$
4. $(\mathbf{A} \times \mathbf{B}) \cdot (\mathbf{C} \times \mathbf{D}) = (\mathbf{A} \cdot \mathbf{C})(\mathbf{B} \cdot \mathbf{D}) - (\mathbf{A} \cdot \mathbf{D})(\mathbf{B} \cdot \mathbf{C})$
5. $\mathbf{A} \times (\mathbf{B} \times (\mathbf{C} \times \mathbf{D})) = (\mathbf{B} \cdot \mathbf{D})(\mathbf{A} \times \mathbf{C}) - (\mathbf{B} \cdot \mathbf{C})(\mathbf{A} \times \mathbf{D})$
6. $(\mathbf{A} \times \mathbf{B}) \cdot ((\mathbf{B} \times \mathbf{C}) \times (\mathbf{C} \times \mathbf{A})) = (\mathbf{A} \cdot (\mathbf{B} \times \mathbf{C}))^2$

B.2 DIFFERENTIAL RELATIONSHIPS

Let a and b denote scalar functions of the independent variables (x_1, x_2, x_3), and **A** and **B** denote vector functions of (x_1, x_2, x_3), all of which have the necessary derivatives defined. Then the following hold:

1. $\nabla(a + b) = \nabla a + \nabla b$
2. $\nabla \cdot (\mathbf{A} + \mathbf{B}) = \nabla \cdot (\mathbf{A}) + \nabla \cdot (\mathbf{B})$
3. $\nabla \times (\mathbf{A} + \mathbf{B}) = \nabla \times (\mathbf{A}) + \nabla \times (\mathbf{B})$
4. $\nabla(ab) = a\nabla(b) + b\nabla(a)$
5. $\nabla(\mathbf{A} \cdot \mathbf{B}) = \mathbf{A} \times \nabla \times \mathbf{B} + \mathbf{B} \times \nabla \times \mathbf{A} + (\mathbf{B} \cdot \nabla)\mathbf{A} + (\mathbf{A} \cdot \nabla)\mathbf{B}$
6. $\nabla \cdot (a\mathbf{A}) = a\nabla \cdot \mathbf{A} + (\nabla a) \cdot \mathbf{A}$
7. $\nabla \cdot (\mathbf{A} \times \mathbf{B}) = \mathbf{B} \cdot \nabla \times \mathbf{A} - \mathbf{A} \cdot \nabla \times \mathbf{B}$
8. $\nabla \times (a\mathbf{A}) = (\nabla a \times \mathbf{A}) + a\nabla \times \mathbf{A}$
9. $\nabla \times (\mathbf{A} \times \mathbf{B}) = \mathbf{A}\nabla \cdot \mathbf{B} - \mathbf{B}\nabla \cdot \mathbf{A} + (\mathbf{B} \cdot \nabla)\mathbf{A} - (\mathbf{B} \cdot \nabla)\mathbf{A}$
10. $\nabla \times \nabla a = 0$
11. $\nabla \cdot \nabla \times \mathbf{A} = 0$
12. $\nabla \times \nabla \times \mathbf{A} = \nabla(\nabla \cdot \mathbf{A}) - \nabla^2 \mathbf{A}$
13. $\nabla \cdot (\nabla a) = \nabla^2 a$

554 APPENDIX B

B.3 INTEGRAL RELATIONSHIPS

Let **A** and f denote a single-valued vector and scalar function, respectively, defined within a volume V and on its bounding surface S. Their derivatives are assumed to be continuous within V, and the outward-pointing unit normal to the surface S is denoted by $\hat{n}$. Then we have the following theorems of Gauss:

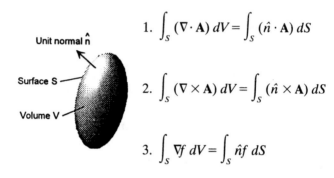

1. $\int_S (\nabla \cdot \mathbf{A}) \, dV = \int_S (\hat{n} \cdot \mathbf{A}) \, dS$

2. $\int_S (\nabla \times \mathbf{A}) \, dV = \int_S (\hat{n} \times \mathbf{A}) \, dS$

3. $\int_S \nabla f \, dV = \int_S \hat{n} f \, dS$

For a two-sided surface S with a boundary contour C and unit normal $\hat{n}$ to the surface defined in the positive direction shown below, the following theorems of Stokes hold:

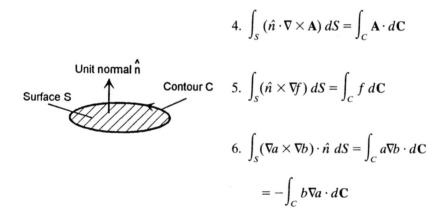

4. $\int_S (\hat{n} \cdot \nabla \times \mathbf{A}) \, dS = \int_C \mathbf{A} \cdot d\mathbf{C}$

5. $\int_S (\hat{n} \times \nabla f) \, dS = \int_C f \, d\mathbf{C}$

6. $\int_S (\nabla a \times \nabla b) \cdot \hat{n} \, dS = \int_C a \nabla b \cdot d\mathbf{C}$

$\qquad = -\int_C b \nabla a \cdot d\mathbf{C}$

B.4 EXPRESSIONS FOR DIFFERENTIAL OPERATORS

Let f represent a scalar function and **A** denote a vector function of the variables x_1, x_2, x_3, with the necessary derivates assumed to exist. Then the following are expressions for the various differential operators:

B.4 EXPRESSIONS FOR DIFFERENTIAL OPERATORS

Rectangular coordinates: $(x_1, x_2, x_3) = (x, y, z)$

1. $\nabla f = \hat{x}\dfrac{\partial f}{\partial x} + \hat{y}\dfrac{\partial f}{\partial y} + \hat{z}\dfrac{\partial f}{\partial z}$

2. $\nabla \cdot \mathbf{A} = \dfrac{\partial A_x}{\partial x} + \dfrac{\partial A_y}{\partial y} + \dfrac{\partial A_z}{\partial z}$

3. $\nabla \times \mathbf{A} = \left(\dfrac{\partial A_z}{\partial y} - \dfrac{\partial A_y}{\partial z}\right)\hat{x} + \left(\dfrac{\partial A_x}{\partial z} - \dfrac{\partial A_z}{\partial x}\right)\hat{y} + \left(\dfrac{\partial A_y}{\partial x} - \dfrac{\partial A_x}{\partial y}\right)\hat{z}$

4. $\nabla^2 f = \dfrac{\partial^2 f}{\partial x^2} + \dfrac{\partial^2 f}{\partial y^2} + \dfrac{\partial^2 f}{\partial z^2}$

5. $\nabla^2 \mathbf{A} = \nabla^2 A_x \hat{x} + \nabla^2 A_y \hat{y} + \nabla^2 A_z \hat{z}$

Cylindrical coordinates: $(x_1, x_2, x_3) = (\rho, \phi, z)$

1. $\nabla f = \hat{\rho}\dfrac{\partial f}{\partial \rho} + \hat{\phi}\dfrac{1}{\rho}\dfrac{\partial f}{\partial \phi} + \hat{z}\dfrac{\partial f}{\partial z}$

2. $\nabla \cdot \mathbf{A} = \dfrac{1}{\rho}\dfrac{\partial(\rho A_\rho)}{\partial \rho} + \dfrac{1}{\rho}\dfrac{\partial A_\phi}{\partial \phi} + \dfrac{\partial A_z}{\partial z}$

3. $\nabla \times \mathbf{A} = \left(\dfrac{1}{\rho}\dfrac{\partial A_z}{\partial \phi} - \dfrac{\partial A_\phi}{\partial z}\right)\hat{\rho} + \left(\dfrac{\partial A_\rho}{\partial z} - \dfrac{\partial A_z}{\partial \rho}\right)\hat{\phi} + \dfrac{1}{\rho}\left(\dfrac{\partial(\rho A_\phi)}{\partial \rho} - \dfrac{\partial A_\rho}{\partial \phi}\right)\hat{z}$

4. $\nabla^2 f = \dfrac{1}{\rho}\dfrac{\partial}{\partial \rho}\left(\rho \dfrac{\partial f}{\partial \rho}\right) + \dfrac{1}{\rho^2}\dfrac{\partial^2 f}{\partial \phi^2} + \dfrac{\partial^2 f}{\partial z^2}$

5. $\nabla^2 \mathbf{A} = \nabla(\nabla \cdot \mathbf{A}) - \nabla \times \nabla \times \mathbf{A}$

Spherical coordinates: $(x_1, x_2, x_3) = (r, \theta, \phi)$

1. $\nabla f = \hat{r}\dfrac{\partial f}{\partial r} + \hat{\theta}\dfrac{1}{r}\dfrac{\partial f}{\partial \theta} + \dfrac{\hat{\phi}}{r\sin\theta}\dfrac{\partial f}{\partial \phi}$

2. $\nabla \cdot \mathbf{A} = \dfrac{1}{r^2}\dfrac{\partial(r^2 A_r)}{\partial r} + \dfrac{1}{r\sin\theta}\dfrac{\partial(\sin\theta A_\theta)}{\partial \theta} + \dfrac{1}{r\sin\theta}\dfrac{\partial A_\phi}{\partial \phi}$

3. $\nabla \times \mathbf{A} = \dfrac{\hat{r}}{r\sin\theta}\left(\dfrac{\partial(\sin\theta A_\phi)}{\partial \theta} - \dfrac{\partial A_\theta}{\partial \phi}\right) + \dfrac{\hat{\theta}}{r}\left(\dfrac{1}{\sin\theta}\dfrac{\partial A_r}{\partial \phi} - \dfrac{\partial(rA_\phi)}{\partial r}\right)$
$\qquad + \dfrac{\hat{\phi}}{r}\left(\dfrac{\partial(rA_\theta)}{\partial r} - \dfrac{\partial A_r}{\partial \theta}\right)$

4. $\nabla^2 f = \dfrac{1}{r^2}\dfrac{\partial}{\partial r}\left(r^2 \dfrac{\partial f}{\partial r}\right) + \dfrac{1}{r^2 \sin\theta}\dfrac{\partial}{\partial \theta}\left(\sin\theta \dfrac{\partial f}{\partial \theta}\right) + \dfrac{1}{r^2 \sin^2\theta}\dfrac{\partial^2 f}{\partial \phi^2}$

5. $\nabla^2 \mathbf{A} = \nabla(\nabla \cdot \mathbf{A}) - \nabla \times \nabla \times \mathbf{A}$

B.5 COORDINATE TRANSFORMATIONS

Rectangular–cylindrical:

$$\begin{bmatrix}\hat{\rho}\\ \hat{\phi}\\ \hat{z}\end{bmatrix} = \begin{bmatrix}\cos\phi & \sin\phi & 0\\ -\sin\phi & \cos\phi & 0\\ 0 & 0 & 1\end{bmatrix}\begin{bmatrix}\hat{x}\\ \hat{y}\\ \hat{z}\end{bmatrix} \qquad \begin{bmatrix}\hat{x}\\ \hat{y}\\ \hat{z}\end{bmatrix} = \begin{bmatrix}\cos\phi & -\sin\phi & 0\\ \sin\phi & \cos\phi & 0\\ 0 & 0 & 1\end{bmatrix}\begin{bmatrix}\hat{\rho}\\ \hat{\phi}\\ \hat{z}\end{bmatrix}$$

Rectangular–spherical:

$$\begin{bmatrix}\hat{r}\\ \hat{\theta}\\ \hat{\phi}\end{bmatrix} = \begin{bmatrix}\sin\theta\cos\phi & \sin\theta\sin\phi & \cos\theta\\ \cos\theta\cos\theta & \cos\theta\sin\theta & -\sin\theta\\ -\sin\phi & \cos\phi & 0\end{bmatrix}\begin{bmatrix}\hat{x}\\ \hat{y}\\ \hat{z}\end{bmatrix}$$

$$\begin{bmatrix}\hat{x}\\ \hat{y}\\ \hat{z}\end{bmatrix} = \begin{bmatrix}\sin\theta\cos\phi & \cos\theta\cos\phi & -\sin\phi\\ \sin\theta\sin\theta & \cos\theta\sin\phi & \cos\phi\\ \cos\theta & -\sin\theta & 0\end{bmatrix}\begin{bmatrix}\hat{r}\\ \hat{\theta}\\ \hat{\phi}\end{bmatrix}$$

Cylindrical–spherical:

$$\begin{bmatrix}\hat{r}\\ \hat{\theta}\\ \hat{\phi}\end{bmatrix} = \begin{bmatrix}\sin\theta & 0 & \cos\theta\\ \cos\theta & 0 & -\sin\theta\\ 0 & 1 & 0\end{bmatrix}\begin{bmatrix}\hat{\rho}\\ \hat{\phi}\\ \hat{z}\end{bmatrix} \qquad \begin{bmatrix}\hat{\rho}\\ \hat{\phi}\\ \hat{z}\end{bmatrix} = \begin{bmatrix}\sin\theta & \cos\theta & 0\\ 0 & 0 & 1\\ \cos\theta & -\sin\theta & 0\end{bmatrix}\begin{bmatrix}\hat{r}\\ \hat{\theta}\\ \hat{\phi}\end{bmatrix}$$

APPENDIX C

Per-Unit-Length Line Parameters

TABLE C.1 Relationships Between the Per-Unit-Length Parameters of a Lossless Two-Conductor Transmission Line

Line Cross-Section

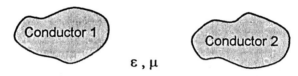

ε, μ

Quantity	Expression
Per-unit-length line	
Inductance	$L' = \mu \mathcal{F}$ (H/m)
Capacitance	$C' = \dfrac{\epsilon}{\mathcal{F}}$ (F/m)
Characteristic line impedance	$Z_c = \sqrt{L'/C'} = \sqrt{\mu/\epsilon}\,\mathcal{F}$ (Ω)
Line propagation speed	$v = 1/\sqrt{L'C'} = 1/\sqrt{\mu\epsilon}$ (m/s)

Notes:
1. The line conductors are assumed to be perfectly conducting.
2. The medium surrounding the conductors has permittivity $\epsilon = \varepsilon_r\epsilon_0$ and permeability $\mu = \mu_r\mu_0$.
3. The function $\mathcal{F}$ depends on the size, shape, and separation of the two conductors comprising the transmission line.

TABLE C.2 Geometrical Parameter $\mathcal{F}$ for Two-Conductor Line Configurations

Configuration	Formula	Diagram
1. Parallel Plate Line	$\mathcal{F} \approx \dfrac{d}{w}$ (for $w \gg d$)	ε, μ; w, d

TABLE C.2 (*Continued*)

2.	Stripline Within a Parallel-Plate	$\mathcal{F} \approx \dfrac{1}{4}\dfrac{d}{w_{\text{eff}} + 0.441d}$ with effective width w_{eff} given by $w_{\text{eff}} = w - \begin{cases} 0 & \text{for } w/d > 0.35 \\ d(0.35 - w/d)^2 & \text{for } w/d < 0.35 \end{cases}$	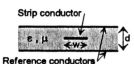 Strip conductor; ε, μ; Reference conductors
3.	Microstrip Line on Dielectric Layer	$\mathcal{F} \approx \begin{cases} \dfrac{1}{2\pi}\ln\!\left(\dfrac{8d}{w} + \dfrac{w}{4d}\right) & \text{for } \dfrac{w}{d} \le 1 \\[6pt] \dfrac{1}{\bigl(w/d + 1.393 + 0.667\ln(w/d + 1.444)\bigr)} & \text{for } \dfrac{w}{d} > 1 \end{cases}$ Use an effective permittivity ε_{eff} $\varepsilon_{\text{eff}} = \dfrac{\varepsilon_r + 1}{2} + \dfrac{\varepsilon_r - 1}{2}\dfrac{1}{\sqrt{1 + 12d/w}}$	Strip conductor; ε_0, μ_0; ε, μ_0; Reference conductor
4.	Coaxial Line	$\mathcal{F} = \dfrac{1}{2\pi}\ln\!\left(\dfrac{d_e}{d_i}\right)$	Diam. d_e; Diam. d_i; ε, μ
5.	Slotted Air Line	$\mathcal{F} \le \dfrac{1}{2\pi}\ln\!\left(\dfrac{d_e}{d_i}\right) + 7.96 \times 10^{-5}\theta^2$ (with θ in radians)	Diam. d_i; ε_0, μ_0; Diam. d_e
6.	Split, Thin-Walled Cylinder	$\mathcal{F} \approx \dfrac{1.269}{\ln\!\left(\cot(\theta/2) + \bigl(\cot^2(\theta/2) - 1\bigr)^{1/2}\right)}$	2θ; D; ε_0, μ_0
7.	Cylindrical Capacitor (with eccentricity)	$\mathcal{F} = \dfrac{1}{2\pi}\operatorname{arccosh}\!\left(\dfrac{d_e^2 + d_i^2 - 4D^2}{2 d_e d_i}\right)$	Diam. d_e; Diam. d_i; ε, μ; D
8.	Two Cylinders	$\mathcal{F} = \dfrac{1}{2\pi}\operatorname{arccosh}\!\left(\dfrac{4D^2 - d_1^2 - d_2^2}{2 d_1 d_2}\right)$ $\mathcal{F} \approx \dfrac{1}{2\pi}\ln\!\left(\dfrac{4D^2}{d_1 d_2}\right)$ (for $D \gg r_1, r_2$)	Diam. d_1; Diam. d_2; ε, μ; D
9.	Two Identical Cylinders	$\mathcal{F} = \dfrac{1}{\pi}\operatorname{arccosh}\!\left(\dfrac{D}{d}\right)$ $\mathcal{F} \approx \dfrac{1}{\pi}\ln\!\left(\dfrac{2D}{d}\right)$ (for $D \gg d$)	Diameters d; ε, μ; D

TABLE C.2 (*Continued*)

10. Cylinder Over Ground	$\mathscr{P} = \dfrac{1}{2\pi}\operatorname{arccosh}\left(\dfrac{2h}{d}\right)$ $\mathscr{P} = \dfrac{1}{2\pi}\ln\left(\dfrac{4h}{d}\right)$ (for $h \gg d$)	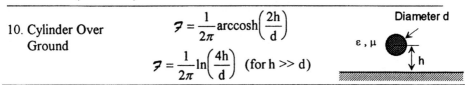

TABLE C.3 Relationships Between the Per-Unit-Length Parameters of a Lossless Multiconductor Transmission Line

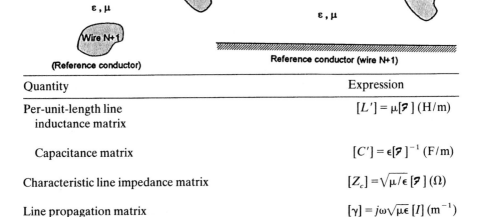

Quantity	Expression
Per-unit-length line inductance matrix	$[L'] = \mu[\mathscr{P}]$ (H/m)
Capacitance matrix	$[C'] = \epsilon[\mathscr{P}]^{-1}$ (F/m)
Characteristic line impedance matrix	$[Z_c] = \sqrt{\mu/\epsilon}\,[\mathscr{P}]$ (Ω)
Line propagation matrix	$[\gamma] = j\omega\sqrt{\mu\epsilon}\,[I]$ (m^{-1})

Notes:
1. The TEM currents in each type of multiconductor line are described in terms of conductor currents I_j for $j = 1 \ldots N+1$, with the sum of all currents being zero. Thus the currents in conductors 1 through N are independent.
2. The TEM voltages are referenced to the $(N+1)$st conductor, so that voltages V_1 through V_n are unique.
3. These lines are described by $N \times N$ per-unit-length inductance and capacitance matrices, plus an impedance and propagation matrix.
4. The line conductors are assumed to be perfectly conducting.
5. The medium surrounding the conductors has permittivity $\epsilon = \epsilon_r\epsilon_0$ and permeability $\mu = \mu_r\mu_0$.
6. The function $[\mathscr{P}]$ depends on the size, shape, and separation of the $N+1$ conductors of the transmission line.

TABLE C.4 Geometrical Parameter Matrix [$\mathcal{P}$] for Lossless Multiconductor Line Configurations

Configuration	Formula	Diagram
1. Isolated N+1-Conductor Line	$[\mathcal{P}] = \dfrac{1}{2\pi}\begin{bmatrix} f_{11} & f_{12} & \cdots & f_{1N} \\ f_{21} & f_{22} & \cdots & f_{2N} \\ \vdots & \vdots & & \vdots \\ f_{N1} & f_{N2} & \cdots & f_{NN} \end{bmatrix}$ $f_{ii} = \ln\left(\dfrac{d_i^2}{a_i a_o}\right)$ $f_{ij} = \ln\left(\dfrac{d_i d_j}{d_{ij} a_o}\right) \; (i \ne j)$	
2. N-Conductor Line Over a Groundplane	$[\mathcal{P}] = \dfrac{1}{2\pi}\begin{bmatrix} f_{11} & f_{12} & \cdots & f_{1N} \\ f_{21} & f_{22} & \cdots & f_{2N} \\ \vdots & \vdots & & \vdots \\ f_{N1} & f_{N2} & \cdots & f_{NN} \end{bmatrix}$ $f_{ii} = \ln\left(\dfrac{2h_i}{a_i}\right)$ $f_{ij} = \ln\left(\dfrac{d'_{ij}}{d_{ij}}\right) \; (i \ne j)$	
3. N-Conductor Line within a Circular Shield	$[\mathcal{P}] = \dfrac{1}{2\pi}\begin{bmatrix} f_{11} & f_{12} & \cdots & f_{1N} \\ f_{21} & f_{22} & \cdots & f_{2N} \\ \vdots & \vdots & & \vdots \\ f_{N1} & f_{N2} & \cdots & f_{NN} \end{bmatrix}$ $f_{ii} = \ln\left(\dfrac{a_o^2 - d_i^2}{a_i a_o}\right)$ $f_{ij} = \ln\sqrt{\dfrac{(d_i d_j / a_o)^2 + a_o^2 - 2 d_i d_j \cos\theta_{ij}}{d_i^2 + d_j^2 - 2 d_i d_j \cos\theta_{ij}}}$ $(i \ne j)$	

TABLE C.5 Geometrical Mean Radius

Case	Configuration	Formula
1	(solid cylinder, diameter $2r$)	$g_{11} = r \cdot \exp(-1/4) = 0.7788 \cdot r$
2	(hollow cylinder, diameter $2r$)	$g_{11} = r$
3	(annular cylinder, inner $2r_1$, outer $2r_2$)	$\ln \dfrac{g_{11}}{r_2} = -\dfrac{r_2^4/4 - r_1^2 r_2^2 + r_1^4 [3/4 + \ln(r_2/r_1)]}{(r_2^2 - r_1^2)^2}$
4	(thin strip, width a, thickness ϵ)	$g_{11} = a \cdot \exp(-3/2) = 0.22313 \cdot a$
5	(solid square, side a)	$g_{11} = 0.44705 \cdot a$
6	(hollow square, side a, thickness ϵ)	$g_{11} = 0.578 a$

TABLE C.6 Geometrical Mean Distances

Case	Configuration	Formula
1	(solid conductor, radius r)	$g_{12} = r$
2	(tube with inner radius r_1, outer radius r_2)	$\ln(g_{12}/r_2) = \dfrac{\ln(r_2/r_1)}{(r_2/r_1)^2 - 1} - 0.5$
3	(point and tube separated by d)	$g_{12} = d$ for any thickness of the tube
4	(two tubes separated by d)	$g_{12} = d$ for any thickness of the two tubes
5	(two perpendicular strips, width a, thickness ϵ, distance d)	$\ln \dfrac{g_{12}}{d} = -1.5 + \dfrac{1}{2}\left(1 - \dfrac{d^2}{a^2}\right)\ln\left(1 + \dfrac{a^2}{d^2}\right) + 2\dfrac{d}{a}\arctan\dfrac{a}{d}$
6	hollow bussbars (width a, thickness ϵ, distance d)	$\ln(g_{12}/d) = 1.5 + (1/2)(1 + d/a)^2 \ln(1 + a/d) +$ $+ (1/2)[(d/a) - 1]^2 \ln[1 - (a/d)]$
7	hollow bussbars (square cross-section, side a, distance d)	$g_{12} \approx d$ for $d/a > 2$ for hollow and solid bussbars $g_{12} = 1.00655d$ for $d/a = 1$

APPENDIX D

Grounding Resistance Parameters

TABLE D.1 Grounding Resistances for Various Electrode Configurations

Configuration	Formula	Diagram
1. Hemisphere at surface	$R = \dfrac{1}{2\pi\sigma r}$	
2. Hemisphere at surface surrounded by a different medium	$R = \dfrac{1}{2\pi\sigma r} + \dfrac{1}{2\pi\sigma_o}\left[\dfrac{1}{r_o} - \dfrac{1}{r}\right]$	
3. Buried sphere	$R = \dfrac{1}{4\pi\sigma}\left[\dfrac{1}{r} + \dfrac{1}{2d}\right] \ (d \gg r)$	
4. Disk on surface	$R = \dfrac{1}{4\sigma r}$	

TABLE D.1 Grounding Resistances for Various Electrode Configurations

Configuration	Formula	Diagram
5. Buried disk	$R = \dfrac{1}{8\sigma r} + \dfrac{1}{8\pi\sigma d}$ $(d > r)$ $R = \dfrac{1}{4\sigma r}\left[1 - \dfrac{4d}{\pi r}\right]$ $(d \ll r)$	
6. Vertical rod	$R = \dfrac{1}{2\pi L \sigma}\left[\ln\left(\dfrac{4L}{a}\right) - 1\right]$ $(L \gg a)$	
7. Two vertical rods	$R = \dfrac{1}{2\pi\sigma L}\left[\ln\left(\dfrac{4L}{\sqrt{ad}}\right) - 1\right] (d \ll L)$ $= \dfrac{1}{4\pi\sigma L}\left[\ln\left(\dfrac{4L}{a}\right) - 1 + \dfrac{L}{d}\right] (d \gg L)$	
8. N vertical rods	$R = \dfrac{1}{2\pi L \sigma}\left[\ln\left(\dfrac{4L}{a'}\right) - 1\right]$ with $a' = \left[Na\left(\dfrac{D}{2}\right)^{N-1}\right]^{1/N}$ $(L \gg a), (D \ll L)$ $R = \dfrac{1}{N 2\pi\sigma L}\left[\ln\left(\dfrac{4L}{a}\right) - 1 + F(N)\right]$ $F(N) = \dfrac{L}{D}\sum_{k=1}^{N-1}\ln\left(\dfrac{1}{\sin(k\pi/N)}\right)$ $(L \gg a), (D \gg L)$	
9. Horizontal rod at surface	$R = \dfrac{1}{\pi\sigma L}\left[\ln\left(\dfrac{2L}{a}\right) - 1\right]$ $(L \gg a)$	

TABLE D.1 *Continued*

Configuration	Formula	Diagram
10. Buried horizontal rod	$R = \dfrac{1}{\pi \sigma L}\left[\ln\left(\dfrac{2L}{\sqrt{2ad}}\right) - 1\right] (d \ll L)$ $= \dfrac{1}{2\pi\sigma L}\left[\ln\left(\dfrac{2L}{a}\right) - 1 + \dfrac{L}{4d}\right] (d \gg L)$	
11. N radial rods at surface	$R = \dfrac{1}{N\pi\sigma L}\left[\ln\left(\dfrac{2L}{a}\right) - 1 + F(N)\right]$ $F(N) = \displaystyle\sum_{k=1}^{N-1} \ln\left(\dfrac{1 + \sin(k\pi/N)}{\sin(k\pi/N)}\right)$ $(L \gg a)$	
12. N buried rods	Replace wire radius a by $a' = \sqrt{2ad}$ in Case 11 (for $d \gg a$)	
13. Ring at surface	$R = \dfrac{1}{2\pi\sigma}\dfrac{1}{\pi r}\ln\left(\dfrac{8r}{a}\right) \; (a \ll r)$	
14. Buried ring	Replace wire radius a by $a' = \sqrt{2ad}$ in Case 12 (for $d \gg a$)	

APPENDIX E

Coaxial Cable and Connector Data

TABLE E.1 Selected Cable Shield Parameters

Cable Type (RG-)[a]	Shield Radius b (cm)	DC Resistance R_0 (mΩ/m)	Aperture Inductance L_a' (nH/m)	Shield Leakage Parameter, S_s (m/μF)
6 (I)	0.24	6.6	0.42	29.0
6 (O)	0.27	7.5	0.36	13.0
11	0.37	4.0	0.25	16.0
22 (I)	0.37	5.5	0.34	19.0
22 (O)	0.40	6.4	0.14	6.0
23	0.50	5.9	0.29	12.0
25 (I)	0.38	4.8	0.46	28.0
25 (O)	0.45	4.4	0.35	19.0
35	0.60	4.3	0.12	5.0
58	0.15	14.2	1.0	66.0
59	0.19	8.6	0.49	32.0
62	0.19	8.7	0.52	34.0
63	0.37	4.0	0.42	27.0
65	0.37	4.0	0.27	18.0
108	0.21	17.6	4.6	250.0
114	0.37	5.1	0.70	44.0
119 (I)	0.43	3.1	0.08	5.0
119 (O)	0.47	6.5	0.65	28.0
122	0.13	16.2	0.37	24.0
130	0.62	2.3	0.33	17.0
142 (I)	0.15	14.1	0.43	27.0
142 (O)	0.18	16.0	0.55	28.0
144	0.37	5.5	0.32	17.0
156 (I)	0.37	6.0	0.11	5.5
156 (O)	0.42	4.7	0.26	13.0
156 (S)	0.52	8.3	0.17	5.1
157 (I)	0.59	7.5	0.12	3.6
157 (O)	0.64	4.1	0.70	31.0
157 (S)	0.74	9.2	0.16	3.4
174	0.08	36.5	2.3	158.0
179	0.08	28.1	0.88	66.0

TABLE E.1 *Continued*

Cable Type (RG-)[a]	Shield Radius b (cm)	DC Resistance R_0 (mΩ/m)	Aperture Inductance L_a' (nH/m)	Shield Leakage Parameter, S_s (m/μF)
181 (I)	0.27	5.7	0.14	9.0
181 (O)	0.62	2.4	0.20	9.5
189 (I)	0.81	1.1	0.008	0.6
189 (O)	0.86	1.4	0.17	10.0
192 (I)	2.19	1.1	0.14	4.9
192 (O)	2.26	1.1	0.25	8.5
192 (S)	2.40	1.2	0.17	5.3
193 (I)	2.19	0.89	0.66	31.0
193 (O)	2.26	1.2	0.39	14.0
193 (S)	2.40	1.7	0.23	5.4
194 (I)	2.19	0.89	0.66	31.0
194 (O)	2.26	1.2	0.11	3.8
210	0.19	8.7	0.52	34.0
211	0.79	1.7	0.09	4.8
212 (I)	0.24	6.6	0.42	29.0
213	0.37	4.0	0.25	16.0
214 (I)	0.37	9.9	0.37	13.0
214 (O)	0.40	8.5	0.13	4.5
217 (I)	0.48	3.2	0.30	19.0
217 (O)	0.51	5.4	0.32	13.2
218	0.88	1.2	0.07	4.4
220	1.17	0.91	0.09	5.3
222 (I)	0.24	6.6	0.42	29.0
222 (O)	0.27	7.5	0.36	19.0
223 (I)	0.15	14.8	0.72	44.0
223 (O)	0.18	16.0	1.3	70.0
225 (I)	0.37	9.8	0.36	13.0
225 (O)	0.40	8.5	0.12	4.4
226 (I)	0.48	5.2	0.03	1.2
226 (O)	0.51	6.6	0.92	37.0
301	0.24	10.0	0.73	44.0
302	0.19	15.0	1.50	84.0
303	0.15	14.1	0.43	27.0
304 (I)	0.24	9.0	0.52	28.0
304 (O)	0.27	7.2	0.40	23.0
316	0.08	26.8	0.88	77.0
326 (I)	0.70	5.9	0.05	2.2
326 (O)	0.72	6.0	0.06	3.0
328 (I)	1.38	1.0	0.14	7.5
328 (O)	1.43	1.7	0.15	6.6
328 (S)	1.56	1.1	0.28	13.0
329 (I)	0.50	2.4	0.38	23.0

TABLE E.1 *Continued*

Cable Type (RG-)[a]	Shield Radius b (cm)	DC Resistance R_0 (mΩ/m)	Aperture Inductance L'_a (nH/m)	Shield Leakage Parameter, S_s (m/μF)
329 (O)	0.55	4.7	0.31	13.0
391	0.39	8.1	0.05	1.7

Source: Adapted from E. F. Vance, *Coupling to Shielded Cables*, R. E. Krieger, Melbourne, FL, 1987. (Copyright © 1977 John Wiley & Sons, Inc. Reprinted by permission from the publisher.)

[a] (I), inner braid; (O), outer braid; (S), shield insulated from braid.

TABLE E.2 Measured Transfer Impedances $Z_t = R_d + j\omega M_{ab}$ for Cable Connectors

Connector	Identification	R_d (Ω)	M_{ab} (H)
Multipin	Burndy NA5-15853	0.0033	5.7×10^{-11}
Aerospace connectors	Deutch 38068-10-5PN	0.15	2.5×10^{-11}
(threaded)	Deutch 38068-18-31SN	0.005	1.6×10^{-10}
	Deutch 38060-22-55SN	0.023	1.1×10^{-10}
	Deutch 38068-14-7SN	0.046	5.0×10^{-11}
	Deutch 38060-14-7SN	0.10	8.2×10^{-11}
	Deutch 38060-14-7SN	0.023	6.7×10^{-11}
	Deutch 38068-12-12SN	0.0033	3.0×10^{-11}
	Deutch 38068-12-12SN	0.012	1.3×10^{-11}
	Deutch 38060-12-12SN	0.012	1.3×10^{-11}
	Deutch 38060-22-12SN	<0.001	2.5×10^{-12}
	Deutch 38068-12-12SN	0.014	3.5×10^{-11}
	AMP	0.0067	1.6×10^{-11}
	AMP	0.0067	1.6×10^{-11}
	AMP	0.0033	1.9×10^{-11}
Type N	UG 21B/U-UG58A/U	[a]	[a]
Type BNC (bayonet)	UG 88C/U-UG1094/U	0.002	$4–8 \times 10^{-11}$
Anodized	MS 24266R-22B-55	5×10^4	$\omega M_{ab} < R_0$ at 20 MHz
Open shell	MS 3126-22-55	0.5–1	$\omega M_{ab} < R_0$ at 20 MHz
Split shell	MS 3100-165-1P MS 3106A	0.001	20×10^{-11}

Source: Adapted from E. F. Vance, *Coupling to Shielded Cables*, Krieger, Melbourne, FL, 1987. (Copyright © 1977 John Wiley & Sons, Inc. Reprinted by permission from the publisher.)

[a] Too small to measure in the test fixture.

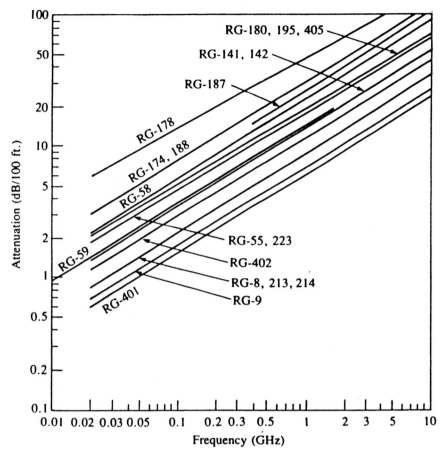

FIGURE E.1 Propagation loss in coaxial cables. (From D. M. Pozar, *Microwave Engineering*, Addison-Wesley, Reading, MA, 1990. © 1990 Addison-Wesley Publishing Co., Inc., reprinted by permission.)

APPENDIX F

Computer Software

F.1 GENERAL COMMENTS ON CODE OPERATION

Four computer programs are provided with this book to aid in understanding transmission line coupling and propagation problems. Several problems in the previous chapters have used these programs. This software is as follows:

- **NULINE:** program to analyze the behavior of an aboveground transmission line excited by either transient or CW lumped voltage or current sources anywhere on the line or by an incident EM field.
- **RISER:** program to compute the voltage and current in a load at one end of a field-excited, aboveground transmission line, modeling the ends (risers) of the line as smaller, vertical transmission lines.
- **LTLINE:** program to compute the load voltages and currents at each end of an aboveground transmission line that is excited by a nearby lightning strike.
- **TOTALFLD:** program to evaluate the total aboveground and belowground **E** and **H**-fields produced by an incident EM plane wave. Both CW and transient results for a number of different user-defined waveforms are presented.

These programs run on IBM-PC class machines under the DOS/Windows 3.1 or 95 operating systems. The NULINE code is a DOS-based, compiled BASIC program having pull-down menus for data entry and a self-contained screen plot utility, NUPLOT. It can be run under Windows in a DOS window, but as such it does not provide the typical graphical interface of a Windows program. A Windows-based driver program, WINULINE, has been written for NULINE and this provides a convenient way of entering the data for NULINE and for plotting calculated results on the screen.

The other three programs are Windows user interface codes for performing the data input, together with compiled FORTRAN-based programs that perform the "number crunching" operations. The latter FORTRAN codes run independently under DOS as long as the necessary input data file is provided to the code.

F.2 DISTRIBUTION OF THE COMPUTER CODES

The four EMC programs are available online as single compressed files from the ftp server at John Wiley and Sons, Inc. The files can be accessed through either a standard ftp program or a Web browser using the ftp protocol. To gain ftp access, type the following at your ftp command prompt:

ftp://ftp.wiley.com

When asked for your name, log in as

anonymous

The files are located in the **emc_code** area of the **public\sci_tech_med** directory. Instructions for installing the programs are included in the README.TXT file. If you need further information about downloading the files, you can reach Wiley's tech support line at 212-850-6194.

F.3 COMPUTER CODE INSTALLATION

Complete instructions for installing the compressed files for each program are included in the README.TXT file in the **emc_code** directory. The compressed program files should be copied to separate directories on your hard drive and uncompressed according to the method described in the instructions. You will then be able to run the Windows setup program for each of the EMC modules.

Extensive tests of these codes with different computers and CPU memory have not been conducted. However, it is believed that these codes will function properly on most PCs capable of supporting the Windows operating system. The fast Fourier transform (FFT) is used to obtain the transient responses for each of these codes, and this can be a source of difficulty to the uninitiated. As a general rule, be sure that the computed waveform has died out sufficiently at the end of the computed transient response (see Problem 6.15d). If this is not the case, try increasing the maximum time of the calculation or changing the loads on the line so that the response damps out more rapidly.

In the following sections each of these codes is discussed in more detail. In addition, consult the help files or other documentation provided on the distribution diskettes for up-to-date information that may not be presented here.

F.4 NULINE TRANSMISSION LINE CODE

F.4.1 Overview

NULINE is a numerical transmission line program for performing either frequency-domain (time-harmonic) or transient analysis of a single-wire transmission line located over a perfectly or imperfectly conducting ground plane (the earth). For this code the frequency-domain transmission line models developed in Chapters 6, 7, and 8 have been implemented into this rapidly running, user-friendly code.

NULINE is a DOS-based program and, as will be described later, it can be run directly from DOS. When the DOS version of NULINE is run for the first time, a welcome message screen will be displayed, followed by a screen asking for the disk drive and directory name where the NULINE code has been placed. At this point, the user should enter the appropriate drive and directory information. Upon doing this, the NULINE.INI file is created, which tells the code where to find its programs and data. In addition, the NULINE.CMD file containing the default data is created and the NULINE code begins executing—ready to perform calculations with the default transmission line geometry and program parameters.

If a mistake in the directory name definition is made, or if it is desired to change the NULINE directory name later, it is possible to edit the NULINE.INI data file with an ASCII text editor. Alternatively, one can simply delete the NULINE.CMD file from the disk file and run NULINE again. The absence of the CMD file will be detected automatically, and the initialization procedure will be reinvoked, allowing the user to enter the new directory.

For ease of use in the Windows graphical environment, an interface program called WINULINE has been developed, and it is provided with NULINE. This program allows the user to easily enter the parameters of the problem to be solved, and then it launches the NULINE program in a DOS window to execute the desired calculations.

After NULINE has terminated, control returns to the WINULINE program, where various screen plots can be made. For single-trace plots of a transient response or a spectral magnitude, a simple plot routine is integrated in WINULINE. However, another DOS-based plot program called NUPLOT has been provided with NULINE. This program also runs in a DOS window, and permits either manual or automatic scaling of the plots, plotting on log or linear scales, and adding a simple, one line title and axes labeling. With NUPLOT, several different responses can be overlaid together on one plot, and for frequency domain data files, it is possible to plot the real, imaginary, magnitude, and phase components of the responses. Moreover, if the responses from the NULINE engine result from a "sweeping" of one of the input parameters from a low value to a high value, the NUPLOT program will plot these results as an overlay of multiple waveforms or spectra.

F.4.2 Problem Geometry

The geometry of the transmission line analyzed in this code is shown in Figure F.1. Each end of the line is terminated by a general series RLC impedance. These loads may be set to a matched open- or short-circuit state, or the individual circuit element parameters may be specified independently.

The line may be excited by either a lumped voltage source or lumped current source located at an arbitrary position on the line. In addition, the excitation of the line by an incident electromagnetic plane wave having an arbitrary angle of incidence and polarization can be treated. For the case of transient excitation of the line, either by a lumped excitation source or by an incident EM field, the user can select a delta-function, step-function, single-exponential, double-exponential, damped-cosine, or damped-sine waveform for the forcing function. User-specified waveforms in an external disk file may also be read in as transient excitation functions. In the case of a frequency-domain analysis, the excitation is taken to be a constant of unit amplitude for all frequencies, thereby giving the delta-function frequency

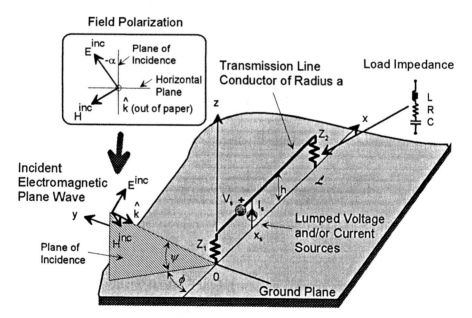

FIGURE F.1 Geometry of the NULINE code.

response. The resulting calculated response is a transfer function, which can later be multiplied by a specified excitation field spectrum to provide an actual spectral response.

Several different line responses, or observables, are available with NULINE, as shown in Figure F.2. All observables are written to user-specified disk files for later processing and possible off-line plotting. One output option permits the calculation of the total load voltage and current at either load 1 (at $x = 0$) or load 2 (at $x = \mathcal{L}$). If the responses at both loads are desired, the code must be run twice. A second option permits calculation of the current and linear charge density at an arbitrary point on the line. A third and fourth output option provides a Norton or Thévenin circuit representation of the line, as seen from an arbitrary point on the line. In the case of an incident field excitation, the Thévenin source constitutes the total voltage response for the line, in that both the scattered field from the line and the incident field components go into determining this voltage.

F.4.3 NULINE Code Operation for DOS

The NULINE code can be run in DOS in several different modes. Normally, the code will be run interactively. In this mode the user can change various transmission line parameters, calculate the line responses, and obtain a screen plot of the responses. NULINE also can be run in batch mode, automatically calculating one or more cases and storing the calculated responses on disk files for future plotting or processing. Such an operational mode is useful for performing a lengthy parametric study of line responses. A third operational mode is the sweep mode, in which a selected parameter in the problem may be swept from a specified minimum value to a maximum value. Outputs from the NULINE code operating in this sweep mode may also be plotted as overlaid line plots on the screen, using the plotting option.

The first time that the code is run on a new system, a default configuration data file named NULINE.CMD is generated and stored on the disk

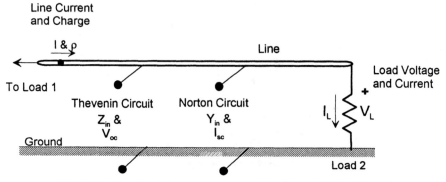

FIGURE F.2 Possible response quantities in the NULINE code.

drive and subdirectory which has been defined during the code initialization procedure. This file contains program parameters and information for defining a default line geometry and calculational parameters. Every time NULINE is executed, this configuration file is read automatically (unless another .CMD file name is specified when NULINE is run), and this sets the initial parameters for the transmission line problem being analyzed. Following is a listing of the default NULINE.CMD data file.

```
! --------------------------------------------------------------------------
!
!
!CONFIGURATION DATA FILE FOR NULINE TRANSMISSION LINE CODE
!
!  FILE WRITTEN ON 01-11-1996 AT 17:59:28
!
!
! --------------------------------------------------------------------------
!
LENGTH=100        !LINE LENGTH (IN M)
HEIGHT=10         !LINE HEIGHT (IN M)
RADIUS=1          !LINE RADIUS (IN CM)


SIGMA=.01         !EARTH CONDUCTIVITY (MHOS/M)
EPSR=10           !RELATIVE EARTH DIELECTRIC CONST.

R1=100            !LOAD #1 RESISTANCE (OHMS)
L1=0              !LOAD #1 INDUCTANCE (MILLI-HENRYS)
C1=1E+10          !LOAD #1 CAPACITANCE (MICRO-FARADS)


R2=100            !LOAD #2 RESISTANCE (OHMS)
L2=0              !LOAD #2 INDUCTANCE (MILLI-HENRYS)
C2=1E+10          !LOAD #2 CAPACITANCE (MICRO-FARADS)

ANALYSIS=TRANSIENT  !TYPE OF ANALYSIS (TRANSIENT OR FREQUENCY)
NTERM=-1          !NO OF TERMS IN DENOMINATOR EXPANSION (<0 FOR NO EXPANSION)
T0REF=0           !FLAG FOR T=0 REFERENCE: =0 FOR T=0 WHEN SOURCE (OR WAVE) FIRST TURNS ON, =1 FOR T=0 AT FIRST RESPONSE

EXTYPE=FIELD      !TYPE OF EXCITATION (FIELD, LUMPED VOLTAGE OR CURRENT)
PSI=45            !VERTICAL (ELEVATION) ANGLE OF INCIDENCE (DEG)
PHI=0             !LINE ORIENTATION ANGLE (DEG) FOR FIELD EXCITATION
SPOS=0            !POSITION OF LUMPED SOURCE (M)

WAVEFORM=DE       !WAVEFORM TYPE (DE, SE, DS, DC, UI, SW)
EPEAK=1           !WAVEFORM MAGNITUDE CONSTANT (V), (A), OR (V/M)
```

APPENDIX F

```
AL=4.76E+08        !DOUBLE EXPONENTIAL ALPHA (1/SEC)
BE=4000000         !DOUBLE EXPONENTIAL BETA (1/SEC)
GAMMA=0            !E-FIELD POLARIZATION ANGLE (DEG), MEASURED FROM VERTICAL
TMAX=6             !MAXIMUM CALCULATION TIME (USEC)
TSTART=0           !STARTING TIME OF EXCITATION WAVEFORM
TWIDTH=0           !WIDTH OF PULSE WAVEFORM IN (US)
FREQ0=1            !OSCILLATION FREQUENCY FOR DAMPED SINE/COSINE WAVEFORM (MHz)
NPT=2048           !NUMBER OF POINTS IN TIME DOMAIN (2^M points)


EVINC=EVINC.TIM    !FILE FOR SPECIFIED VERTICAL INCIDENT FIELD
EHINC=EHINC.TIM    !FILE FOR SPECIFIED HORIZONTAL INCIDENT FIELD
VLUMP=VLUMP.TIM    !FILE FOR SPECIFIED VOLTAGE SOURCE
CLUMP=CLUMP.TIM    !FILE FOR SPECIFIED CURRENT SOURCE


FMIN=.001          !MINIMUM FREQUENCY (MHZ)
FMAX=100           !MAXIMUM FREQUENCY (MHZ)
NPF=1000           !NO. OF POINTS FOR FREQUENCY DOMAIN ANALYSIS
SPACING=LOG        !SPACING TYPE FOR FREQUENCY DOMAIN ANALYSIS (LOG OR LINEAR)


OUTPUT=LOAD        !SPECIFIES OUTPUT QUANTITY (LINE, LOAD, NORTON, THEVENIN)
OBSPT=100          !POSITION OF OUTPUT QUANTITY (M)
FILEV=VOL          !VOLTAGE OUTPUT FILE NAME
FILEC=CUR          !CURRENT OUTPUT FILE NAME


CONTROL=           !BATCH CONTROL PARAMETER (SET TO RUN FOR RUNNING BATCH)
CMDFILE=NULINE.CMD !NAME OF .CMD FILE FOR WRITING NULINE CONFIGURATION DATA


SWEEP=             !VARIABLE TO BE SWEPT IN ANALYSIS
SMIN=0             !MINIMUM VALUE OF SWEEP VARIABLE
SMAX=0             !MAXIMUM VALUE OF SWEEP VARIABLE
NSWEEP=0           !NUMBER OF CALCULATIONS IN SWEEP
RHOW=.000002       !ELECTRICAL CONDUCTIVITY OF WIRE (IN OHM-METERS)
```

One or more of the defined input parameters in this data file may be changed using the F-3 key menus in NULINE, and the resulting new data configuration can then be saved, either as NULINE.CMD or under another name, for future use.

In addition, the first time that NUPLOT is run, a configuration data file named NUPLOT.CMD is generated. This file stores previous plot parameters and allows for a rapid plotting of new results without reentering title, axis, scaling, or data file information. Of course, changes in these plotting parameters can be made easily, as discussed in the code information provided on the installation disk.

F.4.3.1 Interactive Mode. To run NULINE in the interactive mode after the installation has been completed, log onto the directory containing NULINE and at the DOS prompt, type

$$\text{NULINE}\langle\text{Enter}\rangle$$

This will cause the program to execute and will bring up the primary menu screen, which has a menu of F-keys that control the operation of the program.

When the code is first executed, it reads the configuration file NULINE.CMD, which contains previously defined parameters for the transmission line analysis (line length, height, type of analysis, etc). If desired, an alternative .CMD file can be read, one that might have been generated and saved in a previous NULINE session using the F2 key, or may have been modified by an ASCII editor. This is done by starting NULINE with the name of the alternative .CMD file, specified as

$$\text{NULINE CMD} = \textit{filename}\langle\text{enter}\rangle$$

Here, *filename* is the name of the .CMD file, which may be entered without the .CMD suffix, and which resides in the same directory containing the NULINE codes.

Once NULINE is executing, the overall operation of the program is controlled by the F-keys shown below, which provide the following functions:

Key	Function
F-1	Show the present input parameters for the code, as defined from the initial .CMD file and possibly modified by user input.
F-2	Store the current program parameters in a .CMD data file.
F-3	Modify input parameters for the calculations.
F-4	Perform analysis of the selected transmission line responses, using the current program parameters.
F-5	Plot one or more computed response files on the screen.
F-10	Stop and return to DOS.

A more complete description of the use of these functions is presented in the documentation provided on the distribution disk.

F.4.3.2 Batch Mode. NULINE can also be run in a batch mode. This is particularly useful if there is a parametric study of line responses to be conducted. Such calculations can take several hours on a PC and can be accomplished easily using this feature. In addition, commercially available plotting programs can be used in the batch mode to generate plots automatically from the data files produced by NULINE on a printer or plotter.

An example batch file named RUNME.BAT is generated by the INSTALL program. This file shows a number of examples of the use of the batch mode operation. To run NULINE in batch mode, a command of the form

NULINE   CONTROL = RUN

should be included in a batch file (or from the DOS prompt). This will use the *existing* NULINE.CMD data file for defining the parameters of the problem and will run the NULINE code automatically without user prompts. Note that the output data files (perhaps VL1.TIM and IL1.TIM, if these are the output file names defined in the NULINE.CMD file) will be generated automatically when NULINE completes the calculations. The batch file processing can then rename or otherwise manipulate these data files, as required.

If it is desired to use a command file that differs from NULINE.CMD, it is possible to do this in the batch operation. This is done by first generating a different .CMD name when running the NULINE code in interactive mode, by saving a particular set of parameters using the F2 Store option (such as XXX.CMD). Subsequently, when running in the batch mode, the command

NULINE   CMD = XXX   CONTROL = RUN

will load the XXX.CMD command file and begin batch mode excitation.

If the user desires to change several of the data entries from the existing NULINE.CMD file, the batch run can be made with the line

NULINE   LENGTH = 100.   SIGMA = 0.01   CONTROL = RUN

Each change to the data in the .CMD file is separated by a space. Because the default data file NULINE.CMD is being used in this instance, it is not necessary to specify a different .CMD file in the command line. Note that the parameters LENGTH and SIGMA correspond to specific input data quantities defined in NULINE.CMD. Any of the other quantities in this command file can be changed in this manner. As the exact spelling of these parameters is important, the user should print out the NULINE.CMD file to view the correct definition of each program parameter.

F.4.4 NUPLOT Screen Plotting Routine

Data files generated by NULINE, either real-valued transient data or complex frequency-domain data, can be plotted by pressing the F-5 key. This loads the NUPLOT plotting routine, which provides either a CGA OR EGA/VGA screen plot of the data, depending on the hardware available. Although NUPLOT appears to be integrated into NULINE, it is actually a stand-alone screen plot program that can be run independently. It is also menu-driven and operates in a manner similar to NULINE, using a new set of F-key menus.

When it is executed, NUPLOT reads a plotting command file called NUPLOT.CMD, which contains plot configuration data from previous plots. In this manner it is not necessary to reenter file names, axis configurations, and plot titles each time a plot is done, if there is no change in the plotting parameters. This .CMD file is updated and saved automatically each time the plotting routine terminates.

The various F-keys and their functions for the plotting code are listed below. Additional information about the functioning of these keys is presented in the code documentation on the diskette.

Key	Function
F-1	Show the current status of the plotting parameters.
F-3	Define plotting parameters and select the files to be plotted.
F-4	Plot the selected data files.
F-10	Quit plotting and return to NULINE or DOS.

F.5 RISER TRANSMISSION LINE COUPLING CODE

F.5.1 Overview

The RISER code was written to enhance the accuracy of simple transmission line models for computing the coupling of an incident transient plane wave to overhead electrical conductors. An early transmission line model (NULINE) had a very simple treatment of the vertical ends of a long transmission line connected to the earth, and consequently, its accuracy for early-time responses was not adequate for some purposes. As a result, the RISER code was developed, which treated each end of the transmission line as separate transmission line sections, each having a length equivalent to the height of the line over the ground. This, combined with the overall transmission line model for the horizontal line, provided a more accurate calculational model for a physical line (see Problem 7.12).

Important features of this analysis code include the following:

- Excitation provided by a user–definer transient waveform.
- Provision for arbitrary angles of incidence ψ and ϕ, as well as polarization α.
- A large number of time-domain sample points in the waveform ($2^{14} = 16{,}384$) for determining accurate early-time responses.
- Calculated transient responses: the incident excitation waveform, the voltage and current at the load 2 resistance, and the induced current at a user-specified location along riser 2.
- The possibility of "sweeping" over one of the many input parameters, defining the problem so as to perform a parametric study.
- A Windows interface for the code.

This code actually consists of two parts: the Windows-based riser menu program, called RISERDVR.EXE, and a FORTRAN-based analysis engine, called RISER.EXE. The RISERDVR menu program provides a user-friendly interface for defining the data file needed to run the analysis portion of the program. This program defines the excitation waveform, EINC.TIM, and then creates a data file called RISER.DAT, which is read by the RISER.EXE code. The latter code actually performs the coupling analysis. After this is completed, the user is left with the data files named VOLOAD.TIM, CURLOAD.TIM, and CURISER.TIM, which contain the calculated transient response waveforms.

F.5.2 Problem Geometry

The geometry treated in the RISER code is shown in Figure F.3, in which an aboveground transmission line of length $\mathcal{L}$ and height h is located over a lossy earth. The line is excited by an incident transient plane wave and the responses at the $x = \mathcal{L}$ end of the line are computed. For this model the load resistances R_1 and R_2 are assumed to be located at the bottom of the risers of the transmission line, just at the earth–air interface. The calculated responses include the voltage $V_L(t)$ and current $I_L(t)$ at the load at $x = \mathcal{L}$, along with the riser current $I_R(t)$ at location z_0, as indicated in the figure. Note that for the response calculations, the reference $t = 0$ time is chosen to be the time that the incident field first hits the *top* of riser 2 at $x = \mathcal{L}$.

F.5.3 Excitation Field

The incident field is a plane wave with a vertical angle of incidence ψ and an azimuthal angle of incidence ϕ, together with a polarization angle α

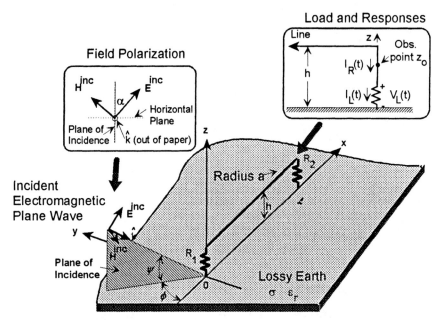

FIGURE F.3 Problem geometry for the RISER code.

measured in the clockwise direction from the vertical plane of incidence, as shown in Figure F.3.

The RISER windows driver program has a menu option for defining six different types of excitation waveforms. These are

- Dirac delta function
- Pulse function
- Single exponential
- Double exponential
- Exponentially damped cosine
- Exponentially damped sine

Equations for each of these functions are provided in the menu and the various parameters can be set to provide a wide variety of different waveshapes. However, if the user wishes to define an alternate excitation file, or perhaps to use a measured waveform, it is possible to simply define an ASCII input file for EINC.TIM as described above.

F.5.4 Input Data File

The input data for the RISER code are contained in the RISER.DAT data file, which is created by the menu program RISERDVR. Following is a list of the input data for a sample run that is provided to the analysis engine:

```
5          ! ℓ Line Length (in m)
.8         ! h Line Height (in m)
.8         ! z₀ Observation Point Location (in m)
.15        ! a Line Radius (in cm)
20         ! R1 Load Resistance (Ohms)
20         ! R2 Load Resistance (Ohms)
45         ! ψ Vertical Angle of Incidence (deg)
0          ! φ Azimuthal Angle of Incidence (deg)
0          ! α E-field Polarization Angle (deg)
1000000    ! σ Earth Conductivity (S/m)
1          ! ε Earth Relative Dielectric Constant
```

Note that the numerical values illustrated in the list are the *default* values, which are initialized each time that a user runs the riser menu program. These values can be changed as desired from within the RISERDVR menu program and can be saved to another data file name for archival purposes.

F.5.5 Code Operation

When properly installed by the setup program, the RISER code is located in a separate program group and may be run by clicking on the RISER program icon. This results in the screen shown in Figure F.4. The user can open a previously saved data file using the top pushbutton or can change any of the program parameters by clicking the mouse on any of the parameters shown on the diagram. The incident field parameters can be edited by clicking on the caption Incident Electromagnetic Plane Wave in the diagram. Alternatively, by selecting the Data option in the menu bar, the user can set the various parameters using a card-file type of input.

After the data are set, they may be saved into a new data file using the Save Data File button. The data can be reset to the default state by pressing the Default button.

The button labeled Compute will begin execution of the RISER.EXE analysis engine. Once the calculations have terminated, the data are in the previously mentioned *.TIM data files. These may be plotted on the screen by using the NUPLOT program provided with NULINE (see Section F.1), or by using another data graphing program of the user's choice. Additional information about running this program is available by using the Help command in the main menu.

F.5.6 Sample Results

As an example of the results from the riser program, and as a means of verifying that the code is functioning properly, the default data discussed above has been used to illustrate the results of the RISER code, with the exception that the value of m has been set to the maximum value of $m = 14$

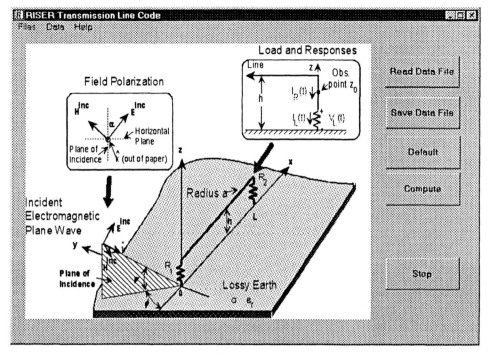

FIGURE F.4 Main menu screen for the RISERDVR program interface for RISER.

to represent more accurately the early-time portion of the response. Figure F.5 illustrates the computed riser current at the top of the riser (at $z = 0.8$ m), as a function of time. Part (*a*) is the complete transient waveform, and part (b) shows the early-time portion. The many oscillations occurring at the early time result from the different times of arrival of the incident field, earth-reflected field, and riser traveling-wave components in the model. Similarly, Figure F.6 illustrates the transient voltage across the load resistance at $x = \mathcal{L}$. Of course, the load current is simply 20 times less than this value and has a waveform similar, but not identical to, the riser current located 0.8 m away.

F.6 LTLINE TRANSMISSION LINE CODE

F.6.1 Overview

LTLINE is a computer program designed for analyzing the load responses of the aboveground line shown in Figure 7.22, which is excited by a nearby lightning strike. This program is a FORTRAN-based computer code that

584 APPENDIX F

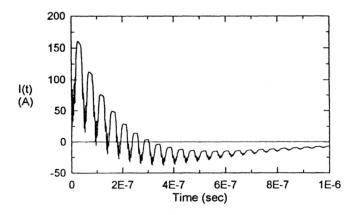

a. Late-time response

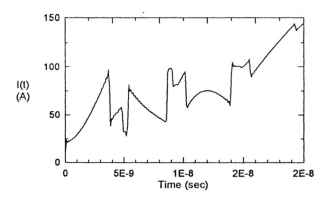

b. Early-time response

FIGURE F.5 Plot of the RISER current for the default data case (with $m = 14$).

can be run independently in DOS or can be controlled by a companion Windows code called the Lightning Coupling Code (LCC). LCC permits the user to input the problem parameters and geometry easily, creates the needed data file, LTLINE.DAT, and then runs the LTLINE code.

The highlights of this analysis package are that it:

- Uses a non-plane-wave excitation of the line.
- Implements a lossy earth model for the line calculations.
- Computes the transient voltage and current responses in both resistive loads at the ends of the line.
- Provides an easy input of the defining parameters using the Windows interface for the code.

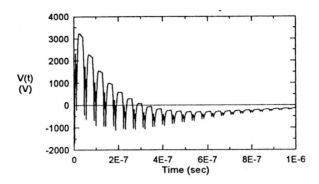

a. Late-time response

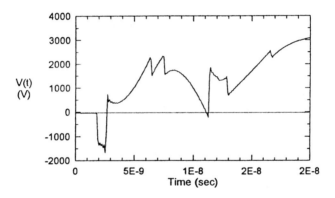

b. Early-time response

FIGURE F.6 Plot of the load 2 voltage for the default data case (with $m = 14$).

F.6.2 Problem Geometry

Referring to Figure 7.22, we note that the transmission line geometry is identical to that of the RISER code but has a different excitation mechanism. The data input for this problem are contained in a file named LTLINE.DAT, which has the following form:

1000.	! $\mathcal{L}$ Line Length (in m)
10.	! h Line Height (in m)
0.5	! a Line Radius (in cm)
−200 −200	! Z_1, Z_2 Load Resistances (Ω) (Neg. for Matched Loads)
500. 50.	! x_c and y_c Coordinates of Lightning Channel (in m)
10. 10000.	! ϵ_r, σ Relative Dielectric Constant and Ground Conductivity (S/m)

10 60   ! m, Tmax(in µS) for the Waveform Definition
10.7 .25 2.1 2   ! C_{I01} in (kA), T_{11} (in (µs), T_{21} in (µs), N_1 for Waveform Contribution #1
6.5 2.5 230 2   ! C_{I02} in (kA), T_{12} (in (µs), T_{22} in (µs), N_2 for Waveform Contribution #2
0.43 1.7   ! Relative Velocity and 1/e Channel Attenuation Length (in km) for the Lightning Channel

Notice that the first three data lines provide the line geometry parameters and the fourth line of data is for the load impedances (resistance only) as Z_1 and Z_2. If these values are *negative*, the impedances are assumed to be *matched* to the characteristic impedance of the line. The eighth and ninth lines define the waveform parameters (as given in Table 7.1). Parameter m in line 7 serves to define the number of points in the transient response, as $NP = 2^m$.

F.6.3 Excitation of the Line by the Lightning Source

The analysis is based on representing the lightning channel current as an upward-traveling current wave having two waveform components, as given in Eqs. (7.65) and (7.66), and in Table 7.1. This leads to nonplanar incident fields which are given by Eq. (7.69) for the case of a perfectly conducting earth.

If the ground has a finite conductivity, the E and H fields produced by the lightning channel will be different from those described in the Section 7.3.5. A rigorous treatment of this problem involves the evaluation of the Sommerfeld integrals [8.1], which provide the fields of a point current element radiating near the ground. Although this is possible to do in the present case, it results in long computer times.

An alternative, yet approximate technique for calculating these fields has been proposed by Rubenstein.[†] This technique is based on use of the wave-tilt formula, which relates the horizontal E-field on the earth surface to the vertical component of the E-field. For many practical cases involving nearby lightning strikes, it has been noted that the vertical component of the E-field is closely modeled by the field component calculated for the perfectly conducting ground.

Using the subscript p to denote the fields for a perfectly conducting ground, the E_z-field component produced by the lightning at a radial distance ρ from the channel shown in Figure 7.24 is approximated by

$$E_z(\rho, z) \approx E_{pz}(\rho, z) \qquad (F.1)$$

The horizontal E-field, E_ρ, is approximated by the sum of the horizontal field for the perfect ground, plus a correction field that arises from the finite

---

† M. Rubenstein, "A Theoretical Study of the Relation Between Vertical and Horizontal Electric Fields at Ground Level from Lightning," Master's Thesis, University of Florida, 1986.

conductivity of the soil:

$$E_\rho(\rho, z) \approx E_{pp}(\rho, z) + E_{cor}(\rho, z) \quad \text{(F.2a)}$$

As developed by Rubenstein, the correction field is approximately independent on the height z and is related to the vertical field at $z = 0$ through the wave-tilt relationship. In this manner, Eq. (F.2a) becomes

$$E_\rho(\rho, z) \approx E_{pp}(\rho, z) + \frac{1}{\sqrt{\epsilon_r + \sigma/j\omega\epsilon}} E_{pz}(\rho, 0) \quad \text{(F.2b)}$$

Thus, in the treatment of the lightning excitation of a long line, Eqs. (F.1) and (F.2b) are used for the excitation fields. To calculate these fields, it is necessary to know the fields for the perfect ground, E_{pp} and E_{pz}, together with the wave-tilt function.

The calculation of the fields produced by the lightning channel and the coupling of these fields to the line are performed in the frequency domain using the BLT equation solution described in the text.

F.6.4 Code Operation

To assist in running the LTLINE, the Windows interface code LCC has been written. This is a graphical data input routine, which permits easy inputting of problem parameters. Upon running the LCC shell program by selecting the lightning icon from the Program Manager, the data entry screen shown in Figure F.7 appears. The operation of this data-entry shell program is very similar to the RISER code discussed previously. Pressing the Default button will set all the parameters to the default values given in Section F.3.2. To change individual parameters, simply click the mouse on one of the parameters in the picture of the line on the menu screen and enter the new value. Once a particular set of parameters has been defined, the data file can be saved under a unique name by pressing the Save Data File button.

After the data have been defined, pressing the Compute button will write the problem to the file LTLINE.DAT and then begin execution of the LTLINE program as a Windows DOS application. When the LTLINE code is executed, this data file is read and the calculation proceeds. As the calculations for each frequency point are completed, an indication as to the status of the calculation is provided in the DOS window.

At this point, this DOS window can be minimized and other tasks undertaken. Keep in mind, however, that the LTLINE program is numerically intensive and requires significant processor resources. Running this calculation as a background job will require a large amount of time for the calculation. If rapid calculations are needed, it is suggested that the LTLINE code be run directly from DOS without Windows. When the calculation is

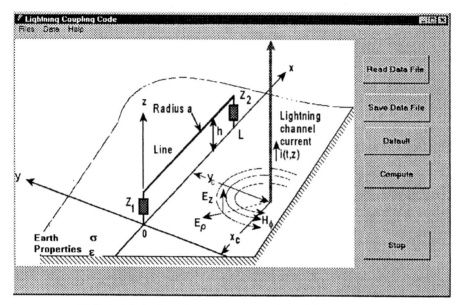

FIGURE F.7 Data input screen for the LCC shell program.

completed, data files V1.TIM, V2.TIM, C1.TIM, and C2.TIM are written to the disk directory containing the code for off-line plotting or processing.

Figure F.8 illustrates typical computation times (in minutes) for the sample data case using a 60-MHz Pentium computer and a 486DX 25-MHz computer. These calculation times are plotted as a function of the number of

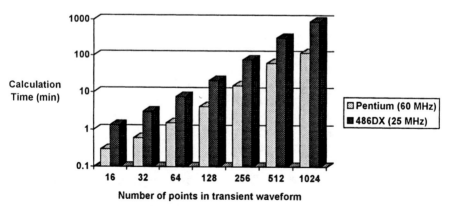

FIGURE F.8 Plot of typical computation times (in minutes) for the LTLINE code (using the Windows95 operating system).

data points in the time-domain waveform given by $NP = 2^m$, where m is the parameter entered in the data file.

F.6.5 Sample Results

As an example of computed lightning-induced load voltages for this lightning channel model, consider the case of a line with the parameters $\mathcal{L} = 1$ km, $h = 10$ m, and $a = 5$ mm. The lightning channel is located a distance $y_c = 50$ m from the midpoint of the line ($x_c = 500$ m). The return-stroke velocity is assumed to be $v = 1.3 \times 10^8$ m/s and the current decay constant in Eq. (7.65) is $\alpha = 0.588$ km^{-1}. For this example, the ground is assumed to be perfectly conducting.

The time-domain behavior of the lightning current at the channel base is that shown in Figure 7.23. Figure F.9 illustrates the early-time behavior of the load voltage as computed using the frequency-domain analysis provided by the LTLINE code. As a check of this result and the numerical method in general, it is useful to compare Figure F.9 with the results of a similar calculation using the direct time-domain method of [7.31]. Figure F.10 illustrates the corresponding load voltage response (solid line) with various components arising from the lumped end sources and the distributed field sources along the line. It should be noted that this result is slightly lower in peak amplitude than the response in Figure F.9.

As a further check of the validity of the analysis in this code, a nonsymmetric lightning strike point ($x_c = -50$ m, $y_c = 50$ m) can be chosen. Using the same line length and lightning parameters, the voltages calculated at the $x = 0$ and $x = 1000$ m ends of the line are illustrated in Figure F.11.

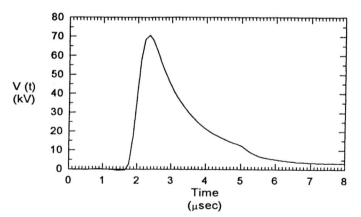

FIGURE F.9 Plot of the induced load voltage for the lightning channel at $x_c = 500$ m, $y_c = 50$ m, as computed using the frequency-domain analysis of LTLINE.

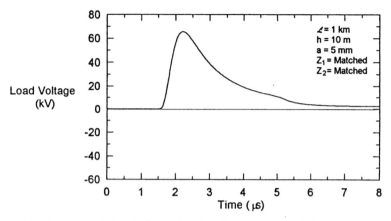

FIGURE F.10 Plot of the induced load voltage for the lightning channel at $x_c = 500$ m, $y_c = 50$ m, as computed using the time-domain approach of [7.31].

The corresponding responses calculated using the direct time-domain approach of [7.31] are shown in Figure F.12, and very good agreement is noted.

F.7 TOTALFLD FIELD CODE

F.7.1 Overview

The TOTALFLD code is designed to compute the total (i.e., incident plus ground-reflected) electromagnetic (EM) fields at a particular height h over a lossy earth, or the EM fields transmitted inside the earth at a burial depth of $-h$. This code assumes that the time-domain behavior of the incident field is one of the same transient functions used by the RISER code. The reflected field from the ground or the field transmitted into the ground is determined by the electrical conductivity of the soil σ and the relative permittivity ϵ_r, as well as the angle of incidence ψ.

As illustrated in Figure F.13, the most general description of the incident field involves both the elevation angle ψ and an azimuthal angle of incidence ϕ, measured positive in the direction indicated in the figure. The present version of this program permits definition of the incident field using both angles of incidence, and thus it is more general in the way that it represents the total fields.

For this code, the following quantities are calculated:

- The time-dependent E- and H-field vector components at the specified observation point for the user-specified excitation waveform.
- The frequency-domain spectrum for the E- and H-field vector com-

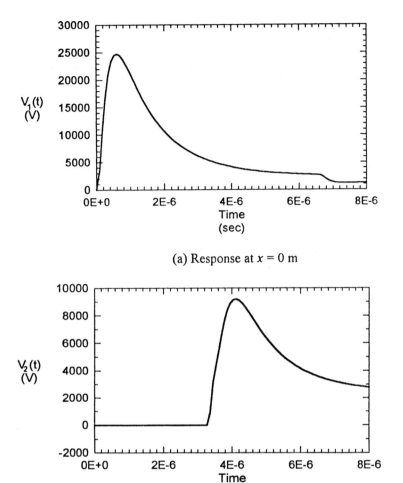

FIGURE F.11 Plot of the induced load voltages for the lightning channel at $x_c = 50$ m, $y_c = -50$ m, as computed the frequency-domain analysis of LTLINE.

ponents at the observation location, assuming an incident E-field of unit amplitude.

F.7.2 Problem Geometry

As illustrated in Figure F.13, the vector direction (i.e., polarization of the incident E-field is described by the angle α, which is measured positive in the clockwise direction from the vertical plane of incidence, looking toward the source. In this manner, an arbitrarily polarized E-field may be decomposed

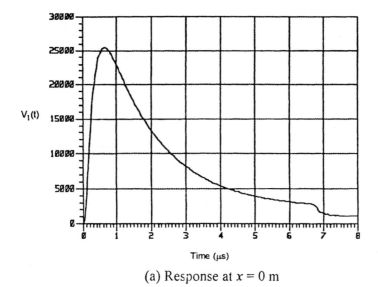

(a) Response at $x = 0$ m

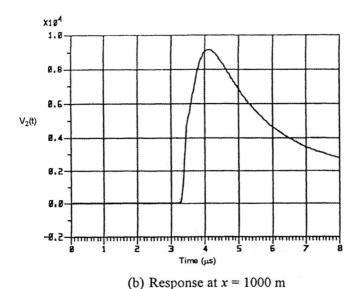

(b) Response at $x = 1000$ m

FIGURE F.12 Plot of the induced load voltages for the lightning channel at $x_c = 50$ m, $y_c = -50$ m, as computed using the direct time-domain analysis of [7.31].

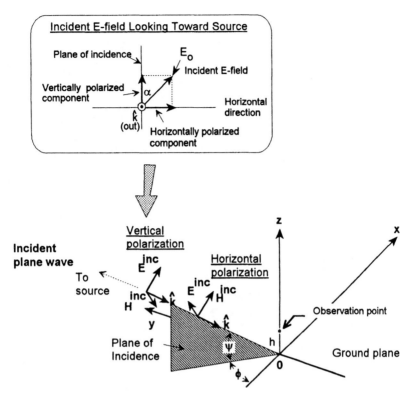

FIGURE F.13 Coordinate system and parameters representing the incident field for the TOTALFLD code.

into a vertically polarized and a horizontally polarized component, as indicated in the figure.

F.7.3 Expressions for the *E* and *H* Field

In the TOTALFLD code, the observation location is assumed to be at position $(x, y, z) = (0, 0, h)$, and the incident field is assumed to arrive at this point at time $t = 0$. Later in time, a reflection from the ground occurs and this modifies the fields at the observation point. The frequency-domain expressions for the incident and reflected *E* and *H* fields at the observation point *h* above the ground, or the transmitted fields into the ground at $-h$, have been developed in Section 8.4. These fields involve the Fresnel reflection and transmission coefficients and are used in the TOTALFLD code for calculating both the frequency-domain fields and transient responses via FFT.

F.7.4 Code Operation

This code operates in a manner similar to the RISER code in that there is a Windows-based graphical interface code called TOTALDVR, which takes the user input data, defines an excitation waveform, creates a data input file called TOTALFL.DAT, and then runs the main analysis program TOTALFLD, which is a compiled FORTRAN program that runs in a DOS window. The latter code can be run independently of the Windows menu program, as long as the input data file is defined.

The default data set provided by the code is as follows:

```
5      ! h Observation location above or below the earth surface (in m)
45     ! ψ Vertical Angle of Incidence (deg)
0      ! φ Azimuthal Angle of Incidence (deg)
0      ! α E-field Polarization Angle (deg)
.01    ! σ Earth Conductivity (S/m)
10     ! ε_r Earth Relative Dielectric Constant
```

Upon entering the TOTFLDVR code, the user can select the various parameters to change in the same manner as in the RISER code: by clicking on the appropriate parameter in the picture in the main menu or by selecting the Define Parameters selection in the menu. After this is done, the analysis code is run in a DOS window and results in the following data files written to disk:

EX.TIM	HX.TIM
EY.TIM	HY.TIM
EZ.TIM	HZ.TIM
EX.FRQ	HX.FRQ
EY.FRQ	HY.FRQ
EZ.FRQ	HZ.FRQ

These contain both the transient and spectral responses for the fields. In addition, the incident field waveform $E^{inc}(t)$ is provided in the data file EINC.TIM.

F.7.5 Samples Results

Calculated results from this code can easily be plotted on the screen with the NUPLOT routine supplied with NULINE. Samples results from this code are shown in Figures 8.6 to 8.9 for several cases of aboveground fields.

INDEX

A priori knowledge, EMF radiation, extended sources center-fed wire antenna, 123
ABCD parameters. *See* Chain parameters
Aboveground transmission line response, lossy ground, 422–428
  angle of incidence variations, 424–428
  earth conductivity variations, 424
  line height variations, 428
Admittance matrices, multiconductor transmission lines, 248–250
Aerial lines, conducted disturbances, mains impedance, 69–70
Agrawal formulation:
  load response for non-plane-wave excitation, 360–363
  single-wire line over perfectly conducting ground plane, 347–350
  transmission line theory, 322
  two-wire transmission line:
    load current and voltage solutions, 339–340
    numerical examples, 336–338
    plane-wave excitation, 340–342
    scattered voltage formulation, 336
Analytical modeling, EMC models, 26–30
  applications, 29–30
  system construction, 28
  system design, 26–28
  verification, 28–29
Angle of incidence variations, lossy ground effects, 424–428
Antenna current:
  approximate analysis of induced current, 169–174
  singularity expansion method (SEM) representation, 166–167
Antenna mode, transmission line response, 226–228, 323
Apertures:
  braided shield cable models:
    multiple apertures, 461–463
    polarizability expressions, 464–465
    single aperture excitation, 460–462
    weave parameters, transfer characteristics, 465–468
  direct time-domain calculations, 216
  EM field penetration, 185–207
  equivalent area, 203–207
  extended antenna radiation, 207–208
  far-field vector field diffraction, 196–199
    rectangular aperture, 198–199
  general vector field diffraction, 192–196
    penetration application, 194–196
  integral equation, 199–203
    field calculation example, 203

595

Apertures (*Continued*)
  low-frequency approximation, 208–211
    dipole moments, 208
    polarizabilities, 209–211
  model overview, 183–184
  scalar diffraction theory, 185–192
    circular aperture, 191–192
    Dirichlet solution, 188
    Kirchhoff approximation, 187–188
    Neumann solution, 188–189
    rectangular aperture, 191
    solution theorems, 189–192
  transient responses, 212–216
  wideband responses, 212–216
Approximate analysis:
  buried cable transmission line, 435–437
  simplifications, 437–438
  electromagnetic reception or scattering
    induced current determination, 155–156
    scattered field determination, 156–157
    thin wire solution, 155–157
  singularity expansion method (SEM), 169–175
    antenna induced current, 169–174
    scattering induced current, 174–175
  wire antenna integral equation, 129–134
    current evaluation, 130–131
    radiated field evaluation, 134
Axial electric field component, shielded cables, 452–453
Axial magnetic field component, braided shield cable model, 479

Babinet's principle, aperture integral equation, 203
Basis function, wire antenna integral equation, 128–129
"Bench" testing, electromagnetic compatibility (EMC) design, 26–28
Bergeron's graphical solution:
  crosstalk models, 15
  time-domain analysis, 275–281
    lumped and distributed parameters, 280–281
    nodal matrix equation, 279–280
    numerical application, 276–279
    principle, 275–276
Bessel functions:
  buried cable, soil impedance, 432–433
  electric dipole in conducting sphere, 148–149
  EMF penetration, scalar diffraction theory, 192
  equivalent area calculations for apertures, 206–207
  finitely conducting wire, telegrapher's equation, 332–333
  lossy ground effects:
    frequency domain representations, 407–409
    plane-wave excitation, transient field evaluation, 422
  solid tubular shield coupling, 456–457
B-field components, transmission line theory, 322
Biot–Savart law, two-wire transmission line, 328–329
Bistatic scattering, electromagnetic reception, 156–157
BLT (Baum–Liu–Tesche) equation:
  braided shield cable models:
    internal load responses, 484
    response calculations, 480
  interrupted cable shields, 490–493, 497–500
  lossy ground effects, 405

INDEX **597**

single-wire transmission line:
  highly resonant structures, 363–367
  load response for non-plane-wave excitation, 362–363
  perfectly conducting ground plane, 353
  resonance matrix expansion, 366–367
  vs. two-wire results, 355–356
transmission line models:
  finite line load response, 243–247
  multiconductor transmission lines, 260, 263–268
  radiation calculations, 373
  transmission network analysis, 380–382
two-wire transmission line:
  frequency-domain responses, 342–343
  load current and voltage solutions, 339–340
  plane-wave excitation, 340–342
  transient response, 343–344
Bounded-wave EMP simulator, interrupted cable shields, 496
Braided shield cable models, 457–479
  aperture polarizabilities, 464–465
  axial magnetic field component, 479
  calculated responses, 480–488
    external transmission line, 480–482
    internal excitation sources, 483
    internal load responses, 483–484
    numerical examples, 484–488
  comparison with measurements, 467–468
  dielectric filling, 467
  EMF penetration and diffraction, 459–460
  geometric configuration, 457–458

  multiple apertures, 462–465
  single aperture excitation, 460–462
  transfer admittance, 466–467
    alternative expression, 479
  transfer impedance, 465–466
    comparison of Demoulin's and Kley's models, 476–479
    Demoulin's model, 472–473
    improved expression, 468–479
    Kley's model, 473, 475–476
    Tyni's model, 469–472
Buried cables, lossy ground effects, 428–443
  cable impedance, 433–434
  current solution, 434–435
  finite length buried lines, 439–443
  infinite buried length response, 438–439
  integral equations, 431–432
  rigorous solution summary, 431–435
  soil impedance, 432–433
  TL solution simplifications, 437–443
  transmission line approximation, 435–437

Cable connectors, interrupted cable shields, 493–494
Cable impedance, buried cable, 433–434
Cable shield parameters, 565–567
Canonical problems, electromagnetic compatibility (EMC) design, 26–28
Canonical shapes, volumetric shields, time-harmonic excitation, 530–534
Capacitance parameters:
  multiconductor transmission lines, 254–257
  time-domain analysis, 274
  transmission line models, 293–310

Capacitance parameters (*Continued*)
  analytical evaluation, 295–296
  finite-element technique,
    partial capacitances, 300–301
  inductance values, 308–310
  integral equation evaluation of per-unit-length matrix, 301–308
  measurement techniques, 294–295
  nonhomogeneous media, 300
  partial capacitances, 296–297
  static capacitances, 297–299
  two-conductor systems, 230–231
Capacitive coupling:
  EMF circuit disturbances:
    electric fields, 83–86
    low-frequency field coupling, 90
    magnetic fields, 73–74
    reduction, 91–94
    reduction techniques, 91–94
  transmission line models, capacitance
    parameter measurement, 294–295
Cartesian coordinates, perfect ground plane, 137–138
Cascaded two-port networks, chain parameters, 57–58
Cavity conductance, electric dipole in, 141–147
Cavity fields, electric dipole in:
  eigen-mode solution, 144–147
  imaging techniques, 142–144
Center-fed wire antenna:
  EMF radiation, extended sources, 123–125
  singularity expansion method (SEM) representation, 171–172
Chain parameters:
  multiconductor transmission lines, 266–267
  multiport circuit models, 62–63

  two-conductor systems, frequency-domain analysis, 233–234
  two-port circuit models, 56–58
Characteristic admittance matrix, capacitance calculations using, 308–310
Chimney effect, braided shield cable transfer impedance, 475–479
Circuit theory, electromagnetic compatibility (EMC) model building, 8–9
Circular aperture:
  EMF penetration, scalar diffraction theory, 191–192
  non-plane-wave shielding, 544–547
Circumferential charge density, per-unit-length capacitance matrix, 302
Closed metallic mesh shield, volumetric shielding, 540–543
Coaxial cable and connector data, 565–567
Compensation theorem, interrupted cable shields, 492–493
Complex variable theory, buried cable current solution, 434–435
Conducted disturbance:
  electrical power systems, 63–72
  harmonic current generation, 64–67
  harmonic current source estimation, 71–72
  mains impedance determination, 67–71
  lumped-parameter circuit models, 47–72
  models for circuit sources, 59–62
  multiport circuits, 62–63
  passive two-port circuits, 50–59
    chain parameters, 56–58
    open-circuit impedance parameters, 51–53

INDEX 599

short-circuit admittance parameters, 53–56
two-port parameter relationships, 58
Thévenin and Norton representations, 48–50
Conducting mesh, plane-wave shielding, 524–527
Conductivities, table of, 552
Conductor charge distribution, per-unit-length capacitance matrix, 302–303
Continuity equation, EMF radiation in frequency domain, 115
Coordinate transformations, 556
Co-polarized field component, aperture integral equation calculations, 203–204
Coupling mechanism:
 EMF circuit disturbances, low-frequency field coupling, 89–90
 generalized shield concept, 506–508
 shielded cables:
  solid tubular shield, 455–457
  transfer admittance and impedance, 455
 singularity expansion method (SEM), 165–166
 scattering current, 167–168
Coupling validation, single-wire line over perfectly conducting ground plane, 356–357
Cross-polarized field component, aperture integral equation calculations, 203–204
Crosstalk
 low-frequency field coupling, 86–90
 models, 15
Current distribution:
 field coupling, 323–326
 multiconductor transmission lines, 256
 wire antennas, 114

Current responses, braided shield cable models, 483–484
Current source excitation, two-wire transmission line, 330–332
Cyclic inductance, conducted disturbances in electrical power systems, mains impedance, 68–69
Cylindrical coordinates, perfect ground plane, 136–138
Cylindrical shield, EMF circuit disturbances, 91–92

Deliberate antenna, defined, 113
Demoulins model:
 braided shield cable transfer impedance, 472–473
 Kley's model compared, 476–479
Diagonalization techniques, multiconductor transmission lines matrix elements, 253–257
[P] and [R] matrices, 251
Diamond-shaped apertures, braided shield cable model polarizabilities, 464–465
Dielectric materials:
 braided shield cable models, 467
 per-unit-length capacitance matrix, 306–307
 static dielectric constants, 551
Differential operator expressions, 554–556
Differential relationships, vector analysis, 553
Dirac delta function:
 braided shield cable models, single aperture excitation, 461–462
 lossy ground effects, plane-wave excitation, transient field evaluation, 422
Dirichlet solution:
 EMF aperture penetration, scalar diffraction theory, 188
 transmission line models, partial

Dirichlet solution (*Continued*)
  capacitance calculations, 300–301
Discontinuous cable shields, 495
Distributed circuit parameters:
  high-frequency problems, 223–226
  shielded cables, transfer admittance and impedance, 454–455
  time-domain analysis, Bergeron's graphical solution, 277–279
  two-conductor systems, 229
    line parameter evaluation, 229–231
Distributed excitations, transmission line models, 224–226

Earth conductivity variations, lossy ground effects, 424
Earth-return coupling, EMF circuit disturbances, 98–100
Eddy current shielding:
  frequency dependence, 527
  general concepts, 515–517
  plane-wave shielding:
    conducting mesh, 524–527
    infinite metal plate, 518–524
    infinite parallel plates, 524
  skin effect and depth, 517–518
E-field shielding effectiveness:
  TOTALFLD code, 592–593
  volumetric shields, time-harmonic excitation, 528, 534
Eigen-mode solution, electric dipole in cavity field, 144–147
Electric dipole, EMF radiation:
  aperture penetration, 208
  cavity conductance, 141–147
  elementary source, 117–120
  extended sources, 122–134
  imperfectly conducting earth, 150–153
  parallel–plate conduction, 138–141
  perfect ground conduction, 135–138
  spherical conduction, 147–150
Electric field coupling, EMF circuit disturbances, 81–86
Electric field integral equation (EFIE):
  thin wire solution in time domain, 161
  time-domain analysis, 158–159
  wire extension over lossy ground, 159–161
Electrical power systems, conducted disturbances, 63–72
  harmonic current generation, 64–67
  harmonic current source estimation, 71–72
  mains impedance determination, 67–71
Electrical shielding, static fields, 508–510
Electrically large circuits, defined, 223
Electrically small circuits, defined, 223
Electromagnetic compatibility (EMC):
  analytical applications, 26–30
    emission testing, 29
    immunity testing, 28–29
    system construction, 28
    system design, 26–28
    verification, 28–29
  analytical errors, 38
  balanced accuracy of analysis, 38–39
  current applications of modeling, 20–21
  limits of modeling, 19–20
  model building, 8–9
  modeling accuracy, 38–39
  modeling concepts, 3–6

modeling examples, 13, 15–16
overview of modeling, 9, 11–12
problem classification, 12–13
signal classification, 16–19
  frequency spectra evaluation, 18–19
system decomposition, 26
topological description, 30–38
  design analysis, 33–34
  design principles, 36–38
  diagram techniques, 31–32
  electromagnetic topology, 30–31
  energy entry points, 32–33
  interaction sequence diagram, 35–36
  system interaction, 34–36
Electromagnetic field (EMF) radiation:
  braided shield cable models, 459–460
  frequency domain:
    electric dipole source, 117–120
    elementary sources, 116–122
    extended sources, 122–134
    magnetic dipole source, 120–122
Electromagnetic interaction, sequence diagram, 35–36
Electromagnetic interference (EMI):
  circuit disturbances, 72–95
    capacitive coupling reduction, 91–94
    electric field coupling, 81–86
    inductive coupling reduction, 94–95
    low-frequency field coupling, 86–90
    low-frequency interference reduction, 90–91
    magnetic field coupling, 72–80
  conducted disturbances in circuits, Thévenin and Norton representations, 48–50
  conducted disturbances in electrical power systems, 63–72
    harmonic current generation, 64–67
    harmonic current source estimation, 71–72
    mains impedance determination, 67–71
  continuous and transient sources, 12, 14
  current applications of modeling, 20–21
  EMC models:
    problem classification, 12–13
    system interaction, 34–36
  emission testing, 29
  ground return disturbances, 95–100
  high-frequency circuit models, 101–103
  limits of modeling, 19–20
  manmade source classification, 12–13
  natural source classification, 12
  noise and frequency levels, 12, 14
  overview of modeling, 9, 11–12
  passive conducted disturbance models, 50–59
  radiated power and field strengths, 12, 14
Electromagnetic penetration, EMC models, 34–36
Electromagnetic propagation, EMC models, 34–36
Electromagnetic pulse (EMP) simulators:
  braided shield cable models, 485–486
  experimental model validation, 7
  single-wire line over perfectly conducting ground plane, 356–357
Electromagnetic reception and scattering, frequency domain radiation, 154–157

Electromagnetic topology, EMC
    models, 30–34
  design parameters, 33–34
  diagram, 31–32
  energy points of entry (POE),
    32–33
Electromagnetic Transients
    Program (EMTP), time-domain
    analysis, 282–285
Electronic components, EMC
    modeling and, 11
Electrostatic discharge (ESD),
    signal classification, 18
Elliptical aperture, braided shield
    cable model polarizabilities,
    464–465
Emission testing, electromagnetic
    compatibility (EMC)
    verification, 29
Equal-velocity assumption,
    multiconductor transmission
    lines, 259–262
Equivalent area calculations for
    apertures, EMF field
    penetration, 203, 205–207
Equivalent-circuit representations:
  conducted disturbances in
    circuits, 49–50
  lossy ground effects, per-unit-
    length parameters, 405–406
Excitation fields:
  braided shield cable models:
    internal excitation sources, 483
    single aperture excitation, 460–
      462
  buried cable, rigorous solution
    summary, 431–435
  lossy ground effects, 398–400
  RISER transmission line coupling
    code, 579–580
  single-wire line over perfectly
    conducting ground plane,
    scattered voltage
    formulation, 348–350
Experimental model validation,
    example, 7

Extended antennas, EMF radiation,
    122–134
  aperture penetration, 207–208
  center-fed wire antenna, 123–126
  wire antenna integral equation,
    126–134
External coupling:
  EMC models, 34
  shielded cable lines, 451
External transmission line, braided
    shield cable models, 480–482

Faraday's cage, static electrical
    shielding, 508–510
Far-field models:
  EMF radiation:
    aperture penetration, vector
      field diffraction, 196–199
    electric dipole source, 118
  perfect ground plane, 136–138
  singularity expansion method
    (SEM) representation, 168
Fast Fourier transform (FFT):
  buried cable, rigorous solution
    summary, 431–435
  computer code installation, 570–
    572
  electric dipole in cavity, 144
  frequency spectra evaluation, 19
  load response for non-plane-wave
    excitation, 360–363
  lossy ground effects:
    plane-wave excitation, transient
      field evaluation, 417–423
    time domain representations,
      410–411
  single-wire line, highly resonant
    structures, 365–366
  transmission line responses:
    analytical transformation from
      frequency to time domain,
      270–271
    numerical transformation from
      frequency to time domain
      analysis, 271–272

two-wire transmission line, transient response, 343–344
volumetric shields, transient electromagnetic fields, 535–540
Ferromagnetic materials:
eddy current shielding, 517
magnetostatic shielding, 510–515
conductivity and permeability, 514
Field attenuation mechanism:
EMC models, 37
magnetostatic shielding, 512–515
non-plane-wave shielding, 545–547
plane-wave shielding, conducting mesh, 526–527
Field propagation techniques, electromagnetic compatibility (EMC) model building, 9–10
Finite buried cable, transmission line solutions, 439–443
Finite element technique, transmission line models, 300–301
Finite line load response, transmission line models BLT equation, 243–244
frequency-domain analysis, 242–247
frequency-domain voltage response, 244–245
validation procedures, 245–247
Finite-difference, time-domain approach (FDTD), wideband aperture calculations, 216
Finite-difference equations:
non-experimental model validation, 8
time-domain analysis, 272–274
Finitely conducting wire, telegrapher's equation modification for, 332–333
Flux measurements, transmission line models, 287–289

Footing resistances, lossy ground effects, 404
Frequency domain analysis:
electromagnetic compatibility (EMC) models, 16, 18–19
EMF radiation, 114–154
dipole radiation with other bodies, 135–153
elementary sources, 116–122
extended sources, 122–134
magnetic field component evaluation, 153–154
overview, 114–116
lossy ground effects:
finite buried cable, 440–442
ground impedance, 406–409
shielding mechanisms, 527
time-domain analysis:
analytical transformation to, 269–271
numerical transformation to, 271–272
transmission line models, 231–268
BLT equation, multiconductor lines, 260–266
chain parameters multiconductor lines, 266–267
finite line load responses, 242–247
general solution, 240–242
lumped-source excitation, 237–239
multiconductor lines, 247–268
Telegrapher's equations for two-conductor line, 232–237
voltage reflection coefficient, 239–240
two-wire transmission line:
field coupling, 342–343
frequency-domain responses, 343
volumetric shields, closed metallic mesh shield, 542–543

Fresnel plane-wave reflection coefficient:
  electric dipole in lossy earth, 151–152
  lossy ground effects:
    plane-wave excitation, 412–417
    transient field evaluation, 420–422
Fresnel transmission coefficient, buried cable current solution, 434–435, 438–439
Fuzzy logic, modeling concepts and, 6

Galerkin's method, wire antenna integral equation, 127–129
"Gedanken" experiments, nonexperimental model validation, 8
Generalized shield, defined, 506–508
Geometrical mean distance (GMD):
  per-unit-length parameters, 562
  transmission line models:
    inductance parameter measurement, 287–289
    mutual and self-inductances, earth as return conductor, 292–293
    mutual inductance per unit length, 289–291
    self-inductance per unit length, 291–292
    between two circuits, 289
Geometrical mean radius (GMR):
  per-unit-length parameters, 561
  transmission line models, inductance parameter measurement, 287–289
Green's function:
  aperture integral equation, 199–203

braided shield cable models, external transmission line calculations, 480–482
buried cable, rigorous solution summary, 431–435
electric dipole in cavity, 141–147
EMF aperture penetration:
  Dirichlet solution, 188
  Neumann solution, 188–189
  scalar diffraction theory, 185–187
  vector field diffraction, 194–196
integrodifferential equation, 158–159
per-unit-length capacitance matrix, 302–303
single-wire line over perfectly conducting ground plane, 351–352
two-wire transmission line, telegrapher's equations, 338–339
Ground impedance, lossy ground effects, 404
  finite buried cable, 440–443
  frequency domain representations, 406–409
  time domain representations, 410–411
Ground returns:
  EMF circuit disturbances, 95–100
  lossy ground effects, plane-wave excitation, 417–422
Grounding resistance parameters, 563–564

Hallén integral, wire antenna radiation model, 130
Hankel functions:
  electric dipole in conducting sphere, 148–149
  lossy ground effects, frequency domain representations, 408–409

INDEX 605

Hard-wired connections, Thévenin and Norton representations, 48–50
Harmonic currents, conducted disturbances:
current generation, 64–67
source estimation, 71–72
Heaviside step function:
singularity expansion method (SEM), 163
two-wire transmission line, telegrapher's equations, 338–339
wire antenna integral equation, 131
Helmholtz wave equation:
EMF aperture penetration, scalar diffraction theory, 185–186
EMF radiation in frequency domain, 115–116
Hertz potentials:
buried cable, soil impedance, 432–433
EMF radiation:
aperture penetration, 208
dipole radiation, 117–120
H-field shielding effectiveness:
TOTALFLD code, 592–593
volumetric shields:
closed metallic mesh shield, 540–543
time-harmonic excitation, 528–534
transient electromagnetic fields, 536–540
High-altitude EMP (HEMP), modeling concepts, 11
High-frequency circuit models:
EMI disturbances, 101–103
distributed parameters, 223–226
transmission line models, 223
Hilbert transform, wideband aperture response, 213–216
Homogeneous media:
transmission line models, static capacitance calculation, 298–299
volumetric shields, 528–540
time-harmonic fields, 528–534
transient electromagnetic fields, 534–540
Horizontal polarization
buried cable current solution, 434–435
lossy ground effects, plane-wave excitation, 413–417
Hustler's alloys, magnetostatic shielding, 510
Hybrid parameters, passive two-port circuits, 58–59

Image theory:
concepts of, 3–6
electric dipole in cavity, 142–144
electric dipole in parallel–plate regions, 138–141
EMF aperture penetration, 209–211
Neumann solution, 188–189
Immunity testing, electromagnetic compatibility (EMC) verification, 28–29
Impedance matrices, multiconductor transmission lines, 248–250
Impulse (δ-function) waveform, signal classification, 18
Inadvertent antenna, defined, 113–114
Incident EM field:
aperture integral equation, 199–203
transmission line radiation, 322, 370–372
two-wire transmission line, 327–328
wire antenna integral equation, 126–127
Induced current:
electromagnetic reception or scattering, 155–156

Induced current (*Continued*)
  singularity expansion method (SEM) of approximate analysis, 169–174
  volumetric shields, closed metallic mesh shield, 543
Inductance parameters:
  capacitance calculations using, 308–310
  multiconductor transmission lines, 254–257
  time-domain analysis, 274
  transmission line theory
    analytical inductance evaluation, 287–289
    determination techniques, 283–293
    geometrical mean distance between circuits, 289
    measurement techniques, 286–287
    mutual and self-inductances, earth return conductor, 292–293
    mutual inductance per unit length, 289–290
    self inductance per unit length, 291–292
  two-conductor systems, 230–231
Inductive coupling:
  EMF circuit disturbances, 94–95
    low-frequency field coupling, 90
    magnetic fields, 73–74
  transmission line models, 286–287
Infinite buried cable:
  current responses, 438–439
  lossy ground effects, 430–431
Infinite metal plates:
  parallel plates, 524
  plane-wave shielding, 517–524
Intake point, generalized shield concept, 508
Integral equations:
  buried cable, rigorous solution summary, 431–432
  electromagnetic reception or scattering, 154–157
    approximate solution for thin wire, 155–157
  EMF aperture penetration, 199–203
  per-unit-length capacitance matrix, 301–308
  time domain equations, 157–161
    EFIE solution for thin wires, 161
    integrodifferential equation, 158–159
    lossy ground extension to wire, 159–161
  vector analysis, 554
  wire antenna radiation, 126–134
    approximate solution, 129–134
    method of moments solution, 127–129
Integrodifferential equations, time-domain analysis, 158–159
Intensity distributions, equivalent area calculations for apertures, 205–207
Interaction sequence diagram, EMC models, 35–37
Intermediate zone locations, EMF radiation, 118
Internal circuits, shielded cable lines, 451–452
Internal excitation sources, braided shield cable models, 483
Internal load responses, braided shield cable models, 483–484
Interrupted cable shields:
  cable connectors, 493–494
  discontinuous shields, 495
  numerical examples, 496–500
  overview, 488–493
  pigtail terminations, 494–495
Inverse hybrid parameters, passive two-port circuits, 58–59
Irradiance patterns, scalar diffraction theory, 189–193

INDEX 607

Isolated circular current loops, EMF circuit disturbances, 80

Kirchhoff approximation:
  EMF aperture penetration, scalar diffraction theory, 187–188
  nonlinear load transmission lines, 383–384
  time-domain analysis, 274
Kley's model:
  braided shield cable transfer impedance, 473, 475–476
  Demoulin's model compared, 476–479

Laplace transform:
  EMF aperture penetration, scalar diffraction theory, 185–186
  lossy ground effects, plane-wave excitation, transient field evaluation, 420–422
  magnetostatic shielding, 511–515
  singularity expansion method (SEM):
    approximate analysis of induced current, 171–174
    mathetmatical expression, 166
    overview, 162–163
Legendre polynomials, electric dipole in conducting sphere, 148–149
Lenz's law, EMF circuit disturbances, 73–80
Lightning EMP (LEMP):
  load response for non-plane-wave excitation, 357–363
  LTLINE transmission line code, 585–586, 591
  models, 15
  single-wire line over perfectly conducting ground plane, 356–357
  topological diagram, 31–32

Line current, two-wire transmission line, 338–339
Line height variations, lossy ground effects, 428–429
Line voltage, telegrapher's equations, 403
Load responses:
  braided shield cable models, 480–488
  external transmission line, 480–482
  internal excitation sources, 483
  internal load responses, 483–484
  numerical examples, 484–488
  single-wire line over perfectly conducting ground plane:
    non-plane-wave excitation, 357–363
    plane-wave excitation, 353–356
    telegrapher's equations for, 353
  transmission line models, finite line, 242–247
Lorentz gauge condition, EMF radiation in frequency domain, 115
Lossless transmission line:
  multiconductor per-unit-length parameters, 559–560
  two-conductor per-unit-length parameters, 557
Lossy ground:
  electric dipole, 150–153
  transmission lines:
    aboveground line responses, 422–428
    buried cables, 428–443
    overview, 395–396
    per-unit-length line parameters, 405–411
    reflected and transmitted plane-wave fields, 411–422
    telegrapher's equations, 396–405
    two-wire transmission line, 333

Lossy ground (*Continued*)
    wire extension, 159–161
Low-frequency approximation, EMF aperture penetration, 208–211
    dipole moments, 208
    lumped circuit elements, 223–226
    polarizabilities, 209–211
    transmission line models, 223
Low-frequency field coupling, EMF circuit disturbances:
    general coupling, 86–90
    interference reduction, 90–91
LTLINE transmission line code, 582–589
    lightning source line excitation, 585–586
    operating principles, 586–588
    problem geometry, 584–585
Lumped circuit parameters:
    braided shield cable models, 481–482
    common ground return disturbances, 95–100
    conducted disturbances, 48–72
        electrical power systems, 63–72
        multiport circuits, 62–63
        passive two-port circuits, 50–59
        Thévenin and Norton representations, 48–50
        two-port models for circuits with sources, 59–62
    electromagnetic field interference (EMI), 72–95
        capacitive coupling reduction, 91–94
        electric field coupling, 81–86
        inductive coupling reduction, 94–95
        low-frequency coupling, 86–90
        low-frequency interference reduction, 90–91
        magnetic field coupling, 72–80
    high-frequency extension, 101–103

interrupted cable shields, 491–493, 498–500
    low-frequency problems, 223–226
    overview, 47–48
    time-domain analysis, Bergeron's graphical solution, 277–279
    transmission line models, 224–226
        frequency-domain analysis, 237–239
    radiation, 372–373

Magnetic dipole, EMF radiation, 120–122
Magnetic effect, modeling of, 4
Magnetic field:
    EMF circuit disturbances, 72–80
        mutual and self-inductance calculations, 77–80
        weak-coupling approximations, 74–77
    frequency domain radiation, 153–154
    Kley's model of transfer impedance, 476
    static relative permeabilities, 550
Magnetic flux density, two-wire transmission line, 328–329
    first telegrapher's equation, 328–329
Magnetostatic shielding, static fields, 510–515
Mains impedance, conducted disturbances
    in electrical power systems, 67–71
Maxwellian capacitance matrix, transmission line models, 297
Maxwell's equations:
    eddy current shielding, 515–517
    electromagnetic compatibility (EMC) model building, 8–9

EMF radiation in frequency domain, 115–116
frequency domain radiation, 153–154
lossy ground effects, 397–400
magnetostatic shielding, 510–515
modeling concepts and, 5
single-wire line over perfectly conducting ground plane:
 scattered voltage formulation, 347–350
 total voltage formulation, 346–347
two-wire transmission line, 329–330
wideband aperture calculations, 216
wire antenna radiation models, 113–114
Mittag–Leffler theorem, singularity expansion method (SEM), 165
Modal equations, multiconductor transmission lines:
 frequency-domain analysis, 252–253
 velocity computations, 255
 voltages and currents, 251–252
Modeling:
 accuracy issues, 38–39
  balanced accuracy, 38–39
  inherent errors, 38
 concepts of, 3–6
 current applications, 20–21
 electromagnetic compatibility (EMC), 8–9
 limits of, 19–20
 history of, 4
 resistor model, 23–25
 sample problems, 23–25
 validation procedures, 6–8
Moment method solution:
 current behavior, 133–134
 wire antenna integral equation, 127–129
Monostatic scattering,

electromagnetic reception, 156–157
Multiconductor lines:
 per-unit-length parameters:
  capacitance matrix, 301–302
  geometrical matrix, 560
  lossless transmission line, 559
 resonant structures, BLT equation, 367–368
 transmission line theory:
  applications, 267–268
  BLT equation, 260–266
  chain parameters, 266–267
  equal-velocity assumption, 259–260
  frequency-domain analysis, 247–268
  impedance and admittance matrices, 248–250
  modal equation solutions, 252–253
  modal voltages and currents, 251–252
  natural propagation modes, 250–251
  open-circuit voltage, semi-infinite line excitation, 258–259
  overview, 226
  [P] and [R] matrices diagonalization, 251
  propagation and diagonalization matrices calculations, 253–257
Multiple apertures, braided shield cable models, 461–463
Multiplicative relationships, 553
Multiport circuits, passive conducted disturbance models, 62–63
Mutual inductance configurations:
 EMF circuit disturbances:
  electric field coupling, 81–86
  magnetic field coupling, 77–80
 transmission line models:

Mutual inductance configurations (*Continued*)
  earth as return conductor, 292–293
  mutual inductance per unit length, 289–291

Natural current mode, singularity expansion method (SEM), 165
Natural frequencies, singularity expansion method (SEM), 164–166
Natural propagation modes, multiconductor transmission lines, 250–251
Near-field shielding, non-plane-wave excitation, 544
Near-zone locations, EMF radiation, 118
Neumann form of mutual inductance, EMF circuit disturbances, 77–80
Neumann solution
  EMF aperture penetration, scalar diffraction theory, 188–189
  transmission line models, 300–301
Neural networks, modeling concepts, 6
Nodal admittance matrix, time-domain analysis, 277–280
Nodal capacitance matrix, transmission line models, 297
Nonexperimental model validation, 7–8
Nonhomogeneous media, static capacitance calculation, 300
Nonlinear loads:
  single transmission line example, 384–388
  Volterra integral equation, 382–384
Non-plane-wave excitation:
  shielding mechanisms, 544–547
  inter-circular loops, 544–547
  near-field shielding, 544
  single-wire line over perfectly conducting ground plane, 357–363
Nonsinusoidal traveling waves, transmission line responses, 268–269
Normalized function, wideband aperture response, 212–216
Norton representation:
  conducted disturbances in circuits, 48–50
  electromagnetic reception or scattering, 155–156
  EMF circuit disturbances, magnetic fields, 75–77
  nonlinear load transmission lines, Volterra integral equation, 383–384
  passive conducted disturbance models, 60–62
  passive two-port circuits, 55–56
  transmission network analysis, 375–380
Nuclear electromagnetic pulse (NEMP):
  modeling, 11, 15
  two-wire transmission line, 343–344
NULINE transmission line code, 571–577
  batch mode, 576–577
  DOS operation, 573–575
  interactive mode, 575–576
  problem geometry, 572–573
Numerical Electromagnetics Code (NEC):
  single-wire line:
    highly resonant structures, 365–367
    vs. two-results, 356
  transmission line theory:
    field coupling, 323–326
    frequency-domain analysis, 246–247

INDEX **611**

radiation calculations, 373
wire antenna integral equation, 129–134
Numerical experiment, defined, 3
Numerical model, scientific method, 5
NUPLOT screen plotting routine, 578
n-Wire multiconductor line, configurations, 226

Obliquity factor, EMF aperture penetration, 189–192
Ohm's law, time-domain analysis, 278–279
Open-circuit impedance parameters, passive conducted disturbance models, 51–53
Open-circuit voltage, semi-infinite multiconductor line, 258–261

Parallel-plate regions, electric dipole in, 138–141
Partial capacitances, transmission line models:
 finite element calculations, 300–301
 measurement techniques, 296–297
 sign of, 297
Perfectly conducting ground plane:
 electric dipoles over, 135–138
 single-wire line, 345–363
  BLT equations, 353
  coupling validation, 356–357
  EM scattering and line excitation, 350–352
  load response solution to telegrapher's equations, 353
  modified telegrapher's equations, 352
  non-plane wave excitation load response, 357–363

plane-wave excitation load responses, 353–356
scattered voltage formulation, 347–350
telegrapher's equations, 345–346
total voltage formulation, 346–347
two-wire results compared, 353–356
static capacitance calculation, homogeneous medium conductors over, 298–299
transmission line models, mutual and self-inductances over, 292–293
Per-unit-length parameters:
 braided shield cable models:
  external transmission line calculations, 482
  multiple apertures, 463–464
 geometrical mean distance (GMD), 562
 geometrical mean radius (GMR), 561
 integral equation evaluation, 301–308
 line parameter tables, 557–562
 lossless multiconductor transmission line, 559–560
 lossless two-conductor transmission line, 557
 lossy ground effects, equivalent line circuits, 405–406
 single-wire line over perfectly conducting ground plane, 347
 two-conductor line geometrical parameter, 557
Physical constant tables, 550–552
π circuit, two-conductor systems, 235–237
Pigtails:
 EMF circuit disturbances, 94
 interrupted cable shields, 488–501

Plane-wave excitation:
  buried cable current solution, 434–435
  eddy current shielding:
    conducting mesh, 524–527
    infinite metal plate, 518–524
    infinite parallel plates, 524
  lossy ground effects, 411–422
    ground reflection of transient field, 417–422
    reflection and transmission from the earth, 412–417
    single-wire line over perfectly conducting ground plane, 353–356
    two-wire results compared, 355–356
  two-wire transmission line, 340–342
Pocklington integrodifferential equation, 127
Point-centered finite-difference technique, time-domain analysis, 272–274
Points of entry (POE):
  electromagnetic-system interaction, 34–36
  EMC models, 32–33
Polarizabilities:
  braided shield cable model aperture expressions, 464–465
  EMF aperture penetration, 209–211
  plane-wave shielding, conducting mesh, 525–527
Porpoising phenomenon, braided shield cable models:
  Demoulin's impedance model, 472–473
  transfer impedance expression, 469
Power cables, conducted disturbances, mains impedance, 69–70

Power density, equivalent area calculations for apertures, 205–207
Poynting vector, EMF radiation, 120
Primary constant table, 550
Printed circuit boards (PCBs):
  low-frequency field coupling, 86–90
  models, 15–16
Propagation matrix, multiconductor transmission lines, 253–257

Quasi static field, two-wire transmission line, 327–328

Radar cross section, electromagnetic reception, 156–157
Radiated disturbance, lumped-parameter circuit models, 47–48
Radiated field evaluation, wire antenna integral equation, 134
Radiated fields, singularity expansion method (SEM) representation, 168
Radiation models:
  electric-field integral equations in time domain, 157–161
    EFIE numerical solution for thin wire, 160
    integrodifferential equation, 158–159
    wire extension over lossy ground, 159–161
  EMF reception and scattering in frequency domain, 154–157
    thin wire approximation, 155–157
  singularity expansion method (SEM), 161–175

antenna current representation, 166–167
approximate antenna analysis, 169–175
mathematical description, 164–166
radiated field representation, 168
scattered field representation, 168–169
scattering current representation, 167–168
transmission lines:
examples, 372–373
line geometry, 370–372
overview, 368
reciprocity theorem, 368–370
wire antennas:
magnetic field components evaluation, 153–154
overview, 113–115
Radiation problem, overview, 113–114
Random transient noise, EMC models, 18
Rayleigh–Ritz variational technique, wire antenna integral equation, 127
Reception (scattering) problem, overview, 114
Reciprocity theory, transmission line radiation, 368–370
Rectangular aperture, EMF penetration:
far-field diffraction, 198–199
scalar diffraction theory, 191
Reflected EM field:
aperture integral equation, 199–203
lossy ground effects, plane-wave excitation, 412–417
Reflection coefficients, wire extension over lossy ground, 159–161
Response coefficients, EMF circuit disturbances, 89–90

Ribbon cables, models, 16
Rigorous solutions, buried cable:
cable impedance, 433–434
current solution, 434–435
integral equations, 431–432
lossy ground effects, 431–435
soil impedance, 432–433
RISER transmission line coupling code, 578–582
excitation field, 579–580
input data file, 580–581
operating principles, 581
problem geometry, 579

Scalar diffraction theory:
EMF aperture penetration, 185–192
circular aperture, 191–192
Dirichlet solution, 188
Kirchhoff approximation, 187–188
Neumann solution, 188–189
rectangular aperture, 191
solution theorems, 189–192
Scattered EM field
defined, 114
electromagnetic reception, 156–157
single-wire line over perfectly conducting ground plane, 350–352
singularity expansion method (SEM) representation, 168–169
transmission line theory, 322–323
two-wire transmission line
first telegrapher's equation, 327–328
wire antenna integral equation, 126–127
Scattered voltage formulation:
lossy ground effects, plane-wave fields, 411
telegrapher's equations, 334–336
lossy ground effects, 398–399, 402–403

614 INDEX

Scattered voltage formulation (*Continued*)
  single-wire line over perfectly conducting ground plane, 347–350
Scattering current:
  induced current analysis, 174–175
  singularity expansion method (SEM) representation, 167–168
Scattering parameters, passive two-port circuits, 58–59
Scientific method, 4–5
Self-capacitance calculations, transmission line models, 301
Self-inductance configurations:
  EMF circuit disturbances:
    electric field coupling, 82–86
    magnetic field coupling, 77–80
  transmission line models:
    mutual and self-inductances, earth as return conductor, 292–293
    self-inductance per unit length, 291–292
Semi-infinite multiconductor line, open-circuit voltage, 258–262
Shielded cables:
  braided shield models, 457–479
    aperture polarizability expressions, 464–465
    axial magnetic field component, 479
    calculated responses, 480–488
    Demoulin's model, 472–473, 476–479
    EMF penetration and diffraction, 459–460
    external transmission line, 480–482
    internal excitation, 483
    internal load, 483–484
    Kley's model, 473, 475–479
    multiple apertures, 462–464
    numerical example, 484–488
    single aperture excitation, 460–462
    transfer admittance, 466–467, 479
    transfer characteristics with weave parameters, 465–469
    transfer impedance, 465–466, 468–469, 474
    Tyni's model, 469–472
  coupling fundamentals, 452–455
    braid optimization, 455
    transfer admittance and impedance, 453–455
  geometry, 452
  interruptions:
    cable connectors, 493–494
    discontinuous shields, 495
    examples, 496–501
    overview, 488–493
    pigtail terminations, 494–495
  overview, 451–452
  tubular shield EM coupling, 455–457
    transfer admittance, 457
    transfer impedance, 455–457
Shielding models, 16
  topology, 31–32
Shielding principles:
  general concepts, 506–508
  mechanisms, 508–527
    static fields, 508–515
    time-varying fields: eddy current shielding, 515–527
  non-plane-wave fields, 544–547
  overview, 505–506
  volumetric shields, 528–543
    closed homogeneous metal shield, 528–540
    closed metallic mesh shield, 540–543
Short-circuit admittance parameters,

passive conducted disturbance models, 53–56
Short-circuited fields, EMF aperture penetration, 209–211
Signals, electromagnetic compatibility (EMC) models, 16–19
Single-source excitation:
  braided shield cable models, 460–462
  transmission line models, 238–239
Single-wire line:
  highly resonant structures, 363–367
    BLT resonance matrix expansion, 366–367
    numerical examples, 364–366
    nonlinear load impedance, 384–388
  perfectly conducting ground plane:
    BLT equations, 353
    coupling validation, 356–357
    EM scattering and line excitation, 350–352
    load response solution to telegrapher's equations, 353
    modified telegrapher's equations, 352
    non-plane-wave excitation load response, 357–363
    plane-wave excitation load responses, 353–356
    scattered voltage formulation, 347–350
    telegrapher's equations, 345–346
    total voltage formulation, 346–347
    two-wire results compared, 355–356
Singularity expansion method (SEM):
  antenna current representation, 166–167
  approximate antenna analysis, 169–175
    antenna induced current problem, 169–174
    scattering induced current problem, 174–175
  mathetmatical description, 164–166
  overview, 161–164
  radiated field representation, 168
  scattered field representation, 168–169
  scattering current representation, 167–168
Skin effect:
  eddy current shielding, 515–518
  shielded cables, 452–453
Slice generator, wire antenna integral equation, 131
Snell's law, lossy ground effects, 412–417
Software directory, 569–593
  code distribution, 570
  code operation, 569–570
  computer code installation, 570–571
  LTLINE transmission line code, 582–589
  NULINE transmission line code, 571–577
  NUPLOT screen plotting routine, 578
  RISER transmission line coupling code, 578–582
  TOTALFLD code, 589–593
Soil impedance, buried cable, 432–433
Solid tubular shield, EM coupling, 455–457
Sommerfeld radiation:
  electric dipole in lossy earth, 150–153

616 INDEX

Sommerfeld radiation (*Continued*)
  EMF aperture penetration, vector field diffraction, 194
Source–victim configuration:
  conducted disturbances in circuits, 48–50
  EMC models:
    orthogonalization, 37–38
    separation, 37–38
  EMF circuit disturbances:
    electric field coupling, 81–86
    ground returns, 97–100
    low-frequency field coupling, 88–90
    weak magnetic field coupling, 76–77
Spectral magnitude:
  braided shield cable models, 486–487
  electromagnetic compatibility (EMC) models, 16–17
Spherical conductance, electric dipole in, 147–150
Static capacitances, transmission line models:
  homogeneous medium over ground, 298–299
  measurement techniques, 297–308
  nonhomogeneous media, 300
Static shielding, 508–515
  electrical shielding, 508–510
Step-function response:
  multiconductor transmission lines, 260, 264–265
  singularity expansion method (SEM), 162–164
    approximate antenna analysis, 173–174
    wideband aperture calculations, 214–216
Stokes' theorem:
  lossy ground effects, total voltage formulation, 397–400
  single-wire line over perfectly conducting ground plane

    scattered voltage formulation, 347–350
  two-wire transmission line:
    first telegrapher's equation, 327–329
    scattered voltage formulation, 335–336
Sunde's formula, lossy ground effects, 407–409
Supermatrix equation, multiport circuit models, 63
System construction, electromagnetic compatibility (EMC) design, 28
System design, electromagnetic compatibility (EMC), 26–28

T equivalent circuit, two-conductor systems, 235–237
Taylor formulation:
  lossy ground effects, 405–406
  transmission line theory, 321
  two-wire transmission line, 336–338
    load current and voltage solutions, 339–340
Telegrapher's equations:
  lossy ground effects, 396–405
    plane-wave fields, 411–422
    scattered voltage formulation, 402–403
    solutions, 404–405
    termination conditions, 403–404
    total voltage formulation, 396–402
  multiconductor transmission lines, 247–260
    equal-velocity assumption, 259–260
    impedance and admittance matrices, 248–250
    modal equation solutions, 252–253

modal voltages and currents, 251–252
natural propagation modes, 250–251
open-circuit voltage, semi-infinite line excitation, 258–259
[P] and [R] matrices diagonalization, 251
propagation and diagonalization matrices, 253–257
single-wire line over perfectly conducting ground plane: derivation, 345–346
total voltage formulation, 346–347
time-domain analysis:
nonsinusoidal traveling waves, 268–269
numerical solutions, 272–274
two-conductor systems:
π and T equivalent two-wire line circuits, 235–237
chain parameter wire line representation, 233–234
frequency-domain solutions, 232–237
overview, 228–231
two-port representation applications, 234–235
two-port representations for two-wire line, 234
two-wire transmission line, 326–344
BLT line current and voltage solutions, 339–340
external excitation derivation, 326–333
finitely conducting wire modification, 332–333
first equation, 327–329
frequency-domain responses, 342–343

line current and voltage solutions, 338–339
lossy medium modification, 333
numerical examples, 336–338
plane-wave excitation load responses, 340–342
scattered voltage formulation, 334–336
second equation, 329–332
total voltage formulation, 333–334
transient responses, 343–344
Termination conditions, lossy ground effects, 403–404
Testing functions, wire antenna integral equation, 128–129
Thévenin representation:
conducted disturbances in circuits, 48–50
EMF circuit disturbances:
electric fields, 82
magnetic fields, 76–77
interrupted cable shields, 489–493
nonlinear load transmission lines, Volterra integral equation, 382–384
passive two-port circuits:
chain parameters, 56–58
open-circuit impedance parameters, 51–53
transmission network analysis, 375–380
Time-domain analysis:
electric-field integral equations, 157–161
EFIE solution for thin wires, 161
integrodifferential equation, 158–159
lossy ground extension to wire, 159–161
electromagnetic compatibility (EMC) models, 16
frequency spectra evaluation, 18–19

Time-domain analysis (*Continued*)
lossy ground effects:
ground impedance, 410–411
plane-wave excitation, transient field evaluation, 420–422
transmission line responses, 268–283
analytical frequency domain transformation to time-domain, 269–271
Bergeron's graphical solution, 275–281
electromagnetic transients program (EMTP), 282–283
inductive and capacitive terminations, 274
nodal matrix equation, 279–280
nonsinusoidal traveling waves, 268–269
numerical frequency domain transformation to time-domain, 271–272
numerical solution to telegrapher's equations, 272–274
time-harmonic excitation, 268
transient circuit after closing two interrupters, 280–281
wideband aperture calculations, 216
Time-domain reflectometer (TDR), multiconductor transmission lines, 249–250
Time-harmonic excitation:
eddy current shielding, 516–517
transmission line responses, 268
analytical transformation from frequency to time domain, 270–271
volumetric shields:
closed homogeneous shields, 528–534
E-field shielding effectiveness, 534

H-field shielding effectiveness, 528–534
limitations, 532–533
Time-varying fields, shielding mechanisms, 515–527
Timotin's expression, lossy ground effects, 411
Topological analysis, EMC models, 30–38
design principles, 36–38
electromagnetic topology, 30–34
system interaction, 34–36
Total voltage formulation:
lossy ground effects, plane-wave fields, 411
telegrapher's equations:
lossy ground effects, 396–402
single-wire line over perfectly conducting ground plane, 346–347
two-wire model, 333–334
TOTALFLD code, 589–593
E- and H-field shielding effectiveness, 592–593
problem geometry, 590, 592
Transfer admittance:
braided shield cable models:
alternative expressions, 479
weave parameters, 466–468
defined, 453–454
shielded cables, 453
solid tubular shield coupling, 457
Transfer impedance:
braided shield cable models:
Demoulin's model, 472–473
improved expressions, 468–479
Kley's model, 473, 475–476
measured and calculated magnitudes, 471–472, 474
Tyni's model, 469–472
weave parameters, 465–466, 468
defined, 453–455

measured impedances for cable connectors, 567
shielded cables, 453
solid tubular shield coupling, 455–456
Transformer circuit models, high and low frequency, 101–103
Transient current pulse measurement, braided shield cable models, 468
Transient electromagnetic fields, volumetric shields, 534–540
Transient response, 211–216
 braided shield cable models, 486–487
 lossy ground effects:
  finite buried cable, 440–442
  plane-wave excitation, 417–422
 multiconductor transmission lines, 260, 262
 time-domain analysis:
  Bergeron's graphical solution, 280–281
  numerical transformation from frequency analysis to, 271–272
 transmission network analysis, 379–380
 two-wire transmission line, 343–344
Transmission line mode, components of, 227, 323
Transmission line theory:
 antenna mode responses, 226–228
 field coupling:
  highly resonant structures, 363–368
  nonlinear loads, 382–388
  overview, 321–326
  radiation from, 368–373
  single line over perfectly conducting ground, 345–363
  transmission networks, 373–382
  two-wire transmission lines, 326–344
 frequency-domain responses, 231–268
  analytical transformation to time-domain, 269–271
  finite line load responses, 242–247
  general solutions, 240–242
  lumped-source excitation, 237–239
  multiconductor line chain parameters, 266–267
  multiconductor line equations, 247–266
  numerical transformation to time-domain, 271–272
  Telegrapher's equations for two-conductor line, 232–237
  voltage reflection coefficient, 239–240
 line capacitance parameters, 293–310
  analytical evaluation, 295–297
  finite element techniques for partial capacitances, 300–301
  homogeneous medium ground conductors, 298–299
  inductance values calculations, 308–310
  integral equations for per-unit-length capacitance matrix, 301–308
  measurement techniques, 294–295
  nonhomogeneous media, 300
  partial capacitance analysis, 296–297
  static capacitance calculations, 298–310
  static capacitance evaluation, 297

Transmission line theory (*Continued*)
line inductance parameters, 283–293
  analytical inductance evaluation, 287–293
  earth return conductor inductances, 292–293
  geometrical mean distance between two circuits, 289
  inductance measurement, 286–287
  mutual inductance per unit length, 289–290
  self inductance per unit length, 291–292
lossy ground effects:
  aboveground line responses, 422–428
  buried cables, 428–443
  overview, 395–396
  per-unit-length line parameters, 405–411
  reflected and transmitted plane-wave fields, 411–422
  telegrapher's equations, 396–405
lumped and distributed circuit parameters, 223–224
lumped and distributed excitations, 224–226
nonlinear loads, 382–388
  single transmission line example, 384–388
  Volterra integral equation, 382–384
overview, 223–231
radiation:
  examples, 372–373
  line geometry, 370–372
  overview, 368
  reciprocity theorem, 368–370
software directory, 569–593
  LTLINE transmission line code, 582–589

NULINE transmission line code, 571–577
NUPLOT screen plotting routine, 578
RISER transmission line coupling code, 578–582
TOTALFLD code, 589–593
telegrapher's equations for two-conductor system, 228–231
time-domain responses, 268–283
  analytical transformation from frequency domain, 269–271
  Bergeron's graphical solution, 275–281
  electromagnetic transients program, 282–283
  inductive and capacitive terminations, 274
  nonsinusoidal traveling waves, 268–269
  numerical transformation from frequency domain, 271–272
  Telegrapher's equation, numerical solution, 272–274
  time-harmonic excitation, 268
two-conductor and multiconductor systems, 226
Transmission networks:
  BLT equation, 380–382
  overview, 373–375
  Thévenin transformations, 375–380
Transmission parameters. *See* Chain parameters
Transverse electromagnetic (TEM) field, 323
  braided shield cable models:
    multiple apertures, 463–464
    single aperture excitation, 460–462
Transverse propagation constant,

INDEX   **621**

buried cable transmission line, 437–438
Two-conductor systems:
  configurations, 226
  frequency-domain analysis, 232–237
  π and T equivalent two-wire line circuits, 235–237
  applications of two-port representation, 234–235
  chain parameter wire line representation, 233–234
  two-port representations for two-wire line, 234
  per-unit-length parameters, 557–558
  Telegrapher's equations, 228–237
Two-port circuit theory:
  passive conducted disturbance models, 50–59
  chain parameters, 56–58
  hybrid, inverse hybrid and scattering parameters, 58–59
  open-circuit impedance parameters, 51–53
  short-circuit admittance parameters, 53–56
  two-port parameter relationships, 58
  two-conductor systems, frequency-domain analysis, 234–235
Two-wire transmission line:
  single-wire line over perfectly conducting ground plane, 355–356
  telegrapher's equations, 326–344
  BLT line current and voltage solutions, 339–340
  external excitation derivation, 326–333
  finitely conducting wire modification, 332–333
  first equation, 327–329
  frequency-domain responses, 342–343
  line current and voltage solutions, 338–339
  lossy medium modification, 333
  numerical examples, 336–338
  plane-wave excitation load responses, 340–342
  scattered voltage formulation, 334–336
  second equation, 329–332
  total voltage formulation, 333–334
  transient responses, 343–344
Tyni's model, braided shield cable transfer impedance, 469–472

Unique observer technique, time-domain analysis, 275–281
Uniqueness theorem, EMF aperture penetration, 189–192

Validation procedures:
  transmission line models, 6–8
  frequency-domain analysis, 245–247
Vance formula, lossy ground effects:
  frequency domain representations, 408–409
  time domain representations, 410–411
Vector analysis and functions, 553–556
Vector field diffraction, EMF aperture penetration, 192–196
  far-field diffraction, 196–199
  fundamental principles, 192, 194
  penetration problem, 194–196
Verification procedures, electromagnetic compatibility (EMC) design, 28–29

622 INDEX

Vertical polarization
 buried cable current solution, 434–435
 lossy ground effects, plane-wave excitation, 413–417
V-I relationships, lossy ground effects, 403–404
Voltage reflection coefficient, transmission line models, 239–242
Voltage response:
 braided shield cable models, 483–484
 interrupted cable shields, 499–500
 transmission line models:
  electromagnetic transients program (EMTP), 282–285
  frequency-domain analysis, 244–245
  multiconductor transmission lines, 257–259
Voltage source excitation:
 single-wire line over perfectly conducting ground plane, 350–352
 transmission line models, frequency-domain analysis, 241–242
 two-wire transmission line:
  second telegrapher's equation, 330–332
  telegrapher's equations, 338–339
Volterra integral equation:
 nonlinear load transmission lines, 382–384
 single-wire nonlinear load impedance, 384–388
Volumetric shields:
 closed, homogeneous metal shield, 528–540
 time-harmonic field shielding, 528–534
 transient electromagnetic fields, 534–540

closed metallic mesh shield, 540–543
 mesh-protected shield induced responses, 543
 overview, 528
 time-harmonic fields:
  E-field shielding effectiveness, 534
  H-field shielding evaluation, 528–532
  H-field shielding examples, 533–534
  limitations, 532–533
 transient electromagnetic fields, 534–540

Waveforms, electromagnetic compatibility (EMC) models, 16–17
Weak-coupling approximations, EMF circuit disturbances:
 electric fields, 82
 magnetic fields, 74–77
Weave parameters, braided shield cable models, 465–468
Wideband aperture response, 211–216
Wire antennas:
 electric-field integral equations in time domain, 157–161
  EFIE numerical solution for thin wire, 161
  integrodifferential equation, 158–159
  wire extension over lossy ground, 159–161
 EMF reception and scattering in frequency domain, 154–157
  thin wire approximation, 155–157
 extended source radiation, 122–134
  center-fed wire antenna, 123–125

integral equation for, 126–134
radiation models:
  magnetic field components evaluation, 153–154
  overview, 113–115
singularity expansion method, 161–175
  antenna current representation, 166–167
  approximate antenna analysis, 169–175
  mathematical description, 164–166
  radiated field representation, 168
  scattered field representation, 168–169
  scattering current representation, 167–168

Y-matrix representation:
  passive conducted disturbance models, 54–56
  two-port models for circuits with sources, 60–62

Z-matrices, passive conducted disturbance models:
  open-circuit impedance parameters, 51–53
  short-circuit admittance parameters, 53–56
  two-port models for circuits with sources, 59–62